CHEMICAL
PROCESS
PRINCIPLES

CHEMICAL PROCESS PRINCIPLES

PART I

MATERIAL AND ENERGY BALANCES

OLAF A. HOUGEN
PROFESSOR OF CHEMICAL ENGINEERING
UNIVERSITY OF WISCONSIN

KENNETH M. WATSON
PROFESSOR OF CHEMICAL ENGINEERING
ILLINOIS INSTITUTE OF TECHNOLOGY

ROLAND A. RAGATZ
PROFESSOR OF CHEMICAL ENGINEERING
UNIVERSITY OF WISCONSIN

Second Edition

JOHN WILEY & SONS, INC.

New York · London · Sydney

ISBN 0 471 41283 X

PRINTED IN THE UNITED STATES OF AMERICA

Preface

"In the following pages certain industrially important principles of chemistry and physics have been selected for detailed study. The significance of each principle is intensively developed and its applicability and limitations scrutinized." Thus reads the preface to the first edition of *Industrial Chemical Calculations*, the precursor of this book. The present book continues to give intensive quantitative training in the practical applications of the principles of physical chemistry to the solution of complicated industrial problems and in methods of predicting missing physicochemical data from generalized principles. In addition, through recent developments in thermodynamics and kinetics, these principles have been integrated into procedures for process design and analysis with the objective of arriving at optimum economic results from a minimum of pilot-plant or test data. The title *Chemical Process Principles* was selected to emphasize the importance of this approach to process design and operation.

The design of a chemical process involves three types of problems, which although closely interrelated depend on quite different technical principles. The first group of problems is encountered in the preparation of the material and energy balances of the process and the establishment of the duties to be performed by the various items of equipment. The second type of problem is the determination of the process specifications of the equipment necessary to perform these duties. Under the third classification are the problems of equipment and materials selection, mechanical design, and the integration of the various units into a coordinated plan.

These three types may be designated as process, unit-operation, and plant-design problems, respectively. In the design of a plant these problems cannot be segregated and each treated individually without consideration of the others. However, in spite of this interdependence in application the three types may advantageously be segregated for

v

study and development because of the different principles involved. Process problems are primarily chemical and physicochemical in nature; unit-operation problems are for the most part physical; the plant-design problems are to a large extent mechanical.

In this book only process problems of a chemical and physicochemical nature are treated, and it has been attempted to avoid overlapping into the fields of unit operations and plant design. The first part deals primarily with the applications of general physical chemistry, thermophysics, thermochemistry, and the first law of thermodynamics. Generalized procedures for estimating vapor pressures, critical constants, and heats of vaporization have been elaborated. New methods are presented for dealing with equilibrium problems in extraction, adsorption, dissolution, and crystallization. The construction and use of enthalpy-concentration charts have been extended to complex systems. The treatment of material balances has been elaborated to include the effects of recycling, by-passing, changes of inventory, and accumulation of inerts.

In the second part the fundamental principles of thermodynamics are presented with particular attention to generalized methods. The applications of these principles to problems in the compression and expansion of fluids, power generation, and refrigeration are discussed. However, it is not attempted to treat the mechanical or equipment problems of such operations.

Considerable attention is devoted to the thermodynamics of solutions with particular emphasis on generalized methods for dealing with deviations from ideal behavior. These principles are applied to the calculation of equilibrium compositions in both physical and chemical processes.

All these principles are combined in the solution of the ultimate problem of the kinetics of industrial reactions. Quantitative treatment of these problems is difficult, and designs generally have been based on extensive pilot-plant operations carried out by a trial-and-error procedure on successively larger scales. However, recent developments of the theory of absolute reaction rates have led to a thermodynamic approach to kinetic problems which is of considerable value in clarifying the subject and reducing it to the point of practical applicability. These principles are developed and their application is discussed for homogeneous, heterogeneous, and catalytic systems. Particular attention is given to the interpretation of pilot-plant data. Economic considerations are emphasized and problems are included in establishing optimum conditions of operation.

In covering so broad a range of subjects, widely varying comprehensibility is encountered. It has been attempted to arrange the material in the order of progressive difficulty. Where the book is used for college instruction in chemical engineering the material of the first part is suitable for second- and third-year undergraduate work. The second part is suitable for third- and fourth-year undergraduate work; the third part for senior and graduate studies.

A few problems were selected from *Chemical Engineering Problems* published by the American Institute of Chemical Engineers (1946), with permission.

The authors wish to acknowledge gratefully the suggestions of Professors Joseph Hirschfelder, R. J. Altpeter, K. A. Kobe, E. N. Lightfoot, R. G. Taecker, and A. L. Lydersen.

OLAF A. HOUGEN
KENNETH M. WATSON
ROLAND A. RAGATZ

Madison, Wisconsin
September 1954

HOUGEN (O. A.), WATSON (K. M.) AND RAGATZ (R. A.)
CHEMICAL PROCESS PRINCIPLES CHARTS. *Second Edition*

HOUGEN (O. A.), WATSON (K. M.) AND RAGATZ (R. A.)
CHEMICAL PROCESS PRINCIPLES

Part 1—Material and Energy Balances. *Second Edition*
Part 2—Thermodynamics. *Second Edition*
Part 3—Kinetics and Catalysis

Contents

Table of Symbols

A	area
A	atomic weight
A	component A
B	component B
B	constant of Calingaert–Davis equation
C	component C
$°C$	degrees centigrade
C_p	heat capacity at constant pressure
C_v	heat capacity at constant volume
C	specific heat
c_p	molal heat capacity at constant pressure
c_v	molal heat capacity at constant volume
D	diameter
d	differential operator
E	energy in general
e	base of natural logarithms
F	force
f	weight fraction
G	mass velocity per unit area
G	specific gravity
(g)	gaseous state
g_c	standard gravitational constant, 32.174 (ft/sec)/sec
H	enthalpy
H	Henry's constant
H	humidity
H_p	percentage humidity
ΔH	change in enthalpy
ΔH_c	heat of combustion
ΔH_f	heat of formation
ΔH_r	heat of reaction

H	enthalpy per mole
$\bar{H}$	partial molal enthalpy
$\Delta\bar{H}$	partial molal enthalpy change
I	inert component
I	integration constant
J	mechanical equivalent of heat
K	characterization factor
°K	degrees Kelvin
K	distribution coefficient
K	equilibrium constant
l	length
(l)	liquid state
ln	natural logarithm
log	logarithm to base 10
M	molecular weight
M_m	mean molecular weight
m	mass
N	Avogadro number $6.024(10^{23})$ molecules per gram-mole
N	mole fraction
n	number of moles
P	pressure (used only in exceptional cases to distinguish pressure of pure components from partial pressures of some component in solution)
p	total pressure
p	partial pressure (with subscript)
Q	heating value of fuel
Q	heat lost from system
q	heat *added* to system
R	gas constant
°R	degrees Rankine
r	radius
S	cross section
S	humid heat
S_p	percentage saturation
S_r	relative saturation
S.C.	standard conditions
(s)	solid state
T	absolute temperature, degrees Rankine or Kelvin
t	temperature, °F or °C
U	internal energy
U	internal energy per mole
u	velocity

$\bar{u}$	average velocity
V	volume
V_r	volume of reactor
$\bar{V}$	partial molal volume
v	volume per mole
w	weight
w	work *done by* system
w_e	work of expansion *done by* system
w_f	electrical work *done by* system
w_s	shaft work *done by* system
x	mole fraction in liquid phase
x	mole fraction of reactant converted
x	quality
y	mole fraction in vapor phase
y^*	mole fraction in vapor, equilibrium value
y_r	relative humidity in percent
y_p	percentage humidity
Z	elevation above datum plane
z	compressibility factor
z	mole fraction in total system

SUBSCRIPTS

A	component A
a	air
B	component B
b	normal boiling point
C	component C
c	critical state
D	component D
e	expansion
f	electrical and radiant
f	formation
f	fusion
G	gas or vapor
H	isenthalpic
i	any component i
j	any component except i
L	liquid
p	constant pressure
r	reduced conditions
r	relative
s	normal boiling point

Table of Symbols

s	saturation
T	isothermal
t	temperature
t	transition
V	constant volume
v	vapor
w	water vapor

GREEK SYMBOLS

(α)	crystal form
(β)	crystal form
(γ)	crystal form
Δ	finite change of a property; positive value indicates an increase
∂	partial differential operator
κ	ratio of heat capacities
Λ	heat of vaporization
λ	heat of vaporization per mole
λ	wave length
λ_f	heat of fusion per mole
μ	viscosity
π	total pressure of mixture, used where necessary to distinguish from p
π	3.1416
ρ	density
Σ	summation
σ	surface tension
τ	time

SUPERSCRIPTS

*	ideal behavior
*	equilibrium state
o	standard state

1

Mathematical Procedures

The principal objective to be gained in the study of this book is the ability to reason accurately and concisely in the application of the principles of physics and chemistry to the solution of industrial problems. It is necessary that each fundamental principle be thoroughly understood, not superficially memorized. However, even though a knowledge of scientific principles is possessed, special training is required to solve the more complex industrial problems. There is a great difference between the mere possession of tools and the ability to handle them skillfully.

Direct and logical methods for the combination and application of certain principles of chemistry and physics are described in the text and indicated by the solution of illustrative problems. These illustrations should be carefully studied and each individual operation justified. However, it is not intended that these illustrations should serve as forms for the solution of other problems by mere substitution of data. Their function is to indicate the organized type of reasoning that will lead to the most direct and clear solutions. In order to test the understanding of these principles and to develop the ability of organized, analytical reasoning, practice in the actual solution of typical problems is indispensable. The problems selected represent, wherever possible, reasonable conditions of actual industrial practice.

The mathematical operations involved in solving the problems of Part One of this text usually do not go beyond arithmetic and algebraic operations, or simple calculus; some problems will require the use of special procedures, which are described in the present chapter.

Solution of Equations by Trial-and-Error Procedures. In solving cubic equations and equations of higher degree, trial-and-error procedures are resorted to because analytical procedures are extremely involved. The details of such a procedure are shown in the following illustration.

Illustration 1. Methyl alcohol can be synthesized by passing a mixture of CO and H_2 over a suitable catalyst. For a feed mixture containing 2 moles of hydrogen to 1 mole of carbon monoxide, x, the number of moles of hydrogen converted at 300° C and 240 atm, is indicated by the following equation:

$$\left(\frac{3-x}{2-x}\right)^2\left(\frac{x}{2-x}\right) - 429 = 0$$

The evaluation of x in this cubic equation can be solved by the following trial-and-error procedure.

The left-hand side of the equation may be designated as $\phi(x)$. The solution to the problem corresponds to that value of x at which $\phi(x)$ reduces to zero. Various

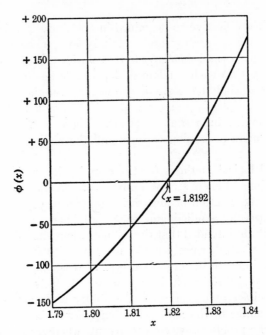

FIG. 1.   Solution of equations by trial-and-error procedures

values for x are assumed, and corresponding values of $\phi(x)$ are calculated, giving the values tabulated below.

x	$\phi(x)$
1.79	−146
1.80	−105
1.81	− 55
1.82	+ 6
1.83	+ 81
1.84	+175

The tabulated values are plotted (Fig. 1). It is observed that the resultant curve cuts the $\phi(x) = 0$ horizontal at an x value of 1.8192 moles. Actually, the accuracy of the equation is such that the value should be rounded off to 1.82.

With equations higher than those of first degree the question of more than one root must be considered.   In general, a quadratic equation always has two roots, a cubic three roots, etc.   However, in most of the situations encountered in chemical engineering calculations, only one root is of real significance.   In the present problem, the only real root is 1.82; the other two have imaginary values of no practical interest.

Solution of Simultaneous Equations.   Problems are frequently encountered in which the values of two or more unknowns are sought. In order to solve problems of this type, it is necessary that as many independent relations be available as there are unknowns.   If the relations are expressed as linear equations, the solution is arrived at most readily by common algebraic procedures.   If, however, one or more of the equations are of higher degree, graphical methods usually are much easier.   For two unknowns the procedure involves plotting the curves for the two independent equations and obtaining the desired values from the point of intersection.

Illustration 2.   In analyzing the equilibrium developed in the outgoing gas when a feed mixture of H_2O and CO in a molal ratio of 5 to 1 is passed through a bed of coke maintained at 1100° K and 10 atm, the following two independent equations are obtained:

$$\frac{(x - y)x}{(1 - x + 2y)(5 - x)} = 0.9442 \tag{a}$$

$$\frac{(1 - x + 2y)^2}{(x - y)(6 + y)} = 1.2201 \tag{b}$$

x = moles of H_2O converted, per 6 moles of feed, according to the water-gas reaction, $H_2O + CO \rightarrow CO_2 + H_2$

y = moles of CO_2 converted, per 6 moles of feed, according to the reaction, $C + CO_2 \rightarrow 2CO$

If the numerical values of x and y can be determined, the equilibrium gas composition may be calculated.   An algebraic solution would be difficult to arrive at, because both of the equations are nonlinear.   A graphical solution is therefore employed, as follows.

Values of x ranging from 0 to 5 are substituted in each of the two equations, and the corresponding values of y are calculated.

Assumed Values of x	y from Equation a	y from Equation b
0	−0.500	−0.092
1	0.117	0.736
2	0.891	1.534
3	1.885	2.316
4	3.198	3.089
5	5.000	3.855

The tabulated values of y are plotted against x in separate curves (Fig. 2).   The point of intersection of these two curves corresponds to the desired values of $x = 3.86$ and $y = 2.98$.

Graphical Integration. Graphical integration must be resorted to where the mathematical relationship between the variables which are required for an analytical integration are not known. Where the curve relating two variables is available, the integration may be carried out graphically.

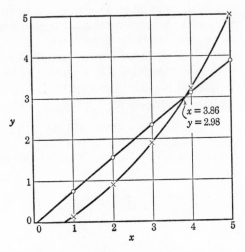

FIG. 2. Graphical solution of simultaneous equations

The process of graphical integration involves measuring the area under a suitable curve between specified limits. In general, $\int_{x_1}^{x_2} f'(x)\ dx$ is the area under the curve when $f'(x)$ values are plotted on the y axis and x values are plotted on the x axis, between the limits x_2 and x_1 using rectangular coordinate paper. The procedure of graphical integration is shown in the following illustration.

Illustration 3. Table 1 gives the atomic heat capacity[1] (atomic specific heat) of graphite at constant pressure for temperatures ranging from 298.16 to 1200° K.

From the data given, determine the heat absorption q when one gram-atom of graphite is heated from a base temperature of 298.16° K to various higher temperatures, up to 1200° K.

$$q = \int_{298.16}^{T_2} c_p\ dT$$

[1] *Selected Values of Chemical Thermodynamic Properties,* as of July 1, 1953, edited by D. D. Wagman, National Bureau of Standards.

TABLE 1. ATOMIC HEAT CAPACITY OF GRAPHITE

Temperature, °K	Heat Capacity, (g-cal)/(g-atom)(K°)	Temperature, °K	Heat Capacity, (g-cal)/(g-atom)(K°)
298.16	2.066	800	4.75
400	2.851	900	4.98
500	3.496	1000	5.14
600	4.03	1100	5.27
700	4.43	1200	5.42

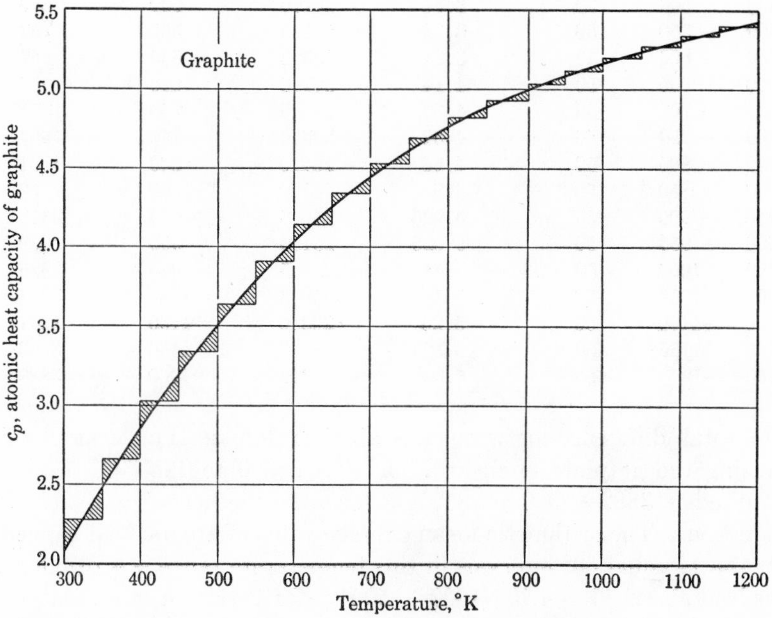

FIG. 3. Graphical integration

A curve is plotted (Fig. 3) with c_p on the y axis and T on the x axis on rectangular cross-section paper. The total temperature interval to be covered is arbitrarily divided into equal increments of 50 K°, except for the first increment, which is 51.84 K°. Over each increment, a mean c_p value is selected, so chosen that the small cross-hatched area above the curve equals the small cross-hatched area below the curve. The heat absorption over each increment then will equal the mean heat capacity for the increment multiplied by the temperature increment. Cumulative values for heat absorption are obtained by adding the heat absorptions over the successive temperature increments as shown in Table 2.

Where an integral number of equal intervals is taken between the upper and lower limits of integration, the area can be obtained from the arithmetic mean of all average ordinate values by multiplying this by

TABLE 2. DATA FOR GRAPHICAL INTEGRATION
Basis: 1 g-atom Graphite

Temperature Interval, °K	ΔT, K°	Mean c_p	Heat Absorbed in Interval, g-cal	Temperature, °K	Heat Absorbed above 298.16° K, g-cal
298.16– 350	51.84	2.28	118.2	350	118
350 – 400	50	2.67	133.5	400	252
400 – 450	50	3.025	151.3	450	403
450 – 500	50	3.34	167.0	500	570
500 – 550	50	3.63	181.5	550	752
550 – 600	50	3.91	195.5	600	947
600 – 650	50	4.14	207.0	650	1154
650 – 700	50	4.33	216.5	700	1371
700 – 750	50	4.52	226.0	750	1597
750 – 800	50	4.68	234.0	800	1831
800 – 850	50	4.82	241.0	850	2072
850 – 900	50	4.925	246.3	900	2318
900 – 950	50	5.025	251.3	950	2569
950 –1000	50	5.11	255.5	1000	2825
1000 –1050	50	5.19	259.5	1050	3084
1050 –1100	50	5.26	263.0	1100	3347
1100 –1150	50	5.325	266.3	1150	3613
1150 –1200	50	5.38	269.0	1200	3882

the total difference between the abscissa limits. For example, in heating the graphite of illustration 3 from 500 to 1200° C, the heat required is $3882 - 570 = 3312$ g-cal by the method of graphical integration. The arithmetic mean average value of atomic heat capacity for the 14 equal 50° intervals in this temperature range is 4.732; hence the heat absorbed $= 4.732(1200 - 500) = 3312$ g-cal in agreement with the given value.

There are many instances where the variable plotted on the y axis is a complex function, but once the points for establishing the curve are determined the integration itself is simple. For example, consider the following equation:

$$(\text{H}^* - \text{H})_T = \int_0^{p_2} \left[T\left(\frac{\partial v}{\partial T}\right)_p - v \right]_T dp$$

In order to carry out a graphical integration, it is necessary to prepare a table relating $[T(\partial v/\partial T)_p - v]_T$ and p. Then a curve is drawn by plotting values of $[T(\partial v/\partial T)_p - v]_T$ on the y axis and values of p on the x axis. The area under the curve is determined graphically from a lower pressure limit of zero to the higher pressure p_2.

Graphical Differentiation. If the equation relating two variables is known, differentiation can usually be carried out analytically with precision, and there is no need to resort to graphical methods. It frequently happens, however, that the equation is not known, but from a table of data a curve relating the two variables is available. In such a case, differentiation may be carried out graphically.

One method of graphical differentiation involves plotting a curve relating the two variables, drawing tangents to the curve at desired locations and then measuring the slopes of these tangents. Although this method is correct in principle, it is difficult to draw tangents accurately, even with special devices developed for that purpose. Accordingly, a graphical procedure, the reverse of graphical integration, is frequently used as illustrated by the following problem.

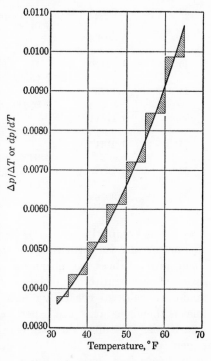

FIG. 4.   Graphical differentiation

Illustration 4. Determine the slope of the vapor-pressure curve of water at 32° F and at intervals of 5 F° from 35 to 60° F.

In Table 3 are tabulated values of the vapor pressure of water[2] against temperature, columns 1 and 2.

The increments in p and T for each interval are tabulated in columns 3 and 4, and the values of $\Delta p/\Delta T$ are tabulated in column 5. In Fig. 4 these values of $\Delta p/\Delta T$ are plotted on rectangular coordinate paper; over each temperature interval a horizontal line is drawn to represent the value of $\Delta p/\Delta T$ for that interval. A smooth curve is then drawn, cutting these horizontal lines in such a manner that the small cross-hatched areas above and below the curve for each increment are equal. Though the curve does not strike the mid-points of the horizontals exactly (except in the case of a straight line), nevertheless the departure from the mid-point is slight unless the curve has a considerable curvature over the interval under consideration. As a first approximation, the curve may be sketched passing through the mid-points of the horizontals, and subsequently any adjustment needed to equalize the areas can be made. The curve thus developed gives the true values of dp/dT, as tabulated in column 6.

[2] J. H. Keenan and F. G. Keyes, *Thermodynamic Properties of Steam*, John Wiley & Sons (1936).

TABLE 3. DATA FOR GRAPHICAL DIFFERENTIATION

Temperature, °F	Vapor pressure, psi	Δp	ΔT	$\Delta p/\Delta T$	dp/dT
1	2	3	4	5	6
32	0.08854				0.00362
		0.01141	3	0.003803	
35	0.09995				0.00398
		0.02175	5	0.004350	
40	0.12170				0.00474
		0.02582	5	0.005164	
45	0.14752				0.00562
		0.03059	5	0.006118	
50	0.17811				0.00663
		0.03599	5	0.007198	
55	0.2141				0.00780
		0.0422	5	0.008440	
60	0.2563				0.00913
		0.0493	5	0.009860	
65	0.3056				

Log-Log Graph Paper. Log-log graph paper is developed by laying off linear scales for logarithms to the base 10 along the x and y axes. Then, next to these two linear scales of logarithms, adjacent nonlinear scales of antilogarithms are laid out. Rather inaccurately, these two nonlinear scales are generally referred to as "log scales." A network of lines is then ruled onto the paper, based on the nonlinear scales of antilogarithms. Usually, on the commercially available log-log graph papers, the uniform scale of logarithms is omitted, and only the network of lines based on the nonuniform scale of antilogarithms remains.

If logarithms to the base 10 increase by increments of 1, the corresponding successive antilogarithms have a ratio of 10 to 1 to each other. One so-called cycle covers an increment of 1 on the scale of logarithms, and the corresponding antilogarithms for this increment show a ratio of 10 to 1 to each other. Each cycle on the scale of antilogarithms is divided into 9 nonuniform major divisions, with the ratios of the successive antilogarithms to the antilogarithm at the beginning of the cycle being 2, 3, 4, 5, 6, 7, 8, 9, 10. Usually, the antilogarithm assigned to the beginning of the cycle is a power of 10, as 0.1, or 1, or 10, etc. However, it is not necessary to adhere to this convention. The cycle may start at any desired number, and end with a number 10 times as great as the selected starting number. For example, if the cycle starts at 50, the succeeding subdivisions of the cycle will correspond to 2×50, 3×50, 4×50, etc., with the last one corresponding to 500.

If an equation has the general form, $y = ax^b$, where a and b are constants, logarithms may be taken of both sides of the equation, resulting in the following:

$$\log y = b \log x + \log a$$

This indicates that a straight line will be developed if values of $\log y$ are plotted on the linear scale of logarithms on the y axis (or values of y on the nonlinear scale of antilogarithms) and if values of $\log x$ are plotted on the linear scale of logarithms on the x axis (or values of x on the nonlinear scale of antilogarithms).

The slope of such a line, referred to the linear scales of logarithms, is b, while the y intercept, referred to the linear scale of logarithms on the y axis, is equal to $\log a$.

Illustration 5. If air is expanded reversibly and adiabatically, the following two equations give the relation among pressure, volume, and absolute temperature, where C_1 and C_2 are constants which depend on the mass and initial state of the air, and on the units for p, V, and T.

$$pV^{1.40} = C_1 \quad \text{or} \quad \log p = -1.40 \log V + \log C_1$$

$$pT^{-3.50} = C_2 \quad \text{or} \quad \log p = 3.50 \log T + \log C_2$$

If one pound-mole of air (29.0 lb) is originally at 800° R (absolute Fahrenheit temperature) and 1 atm, draw curves showing how volume and temperature will change if the air is expanded reversibly and adiabatically. The initial volume of the air, as calculated from the ideal-gas law, is 583 cu ft.

On a sheet of log-log paper (Fig. 5), the initial state points ($p = 1$ atm, $V = 583$ cu ft), ($p = 1$ atm, $T = 800°$ R) are located. Straight lines having respective slopes of -1.40 and 3.50 with reference to linear coordinates are drawn through these points. These two straight lines then represent the pressure-volume and the pressure-temperature relationships for the particular set of initial conditions specified for this problem. If the initial state of the air is different from that specified in this problem, the same procedure is employed, except that the lines having the respective slopes of -1.40 and 3.50 are drawn through the new initial state points. The exponents 1.40 and 3.50 in the equations are the same; hence the slopes will be the same as before.

Frequently it is known that the law relating two variables is of the form $y = ax^b$, and it is required to evaluate a and b from experimental data. This involves plotting the experimental values of x and y on log-log paper and then drawing the best straight line through the plotted points. The slope is then evaluated with reference to linear coordinates, and is equal to the exponent b. The intercept on the y axis of the linear-coordinate system gives $\log a$ if read on the linear scale of logarithms, or a itself if read on the antilogarithm scale. It must be kept in mind that the origin $(0, 0)$ of the system of linear coordinates for logarithms corresponds with the point (1, 1) located on the antilogarithm scale of coordinates.

Semilogarithmic Graph Paper. Semilogarithmic paper has a non-uniform logarithmic scale on the y axis (an auxiliary uniform scale of logarithms may or may not be shown) and a linear scale on the x axis. It is particularly adapted to plotting graphs of equations of the type

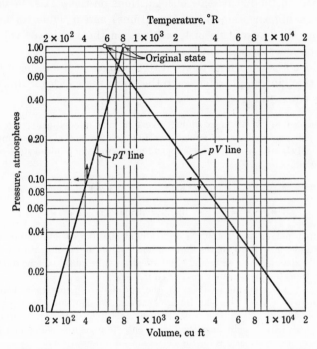

Fig. 5.   Use of log-log paper to show pV and pT lines in the reversible adiabatic expansion of air

$y = ab^{cx}$, where a, b, and c are constants. Usually b is made equal to 10 or to e, the base of natural logarithms. Taking logarithms of both sides gives

$$\log y = (c \log b)(x) + \log a$$

Inspection of the foregoing equation indicates that, if values of y are plotted on the nonlinear logarithmic scale on the y axis, and if values of x are plotted on the linear scale on the x axis, a straight line will result. The slope of the line, referred to the auxiliary linear scale on the y axis, will equal $c \log b$, while the y intercept on the auxiliary linear scale will equal $\log a$.

Illustration 6. The Arrhenius equation which relates rate of reaction with temperature is as follows:

$$k = Ae^{-E/RT}$$

k = specific reaction rate, $\sec^{-1}$
A = frequency factor, $\sec^{-1}$
e = base of natural logarithms
E = energy of activation, g-cal per g-mole
R = gas constant, g-cal/(g-mole)(K°)
T = absolute temperature, °K

In investigating a certain chemical reaction, the following data were obtained:

Temperature			
°C	°K	$1/T$	k
100	273.16	3.661×10^{-3}	1.055×10^{-16}
110	283.16	3.532×10^{-3}	1.070×10^{-15}
120	293.16	3.411×10^{-3}	9.25×10^{-15}
130	303.16	3.299×10^{-3}	6.94×10^{-14}
140	313.16	3.193×10^{-3}	4.58×10^{-13}
150	323.16	3.085×10^{-3}	3.19×10^{-12}

Using the data in the table, evaluate E and A of the Arrhenius equation.

Logarithms are taken of both sides of the Arrhenius equation, giving the following:

$$\log k = \left(-\frac{E \log e}{R} \right) \frac{1}{T} + \log A$$

Values of k and $1/T$ are plotted on semilogarithmic paper (Fig. 6) with k on the log scale and $1/T$ on the linear scale. A straight line is drawn through the plotted points, and the slope is determined to be $-(5)/(0.646 \times 10^{-3})$.

$$\text{Slope} = -\frac{E \log e}{R} = -\frac{(E)(0.4343)}{1.987} = -\frac{5}{0.606 \times 10^{-3}}$$

$$E = 35{,}400 \text{ g-cal per g-mole}$$

The y intercept must be computed using the value for the slope, as the curve is far removed from the origin of linear coordinates.

$$y \text{ intercept} = 12.375 = \log A$$

$$A = 2.36 \times 10^{12}$$

Triangular Diagrams. In dealing with three-component systems, compositions may be represented by points within a triangle, each point representing one unique composition. While an equilateral triangle is frequently used for this purpose, a triangle of any shape may be employed. To demonstrate the fact that the principles developed in this section may be applied to any triangle and are not restricted to the equilateral type, a triangle having three unequal sides is used.

In Fig. 7, apexes A, B, and C represent the respective pure compo-

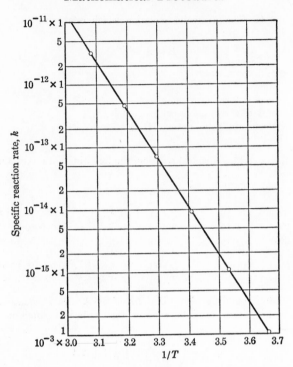

Specific reaction rate, k

$1/T$

FIG. 6.   Use of semilog paper to determine constants of Arrhenius equation

nents.   Assume a ternary system P of known composition, and let x_{AP}, x_{BP}, and x_{CP} represent the respective weight fractions of A, B, and C in the system.

On Fig. 7, points D and I are so located on the base BC that $BD/BC = x_{CP}$ and $IC/BC = x_{BP}$.   Then DI/BC will equal x_{AP}.   Through points D and I, lines DE and HI are drawn parallel to the respective sides BA and AC, thus defining point P.   Through point P, line FG is drawn parallel to the base BC.   From similar triangles,

$$x_{AP} = \frac{DI}{BC} = \frac{BF}{BA} = \frac{GC}{AC}$$

$$x_{BP} = \frac{IC}{BC} = \frac{HA}{BA} = \frac{EG}{AC}$$

$$x_{CP} = \frac{BD}{BC} = \frac{FH}{BA} = \frac{AE}{AC}$$

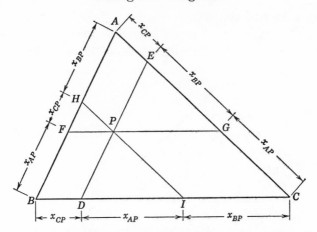

FIG. 7.   Compositions on a triangular diagram of unequal scales

If each side of the triangle is provided with a linear scale ranging from 0 to 1.0,

$$x_{AP} = DI = BF = GC$$

$$x_{BP} = IC = HA = EG$$

$$x_{CP} = BD = FH = AE$$

Where all sides of the triangle are of unequal length, as in Fig. 7, the scale units will be of different length on each of the three sides. Only for a triangle of the equilateral type will the length of the scale units on the three sides be equal. Also, for equilateral triangles, it is possible to express the weight fractions of the three components corresponding to any point P as the perpendicular distances from the point to the three sides of the triangle, if the perpendicular distance from any apex to the opposite side is taken as unity. Only for equilateral triangles will the ratio of the weight fractions to one another be the same as the ratio of the respective perpendicular distances to the sides of the triangle.

Figure 8 shows a triangular composition diagram with scales for weight fraction of B and C on base line BC and a scale for the weight fraction of A on side BA. To facilitate reading the composition of any point within the diagram by projection to these scales, an equally spaced network of three sets of lines is ruled in, parallel, respectively, to each of the three sides of the triangle. The composition corresponding to point Q can then be read directly from Fig. 8 as $x_{AQ} = 0.3$, $x_{BQ} = 0.2$, $x_{CQ} = 0.5$.

The use of a *right triangle*, not necessarily of the isosceles type, with only two scales and two sets of coordinate lines has much to recommend it. No special graph paper is needed, there is no confusion in projecting to the proper composition scale, and scales may be changed

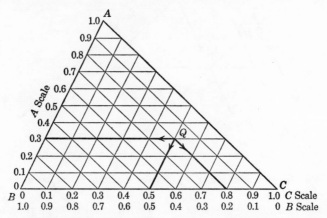

Fig. 8. Network of composition lines on a triangular diagram of unequal scales

without the need for ruling a new triangle to secure the correct projection lines. A fact not fully appreciated is that, when standard equilateral triangular paper is used, the adoption of scales that are not equal invalidates one of the three sets of projection lines.

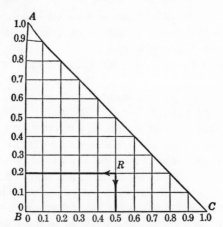

Fig. 9.   Use of right-triangular diagram

In the right triangle of Fig. 9, point R indicates a ternary system containing 0.5 weight fraction of C and 0.2 weight fraction of A. The concentration of B is not read from any scale but is obtained by difference and is 0.3 weight fraction in this instance.

A principle of interest is indicated by means of Fig. 10. In systems having composition points that fall on a straight line passing through an apex corresponding to one pure component, the other two components are present in fixed ratio. For example, in Fig. 10, consider the points P, Q, and R, all of which lie on the same straight line drawn

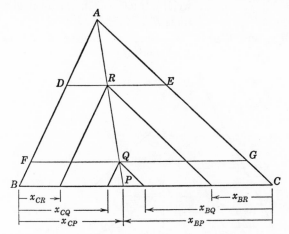

FIG. 10.   Fixed ratio of two components of a ternary system

through the apex A. Because the lines DRE, FQG, and BPC are parallel, and from the principle of similar triangles,

$$\frac{DR}{RE} = \frac{FQ}{QG} = \frac{BP}{PC}$$

Therefore, with reference to Fig. 10,

$$\frac{x_{CR}}{x_{BR}} = \frac{x_{CQ}}{x_{BQ}} = \frac{x_{CP}}{x_{BP}}$$

While the absolute values of x_B and x_C change, their ratio is fixed in all three systems.

Mass Relationships in Triangular Diagrams. Assume that two ternary systems R and S, each composed of components A, B, and C, are mixed to form a new ternary system M. If the composition points for the two original systems R and S are located on a triangular diagram and a straight line is drawn between them, the composition point of the new system M will lie on this straight line, as shown in Fig. 11. Furthermore, the relative masses of the two systems that are mixed together are given by opposite line segments, thus,

$$\frac{W_S}{W_R} = \frac{RM}{MS}$$

These relations may be proved as follows. Let

x_{AR}, x_{AS}, and x_{AM} = the concentrations of component A, as weight fractions, in the respective systems R, S, and M

x_{CR}, x_{CS}, and x_{CM} = the concentrations of component C, as weight fractions, in the respective systems R, S, and M

W_R, W_S, and W_M = the respective masses of the systems R, S, and M

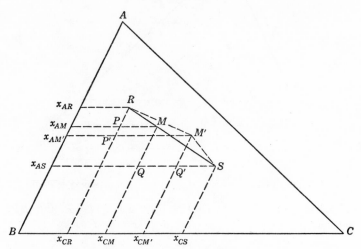

Fig. 11. Composition relations in mixing two ternary systems

Over-all Material Balance

$$W_M = W_R + W_S \qquad (1)$$

Material Balance for Component A

$$W_M x_{AM} = W_R x_{AR} + W_S x_{AS} \qquad (2)$$

Material Balance for Component C

$$W_M x_{CM} = W_R x_{CR} + W_S x_{CS} \qquad (3)$$

By substituting for W_M in equation 2, the following is obtained:

$$\frac{W_S}{W_R} = \frac{x_{AR} - x_{AM}}{x_{AM} - x_{AS}} \qquad (4)$$

Substituting for W_M in equation 3 gives

$$\frac{W_S}{W_R} = \frac{x_{CM} - x_{CR}}{x_{CS} - x_{CM}} \qquad (5)$$

Equating 4 and 5 yields

$$\frac{x_{AR} - x_{AM}}{x_{AM} - x_{AS}} = \frac{x_{CM} - x_{CR}}{x_{CS} - x_{CM}} \tag{6}$$

It should be noted that the foregoing equation was derived solely from material balances, independent of any reference to the triangular diagram. That being the case, point M must be so located in the diagram that equation 6 is satisfied.

If point M falls on a straight line connecting R and S, the triangles RPM and MQS are similar and equation 6 is satisfied. On the other hand, if the point representing the composition of the mixture is at some point not on the line RS, such as at M', the triangles $RP'M'$ and $M'Q'S$ are not similar and equation 6 is not satisfied. It is therefore concluded that the composition point for the system M must fall on the straight line that connects R and S.

From the geometry of the diagram,

$$\frac{x_{AR} - x_{AM}}{x_{AM} - x_{AS}} = \frac{RM}{MS} = \frac{x_{CM} - x_{CR}}{x_{CS} - x_{CM}} \tag{7}$$

Then, substituting in either equation 4 or 5 gives

$$\frac{W_S}{W_R} = \frac{RM}{MS} \tag{8}$$

The above derivation was made with reference to a triangle having three unequal sides. It is therefore evident that the line-segment principle is valid, regardless of the shape of the triangle.

Although the foregoing discussion was based on a process involving the mixing of two systems to form a third one, it should be apparent that the discussion applies equally well to the reverse, where an original system of composition M is separated into two systems of composition R and S.

If three ternary systems are mixed to form a fourth, the relative weights can be readily solved by an extension of the principles previously developed. In Fig. 12, points A', B', and C' represent the compositions of three specific ternary systems containing components A, B, and C. It is required that a system of composition R be made by mixing systems A', B', and C' in suitable proportions. Points A', B', and C' are connected with straight lines, forming the internal triangle $A'B'C'$. The general principles previously developed may be applied to this internal triangle. Through point R, lines DE, FG, and HI are drawn parallel, respectively, to the sides of the inner triangle. The respective weight fractions $x_{A'R}$, $x_{B'R}$, and $x_{C'R}$ of A', B', and C'

that must be taken to form the mixture of composition R are indicated by the following equations:

$$x_{A'R} = \frac{B'F}{B'A'} = \frac{GC'}{A'C'} = \frac{DI}{B'C'}$$

$$x_{B'R} = \frac{HA'}{B'A'} = \frac{EG}{A'C'} = \frac{IC'}{B'C'}$$

$$x_{C'R} = \frac{FH}{B'A'} = \frac{A'E}{A'C'} = \frac{B'D}{B'C'}$$

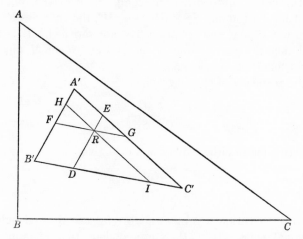

FIG. 12. Composition relations in mixing three ternary systems

It should be evident that, if A', B', and C' are the only materials available, and if operations are restricted to simple mixing, it is impossible to form a ternary system having a composition falling outside the triangle $A'B'C'$.

The entire foregoing discussion was based on the assumption that compositions were expressed in weight fraction. If the composition scales are graduated uniformly in mole fraction rather than in weight fraction, the opposite line-segment principle must be modified by replacing weights with moles. For example, if the composition scales of Fig. 11 were graduated in mole fractions,

$$\frac{RM}{MS} \text{ would equal } \frac{n_S}{n_R}$$

where n_S and n_R equal the respective number of moles of S and R.

Conversion of Units

Conversion of Numerical Values. One of the common problems encountered in engineering calculations is the conversion of numerical values based on one set of units into numerical values based on a different set of units. The recommended procedure is demonstrated in the following illustration.

Illustration 7. Stainless steel, type 304 (18% Cr, 8% Ni, 0.08% C max.), has a thermal conductivity k of 16.2 $\dfrac{\text{Btu}}{(\text{hr})(\text{ft})(\text{F}°)}$. Convert this value of thermal conductivity into $\dfrac{\text{g-cal}}{(\text{sec})(\text{cm})(\text{C}°)}$.

The following conversion factors are required:

$$1 \text{ Btu} = 252 \text{ g-cal} \qquad 1 \text{ ft} = 30.48 \text{ cm}$$

$$1 \text{ hr} = 3600 \text{ sec} \qquad 1 \text{ F}° = (5/9) \text{ C}°$$

These are substituted in the expression for k, thus:

$$k = 16.2 \frac{\text{Btu}}{(\text{hr})(\text{ft})(\text{F}°)} = (16.2) \frac{(252)(\text{g-cal})}{(3600)(\text{sec})(30.48)(\text{cm})(5/9)(\text{C}°)}$$

$$= 0.0670 \frac{\text{g-cal}}{(\text{sec})(\text{cm})(\text{C}°)}$$

Conversion of Equations. Another problem frequently encountered in engineering work is the modification of an equation if the units for one or more of the members of the equation are altered. The procedures that may be employed are shown in the following illustration.

Illustration 8. The equation for the heat transfer to or from a stream of gas flowing in turbulent motion is as follows:[3]

$$h = \alpha \frac{c_p G^{0.8}}{D^{0.2}} = 16.6 \frac{c_p G^{0.8}}{D^{0.2}}$$

c_p = heat capacity, as $\dfrac{\text{Btu}}{(\text{lb})(\text{F}°)}$

D = internal diameter of pipe, as in.

G = mass velocity, as $\dfrac{\text{lb}}{(\text{sec})(\text{ft}^2)}$

h = heat-transfer coefficient, as $\dfrac{\text{Btu}}{(\text{hr})(\text{ft})^2(\text{F}°)}$

[3] John H. Perry, *Chemical Engineers' Handbook*, 3d ed., McGraw-Hill Book Co., p. 467 (1950).

It is desired to transform the equation into a new form:

$$h' = \alpha' \frac{c'_p (G')^{0.8}}{(D')^{0.2}}$$

c'_p = heat capacity, as $\dfrac{\text{kcal}}{(\text{kg})(\text{C}°)}$

D' = internal diameter of pipe, as　cm

G' = mass velocity, as $\dfrac{(\text{sec})(\text{m})^2}{\text{kg}}$

h' = heat-transfer coefficient, as $\dfrac{\text{kcal}}{(\text{hr})(\text{m})^2(\text{C}°)}$

The procedure recommended is to treat each symbol as if it were a numerical value, and to convert the units according to the procedure followed in illustration 7.　The conversion factors needed are as follows:

$$1 \text{ kcal} = 3.966 \text{ Btu} \qquad 1 \text{ meter} = 3.281 \text{ ft}$$

$$1 \text{ cm} \ \ = 0.3937 \text{ in.} \qquad 1 \text{ kg} \ \ \ = 2.205 \text{ lb}$$

$$1 \text{ C}° \ \ = (9/5) \text{ F}°$$

Heat capacity　　$= c_p \dfrac{\text{Btu}}{(\text{lb})(\text{F}°)} = c'_p \dfrac{\text{kcal}}{(\text{kg})(\text{C}°)} = c'_p \dfrac{(3.966)(\text{Btu})}{(2.205)(\text{lb})(9/5)(\text{F}°)}$

$$c_p = \frac{3.966}{(2.205)(9/5)} c'_p = c'_p$$

Mass velocity　　$= G \dfrac{\text{lb}}{(\text{sec})(\text{ft})^2} = G' \dfrac{\text{kg}}{(\text{sec})(\text{m})^2} = G' \dfrac{2.205 \text{ lb}}{(\text{sec})(3.281)^2 (\text{ft})^2}$

$$G = \frac{2.205}{(3.281)^2} G'$$

$$G^{0.8} = \frac{(2.205)^{0.8}}{(3.281)^{1.6}} (G')^{0.8} = 0.2813 (G')^{0.8}$$

Internal diameter $= D(\text{in.}) = D'(\text{cm}) = D'(0.3937)(\text{in.})$

$$D = 0.3937 D'$$

$$D^{0.2} = (0.3937)^{0.2} (D')^{0.2} = 0.8299 (D')^{0.2}$$

Heat-transfer coefficient　$= h \dfrac{\text{Btu}}{(\text{hr})(\text{ft})^2(\text{F}°)} = h' \dfrac{\text{kcal}}{(\text{hr})(\text{m})^2(\text{C}°)}$

$$= h' \frac{3.966 \text{ Btu}}{(\text{hr})(3.281)^2(\text{ft})^2(9/5)(\text{F}°)}$$

$$h = \frac{3.966}{(3.281)^2(9/5)} h' = 0.2047 h'$$

Substituting in the original equation gives

$$0.2047h' = 16.6 \frac{(c'_p)(0.2813)(G')^{0.8}}{(0.8299)(D')^{0.2}}$$

$$h' = 16.6 \frac{0.2813}{(0.2047)(0.8299)} \frac{c'_p(G')^{0.8}}{(D')^{0.2}}$$

$$= 27.5 \frac{c'_p(G')^{0.8}}{(D')^{0.2}}$$

An alternate procedure is to deduce the effective units of the numerical constant, and then convert this to a new numerical constant in the desired net units, employing the procedures demonstrated in illustration 7. The net units are deduced by solving the equation for α and then writing down the units pertaining to the variables on the other side of the equation. It proves possible to cancel and combine certain of the units.

$$\alpha = 16.6 \frac{(\text{in.})^{0.2}(\text{lb})^{0.2}(\text{sec})^{0.8}}{(\text{ft})^{0.4}(\text{hr})}$$

The net units for α' when heat capacity, internal diameter, mass velocity, and heat-transfer coefficient are expressed in the new set of units are $\dfrac{(\text{cm})^{0.2}(\text{kg})^{0.2}(\text{sec})^{0.8}}{(\text{m})^{0.4}(\text{hr})}$.
The appropriate conversion factors, when inserted into the above equation, yield the following:

$$\alpha' = 16.6 \frac{(2.540)^{0.2}(\text{cm})^{0.2}(0.4536)^{0.2}(\text{kg})^{0.2}(\text{sec})^{0.8}}{(0.3048)^{0.4}(\text{m})^{0.4}(\text{hr})}$$

$$= 16.6 \times 1.655 \frac{(\text{cm})^{0.2}(\text{kg})^{0.2}(\text{sec})^{0.8}}{(\text{m})^{0.4}(\text{hr})}$$

$$= 27.5 \frac{(\text{cm})^{0.2}(\text{kg})^{0.2}(\text{sec})^{0.8}}{(\text{m})^{0.4}(\text{hr})}$$

Illustration 9. The heat-capacity equation for carbon dioxide is as follows:

$$c_p = 6.339 + 10.14 \times 10^{-3}T_K - 3.415 \times 10^{-6}T^2_K$$

where c_p is the heat capacity as $\dfrac{\text{g-cal}}{(\text{g-mole})(\text{K}°)}$ and T_K is the temperature in degrees Kelvin. It is desired to transform the equation into a new form, wherein c'_p is the heat capacity as $\dfrac{\text{Btu}}{(\text{lb})(\text{F}°)}$ and temperature t_f is on the Fahrenheit scale. The essential conversion factors and relations are as follows:

Molecular weight $CO_2 = 44.01$

$$1 \text{ lb } CO_2 = (1/44.01) \text{ lb-mole } CO_2 = (453.6/44.01) \text{ g-mole } CO_2$$

$$1 \text{ Btu} = 252 \text{ g-cal}; \qquad 1 \text{ C}° = 1 \text{ K}°; \qquad 1 \text{ F}° = (\tfrac{5}{9}) \text{ C}°$$

$$T_K = (t_c + 273.16) = [(t_f - 32)(\tfrac{5}{9}) + 273.16] = \frac{t_f + 459.69}{1.8}$$

$$\text{Heat capacity} = c_p \frac{\text{g-cal}}{(\text{g-mole})(\text{C}°)} = c'_p \frac{\text{Btu}}{(\text{lb})(\text{F}°)}$$

$$= c'_p \frac{(252)(\text{g-cal})}{(453.6/44.01)(\text{g-mole})(\frac{5}{9})(\text{C}°)}$$

$$c_p = 44.01 c'_p$$

By making substitutions in the original equation, the following is obtained:

$$c'_p = \frac{1}{44.01}\left[6.339 + 10.14 \times 10^{-3} \frac{(t_f + 459.69)}{1.8} - 3.416 \times 10^{-6} \frac{(t_f + 459.69)^2}{1.8^2}\right]$$

$$= 0.19782 + 1.059 \times 10^{-4} t_f - 2.395 \times 10^{-8} t_f{}^2$$

Dimensionless Groups and Constants. Certain fundamental scientific equations include variables collected into groups which are dimensionless provided a set of consistent units is used for each variable of the group. If an equation has all of the variables collected into dimensionless groups, any numerical constants appearing outside of the groups will also be dimensionless and will have fixed numerical values, regardless of the units that may be employed for the individual variables within the different groups. An equation[4] for heat transfer, which is an example of this type, is as follows:

$$\left(\frac{hD}{k}\right) = 0.023 \left(\frac{DG}{\mu}\right)^{0.8} \left(\frac{C_p \mu}{k}\right)^{\frac{1}{3}}$$

TABLE 4

Symbol	Variable	Units	
C_p	Heat capacity	$\dfrac{\text{Btu}}{(\text{lb})(\text{F}°)}$	$\dfrac{\text{g-cal}}{(\text{g})(\text{C}°)}$
D	Internal pipe diameter	ft	cm
G	Mass velocity	$\dfrac{\text{lb}}{(\text{sec})(\text{ft})^2}$	$\dfrac{\text{g}}{(\text{sec})(\text{cm})^2}$
h	Heat-transfer coefficient	$\dfrac{\text{Btu}}{(\text{sec})(\text{ft})^2(\text{F}°)}$	$\dfrac{\text{g-cal}}{(\text{sec})(\text{cm})^2(\text{C}°)}$
k	Thermal conductivity	$\dfrac{\text{Btu}}{(\text{sec})(\text{ft})(\text{F}°)}$	$\dfrac{\text{g-cal}}{(\text{sec})(\text{cm})(\text{C}°)}$
μ	Viscosity	$\dfrac{\text{lb}}{(\text{ft})(\text{sec})}$	$\dfrac{\text{g}}{(\text{cm})(\text{sec})}$

Each of the three groups of variables enclosed in parentheses is dimensionless, provided a proper choice of units for the individual members

[4] A. P. Colburn, *Trans. Am. Inst. Chem. Engrs.*, **29**, 174 (1933).

of the group is made. To illustrate this fact, the Reynolds group DG/μ will be considered, with respect to the two sets of units indicated in Table 4.

Employing the first set of units for the Reynolds group gives

$$\frac{(\text{ft})(\text{lb})/(\text{sec})(\text{ft})^2}{(\text{lb})/(\text{ft})(\text{sec})} = 1$$

With the second set of units,

$$\frac{(\text{cm})(\text{g})/(\text{sec})(\text{cm})^2}{(\text{g})/(\text{cm})(\text{sec})} = 1$$

It should be apparent from the foregoing that other units of length, mass, and time could be used, provided they are used consistently within the group.

A similar analysis made of the other two groups of variables of the above equation shows them to be dimensionless.

2

Stoichiometric and
Composition Relationships

In this chapter the mass and gas volumetric relationships in chemical reactions are presented together with the basic units for expressing changes in mass and composition. These relationships are designated as the principles of *stoichiometry*.

Conservation of Mass. A *system* refers to a substance or a group of substances under consideration and a *process* to the changes taking place within that system. Thus, hydrogen, oxygen, and water may constitute a system, and the combustion of hydrogen to form water, the process. A system may be a mass of material contained within a single vessel and completely isolated from the surroundings, it may include the mass of material in this vessel and its association with the surroundings, or it may include all the mass and energy included in a complex chemical process contained in many vessels and connecting lines and in association with the surroundings. In an *isolated system* the boundaries of the system are limited by a mass of material, and its energy content is completely detached from all other matter and energy. Within a given isolated system the mass of the system remains constant, regardless of the changes taking place within the system. This statement is known as the *law of conservation of mass* and is the basis of the so-called *material balance* of a process.

The state of a system is defined by numerous properties that are classified as *extensive* if they are dependent on the mass under consideration and *intensive* if they are independent of mass. For example, volume is an extensive property, whereas density and temperature are intensive properties.

In the system of hydrogen, oxygen, and water undergoing the process of combustion the total mass in the isolated system remains the same. If the reaction takes place in a vessel and hydrogen and oxygen are fed to the vessel and products are withdrawn, then the incoming and outgoing streams must be included as part of the system in applying the law of conservation of mass or in establishing a material balance.

The law of conservation of mass may be extended and applied to the mass of each element in a system. Thus, in the isolated system of hydrogen, oxygen, and water undergoing the process of combustion the mass of hydrogen in its molecular, atomic, and combined forms remains constant. The same is true for oxygen.

In a universal sense the conservation law applies to the total energy content of a closed system and not to its mass. By the emission of radiant energy and by the transmutation of the elements the mass of a system is partly converted into energy. In processes of nuclear fission and condensation the law of conservation of mass becomes invalid, but in ordinary industrial processes it is accepted as rigorous.

Since the word *weight* is entrenched in engineering literature as synonymous with *mass*, the common practice will be followed in frequently referring to weights of material instead of using the more exact term *mass* as a measure of quantity. Weights and masses are numerically equal only at a location where the gravitational constant has the standard value of 32.174 ft per sec^2. The variation in the weight of a given mass over the earth's surface is negligible for ordinary engineering work.

Stoichiometric Relations

Nature of Chemical Compounds. According to generally accepted theory, the chemical elements are composed of submicroscopic particles which are known as atoms. Further, it is postulated that all the atoms of a given element have the same mass, but that the atoms of different elements have characteristically different masses.

Because of the existence of isotopes, it is recognized that the individual atoms of the various elements vary in mass, and that the so-called atomic weight of an element is, in reality, the weighted average of the atomic weights of the isotopes. In nature the various isotopes of a given element are found in the same proportions; hence in computational work it is permissible to use the weighted average atomic weight as though all atoms actually possessed this average atomic weight.

When the atoms of the elements unite to form a particular compound, it is observed that the composition of the compound is fixed and definite rather than variable. For example, when various samples of carefully purified sodium chloride are analyzed, they all are found to contain 60.6% chlorine and 39.4% sodium. Since the sodium chloride is composed of sodium atoms, each of which has the same mass, and of chlorine atoms, each of which also has the same mass (but a mass different from that of the sodium atoms), it is concluded that in the compound sodium chloride the atoms of sodium and chlorine have combined according to some fixed and definite integral ratio.

By making a careful study of the relative weights by which the chemical elements unite to form various compounds, the relative weights of the atoms have been computed. Work of this type occupied the attention of many of the early leaders in chemical research and has continued to the present day. This work has resulted in the familiar table of international atomic weights, which is still subject to periodic revision and refinement. In this table, the numbers, which are known as atomic weights, give the relative masses of the atoms of the various chemical elements, all referred to the arbitrarily assigned value of exactly 16 for the oxygen atom.

A large amount of work has been done to determine the composition of chemical compounds. As a result of this work, the composition of a great variety of chemical compounds can now be expressed by formulas which indicate the elements that comprise the compound and the relative number of the atoms of the various elements present.

It should be pointed out that the formula of the compound as ordinarily written does not necessarily indicate the exact nature of the atomic aggregates that comprise the compound. For example, the formula for water is written as H_2O, which indicates that, when hydrogen and oxygen unite to form water, the union of the atoms is in the ratio of 2 atoms of hydrogen to 1 atom of oxygen. If this compound exists as steam, there are two atoms of hydrogen permanently united to one atom of oxygen, forming a simple aggregate termed a molecule. Each molecule is in a state of random motion and has no permanent association with other similar molecules to form aggregates of larger size.

However, when this same substance is condensed to the liquid state, there is good evidence to indicate that the individual molecules become associated to form aggregates of larger size, $(H_2O)_x$, where x is an integral number. With respect to solid substances, it may be said that the formula as written merely indicates the relative number of atoms present in the compound and has no further significance. For example, the formula for cellulose is written $C_6H_{10}O_5$, but it should not be therefore concluded that individual molecules, each of which contains only 6 atoms of carbon, 10 atoms of hydrogen, and 5 atoms of oxygen exist. There is much evidence to indicate that aggregates of the nature of $(C_6H_{10}O_5)_x$ are formed where x is a large number.

It is general practice wherever possible to write the formula of a chemical compound to correspond to the number of atoms making up one molecule in the gaseous state. If the degree of association in the gaseous state is unknown, the formula is written to correspond to the lowest possible number of integral atoms that might make up the molecule. However, where the actual size of the molecule is important care must be exercised in determining the degree of association of a compound even in the gaseous state. For example, hydrogen fluoride is commonly designated by the formula HF and at high temperatures and low pressures exists in the gaseous state in molecules, each comprising one atom of fluorine and one atom of hydrogen. However, at high pressures and low temperatures even the gaseous molecules undergo association, and the compound behaves in accordance with the formula $(HF)_x$, with x a function of temperature and pressure.

Mass Relations in Chemical Reactions. In stoichiometric calculations, the mass relations existing between the reactants and products of a chemical reaction are of primary interest. Such information may be deduced from a correctly written reaction equation, used in conjunction with atomic-weight values selected from a table of atomic weights. As a typical example of the procedures followed, the reaction between iron and steam, resulting in the production of hydrogen and the magnetic oxide of iron, Fe_3O_4, may be considered. The first requisite is a correctly written reaction equation. The formulas of the various reactants are set down on the left side of the equation, and the formulas of the products are set down on the right side, care being taken to indicate correctly the formula of each substance involved in the reaction. Next, the equation must be balanced by inserting before each formula a coefficient such that for each element the total number of atoms in the reactants will exactly equal the total number of atoms present in the products. For the reaction under consideration the following equation may be written:

$$3Fe + 4H_2O \rightarrow Fe_3O_4 + 4H_2$$

The next step is to ascertain the atomic weight of each element involved in the reaction by consulting a table of atomic weights. From these atomic weights the respective molecular weights of the various compounds may be calculated.

Atomic Weights:

Iron	55.84
Hydrogen	1.008
Oxygen	16.00

Molecular Weights:

H_2O	$(2 \times 1.008) + 16.00$	$= 18.02$
Fe_3O_4	$(3 \times 55.84) + (4 \times 16.00)$	$= 231.5$
H_2	(2×1.008)	$= 2.016$

The respective relative weights of the reactants and products may be determined by multiplying the respective atomic or molecular weights by the coefficients that precede the formulas of the reaction equation. These figures may conveniently be inserted directly below the reaction equation, thus:

$$3\text{Fe} \quad + \quad 4\text{H}_2\text{O} \quad \rightarrow \text{Fe}_3\text{O}_4 \quad + \quad 4\text{H}_2$$

$$
\begin{array}{cccc}
(3 \times 55.84) & (4 \times 18.02) & 231.5 & (4 \times 2.016) \\
167.52 & 72.08 & 231.5 & 8.064
\end{array}
$$

Thus, 167.52 parts by weight of iron react with 72.08 parts by weight of steam, to form 231.5 parts by weight of the magnetic oxide of iron and 8.064 parts by weight of hydrogen. By the use of these relative weights it is possible to work out the particular weights desired in a given problem. For example, if it is required to compute the weight of iron and of steam required to produce 100 lb of hydrogen, and the weight of the resulting oxide of iron formed, the procedure would be as follows:

Reactants:

Weight of iron $= 100 \times (167.52/8.064)$ $= 2075$ lb
Weight of steam $= 100 \times (72.08/8.064)$ $=\ \ 894$
Total   2969 lb

Products:

Weight of iron oxide $= 100 \times (231.5/8.064) = 2869$ lb
Weight of hydrogen $=\ \ 100$
Total   2969 lb

Volume Relations in Chemical Reactions. A correctly written reaction equation will indicate not only the relative weights involved in a chemical reaction but also the relative volumes of those reactants and products that are in the gaseous state. The coefficients preceding the molecular formulas of the gaseous reactants and products indicate the relative volumes of the different substances. Thus, for the reaction under consideration, for every 4 volumes of steam, 4 volumes of hydrogen are produced, when both materials are reduced to the same temperature and pressure. This volumetric relation follows from Avogadro's law, which states that equal volumes of gas at the same conditions of temperature and pressure contain the same number of molecules, regardless of the nature of the gas. That being the case, and since 4 molecules of steam produce 4 molecules of hydrogen, it may be concluded that 4 volumes of steam will produce 4 volumes of hydrogen. It cannot be emphasized too strongly that this volumetric relation holds only for ideal gaseous substances, and must not be applied to liquid or to solid substances.

The Gram-Atom and the Pound-Atom.   The numbers appearing in a table of atomic weights give the relative weights of the atoms of the various elements.   It therefore follows that, if masses of different elements are taken in such proportion that they bear the same ratio to one another as do the respective atomic weight numbers, these masses will contain the same number of atoms.   For example, if 35.46 grams of chlorine, which has an atomic weight of 35.46, are taken, and if 55.84 grams of iron, which has an atomic weight of 55.84, are taken, there will be exactly the same number of chlorine atoms as of iron atoms in these respective masses of material.

The mass in grams of a given element that is equal numerically to its atomic weight is termed a *gram-atom*.   Similarly, the mass in pounds of a given element that is numerically equal to its atomic weight is termed a *pound-atom*.   From these definitions, the following equations may be written:

$$\text{G-atoms of an elementary substance} = \frac{\text{mass in grams}}{\text{atomic weight}}$$

$$\text{Grams of an elementary substance} = \text{g-atoms} \times \text{atomic weight}$$

$$\text{Lb-atoms of an elementary substance} = \frac{\text{mass in lb}}{\text{atomic weight}}$$

$$\text{Lb of an elementary substance} = \text{lb-atoms} \times \text{atomic weight}$$

The actual number of atoms in one gram-atom of an elementary substance has been determined by several methods, the average result being 6.024×10^{23}.   This number, known as the Avogadro number, is of considerable theoretical importance.

The Gram-Mole and the Pound-Mole.   It has been pointed out that the formula of a chemical compound indicates the relative numbers and the kinds of atoms that unite to form a compound.   For example, the formula NaCl indicates that sodium and chlorine atoms are present in the compound in a 1:1 ratio.   Since the gram-atom as above defined contains a definite number of atoms, which is the same for all elementary substances, it follows that gram-atoms will unite to form a compound in exactly the same ratio as the atoms themselves, forming what may be termed a gram-mole of the compound.   For the case under consideration, it may be said that one gram-atom of sodium unites with one gram-atom of chlorine to form one gram-mole of sodium chloride.

One gram-mole represents the weight in grams of all the gram-atoms which, in the formation of the compound, combine in the same ratio as the atoms themselves.   Similarly, one pound-mole represents the weight in pounds of all of the pound-atoms which, in the formation of the compound, combine in the same ratio as the atoms themselves. From these definitions, the following equations may be written:

$$\text{G-moles of a substance} = \frac{\text{mass in grams}}{\text{molecular weight}}$$

$$\text{Grams of a substance} = \text{g-moles} \times \text{molecular weight}$$

$$\text{Lb-moles of a substance} = \frac{\text{mass in lb}}{\text{molecular weight}}$$

$$\text{Lb of a substance} = \text{lb-moles} \times \text{molecular weight}$$

The value of these concepts may be demonstrated by consideration of the reaction equation for the production of hydrogen by passing steam over iron. The reaction equation as written indicates that 3 atoms of iron unite with 4 molecules of steam to form 1 molecule of magnetic oxide of iron and 4 molecules of hydrogen. It may also be interpreted as saying that 3 g-atoms of iron unite with 4 g-moles of steam, to form 1 g-mole of Fe_3O_4 and 4 g-moles of H_2. In other words, the coefficients preceding the chemical symbols represent not only the relative number of molecules (and atoms for elementary substances that are not in the gaseous state) but also the relative number of gram-moles (and of gram-atoms for elementary substances not in the gaseous state).

Relation between Mass and Volume for Gaseous Substances. Laboratory measurements have shown that, *for all substances in the ideal gaseous state, 1.0 g-mole of material at standard conditions (0° C, 760 mm Hg) occupies 22.414 liters.*[1] Likewise, *if 1.0 lb-mole of the gaseous material is at standard conditions, it will occupy a volume of 359.0 cu ft.*

Accordingly, with respect to the reaction equation previously discussed, it may be said that 167.52 grams of iron (3 g-atoms) will form 4 g-moles of hydrogen, which will, when brought to standard conditions, occupy a volume of 4 × 22.4 liters, or 89.6 liters. Or, if English units are to be used, it may be said that 167.52 lb of iron (3 lb-atoms) will form 4 lb-moles of hydrogen, which will occupy a volume of 4 × 359 cu ft (1436 cu ft) at standard conditions.

Illustration 1. A cylinder contains 25 lb of liquid chlorine. What volume in cubic feet will the chlorine occupy if it is released and brought to standard conditions? *Basis of Calculation:* 25 lb of chlorine.

[1] The actual volume corresponding to 1 g-mole of gas at standard conditions will show some variation from gas to gas owing to various degrees of departure from ideal behavior.

Liquid chlorine, when vaporized, forms a gas composed of diatomic molecules, Cl_2.

Molecular weight of chlorine gas $= (2 \times 35.46)$ $= 70.92$
Lb-moles of chlorine gas $= (25/70.92)$ $= 0.3525$
Volume at standard conditions $= (0.3525 \times 359) = 126.7$ cu ft

Illustration 2. Gaseous propane, C_3H_8, is to be liquefied for storage in steel cylinders. How many grams of liquid propane will be formed by the liquefaction of 500 liters of the gas, the volume being measured at standard conditions?
Basis of Calculation: 500 liters of propane at standard conditions.

Molecular weight of propane $= 44.06$
Gram-moles of propane $= (500/22.4)$ $= 22.32$
Weight of propane $= 22.32 \times 44.06 = 985$ grams

The Use of Molal Units in Computations. The desirability of using molal units for the expression of quantities of chemical compounds cannot be overemphasized. Since one molal unit of one compound will always react with a simple multiple number of molal units of another, calculations of weight relationships in chemical reactions are greatly simplified if the quantities of the reacting compounds and products are expressed throughout in molal units. This simplification is not important in very simple calculations, centered about a single compound or element. Such problems are readily solved by means of the combining weight ratios, which are commonly used as the desirable method for making such calculations as may arise in quantitative analyses.

However, in an industrial process successive reactions may take place with varying degrees of completion, and it may be desired to calculate the weight relationships of *all* the materials present at the various stages of the process. In such problems the use of ordinary weight units with combining weight ratios will lead to great confusion and opportunity for arithmetical error. The use of molal units, on the other hand, will give a more direct and simple solution in a form that may be easily verified. It is highly advisable that familiarity with molal units be gained through use of them in *all* calculations of weight relationships in chemical compounds and reactions.

A still more important argument for the use of molal units is that many of the physicochemical properties of materials are expressed by simple laws when these properties are on the basis of a molal unit quantity.

The molal method of computation is shown by the following illustrative problem which deals with the reaction considered earlier in this section, namely, that between iron and steam to form hydrogen and the magnetic oxide of iron.

Illustration 3. (*a*) Calculate the weight of iron and of steam required to produce 100 lb of hydrogen, and the weight of the Fe_3O_4 formed. (*b*) What volume will the hydrogen occupy at standard conditions?

Reaction Equation

$$3Fe + 4H_2O \rightarrow Fe_3O_4 + 4H_2$$

Basis of Calculation: 100 lb of hydrogen.

Molecular and atomic weights:

Fe	55.84
H_2O	18.02
Fe_3O_4	231.5
H_2	2.016

	Hydrogen produced	= 100/2.016	=	49.6	lb-moles
	Iron required	= 49.6 × 3/4	=	37.2	lb-atoms
		37.2 × 55.84	= 2075	lb	
	Steam required	= 49.6 × 4/4	=	49.6	lb-moles
or		49.6 × 18.02	= 894	lb	
	Fe_3O_4 formed	= 49.6 × 1/4	=	12.4	lb-moles
or		12.4 × 231.5	= 2870	lb	
	Total input	= 2075 + 894	= 2969	lb	
	Total output	= 2870 + 100	= 2970	lb	

Volume of hydrogen at standard conditions = 49.6 × 359 = 17,820 cu ft

In this simple problem the full value of the molal method of calculation is not apparent; as a matter of fact, the method seems somewhat more involved than the solution which was presented earlier in this section, and which was based on the simple rules of ratio and proportion. It is in the more complex problems pertaining to industrial operations that the full benefits of the molal method of calculation are realized.

Excess Reactants. In most chemical reactions carried out in industry, the quantities of reactants supplied usually are not in the exact proportions demanded by the reaction equation. It is generally desirable that some of the reacting materials be present in excess of the amounts theoretically required for combination with the others. Under such conditions the products obtained will contain some of the uncombined reactants. The quantities of the desired compounds which are formed in the reaction will be determined by the quantity of the *limiting reactant*, that is, the material that is not present in excess of that required to combine with any of the other reacting materials. The amount by which any reactant is present in excess of that required to combine with the limiting reactant is usually expressed as its *percentage excess*. The percentage excess of any reactant is defined as the percentage ratio of the excess to the amount theoretically required by the stoichiometric equation for combination with the limiting reactant.

The definition of the amount of reactant theoretically required is sometimes arbitrarily established to comply with particular requirements. For example, in combustion calculations, the fuel itself may contain some oxygen, and the normal procedure then is to give a figure for percentage of excess oxygen supplied by the air which is based on the *net* oxygen demand, which is the total oxygen demanded for complete oxidation of the combustible components minus the oxygen in the fuel.

Degree of Completion. Even though certain of the reacting materials may be present in excess, many industrial reactions do not proceed to the extent that would result from the complete reaction of the limiting material. Such partial completion may result from the establishment of an equilibrium in the reacting mass or from insufficient time or opportunity for completion to the theoretically possible equilibrium. The *degree of completion* of a reaction is ordinarily expressed as the percentage of the limiting reacting material which is converted or decomposed into other products. In processes in which two or more successive reactions of the same materials take place, the degree of completion of each step may be separately expressed.

Where excess reactants are present and the degree of completion is 100%, the material leaving the process will contain not only the direct products of the chemical reaction but also the excess reactants. Where the degree of completion is below 100%, the material leaving the process will contain some of each of the reactants, as well as the direct products of the chemical reactions that took place.

Basis of Calculation

Normally, all the calculations connected with a given problem are presented with respect to some specific quantity of one of the streams of material entering or leaving the process. This quantity of material is designated as the basis of calculation, and should always be specifically stated as the initial step in presenting the solution to the problem. Frequently the statement of the problem makes the choice of a basis of calculation quite obvious. For example, in illustration 3, the weights of iron, steam, and magnetic oxide of iron involved in the production of 100 lb of hydrogen are to be computed. The simplest procedure obviously is to choose 100 lb of hydrogen as the basis of calculation, rather than to select some other basis, such as 100 lb of iron oxide, for example, and finally convert all the weights thus computed to the basis of 100 lb of hydrogen produced.

In some instances, considerable simplification results if 100 units of one of the streams of material that enter or leave the process is selected as the basis of computation, even though the final result desired may be with reference to some other quantity of material. If the compositions are given in weight per cent, 100 lb or 100 grams of one of the entering or leaving streams of material may be chosen as the basis of calculations, and, at the close of the solution, the values that were computed with respect to this basis can be converted to any other basis that the statement of the problem may demand.

For example, if it were required to compute the weight of CaO, MgO, and CO_2 that can be obtained from the calcination of 2500 lb of limestone containing 90% $CaCO_3$, 5% $MgCO_3$, 3% inerts, and 2% H_2O, one procedure would be to select 2500 lb of the limestone as the basis of calculation, and, if this choice is made, the final figures will represent the desired result. An alternative procedure is to select 100 lb of limestone as the basis of calculation, and then, at the close of the computation, convert the weights computed to the desired basis of 2500 lb of limestone. In this simple illustration, there is little choice between the two procedures, but in complex problems, where several streams of material are involved in the process and where several of the streams are composed of several components, simplification will result if the second procedure is adopted. It should be added that, if the compositions are given in mole per cent, it will prove advantageous to choose 100 lb-moles or 100 g-moles as the basis of calculation.

In presenting the solutions to the short illustrative problems of this chapter, it may have appeared superfluous to make a definite statement concerning the basis of calculation. However, since such a statement is of extreme importance in working out complex problems, it is considered desirable to follow the rule of *always* stating the basis of calculation at the beginning of the solution, even though the problem may be relatively simple.

Methods of Expressing the Composition
of Mixtures and Solutions

Various methods are employed for expressing the composition of mixtures and solutions. The different methods that are in common use may be illustrated by considering a binary system, composed of components which will be designated as A and B. The following symbols will be used in this discussion:

w = total weight of the system

w_A and w_B = the respective weights of components A and B

M_A and M_B = the respective molecular weights of components A and B if they are compounds

A_A and A_B = the respective atomic weights of components A and B if they are elementary substances

V = volume of the system at a particular temperature and pressure

V_A and V_B = the respective pure-component volumes of components A and B. (The pure-component volume is defined as the volume occupied by a particular component if it is separated from the mixture, and brought to the same temperature and pressure as the original mixture.)

Weight Per Cent. The weight percentage of each component is found by dividing its respective weight by the total weight of th system, and then multiplying by 100:

$$\text{Weight \% of } A = \frac{w_A}{w} \times 100 \tag{1}$$

This method of expressing compositions is commonly employed for solid systems and also for liquid systems. It is not used ordinarily for gaseous systems. Percentage figures applying to a solid or to a liquid system may be assumed to be in weight per cent, if there is no definite specification to the contrary. One advantage of expressing composition on the basis of weight per cent is that the composition values do not change if the temperature of the system is varied (assuming there is no loss of material through volatilization or crystallization, and that no chemical reactions occur). The summation of all the weight percentages for a given system, of necessity, totals exactly 100.

Volumetric Per Cent. The per cent by volume of each component is found by dividing its pure-component volume by the total volume of the system, and then multiplying by 100.

$$\text{Volumetric \% of } A = \left(\frac{V_A}{V}\right) \times 100 \tag{2}$$

This method of expressing compositions is almost always used for gases at low pressures, occasionally for liquids (particularly for the ethyl alcohol-water system), but very seldom for solids.

The analysis of gases is carried out at room temperature and atmospheric pressure. Under these conditions, the behavior of the mixture and of the individual gaseous components is nearly ideal, and the sum of the pure-component volumes will equal the total volume. That is, $V_A + V_B + \cdots = V$. This being the case, the percentages total exactly 100. Furthermore, since changes of temperature produce the same relative changes in the respective partial volumes as in the total volume, the volumetric composition of the gas is unaltered by changes in temperature. Compositions of gases are so commonly given on the basis of volumetric percentages that, if percentage figures are given with no specification to the contrary, it may be assumed that they are per cent by volume.

With liquid solutions, it is common to observe that when the pure components are mixed a shrinkage or expansion occurs. In other words, the sum of the pure-component volumes does not equal total volume. In such instances, the percentages will not be equal to exactly 100. Furthermore, the expansion characteristics of the pure components usually are not the same, and are usually different from those of the mixture. This being the case, the volumetric composition of a liquid solution will change with the temperature. Accordingly, a figure for volumetric per cent as applied to a liquid solution should be accompanied by a statement concerning the temperature. For the alcohol-water system, the volumetric percentages are normally given with respect to a temperature of 60° F. If the actual determination is made at a temperature other than 60° F, a suitable correction is applied.

Mole Fraction and Mole Per Cent. If the components A and B are compounds, the system is a mixture of two kinds of molecules. The total number of A molecules or moles present divided by the sum of the A and the B molecules or moles represents the mole fraction of A in the system. By multiplying the mole fraction by 100, the mole per cent of A in the system is obtained. Thus,

$$\text{Mole fraction of } A = \frac{w_A/M_A}{w_A/M_A + w_B/M_B} \tag{3}$$

$$\text{Mole } \% \text{ of } A = \text{mole fraction} \times 100 \tag{4}$$

The summation of all the mole percentages for a given system totals exactly 100. The composition of a system expressed in mole per cent will not vary with the temperature, assuming there is no loss of material from the system, and that no chemical reactions or associations occur.

Illustration 4. An aqueous solution contains 40% Na_2CO_3 by weight. Express the composition in mole per cent.

Basis of Calculation: 100 grams of solution.

Molecular Weights:

$Na_2CO_3 = 106.0$, $H_2O = 18.02$

Na_2CO_3 present $= 40$ g, or $40/106$ $= 0.377$ g-moles
H_2O present $= 60$ g, or $60/18.02 = \underline{3.33}$
    Total $\overline{3.71}$ g-moles

Mole $\%$ $Na_2CO_3 = (0.377/3.71) \times 100 = 10.16$
Mole $\%$ H_2O $= (3.33/3.71) \times 100 = \underline{89.8}$
                                          $\overline{100.0}$

Illustration 5. A solution of naphthalene, $C_{10}H_8$, in benzene, C_6H_6, contains 25 mole per cent of naphthalene. Express the composition of the solution in weight per cent.

Basis of Calculation: 100 g-moles of solution.

Molecular Weights:

$C_{10}H_8 = 128.1$, $C_6H_6 = 78.1$

$C_{10}H_8$ present = 25 g-moles, or $25 \times 128.1 = 3200$ grams
C_6H_6 present = 75 g-moles, or $75 \times 78.1 = 5860$
         Total $\overline{9060}$ grams

Weight % of $C_{10}H_8 = (3200/9060) \times 100 = 35.3$
Weight % of $C_6H_6 = (5860/9060) \times 100 = \underline{64.7}$
                                              100.0

For ideal gases, the composition in mole per cent is exactly the same as the composition in volumetric per cent. This deduction follows from a consideration of Avogadro's law. It should be emphasized that this relation holds only for gases, and does not apply to liquid or to solid systems.

Illustration 6. A natural gas has the following composition, all figures being in volumetric per cent:

Methane, CH_4	83.5%
Ethane, C_2H_6	12.5
Nitrogen, N_2	4.0
	100.0%

Calculate:

(a) Composition in mole per cent.
(b) Composition in weight per cent.
(c) Average molecular weight.
(d) Density at standard conditions, as pounds per cubic foot.

(a) It has been pointed out that, for gaseous substances, the composition in mole per cent is identical with the composition in volumetric per cent. Accordingly, the above figures give the respective mole per cents directly.

(b) Calculation of composition in weight per cent:

Basis of Calculation: 100 lb-moles of gas.

	Lb-moles	Molecular Weight	Weight, lb	Weight %
CH_4	83.5	16.03	$83.5 \times 16.03 = 1339$	$(1339/1827) \times 100 = 73.3$
C_2H_6	12.5	30.05	$12.5 \times 30.05 = 376$	$(376/1827) \times 100 = 20.6$
N_2	4.0	28.02	$4.0 \times 28.02 = 112$	$(112/1827) \times 100 = 6.1$
	100.0		1827	100.0

(c) The molecular weight of a gas is numerically the same as the weight in pounds of one pound-mole. Therefore, the molecular weight equals 1827/100, or 18.27.

(d) Density at standard conditions, as pounds per cubic foot:

Volume at standard conditions = 100 × 359 = 35,900 cu ft

Density at standard conditions = 1827/35,900 = 0.0509 lb per cu ft

Atomic Fraction and Atomic Per Cent. The general significance of these terms is the same as for mole fraction and mole per cent, except that the atom is the unit under consideration rather than the molecule. Thus,

$$\text{Atomic fraction of } A = \frac{(w_A/A_A)}{(w_A/A_A) + (w_B/A_B)} \tag{5}$$

$$\text{Atomic \% of } A = \text{atomic fraction} \times 100 \tag{6}$$

The summation of all of the atomic percentages for a given system is exactly 100. The composition, expressed in atomic per cent, will not vary with temperature, provided that no loss of material occurs. The composition of a system expressed in atomic per cent will remain the same, regardless of whether or not reactions occur within the system.

Mass of Material per Unit Volume of the System. Various units are employed for mass and for volume. Masses are commonly expressed in grams or pounds and the corresponding gram-moles or pound-moles. For volume, the common units are liters, cubic feet, and U. S. gallons. Some common combinations for expression of compositions are grams per liter, gram-moles per liter, pounds per U. S. gallon, and pound-moles per U. S. gallon.

This general method of indicating compositions finds its widest application in dealing with liquid solutions, both in the laboratory and in plant work. This is primarily due to the ease with which liquid volumes may be measured.

Mass of Material per Unit Mass of Reference Substance. One component of the system may be arbitrarily chosen as a reference material, and the composition of the system indicated by stating the mass of each component associated with unit mass of this reference material. For example, in dealing with binary liquid systems, compositions may be expressed as mass of solute per fixed mass of solvent. Some of the common units employed are: (1) pounds of solute per pound of solvent, (2) pound-moles of solute per pound-mole of solvent, (3) pound-moles of solute per 1000 lb of solvent. The concentration of a solution expressed in the last unit is termed its *molality*.

In dealing with problems involving the drying of solids, the moisture content is frequently indicated as pounds of water per pound of moisture-free material. In dealing with mixtures of condensable vapors and so-called permanent gases, the concentration of the condensable vapor may be indicated as pounds of vapor per pound of vapor-free gas, or as pound-moles of vapor per pound-mole of vapor-free gas.

In all the instances cited, the figure that indicates the composition is, in reality, a dimensionless ratio; hence the metric equivalents have the same numerical value as when the above-specified English units are employed.

For processes involving gain or loss of material, calculations are simplified if the compositions are expressed in this manner. In instances of this kind, the reference component chosen is one that passes through the process unaltered in mass. Compositions expressed in these terms are independent of temperature and pressure.

Illustration 7. A solution of sodium chloride in water contains 230 grams of NaCl per liter at 20° C. The density of the solution at this temperature is 1.148 grams per cubic centimeter. Calculate the following items:

(a) Composition in weight per cent.
(b) Volumetric per cent of water.
(c) Composition in mole per cent.
(d) Composition in atomic per cent.
(e) Molality.
(f) Pounds NaCl per pound H_2O.

Basis of Calculation: 1000 cc of solution.

Total weight $= 1000 \times 1.148 = 1148$ grams

NaCl	$= 230$ g, or $230/58.5$	$=$	3.93 g-moles
H_2O	$= 1148 - 230 = 918$ grams, or $918/18.02$	$= 50.9$	
Total			54.8 g-moles

(a) Composition in weight per cent:
Weight % NaCl $= (230/1148) \times 100 = $ 20.0
Weight % H_2O $= (918/1148) \times 100 = $ 80.0
100.0

(b) Volumetric per cent water:
Density of pure water at 20° C $= 0.998$ grams per cubic centimeter
Volume of pure water $= 918/0.998 = 920$ cc
Volumetric % of water $= (920/1000) \times 100 = 92.0$

(c) Composition in mole per cent:
Mole % NaCl $= (3.93/54.8) \times 100 = $ 7.17
Mole % H_2O $= (50.9/54.8) \times 100 = $ 92.8
100.0

(d) Composition in atomic per cent:
G-atoms of sodium $=$ 3.93
G-atoms of chlorine $=$ 3.93
G-atoms of hydrogen $= 2 \times 50.9 = $ 101.8
G-atoms of oxygen $=$ 50.9
Total 160.6

Atomic % of sodium   =   (3.93/160.6) × 100 =   2.45
Atomic % of chlorine   =   (3.93/160.6) × 100 =   2.45
Atomic % of hydrogen = (101.8/160.6) × 100 =   63.4
Atomic % of oxygen   =   (50.9/160.6) × 100 =   31.7
                                                                  ‾‾‾‾‾
                                                                  100.0

(e) Molality = 3.93 × (1000/918) = 4.28 lb-moles of NaCl per 1000 lb H_2O
(f) Lb NaCl per lb H_2O = 230/918 = 0.251

Density and Specific Gravity

Density is defined as mass per unit volume. Density values are commonly expressed as grams per cubic centimeter or as pounds per cubic foot. The density of water at 4° C is 1.0000 gram per cubic centimeter or 62.43 lb per cu ft.

The specific gravity of a solid or liquid is the ratio of its density to the density of water at some specified reference temperature. The temperatures corresponding to a value of specific gravity are generally symbolized by a fraction, the numerator of which is the temperature of the liquid in question, and the denominator the temperature of the water that serves as the reference. Thus the term sp. gr. 70°/60° F indicates the specific gravity of a liquid at 70° F referred to water at 60° F, or the ratio of the density of the liquid at 70° F to that of water at 60° F. It is apparent that if specific gravities are referred to water at 4° C (39.2° F) they will be numerically equal to densities in grams per cubic centimeter.

The densities of solutions are functions of both concentration and temperature. The relationships between these three properties have been determined for a majority of the common systems. Standard chemical handbooks contain extensive tabulations giving the densities of solutions of varying concentrations at specified temperatures. These data are most conveniently used in graphical form in which density is plotted against concentration. Each curve on such a chart will correspond to a specified, constant temperature. The density of a solution of any concentration at any temperature may be readily estimated by interpolation between these curves. In Fig. 13 are plotted the densities of solutions of sodium chloride at various temperatures.

For a given system of solute and solvent the density or specific gravity at a specified temperature may serve as an index to the concentration. This method is useful only when there is a large difference between the densities of the solutions and of the pure solvent. In several industries specific gravities have become the universally accepted means of indicating concentrations, and products are purchased and sold on the basis of specific gravity specifications. Sulfuric acid, for example, is marketed

almost entirely on this basis.   Specific gravities are also made the basis
for the control of many industrial processes in which solutions are involved.   To meet the needs of such industries, special means of numerically designating specific gravities have been developed.   Several scales
are in use in which specific gravities are expressed in terms of *degrees*

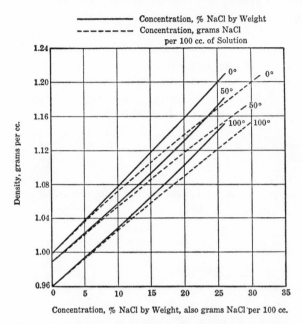

FIG. 13.   Densities of aqueous sodium chloride solutions
Temperatures are in degrees centigrade

which are related to specific gravities and densities by arbitrary mathematical definitions.

Baumé Gravity Scale.   Two so-called Baumé gravity scales are in
common use, one for liquids lighter and the other for liquids heavier
than water.   The former is defined by the following expression:

$$\text{Degrees Baumé} = \frac{140}{G} - 130 \tag{7}$$

where G is the specific gravity at 60°/60° F.   It is apparent from this
definition that a liquid having the density of water at 60° F ($G = 1.0$)
will have a gravity of 10° Bé.   Lighter liquids will have *higher* gravities
on the Baumé scale.   Thus, a material having a specific gravity of 0.60
will have a gravity of 103° Bé.

The Baumé scale for liquids heavier than water is defined as follows:

$$\text{Degrees Baumé} = 145 - \frac{145}{G} \qquad (8)$$

Gravities expressed on this scale increase with increasing density. Thus, a specific gravity of 1.0 at 60°/60° F corresponds to 0.0° Bé, and a specific gravity of 1.80 corresponds to 64.44° Bé. It will be noted that both the Baumé scales compare the densities of liquids at 60° F. In order to determine the Baumé gravity, the specific gravity at 60°/60° F must either be directly measured or estimated from specific-gravity measurements at other conditions. The Baumé gravity of a liquid is thus independent of its temperature. Readings of Baumé hydrometers at temperatures other than 60° F must be corrected for temperature so as to give the value at 60° F.

API Scale. As a result of confusion of standards by instrument manufacturers, a special gravity scale has been adopted by the American Petroleum Institute for expression of the gravities of petroleum products. This scale is similar to the Baumé scale for liquids lighter than water as indicated by the following definition:

$$\text{Degrees API} = \frac{141.5}{G} - 131.5 \qquad (9)$$

As on the Baumé scale, a liquid having a specific gravity of 1.0 at 60°/60° F has a gravity of 10°. However, a liquid having a specific gravity of 0.60 has an API gravity of 104.3 as compared with a Baumé gravity of 103.3. The gravity of a liquid in degrees API is determined by its density at 60° F and is independent of temperature. Readings of API hydrometers at temperatures other than 60° F must be corrected for temperature so as to give the value at 60° F. API gravities are readily converted to specific gravities 60°/60° F by Fig. B in the appendix.

Twaddell Scale. The Twaddell scale is used only for liquids heavier than water. Its definition is as follows:

$$\text{Degrees Twaddell} = 200(G - 1.0) \qquad (10)$$

This scale has the advantage of a simple relationship to specific gravities.

Numerous other scales have been adopted for special industrial uses; for example, the Brix scale measures directly the concentration of sugar solutions.

If the stem of a hydrometer graduated in specific-gravity units is examined, it is observed that the scale divisions are not uniform. The scale becomes compressed and crowded together at the lower end. On the other hand, a Baumé or API hydrometer will have uniform scale graduations over the entire length of the stem.

For gases, water is unsatisfactory as a reference material for expressing specific-gravity values because of its high density in comparison with the density of gas. Therefore, it is conventional to express specific-gravity values with reference to dry air at the same temperature and pressure as those of the gas.

Problems

Tables of Common Atomic Weights aud Conversion Factors will be found in the Appendix.

Although the simple stoichiometric relations included in the following group of problems may easily be solved by the rules of ratio and proportion, it is nevertheless recommended that the molal method of calculation be adhered to as a preparation for the more complex problems to be encountered in succeeding chapters.

In all instances, the basis of calculation should be stated definitely at the start of the solution.

1. $BaCl_2 + Na_2SO_4 = 2NaCl + BaSO_4$.

(a) How many grams of barium chloride will be required to react with 5.0 grams of sodium sulfate? **Ans.** 7.33 grams.

(b) How many grams of barium chloride are required for the precipitation of 5.0 grams of barium sulfate? **Ans.** 4.46 grams.

(c) How many grams of barium chloride are equivalent to 5.0 grams of sodium chloride? **Ans.** 8.91 grams.

(d) How many grams of sodium sulfate are necessary for the precipitation of the barium of 5.0 grams of barium chloride? **Ans.** 3.41 grams.

(e) How many grams of sodium sulfate have been added to barium chloride if 5.0 grams of barium sulfate are precipitated? **Ans.** 3.04 grams.

(f) How many pounds of sodium sulfate are equivalent to 5.0 lb of sodium chloride? **Ans.** 6.08 lb.

(g) How many pounds of barium sulfate are precipitated by 5.0 lb of barium chloride? **Ans.** 5.60 lb.

(h) How many pounds of barium sulfate are precipitated by 5.0 lb of sodium sulfate? **Ans.** 8.22 lb.

(i) How many pounds of barium sulfate are equivalent to 5.0 lb of sodium chloride? **Ans.** 9.98 lb.

2. How many grams of chromic sulfide will be formed from 0.928 gram of chromic oxide according to the equation

$$2Cr_2O_3 + 3CS_2 = 2Cr_2S_3 + 3CO_2$$

3. How much charcoal is required to reduce 1.5 lb of arsenic trioxide?

$$As_2O_3 + 3C = 3CO + 2As$$

Ans. 0.273 lb.

4. Oxygen is prepared according to the equation $2KClO_3 = 2KCl + 3O_2$. What is the yield of oxygen when 9.12 grams of potassium chlorate are decomposed? How many grams of potassium chlorate must be decomposed to liberate 2.5 grams of oxygen?

5. Sulfur dioxide may be produced by the reaction:

$$Cu + 2H_2SO_4 = CuSO_4 + 2H_2O + SO_2$$

(a) How much copper, and (b) how much sulfuric acid (94 weight % H_2SO_4) must be used to obtain 32 lb of sulfur dioxide? **Ans.** (a) 31.8 lb, (b) 104.2 lb.

6. A limestone analyzes $CaCO_3$ 94.52%, $MgCO_3$ 4.16%, and insoluble matter 1.32%.

(a) How many pounds of calcium oxide could be obtained from 4 tons of the limestone?

(b) How many pounds of carbon dioxide are given off per pound of this limestone?

7. How much superphosphate fertilizer can be made from one ton of calcium phosphate, 93.5% pure? The reaction is

$$Ca_3(PO_4)_2 + 2H_2SO_4 = CaH_4(PO_4)_2 + 2CaSO_4$$

Ans. 1.591 tons.

8. How much potassium chlorate must be taken to produce the same amount of oxygen as will be produced by 2.3 grams of mercuric oxide?

9. Ammonium phosphomolybdate, $(NH_4)_3PO_4 \cdot 12MoO_3 \cdot 3H_2O$, is made up of the radicals NH_3, H_2O, P_2O_5, and MoO_3. What is the percentage composition of the molecule with respect to these radicals? **Ans.** 2.64%, 4.20%, 89.48%, 3.68%.

10. How many pounds of salt are required to make 2500 lb of salt cake (Na_2SO_4)? How many pounds of Glauber's salt ($Na_2SO_4 \cdot 10H_2O$) will this amount of salt cake make?

11. In the reactions

$$2KMnO_4 + 8H_2SO_4 + 10FeSO_4 = 5Fe_2(SO_4)_3 + K_2SO_4 + 2MnSO_4 + 8H_2O$$

$$K_2Cr_2O_7 + 7H_2SO_4 + 6FeSO_4 = 3Fe_2(SO_4)_3 + K_2SO_4 + Cr_2(SO_4)_3 + 7H_2O$$

(a) how many grams of potassium dichromate are equivalent to 5.0 grams of potassium permanganate? (b) how many grams of potassium permanganate are equivalent to 3.0 grams of potassium dichromate? **Ans.** (a) 7.76 grams, (b) 1.934 grams.

12. A compound whose molecular weight is 103 analyzes C = 81.5%, H = 4.9%, N = 13.6%. What is its formula?

13. What is the weight of one liter of methane under standard conditions? **Ans.** 0.715 gram.

14. If 45 grams of iron react thus, $Fe + H_2SO_4 = FeSO_4 + H_2$, how many liters of hydrogen are liberated at standard conditions?

15. A solution of sodium chloride in water contains 23.0% NaCl by weight. Express the concentration of this solution in the following terms, using data from Fig. 13:

(a) Gram-moles of NaCl per 1000 grams of water (molality). **Ans.** 5.11 g-moles.

(b) Mole fraction of NaCl. **Ans.** 0.0843 mole fraction.

(c) Gram-moles of NaCl per liter of solution at 30° C. **Ans.** 4.58 g-moles.

(d) Pounds of NaCl per U. S. gallon of solution at 40° C. **Ans.** 2.22 lb.

16. In the following table are listed various aqueous solutions, with the density at 20 and 80° C given, as well as the composition. For one or more of the solutions assigned from the table, report the composition expressed in the following ways ·

(a) Weight per cent.

(b) Mole per cent.

(c) Pounds of solute per pound of solvent.

(d) Pound-moles of solute per pound of solvent.
(e) Grams solute per 100 ml of solution at 80° C.
(f) Grams solute per 100 ml of solution at 20° C.
(g) Gram-moles solute per liter of solution at 20° C.
(h) Pounds of solute per U. S. gallon of solution at 68° F (20° C).
(i) Pound-moles of solute per U. S. gallon of solution at 68° F (20° C).
(j) Molality.
(k) Normality.

Solute	Composition of Solution	Density, g/ml 20° C	80° C
HCl	Weight % HCl = 30	1.149	1.115
H_2SO_4	Mole % H_2SO_4 = 36.8	1.681	1.625
HNO_3	Lb HNO_3/lb H_2O = 1.704	1.382	1.296
NH_4NO_3	Lb-mole NH_4NO_3/lb H_2O = 0.01250	1.226	1.187
$ZnBr_2$	Grams $ZnBr_2$/100 ml of solution at 20° C = 130.0	2.002	1.924
$CdCl_2$	Grams $CdCl_2$/100 ml of solution at 80° C = 68.3	1.575	1.519
$MgCl_2$	G-mole $MgCl_2$/1 of solution at 20° C = 3.99	1.269	1.245
$CaBr_2$	Lb $CaBr_2$/U. S. gal at 68° F (20° C) = 4.03	1.381	1.343
$SrCl_2$	Lb-mole $SrCl_2$/U. S. gal at 68° F (20° C) = 0.02575	1.396	1.358
LiCl	Molality = 10.13	1.179	1.158
KCl	Normality = 2.70	1.118	1.088

17. An aqueous solution of sodium chloride contains 28 grams of NaCl per 100 cc of solution at 20° C. Express the concentration of this solution in the following terms, using data from Fig. 13:

(a) Percentage NaCl by weight. **Ans. 23.7%.**
(b) Mole fraction of NaCl. **Ans. 0.0871.**
(c) Pound-moles of NaCl per 1000 lb of water (molality). **Ans. 5.31.**
(d) Pound-moles of NaCl per U. S. gallon of solution at 0° C. **Ans. 0.0404.**

18. It is desired to prepare a solution of sodium chloride in water, having a molality of 1.80. Calculate the weight of sodium chloride that should be placed in a 1000-cc volumetric flask in order that the desired concentration will be obtained by subsequently filling the flask with water, keeping the temperature of the solution at 30° C.

19. For the operation of a refrigeration plant it is desired to prepare a solution of sodium chloride containing 20% by weight of the anhydrous salt.

(a) Calculate the weight of sodium chloride that should be added to one gallon of water at 30° C in order to prepare this solution. **Ans. 2.070 lb.**

(b) Calculate the volume of solution formed per gallon of water used, keeping the temperature at 30° C. **Ans. 1.090 gal.**

20. (a) A solution has a gravity of 100° Twaddell. Calculate its specific gravity and its gravity in degrees Baumé.

(b) An oil has a specific gravity at 60°/60° F of 0.790. Calculate its gravity in degrees API and degrees Baumé.

3

Behavior of Ideal Gases

In the preceding chapter consideration was given to problems pertaining to the transformation of matter from one physical or chemical state to another. It has been pointed out that matter is essentially indestructible despite all the transformations it may undergo and that the mass of a given system remains unaltered. However, in order to complete the quantitative study of a system it is necessary to consider also its energy content.

Energy. The properties of a moving ball, a swinging pendulum, or a rotating flywheel are different from those of the same objects at rest. The differences lie in the motions of the bodies and in the ability of the moving objects to perform *work*, which is defined as the action of a force moving under restraint through a distance. Likewise, the properties of a red-hot metal bar are different from those of the same metal bar when cold. The red-hot bar produces effects on the eye and the touch very different from those of the cold bar.

Under the classification of *potential energy* are included all forms not associated with motion but resulting from the position and arrangement of matter. The energy possessed by an elevated weight, a compressed spring, a charged storage battery, a tank of gasoline, or a lump of coal is potential energy. Similarly, potential energy is stored within an atom as the result of forces of attraction among its subatomic parts. Thus potential energy can be further classified as *external potential* energy, which is inherent in matter as a result of its position relative to the earth, or as *internal potential energy*, which resides within the structure of matter.

In contrast, energy associated with motion is referred to as *kinetic energy*. The energy represented by the flow of a river, the flight of a bullet, or the rotation of a flywheel is kinetic energy. Also, individual molecules possess kinetic energy by virtue of their translational, rotational, and vibrational motions. Like potential energy, kinetic energy is subclassified as *internal kinetic energy*, such as that associated with

46

molecular and atomic structure, and as *external kinetic energy*, such as that associated with external motion.

In addition to the forms of energy associated with composition, position, or motion of matter, energy exists in the forms of electricity, magnetism, and radiation, which are associated with electronic phenomena.

The science pertaining to the transformation of one form of energy to another is termed *thermodynamics*. Early studies of the transformation of energy lead to the realization that, although energy can be transformed from one form to another, it can never be destroyed and that the total energy of the universe is constant. This principle of the *conservation of energy* is referred to as the *first law of thermodynamics*. Many experimental verifications have served to establish the validity of this law.

Temperature and Heat. Energy may be transferred not only from one form to another but also from one aggregation of matter to another without change of form. The transformation of energy from one form to another or the transfer of energy from one body to another always requires the influence of some *driving force*. As an example, if a hot metal bar is placed in contact with a cold one, the former will be cooled and the latter warmed. The sense of "hotness" is an indication of the internal energy of matter. The driving force which produces a transfer of internal energy is termed *temperature* and that form of energy which is transferred from one body to another as a result of a difference in temperature is termed *heat*.

The Kinetic Theory of Gases. A gas is believed to be composed of molecules, each of which is a material particle and separate from all others. These particles are free to move about in space according to Newton's laws of the motion of material bodies. It is furthermore assumed that each particle behaves as a perfectly elastic sphere. As a consequence of this assumption, there is no change in total kinetic energy when two particles collide or when a particle strikes an obstructing or confining surface. On the basis of these assumptions it is possible to explain many physical phenomena by considering that *each particle of matter is endowed with a certain inherent kinetic energy of translation*. As a result of this energy, the particles will be in constant motion, striking against and rebounding from one another and from obstructing surfaces.

The energy that is represented by the sum of the energies of the component particles of matter is termed the total internal energy. When heat is added to a gas, additional kinetic energy is imparted to its component particles. The average quantity of kinetic energy of *translation* that is possessed by the particles of a gas determines its temperature.

At any specified temperature the particles of a gas possess definitely fixed, average kinetic energies of translation which may be varied only by a change in temperature resulting from the addition or removal of heat. Thus, an increase in temperature signifies an increase in average kinetic energy of translation, which in turn is accompanied by increased speeds of translation of the particles. Conversely, when, by any means, the kinetic energies of translation of the particles of a gas are increased, the temperature is raised.

The theory outlined above accounts for the pressure that is exerted by a gas against the walls of a confining vessel. The translational motion of the particles is assumed to be entirely random, in every direction, and it may be assumed for ordinary cases that the number of particles per unit volume will be constant throughout the space. These assumptions are justified when the number of particles per unit volume is very large. Then, each element of area of confining surface will be subjected to continual bombardment by the particles adjacent to it. Each impact will be accompanied by an elastic rebound and will exert a pressure due to the change of momentum involved. In a pure, undissociated gas all particles may be considered to be of the same size and mass. On the basis of these assumptions the following expression for pressure may be derived from the principles of mechanics;[1]

$$p = \tfrac{2}{3} \frac{\nu}{V} \left(\tfrac{1}{2}mu^2 \right) \tag{1}$$

where ν = number of molecules under consideration
V = volume in which ν molecules are contained
m = mass of each molecule
u = average translational velocity of the molecules

From the definition of the molal units of quantity it was pointed out that one mole of a substance will contain a definite number of single molecules, the same for all substances. Then

$$\nu = nN \tag{2}$$

where n = number of moles in volume V
N = number of molecules in a mole, a universal constant equal to 6.024×10^{23} for the gram-mole

Combining equations 1 and 2 yields

$$pV = n\tfrac{2}{3}N\left(\tfrac{1}{2}mu^2\right) = n\tfrac{2}{3}\mathrm{U}_t \tag{3}$$

[1] This derivation may be found in simplified form in any good physics or physical chemistry text or in more rigorous form in the more advanced books dealing with kinetic theory.

where u_t represents the total translational kinetic energy possessed by one mole of gas.

From extensive experimental investigations the ideal-gas law has been established. The definition of the absolute scale of temperature is based on this relationship:

$$pv = RT \tag{4}$$

or

$$pV = nRT \tag{5}$$

where R = a proportionality factor
T = absolute temperature
v = volume of one mole of gas
n = number of moles of gas
V = volume of n moles of gas

Rearranging equation 4 gives

$$R = \frac{pv}{T} \tag{6}$$

Assuming the validity of the Avogadro principle that equimolal quantities of all gases occupy the same volume at the same conditions of temperature and pressure, it follows from equation 6 that the gas-law factor R is a universal constant. The Avogadro principle and the ideal gas-law have been experimentally shown to approach perfect validity for all gases under conditions of extreme rarefaction, that is, where the number of molecules per unit volume is very small. The constant R may be evaluated from a single measurement of the volume occupied by a known molal quantity of any gas at a known temperature and at a known low pressure.

Combining equations 3 and 5 gives

$$\tfrac{2}{3}u_t = RT \tag{7}$$

or

$$\tfrac{1}{2}mu^2 = \tfrac{3}{2}\frac{R}{N}T \tag{8}$$

Equation 7 states that the average kinetic energy of translation of a molecule in the gaseous state is directly proportional to the absolute temperature. The absolute zero is the temperature at which the kinetic energies of all molecules become zero and molecular motion ceases. From the fact that R, the gas-law constant, and N, the Avogadro number, are universal constants, it follows that equation 7 must apply to all gases. In other words, *the average translational kinetic energy with which a gas molecule is endowed is dependent only on the absolute temperature and is independent of its nature and size.* This conclusion is of far-reaching significance. It follows that a molecule of hydrogen possesses the same average translational kinetic energy as a molecule of bromine at the same temperature. Since the bromine molecule has 80 times the mass of the hydrogen molecule, the latter must move at a correspondingly higher velocity of translation. If the temperature increases, the squares of the velocities of translation of both molecules will be increased in the same proportion.

Extension of the Kinetic Theory. Although the kinetic theory was originally developed to explain the behavior of gases, it has been extended and found to apply with good approximation wherever small particles of matter are permitted to move freely in space. It has been shown that all such particles *may be considered as endowed with the same kinetic energy of translation when at the same temperature, regardless of composition or size.* This principle is believed to apply not only to the molecules of all gases but also to the molecules of all liquids and of substances that are dissolved in liquids. It has been extended still further and shown to apply also to particles of solid matter of considerable size suspended in gases or liquids. Thus, at any selected temperature, a molecule of hydrogen gas, a molecule of iodine vapor, a molecule of liquid water, and a molecule of liquid mercury all are supposed to possess the same translational kinetic energy, indicated by equation 3. Furthermore, this same energy is possessed by a molecule of sulfuric acid in solution in water and by each of the ions formed by the dissociation of such a molecule. A colloidal particle of gold, containing hundreds of atoms, or a speck of dust suspended in air each presumably has the same translational kinetic energy as a molecule of hydrogen gas at the same temperature. The larger particles must therefore exhibit correspondingly slower velocities of translational motion. This generalization is of the greatest importance in the explanation of such phenomena as diffusion, heat conduction, osmotic pressure, and the general behavior of colloidal systems.

The Gas-Law Units and Constants. In the use of the gas-law equations great care must be exercised that consistent units are employed for the expression of both the variable and constant terms. Temperature must always be expressed on an absolute scale. Two such scales are in common use. The Kelvin scale corresponds, in the size of its unit degree, to the centigrade scale. The zero of the centigrade scale corresponds to 273.16° on the Kelvin scale. Thus:

$$t^\circ_C = T^\circ_K - 273.16 \qquad (9)$$

The Rankine scale of absolute temperature corresponds, in the size of its unit degree, to the Fahrenheit scale. The zero of the Fahrenheit scale corresponds to 459.69° on the Rankine scale. Thus:

$$t°_F = T°_R - 459.69 \tag{10}$$

The Avogadro number N denoting the number of molecules in a mole is one of the most important of physical constants and has been carefully determined by a variety of methods. The accepted value is 6.024 $\times 10^{23}$ for the number of molecules in one gram-mole, or 2.73 $\times 10^{26}$ molecules per pound-mole.

From equation 7:

$$R = \tfrac{2}{3}\frac{U_t}{T} \tag{11}$$

From equation 11 it is seen that R represents two thirds of the translational kinetic energy possessed by one mole per degree of absolute temperature. The numerical value of R has been carefully determined and may be expressed in any desired energy units. Following are values corresponding to various systems of units:

Units of Pressure	Units of Volume	R
Per gram-mole (Temperatures: Kelvin)		
Atmospheres	Cubic centimeters	82.06
Per pound-mole (Temperatures: Rankine)		
Pounds per square inch	Cubic inches	18,510
Pounds per square inch	Cubic feet	10.731
Atmospheres	Cubic feet	0.7302

Applications of the Ideal-Gas Law

When substances exist in the gaseous state two general types of problems arise in determining the relationships among mass, pressure, temperature, and volume. The first type involves only the last three variables; pressure, temperature, and volume. For example, a specified volume of gas is initially at a specified temperature and pressure. The conditions are changed, two of the variables in the final state being specified, and it is desired to calculate the third. For such calculations it is not required to know the weight of the gas. The second, more general type of problem involves the weight of the gas. A specified weight of substance exists in the gaseous state under conditions two of which are specified and the third is to be calculated. Or, conversely, it is desired to calculate the weight of a given quantity of gas existing at specified conditions of temperature, pressure, and volume.

Problems of the first type, in which weights are not involved, may be readily solved by means of the proportionality indicated by the gas law.

Equation 5 may be applied to n moles of gas at conditions p_1, V_1, T_1 and also at conditions p_2, V_2, T_2.

$$p_1 V_1 = nRT_1$$
$$p_2 V_2 = nRT_2$$

Combining gives:

$$\frac{p_1 V_1}{p_2 V_2} = \frac{T_1}{T_2} \qquad (12)$$

This equation may be applied directly to any quantity of gas. If the three conditions of state 1 are known, any one of those of state 2 may be calculated to correspond to specified values of the other two. Any units of pressure, volume, or absolute temperature may be used, the only requirement being that the units in both initial and final states be the same.

Equation 5 is in a form that permits direct solution of problems of the second type, that involve both weights and volumes of gases. With weights expressed in molal units the equation may be solved for any one of the four variables if the other three are known. However, this calculation requires a value of the constant R expressed in units that correspond to those used in expressing the four variable quantities. So many units of expression are in common use for each variable quantity that a large table of values of R would be required or else the variable quantities would have to be converted into standard units. Either method is inconvenient.

It proves much more desirable to separate such calculations into two steps. As a primary constant, the *normal molal volume* is used instead of R. The normal molal volume is the volume occupied by one mole of a gas at arbitrarily selected *standard conditions*, assuming that the ideal-gas law is obeyed. The normal molal volume at any one set of standard conditions, if the validity of equation 5 is assumed, must be a universal constant, the same for all gases. The volume, at the standard conditions, of any weight of gas is the product of the number of moles present and the normal molal volume. The general type of problem involving weights and volumes at any desired conditions may then be solved in two steps. In one step the differences between the properties of the gas at standard conditions and at those conditions specified in the problem are determined by equation 12. In the other step the relationship between volume at standard conditions and weight is determined by means of the normal molal volume constant.

Standard Conditions. An arbitrarily specified standard state of temperature and pressure serves two purposes. It establishes the normal molal volume constant required in the system of calculation described in the preceding section. It also furnishes convenient specifications under which quantities of gases may be compared when expressed in terms of volumes. Some such specification is necessary because of the fact that the volume of a gas depends not only on the quantity but on the temperature and pressure as well.

Several specifications of standard conditions are in more or less common use, but the one most universally adopted is a temperature of 0° C and a pressure of one atmosphere. It is recommended that these conditions be adopted as the standard for all calculations. Under these conditions the normal molal volumes are as follows (the abbreviation S.C. is used to designate the standard conditions):

Volume of 1 g-mole (S.C.) = 22.4140 liters
Volume of 1 lb-mole (S.C.) = 359.046 cu ft
Volume of 1 kg-mole (S.C.) = 22.4146 cubic meters

These important constants should be memorized. The conditions of the standard state may be expressed in any desired units as in the following table:

STANDARD CONDITIONS

Temperature	Pressure
0° C	1 atm
273.16° K	760 mm Hg
32° F	29.92 in. Hg
491.69° R	14.70 psi

There are many substances that cannot actually exist in the gaseous state at these specified conditions. For example, at a temperature of 0° C water cannot exist in a stable gaseous form at a pressure greater than 4.6 mm of mercury. Higher pressures cause condensation. Yet it is convenient to refer to the hypothetical volume occupied by water vapor at standard conditions. In such a case the volume at standard conditions indicates the hypothetical volume that would be occupied by the substance if it could exist in the vapor state at these conditions and if it obeyed the ideal-gas law.

Gage Pressure. All ordinary pressure gages indicate the magnitude of pressure above or below that of the atmosphere. In order to obtain the *absolute pressure* which must be used in the gas law, the pressure of the atmosphere must be added to the *gage pressure*. The average atmospheric pressure at sea level is 14.70 psi or 29.92 in. of mercury.

Gas Densities and Specific Gravities. The density of a gas is ordinarily expressed as the weight in grams of one liter or the weight in pounds of one cubic foot. Unless otherwise specified the volumes are at the standard conditions of 0° C and a pressure of 1.0 atm. On this basis air has a normal density of 1.293 grams per liter or of 0.0807 lb per cu ft.

The *specific gravity* of a gas is usually defined as the ratio of its density to that of air at the same conditions of temperature and pressure.

The gas law expresses the relationship among four properties of a gas: mass, volume, pressure, and temperature. In order to calculate any one of these properties the others must be known or specified. Four different types of problems arise, classified according to the property being sought. The following illustrations show the application of the recommended method of calculation to each of these types of problems.

For establishment of correct ratios to account for the effects of pressure and temperature a simple rule may be followed which offers less opportunity for error than attempting to recall equation 12. *The ratio of pressures or temperatures should be greater than unity when the changes in pressure or temperature are such as to cause increase in volume. The ratios should be less than unity when the changes are such as to cause decrease in volume.*

Illustration 1 (*Volume Unknown*). Calculate the volume occupied by 30 lb of chlorine at a pressure 743 mm Hg and 70° F.

Basis: 30 lb of chlorine or $30/71 = 0.423$ lb-mole

Volume at S.C. $= 0.423 \times 359 = 152$ cu ft

$$V_2 = V_1 \frac{p_1}{p_2} \times \frac{T_2}{T_1}$$

$70° \text{ F} = 530° \text{ R}$

Volume at 743 mm Hg, 70° F $= 152 \times \dfrac{760}{743} \times \dfrac{530}{492} = 167$ cu ft

Illustration 2 (*Weight Unknown*). Calculate the weight of 100 cu ft of water vapor, measured at a pressure of 15.5 mm Hg and 23° C.

Basis: 100 cu ft of water vapor at 15.5 mm Hg, 23° C.

Volume at S.C. $= 100 \dfrac{15.5}{760} \times \dfrac{273}{296} = 1.88$ cu ft

Moles of H_2O $= 1.88 \div 359 = 0.00523$ lb-mole

Weight of H_2O $= 0.00523 \times 18 = 0.0942$ lb

Illustration 3 (*Pressure Unknown*). It is desired to compress 10 lb of carbon dioxide to a volume of 20 cu ft. Calculate the pressure in pounds per square inch that is required at a temperature of 30° C, assuming the applicability of the ideal-gas law.

Basis: 10 lb of CO_2 or $10/44 = 0.228$ lb-mole

Volume at S.C. $= 0.228 \times 359 = 81.7$ cu ft

From equation 12:

$$p_2 = p_1 \frac{V_1}{V_2} \times \frac{T_2}{T_1}$$

30° C   = 303° K

Pressure at 20 cu ft, 30° C $= 14.7 \frac{81.7}{20} \times \frac{303}{273} = 66.6$ psi

Illustration 4 (*Temperature Unknown*). Assuming the applicability of the ideal-gas law, calculate the maximum temperature to which 10 lb of nitrogen, enclosed in a 30-cu-ft chamber, may be heated without the pressure exceeding 150 psi.

Basis: 10 lb of nitrogen or $10/28 = 0.357$ lb-mole

Volume at S.C. $= 0.357 \times 359 = 128.1$ cu ft

$$T_2 = T_1 \frac{p_2}{p_1} \times \frac{V_2}{V_1}$$

Temperature at 30 cu ft, 150 psi $= 273 \frac{150}{14.7} \times \frac{30}{128.1} = 652°$ K or 379° C

Dissociating Gases. Certain chemical compounds when in the gaseous state apparently do not even approximately follow the relationships deduced above. The tendency of hydrogen fluoride to associate into large molecules was mentioned in Chapter 2. Ammonium chloride, nitrogen peroxide, and phosphorus pentachloride exhibit an abnormality opposite in effect which has been definitely proved to result from dissociation of the molecules into mixtures containing two or more other compounds. Ammonium chloride molecules in the vapor state separate into molecules of hydrogen chloride and ammonia:

$$NH_4Cl = NH_3 + HCl$$

Thus gaseous ammonium chloride is not a pure gas but a mixture of three gases, NH_4Cl, HCl, and NH_3. By decomposition, two gas particles are produced from one, and the pressure or volume of the gas increases above that which would exist had no decomposition taken place. For this reason, when one gram-mole of ammonium chloride is vaporized the volume occupied will be much greater than that indicated by equation 5. However, when proper account is taken of the fact that in the gaseous state actually more than one gram-mole is present, it is found that the ideal-gas law applies. Conversely, from the apparent deviation from the gas law the percentage of dissociation can be calculated if the chemical reaction involved is known.

Illustration 5. When heated to 100° C and 720 mm pressure 17.2 grams of N_2O_4 gas occupy a volume of 11,450 cc. Assuming that the ideal-gas law applies, calculate the percentage dissociation of N_2O_4 to NO_2.

$$\text{G-moles of } N_2O_4 \text{ initially present} = \frac{17.2}{92} = 0.187$$

Let x = gram-moles of N_2O_4 dissociated. Then $2x$ = gram-moles of NO_2 formed.
Total g-moles present after dissociation = $0.187 - x + 2x$

$$= \frac{11,450}{22,400} \times \frac{273}{373} \times \frac{720}{760} = 0.355$$

Solving, $x = 0.168$.

$$\% \text{ dissociation} = \frac{0.168}{0.187} \times 100 = 90\%$$

Gaseous Mixtures

In a mixture of different gases the molecules of each component gas are distributed throughout the entire volume of the containing vessel, and the molecules of each component gas contribute by their impacts to the total pressure exerted by the entire mixture. The total pressure is equal to the sum of the pressures exerted by the molecules of each component gas. These statements apply to all gases, whether or not their behavior is ideal. In a mixture of ideal gases the molecules of each component gas behave independently as though they alone were present in the container. Before considering the actual behavior of gaseous mixtures it will be necessary to define two terms commonly employed, namely, *partial pressure* and *pure-component volume*. By definition, the *partial pressure of a component gas that is present in a mixture of gases is the pressure that would be exerted by that component gas if it alone were present in the same volume and at the same temperature as the mixture.* By definition, *the pure-component volume of a component gas that is present in a mixture of gases is the volume that would be occupied by that component gas if it alone were present at the same pressure and temperature as the mixture.*

The partial pressure as defined above does not represent the actual pressure exerted by the molecules of the component gas when it is present in the mixture except under certain limiting conditions. Also, the pure-component volume does not represent the volume occupied by the molecules of the component gas when it is present in the mixtures, for obviously the molecules are distributed uniformly throughout the entire volume of the mixture.

The pure-component volume has also been termed *partial volume*. However, the latter term is currently used to designate the differential increase in volume when a component is added to a mixture. Pure-component volumes and partial volumes are not necessarily the same except under ideal conditions.

Laws of Dalton and Amagat. From the simple kinetic theory of the constitution of gases it would be expected that many properties of gaseous mixtures would be additive. The additive nature of partial pressures is expressed by *Dalton's law*, which states that *the total pressure exerted by a gaseous mixture is equal to the sum of the partial pressures*; that is:

$$p = p_A + p_B + p_C + \cdots \tag{13}$$

where p is the total pressure of the mixture and p_A, p_B, p_C, etc., are the partial pressures of the component gases as defined above.

Similarly, the additive nature of pure-component volumes is given by the *law of Amagat*, or Leduc's law, which states that the *total volume occupied by a gaseous mixture is equal to the sum of the pure-component volumes;* that is:

$$V = V_A + V_B + V_C + \cdots \tag{14}$$

where V is the total volume of the mixture and V_A, V_B, V_C, etc., are the pure-component volumes of the component gases as defined above. It will be shown later that each of these laws is correct where conditions are such that the mixture and each of the components obey the ideal-gas law.

Where small molal volumes are encountered, such that the ideal-gas law does not apply, either Dalton's or Amagat's law may apply, but both laws apply simultaneously only for ideal gases. Under such conditions pressures may not be additive, because the introduction of additional molecules into a gas-filled container may appreciably affect the pressure exerted by those already there. The presence of new molecules will reduce the space available for the free motion of those originally present and will exert attractive forces on them. Similarly, if quantities of two gases at the same pressure are allowed to mix at that same pressure, the like molecules of each gas will be separated by greater distances and will be in the presence of unlike molecules, which condition may alter the order of attractive forces existing among them. As a result, the volume of the mixture may be quite different from the sum of the original volumes. These effects are negligible under conditions of large molal volumes.

Where conditions are such that the ideal-gas law is applicable,

$$p_A = \frac{n_A R T}{V} \tag{15}$$

where V = total volume of mixture

n_A = number of moles of component A in mixture

Similar equations represent the partial pressures of components B, C, etc. Combining these equations with Dalton's law (equation 13) gives

$$p = (n_A + n_B + n_C + \cdots)\frac{RT}{V} \qquad (16)$$

This equation relates the pressure, temperature, volume, and molal quantity of any gaseous mixture under such conditions that the mixture and each of the components follow the ideal-gas law and Dalton's law. By combining equations 15 and 16 a useful relationship between total and partial pressure is obtained:

$$p_A = \frac{n_A}{n_A + n_B + n_C \cdots}\, p = N_A p \qquad (17)$$

The quantity $N_A = n_A/(n_A + n_B + n_C + \cdots)$ is the mole fraction of component A. Equation 17 then signifies that, *where the ideal-gas law may be applied, the partial pressure of a component of a mixture is equal to the product of the total pressure and the mole fraction of that component.*

When conditions are such that the ideal-gas law is applicable,

$$pV_A = n_A RT \qquad (18)$$
$$pV_B = n_B RT \qquad (19)$$

where V_A, V_B, etc., are the pure-component volumes as defined above.

Adding these equations gives

$$p(V_A + V_B + \cdots) = (n_A + n_B + \cdots)RT \qquad (20)$$

Combining equations 18 and 19 with Amagat's law (equation 14) yields

$$\frac{V_A}{V} = \frac{n_A}{n_A + n_B + \cdots} \qquad (21)$$

or

$$V_A = N_A V \qquad (22)$$

Equation 22 signifies that, *where the ideal-gas law may be applied, the pure-component volume of a component of a gaseous mixture is equal to the product of the total volume and the mole fraction of that component.*

From equations 16 and 20 it is evident that, when the ideal-gas law is valid, both Amagat's and Dalton's laws apply; that is, both pure-component volumes and partial pressures are additive.

Average Molecular Weight of a Gaseous Mixture. A certain group of components of a mixture of gases may in many cases pass through a process without being changed in composition or weight. For example, in a drying process, dry air merely serves as a carrier for the vapor being removed and undergoes no change in composition or in weight. It is frequently convenient to treat such a mixture as though it were a single gas and assign to it an average molecular weight which may be used for calculation of its weight and volume relationships. The average molecular weight is calculated by adopting a unit molal quantity of the mixture as the basis of calculation. The weight of this molal quantity is then calculated and represents the average molecular weight. By this method the average molecular weight of air is found to be 29.0.

Illustration 6. Calculate the average molecular weight of a flue gas having the following composition by volume:

$$
\begin{array}{ll}
CO_2 & 13.1\% \\
O_2 & 7.7 \\
N_2 & \underline{79.2} \\
& 100.0\%
\end{array}
$$

Basis: 1 g-mole of the mixture.

$CO_2 = 0.131$ g-mole or   5.76 grams
$O_2 \ \ = 0.077$ g-mole or   2.46
$N_2 \ \ = 0.792$ g-mole or $\underline{22.18}$

Weight of 1 g-mole $= \overline{30.40}$ grams, which is the average molecular weight

Densities of Gaseous Mixtures. If the composition of a gas mixture is expressed in molal or weight units, the density is readily determined by selecting a unit molal quantity or weight as the basis and calculating its volume at the specified conditions of temperature and pressure through the use of the ideal-gas law.

Illustration 7. Calculate the density in pounds per cubic foot at 29 in. Hg and 30° C of a mixture of hydrogen and oxygen that contains 11.1% H_2 by weight.

Basis: 1 lb of mixture.

$H_2 = 0.111$ lb or	0.0555 lb-mole
$O_2 = 0.889$ lb or	0.0278 lb-mole
Total molal quantity	$=$ 0.0833 lb-mole
Volume at S.C. $= 0.0833 \times 359$	$=$ 29.9 cu ft

Volume at 29 in. Hg, 30° C $= 29.9 \times \dfrac{29.92 \times 303}{29.0 \times 273} = 34.2$ cu ft

Density at 29 in. Hg, 30° C $= \dfrac{1}{34.2}$ $= 0.0292$ lb per cu ft

If the composition of a mixture of gases is expressed in volume units, the ideal-gas law is ordinarily applicable. In this case the volume analysis is the same as the molal analysis, and the density is readily calculated on the basis of a unit molal quantity of the mixture. The weight of the basic quantity is first calculated and then its volume at the specified conditions.

Illustration 8. Air is assumed to contain 79.0% nitrogen and 21.0% oxygen by volume. Calculate its density in grams per liter at a temperature of 70° F and a pressure of 741 mm Hg.

Basis: 1.0 g-mole of air.

O_2 = 0.210 g-mole or	6.72 grams
N_2 = 0.790 g-mole or	22.10 grams
Total weight	= 28.82 grams
Volume at S.C.	= 22.41 liters

$$\text{Volume, 741 mm Hg, 70° F} = 22.41 \times \frac{760 \times 530}{741 \times 492} = 24.8 \text{ liters}$$

$$\text{Density} = \frac{28.82}{24.8} = 1.162 \text{ grams per liter (741 mm Hg, 70° F)}$$

The actual density of the atmosphere is slightly higher, owing to the presence of about 1% of argon which is classed as nitrogen in the above problem. The mixture of nitrogen and inert gases in the atmosphere may be termed *atmospheric nitrogen*. The average molecular weight of this mixture is 28.2.

Composition of Gases on Dry or Wet Basis. In reporting the volumetric analysis of a gas it is customary to state percentage compositions on a moisture-free basis. This procedure is followed because the water-vapor content of a gas is subject to fluctuations independent of the composition of the dry gas. Also, volumetric gas analyses are usually made over aqueous solutions which maintain the gases saturated with water vapor. Each component removed in analysis and the remainder of the sample are thus measured at the same total pressure and at the same water-vapor pressure. If the moisture content of a sample of gas is desired, it is always determined by a separate procedure, as, for example, by measuring its dew point.

Volume Changes with Change in Composition

Such operations as gas absorption, drying, and some types of evaporation involve changes in the compositions of gaseous mixtures, owing to the addition or removal of certain components. In a drying operation a stream of air takes on water vapor. In the scrubbing of coal gas, ammonia is removed from the mixture. It is of interest to calculate the relationships existing between the initial and final volumes of the mixture and the volume of the material removed or added to the mixture in such a process. The situation is ordinarily complicated by changes of temperature and pressure concurrent with the composition changes.

Solution may be carried out by the methods of Chapter 2 if the quantities specified in the problem are first converted to weight or molal units.   The quantities that are unknown may then be calculated in these same units.   The last step will then be the conversion of the results from molal or weight units into volumes at the specified conditions of temperature and pressure.   The relationships between molal units and volumes under any conditions are expressed by equations 16 to 22.   This method of solution may be applied by the use of either the ideal-gas law or more nearly accurate equations.   The following illustration demonstrates the method for a case in which the ideal-gas law is applicable.   As in the problems of Chapter 2, the calculations must be based on a definite quantity of a component which passes through the process unchanged.

Illustration 9.   Combustion gases having the following molal composition are passed into an evaporator at a temperature of 200° C and a pressure of 743 mm Hg.

Nitrogen	79.2%
Oxygen	7.2
Carbon dioxide	13.6
	100.0%

Water is evaporated, the gases leaving at a temperature of 85° C and a pressure of 740 mm Hg with the following molal composition:

Nitrogen	48.3%
Oxygen	4.4
Carbon dioxide	8.3
Water	39.0
	100.0%

Calculate:

(a) Volume of gases leaving the evaporator per 100 cu ft entering.

(b) Weight of water evaporated per 100 cu ft of gas entering.

Solution

Basis: 1 g-mole of the entering gas.

N_2	0.792 g-mole
O_2	0.072 g-mole
CO_2	0.136 g-mole

Total volume (743 mm Hg, 200° C) calculated from equations 14 and 20:

$$p = 743/760 \text{ or } 0.978 \text{ atm}$$
$$T = 473° \text{ K}$$
$$R = 82.1 \text{ cc-atm per K°}$$

$$V = \frac{(n_A + n_B + n_C)RT}{p} = \frac{(0.792 + 0.072 + 0.136) \, 82.1 \times 473}{0.978}$$

$$= \frac{1.0 \times 82.1 \times 473}{0.978} = 39,750 \text{ cc or } 1.40 \text{ cu ft}$$
$$(743 \text{ mm Hg, } 200° \text{ C})$$

This 1.0 g-mole of gas entering forms 61% by volume of the gases leaving the evaporator.

Gases leaving = 1.0/0.61 = 1.64 g-moles
Water leaving = 1.64 − 1.0 = 0.64 g-mole
Volume of gas leaving, from equations 14 and 20:

$$p = 740/760 = 0.973 \text{ atm}$$
$$T = 358° \text{ K}$$
$$R = 82.1 \text{ cc-atm per K°}$$

$$V = \frac{(0.792 + 0.072 + 0.136 + 0.64) \times 82.1 \times 358}{0.973}$$

$$= \frac{1.64 \times 82.1 \times 358}{0.973} = 49,500 \text{ cc or } 1.75 \text{ cu ft}$$

Volume of gas leaving per 100 cu ft entering,

$$\frac{1.75 \times 100}{1.40} = 125 \text{ cu ft } (740 \text{ mm Hg, } 85° \text{ C})$$

Weight of water leaving evaporator = 0.64 × 18 = 11.5 grams or 0.0254 lb

Weight of water evaporated per 100 cu ft of gas entering   =

$$\frac{0.0254 \times 100}{1.40} = 1.81 \text{ lb}$$

Pure-Component Volume Method. Another method of calculation involves the use of pure-component volumes. The volume of any ideal mixture may be obtained by adding together the pure-component volumes of its components. Similarly, the removal of a component from a mixture will decrease the total volume by its pure-component volume. Care must be taken in the use of this method that all volumes which are added or subtracted are expressed at the same conditions of temperature and pressure. A process involving changes in temperature and pressure as well as in composition is best considered as taking place in two steps. First the change in composition at the initial conditions of temperature and pressure is evaluated. Again the entire calculation must be based on a definite quantity of a component which passes through the process without change in mass. This procedure is indicated in the following illustration.

Illustration 10. In the manufacture of hydrochloric acid a gas is obtained that contains 25% HCl and 75% air by volume. This gas is passed through an absorption system in which 98% of the HCl is removed. The gas enters the system at a temperature of 120° F and a pressure of 743 mm Hg and leaves at a temperature of 80° F and a pressure of 738 mm Hg.

(a) Calculate the volume of gas leaving per 100 cu ft entering the absorption apparatus.

(*b*) Calculate the percentage composition by volume of the gases leaving the absorption apparatus.

(*c*) Calculate the weight of HCl removed per 100 cu ft of gas entering the absorption apparatus.

Solution

Basis: 100 cu ft of entering gas (743 mm Hg, 120° F) containing 75 cu ft of air which will be unchanged in quantity.

Pure-component volume of HCl   = 25   cu ft
Pure-component volume of HCl absorbed   = 24.5   cu ft
Pure-component volume of HCl remaining =   0.50 cu ft
Volume of gas remaining = 75 + 0.50   = 75.5   cu ft (743 mm, 120° F)

$$\text{Volume of gas leaving} = 75.5 \times \frac{743}{738} \times \frac{540}{580} = 70.8 \quad \text{cu ft (738 mm, 80° F)}$$

Composition of gases leaving:

HCl, 0.5/75.5   =   0.66%
Air   = 99.34%

Volume at S.C. of HCl absorbed =

$$24.5 \times \frac{743}{760} \times \frac{492}{580} = 20.3 \text{ cu ft}$$

HCl absorbed = 20.3/359 =
                    0.0565 lb-mole or =   2.07 lb

Partial-Pressure Method. In certain types of work, especially where condensable vapors are involved, it is convenient to express the compositions of gaseous mixtures in terms of the partial pressures of the various components. Where data are presented in this form, problems of the type discussed above may be more conveniently solved by considering only the pressure changes resulting from the changes in composition. The addition or removal of a component of a mixture may be considered as producing only a change in the partial pressure of all of the other components. The *actual* volume occupied by each of these components will always be exactly the same as that of the entire mixture. The volume of the mixture may then always be determined by application of the gas law to any components that pass through the process unchanged in quantity and whose partial pressures are known at both the initial and final conditions. The use of this method is shown in the following illustration.

Illustration 11.   Calcium hypochlorite is produced by absorbing chlorine in milk of lime.   A gas produced by the Deacon chlorine process enters the absorption apparatus at a pressure of 740 mm Hg and a temperature of 75° F.   The partial pressure of the chlorine is 59 mm Hg, the remainder being inert gases.   The gas leaves the absorption apparatus at a temperature of 80° F and a pressure of 743 mm Hg with a partial pressure of chlorine of 0.5 mm Hg.   Calculate:

(a) Volume of gases leaving the apparatus per 100 cu ft entering.
(b) Weight of chlorine absorbed per 100 cu ft of gas entering.

Solution

Basis: 100 cu ft of gas entering (740 mm Hg, 75° F).

Partial pressure of inert gases entering $= 740 - 59$ $= 681$ mm Hg
Partial pressure of inert gases leaving $= 743 - 0.5$ $= 742.5$ mm Hg
Actual volume of inert gas entering $= 100$ cu ft

$$\text{Actual volume of inert gases leaving} = 100 \times \frac{681}{742.5} \times \frac{540}{535} = 92.5 \text{ cu ft.}$$

This is also the total volume of gases leaving (743 mm Hg, 80° F).

The actual volumes of chlorine entering and leaving are also 100 and 92.5 cu ft, respectively.

$$\text{Volume at S.C. of chlorine entering} = 100 \frac{59 \times 492}{760 \times 535} = 7.14 \text{ cu ft}$$

$$\text{Volume at S.C. of chlorine leaving} = 92.5 \frac{0.5 \times 492}{760 \times 540} = 0.055 \text{ cu ft}$$

$$\text{Volume at S.C. of chlorine absorbed} = 7.14 - 0.055 = 7.08 \text{ cu ft}$$

$$\text{Chlorine absorbed} = \frac{7.08}{359} = 0.0197 \text{ lb-mole or} \qquad 1.40 \text{ lb}$$

Gases in Chemical Reactions

In a great many chemical and metallurgical reactions gases are present, either in the reacting materials or the products or in both. Quantities of gases are ordinarily expressed in volume units because the common methods of measurement give results directly on this basis. The general types of reaction calculations must, therefore, include the complications introduced by the expression of gaseous quantities and compositions in volume units.

In Chapter 2 methods are demonstrated for the solution of reaction calculations through the use of molal units for the expression of quantities of reactants and products. Where this is the scheme of calculation, the introduction of volumetric data adds no complications. By the use of the normal molal volume constants combined with the proportions of the ideal-gas law it is easy to convert from molal to volume units, and the reverse. The methods of conversion have been explained in the preceding sections.

The same general methods of solution are followed as were described in Chapter 2. All quantities of active materials, whether gaseous, solid, or liquid, are expressed in molal units and the calculations carried out on this basis. Results are thus obtained in molal units which may readily be converted to volumes at any desired conditions. The most convenient choice of a quantity of material to serve as the basis of calculation is determined by the manner of presentation of the data. In general, if the data regarding the basic material are in weight units, a unit weight is the best basis of calculation.   If the data are in volume units, a unit molal quantity is ordinarily the most desirable basis.

Illustration 12.   Nitric acid is produced in the Ostwald process by the oxidation of ammonia with air.   In the first step of the process ammonia and air are mixed together and passed over a catalyst at a temperature of 700° C.   The following reaction takes place:

$$4NH_3 + 5O_2 = 6H_2O + 4NO$$

The gases from this process are passed into towers where they are cooled, and the oxidation is completed according to the following theoretical reactions:

$$2NO + O_2 = 2NO_2$$

$$3NO_2 + H_2O = 2HNO_3 + NO$$

The NO liberated is in part reoxidized and forms more nitric acid in successive repetitions of the above reactions.   The ammonia and air enter the process at a temperature of 20° C and a pressure of 755 mm Hg.   The air is present in such proportion that the oxygen will be 20% in excess of that required for complete oxidation of the ammonia to nitric acid and water.   The gases leave the catalyzer at a pressure of 743 mm Hg and a temperature of 700° C.

(*a*) Calculate the volume of air to be used per 100 cu ft of ammonia entering the process.

(*b*) Calculate the percentage composition by volume of the gases entering the catalyzer.

(*c*) Calculate the percentage composition by volume of the gases leaving the catalyzer, assuming that the degree of completion of the reaction is 85% and that no other decompositions take place.

(*d*) Calculate the volume of gases leaving the catalyzer per 100 cu ft of ammonia entering the process.

(*e*) Calculate the weight of nitric acid produced per 100 cu ft of ammonia entering the process, assuming that 90% of the nitric oxide entering the tower is oxidized to nitric acid.

Basis of Calculation: 1.0 lb-mole of NH_3.

$$NH_3 + 2O_2 = HNO_3 + H_2O$$

(*a*) O_2 required   =   2.0 lb-moles
        O_2 supplied   = 2.0 × 1.2 =   2.4 lb-moles

Air supplied $= \dfrac{2.4}{0.210}$ = 11.42 lb-moles

Therefore:

Volume of air = 11.42 × (volume of ammonia at same conditions)

or

Volume of NH_3 = 359 × $\dfrac{293 \times 760}{273 \times 755}$ =   388 cu ft (20° C, 755 mm Hg)

Volume of air   = 11.42 × 388   = 4440 cu ft (20° C, 755 mm Hg)

Volume of air per 100 cu ft of NH₃ =

$$\frac{4440 \times 100}{388} = 1142 \text{ cu ft}$$

(b) Gases entering process, N_2, O_2, NH_3.

N₂ present in air = 0.790 × 11.42   = 9.02 lb-moles

Total quantity of gas entering catalyzer = 11.42 + 1 = 12.42 lb-moles

Composition by volume:

$$\begin{aligned}
NH_3 &= 1.0/12.42 &=&\quad 8.0\% \\
O_2 &= 2.4/12.42 &=&\quad 19.3 \\
N_2 &= 9.02/12.42 &=&\quad \underline{72.7} \\
& & &\quad 100.0\%
\end{aligned}$$

(c) Gases leaving catalyzer, N_2, NH_3, O_2, NO, and H_2O.

NH₃ oxidized in catalyzer	=	0.85 lb-mole
NH₃ leaving catalyzer	=	0.15 lb-mole
O₂ consumed in catalyzer = (5/4) × 0.85	=	1.06 lb-moles
O₂ leaving catalyzer = 2.40 − 1.06	=	1.34 lb-moles
NO formed in catalyzer	=	0.85 lb-mole
H₂O formed in catalyzer = (6/4) × 0.85	=	1.275 lb-moles
Total quantity of gas leaving catalyzer = 9.02 + 0.15		
+ 1.34 + 0.85 + 1.275	=	12.64 lb-moles

Composition by volume:

$$\begin{aligned}
NO &= 0.85/12.64 &=&\quad 6.7\% \\
H_2O &= 1.275/12.64 &=&\quad 10.1 \\
NH_3 &= 0.15/12.64 &=&\quad 1.2 \\
O_2 &= 1.34/12.64 &=&\quad 10.6 \\
N_2 &= 9.02/12.64 &=&\quad \underline{71.4} \\
& & &\quad 100 \ \%
\end{aligned}$$

Basis of Calculation: 100 cu ft of NH₃ entering the process.

(d) Moles of NH₃ = $\dfrac{1.0 \times 100}{388}$ =   0.258 lb-mole

Moles of gas leaving catalyzer = 0.258 × 12.64   =   3.26 lb-moles
Volume at S.C. of gas leaving catalyzer = 3.26 × 359 = 1170 cu ft

Volume of gas leaving catalyzer = $1170 \times \dfrac{760 \times 973}{743 \times 273} = 4270$ cu ft

(700° C, 743 mm Hg) per 100 cu ft of NH₃ entering.

(e) NO produced in catalyzer = 0.258 × 0.85   =   0.219 lb-mole
NO oxidized in tower = 0.219 × 0.90   =   0.197 lb-mole
HNO₃ formed = 0.197 lb-mole or 0.197 × 63 = 12.4   lb

Range of Applicability of the Ideal-Gas Law.   The ideal-gas law is applicable only at conditions of low pressure and high temperature corresponding to large molal volumes.   At conditions resulting in small molal volumes the attractive forces among the molecules become significant and volumes calculated from the ideal-gas law tend to be too large.   In extreme cases the volume calculated from the ideal-gas law may be five times the actual volume.

If an error of 1% is permissible, the ideal-gas law may be used for diatomic gases where gram-molal volumes are as low as 5 liters (80 cu ft per lb-mole) and for gases of more complex molecular structure such as carbon dioxide, acetylene, ammonia, and the lighter hydrocarbon vapors, where gram-molal volumes exceed 20 liters (320 cu ft per lb-mole).

The actual behavior of gases at high-pressures is discussed in Chapter 14 where rigorous methods of calculation are presented.

Problems

Pressures are absolute unless otherwise stated

1. It is desired to market oxygen in small cylinders having volumes of 0.5 cu ft and each containing 1.0 lb of oxygen.   If the cylinders may be subjected to a maximum temperature of 120° F, calculate the pressure for which they must be designed, assuming the applicability of the ideal-gas law. **Ans.   389 psi.**

2. Calculate the number of cubic feet of hydrogen sulfide, measured at a temperature of 50° C and a pressure of 29.5 in. Hg, which may be produced from 7 lb of iron sulfide (FeS).

3. An automobile tire is inflated to a gage pressure of 35 psi at a temperature of 0° F.   Calculate the maximum temperature to which the tire may be heated without the gage pressure exceeding 50 psi.   (Assume that the volume of the tire does not change.) **Ans.   139° F.**

4. Calculate the densities in pounds per cubic foot at standard conditions and the specific gravities of the following gases: (a) ethane, (b) sulfur dioxide, (c) hydrogen sulfide, (d) chlorine.

5. The gas acetylene is produced according to the following reaction by treating calcium carbide with water:

$$CaC_2 + 2H_2O = C_2H_2 + Ca(OH)_2$$

Calculate the number of hours of service that can be derived from 1.0 lb of carbide in an acetylene lamp burning 2 cu ft of gas per hour at a temperature of 75° F and a pressure of 743 mm Hg. **Ans.   3.11 hr.**

6. A natural gas has the following composition by volume:

CH$_4$	94.1%
C$_2$H$_6$	3.0
N$_2$	2.9
	100.0%

This gas is piped from the well at a temperature of 80° F and an absolute pressure of 50 psi.   It may be assumed that the ideal-gas law is applicable.   Calculate:

(a) Partial pressure of the nitrogen.

(b) Pure-component volume of nitrogen per 100 cu ft of gas.

(c) Density of the mixture in pounds per cubic foot at the existing conditions.

7. A gas mixture contains 0.274 lb-mole of HCl, 0.337 lb-mole of nitrogen, and 0.089 lb-mole of oxygen. Calculate (a) the volume occupied by this mixture and (b) its density in pounds per cubic foot at a pressure of 40 psi and a temperature of 30° C. **Ans.** (a) 102.5 cu ft, (b) 0.2174 lb per cu ft.

8. A chimney gas has the following composition by volume:

CO_2	9.5%
CO	0.2
O_2	9.6
N_2	80.7
	100 %

Using the ideal-gas law, calculate:

(a) Its composition by weight.
(b) Volume occupied by 1.0 lb of the gas at 80° F and 29.5 in. Hg pressure.
(c) Density of the gas in pounds per cubic foot at the conditions of part b.
(d) Specific gravity of the mixture.

9. By electrolyzing a mixed brine a mixture of gases is obtained at the cathode having the following composition by weight:

Cl_2	67%
Br_2	28
O_2	5
	100%

Using the ideal-gas law, calculate:

(a) Composition of the gas by volume. **Ans.** 74.03% Cl_2, 13.73% Br_2, 12.24% O_2.
(b) Density of the mixture in grams per liter at 25° C and 740 mm of Hg pressure. **Ans.** 3.12 grams per liter.
(c) Specific gravity of the mixture. **Ans.** 2.70 (air = 1.0).

10. A mixture of ammonia and air at a pressure of 730 mm Hg and a temperature of 30° C contains 5.1% NH_3 by volume. The gas is passed at a rate of 100 cu ft per min through an absorption tower in which only ammonia is removed. The gases leave the tower at a pressure of 725 mm Hg, a temperature of 20° C, and contain 0.05% NH_3 by volume. Using the ideal-gas law, calculate:

(a) Rate of flow of gas leaving the tower in cubic feet per minute.
(b) Weight of ammonia absorbed in the tower per minute.

11. A volume of moist air of 1000 cu ft at a total pressure of 740 mm Hg and a temperature of 30° C contains water vapor in such proportions that its partial pressure is 22.0 mm Hg. Without the total pressure being changed, the temperature is reduced to 15° C and some of the water vapor removed by condensation. After cooling it is found that the partial pressure of the water vapor is 12.7 mm Hg. Using the partial pressure method, calculate:

(a) Volume of the gas after cooling. **Ans.** 938 cu ft.
(b) Weight of water removed. **Ans.** 0.563 lb.

12. In producing activated charcoal for adsorption purposes, coconut shells are destructively distilled and treated with superheated steam. This results in the evolution of gases which leave the processing equipment mixed with the steam. In a plant employing such a process, the gas yield is 100,000 cu ft per hr at 850° F and 15 psi absolute, with the following volumetric analysis:

N_2	10.0%	H_2	20.0%
CO_2	5.0	CO	25.0
H_2O	15.0	NH_3	20.0
		C_2H_4	5.0
			100.0%

Before absorption of the NH_3 in H_2SO_4 the gases are cooled to 150° F in a heat exchanger. This results in the removal of 90% of the water by condensation. The NH_3 absorbed by the hot condensed water is negligible. Absorption of the NH_3 and removal of practically all of the residual H_2O vapor is accomplished by further cooling and absorption in sulfuric acid solution. Calculate:

(a) Average molecular weight of the original gas mixture (including the steam).

(b) Specific volume of the gases leaving the processing equipment at 850° F and 15 psi expressed as cubic feet per pound.

(c) Weight of gases (including steam) leaving the processing equipment expressed as pounds per hour.

(d) Weight of H_2O condensed in the heat exchanger, pounds per hour.

(e) Volume of gases (including remaining steam) leaving the heat exchanger at 150° F and 15 psi expressed as cubic feet per hour.

(f) Weight of $(NH_4)_2SO_4$ produced, pounds per hour.

(g) Volumetric analysis of the gases leaving the H_2SO_4 absorber.

13. A producer gas has the following composition by volume:

CO	23.0%
CO_2	4.4
O_2	2.6
N_2	70.0
	100 %

(a) Calculate the cubic feet of gas, at 70° F and 750 mm Hg pressure, per pound of carbon present. **Ans.** 119.2 cu ft per lb.

(b) Calculate the volume of air, at the conditions of part a, required for the combustion of 100 cu ft of the gas at the same conditions if it is desired that the total oxygen present before combustion shall be 20% in excess of that theoretically required. **Ans.** 53.3 cu ft.

(c) Calculate the percentage composition by volume of the gases leaving the burner of part b, assuming complete combustion. **Ans.** 19.32% CO_2; 1.62% O_2, 79.06% N_2.

(d) Calculate the volume of the gases leaving the combustion in parts b and c at a temperature of 600° F and a pressure of 750 mm Hg per 100 cu ft of gas burned. **Ans.** 283.8 cu ft.

14. The gas from a sulfur burner has the following composition by volume:

SO_3	0.8%
SO_2	7.8
O_2	12.2
N_2	79.2
	100 %

(a) Calculate the volume of the gas at 600° F and 29.2 in. Hg formed per pound of sulfur burned.

(b) Calculate the percentage excess oxygen supplied for the combustion above that required for complete oxidation to SO_3.

(c) Calculate the volume of air at 70° F and 29.2 in. Hg supplied for the combustion per pound of sulfur burned.

15. A furnace is to be designed to burn coke at the rate of 200 lb per hr. The coke has the following composition:

$$\begin{array}{ll} \text{Carbon} & 89.1\% \\ \text{Ash} & 10.9\% \end{array}$$

The grate efficiency of the furnace is such that 90% of the carbon present in the coke charged is burned. Air is supplied in 30% excess of that required for the complete combustion of all the carbon charged. It may be assumed that 97% of the carbon burned is oxidized to the dioxide, the remainder forming monoxide.

(a) Calculate the composition, by volume, of the flue gases leaving the furnace. **Ans.** CO_2 14.08%, CO 0.43%, O_2 6.66%, N_2 78.83%.

(b) If the flue gases leave the furnace at a temperature of 550° F and a pressure of 743 mm Hg, calculate the rate of flow of gases, in cubic feet per minute, for which the stack must be designed. **Ans.** 1157 cu ft per min.

16. A by-product coke oven produces one million cubic feet of coal gas per hour having the following analysis by volume:

$$\begin{array}{llll} C_6H_6 & 5.0\% & CO & 7.0\% \\ C_7H_8 & 5.0 & H_2 & 35.0 \\ CH_4 & 40.0 & CO_2 & 5.0 \\ & & N_2 & \underline{3.0} \\ & & & 100.0\% \end{array}$$

The gas leaves the oven at 20.0 psi absolute pressure and 740° F. After cooling to 100° F the benzene and toluene are completely removed by absorption.
Calculate:

(a) Average molecular weight of the gas leaving the oven and the absorber.

(b) Weight of gas leaving the oven and the absorber.

(c) Volumetric composition of the gas leaving the absorber.

(d) Weights of benzene and toluene absorbed.

17. In the fixation of nitrogen by the arc process, air is passed through a magnetically flattened electric arc. Some of the nitrogen is oxidized to NO, which on cooling oxidizes to NO_2. Of the NO_2 formed, 66% will be associated to N_2O_4 at 26° C. The gases are then passed into water-washed absorption towers where nitric acid is formed by the following reaction:

$$H_2O + 3NO_2 = NO + 2HNO_3$$

The NO liberated in this reaction will be reoxidized in part and form more nitric acid.

In the operation of such a plant it is found possible to produce gases from the arc furnace in which the nitric oxide is 2% by volume, while hot. The gases are cooled to 26° C at a pressure of 750 mm Hg before entering the absorption apparatus.

(a) Calculate the complete analysis by volume of the hot gases leaving the furnace, assuming that the air entering the furnace was of average atmospheric composition.   **Ans.**   78.0% N_2, 20.0% O_2, 2.0% NO.

(b) Calculate the partial pressures of the NO_2 and N_2O_4 in the gas entering the absorption apparatus.   **Ans.**   5.18 mm, 5.03 mm.

(c) Calculate the weight of HNO_3 formed per 1000 cu ft of gas entering the absorption system if the conversion to nitric acid of the combined nitrogen in the furnace gases is 85% complete.   **Ans.**   2.73 lb.

18. The gas leaving a gasoline stabilizer has the following analysis by volume:

C_3H_8	8.0%
CH_4	78.0
C_2H_6	10.0
C_4H_{10}	4.0
	100.0%

This gas leaving at 90° F and 16 psi absolute at a rate of 70,000 cu ft per hr is fed to a gas re-forming plant where the following reactions take place:

$$C_nH_{2n+2} + nH_2O = nCO + (2n + 1)H_2$$

$$CO + H_2O = CO_2 + H_2$$

The C_nH_{2n+2} gases are 95% converted and the resulting CO 90% converted by these reactions.   Calculate:

(a) Average molecular weight of the gas leaving the stabilizer.

(b) Weight of gas feed to the re-forming plant pounds per hour.

(c) Weight of hydrogen leaving the re-forming plant, pounds per hour.

Vapor Pressures

Liquefaction and the Liquid State. Molecules in the gaseous state of aggregation exhibit opposing tendencies. The translational kinetic energy possessed by each molecule represents a continual, random motion which tends to separate the molecules from one another and to cause them to be uniformly distributed throughout the entire available space. On the other hand, the attractive forces between

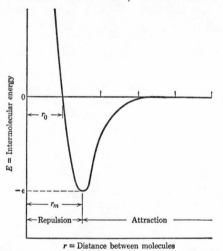

FIG. 14. Attractive force between molecules

the molecules tend to draw them together. If the molecules are pressed so close together as to cause distortion they tend to repel each other.

The first tendency, dispersion, is dependent on the temperature. An increase in temperature increases the translational kinetic energy of each molecule and therefore increases its ability to overcome the forces tending to draw it toward other molecules. The second tendency, aggregation, is determined by the magnitudes of the attractive forces between the molecules and by their proximity to one another. The intermolecular attractive forces increase to definite maxima as the distances between molecules are diminished. This behavior is shown in Fig. 14 in which are plotted intermolecular energy as ordinates and distances of separation between two molecules as abscissas. The greatest attractive force between the two molecules exists when they are separated by a distance r_m corresponding to a minimum value of intermolecular energy. If the distance of separation is diminished below r_m the potential energy rapidly changes to a value of zero at distance r_0. Any attempt made to bring the two molecules closer together than r_m will result in repulsion and an increase in the intermolecular energy.

An empirical equation commonly used to express the relation is given by Lennard-Jones as

$$E = 4\epsilon \left[\left(\frac{r_0}{r} \right)^{12} - \left(\frac{r_0}{r} \right)^6 \right] \tag{1}$$

where E = intermolecular energy between two colliding molecules
  r = distance between centers of colliding molecules
  r_0 = distance of separation when energies of attraction and repulsion are balanced
  ϵ = minimum intermolecular energy

According to equation 1 the energy of repulsion varies inversely with the twelfth power of the intermolecular distance, and the energy of attraction varies inversely with the sixth power of this distance. The total intermolecular energy is equal to the energy of repulsion minus the energy of attraction.

When the energy of attraction of one molecule for another exceeds its kinetic energy of translation, the molecules will form a dense aggregation which is termed a liquid. The characteristic that differentiates a liquid from a gas is the fact that a liquid does not necessarily occupy the entire available void space. The individual molecules of the liquid are in motion, owing to their internal kinetic energies, but this motion takes the form of vibrations, alternately increasing and decreasing the distances of separation.

Vaporization. As pointed out above, the liquid state results when conditions are such that the potential energies of attraction between molecules exceed their kinetic energies of translation. These conditions are brought about when the temperature of a substance is lowered, decreasing the kinetic energies of translation, or when the molecules are crowded close together, increasing the energies of attraction. On the basis of this theory the surface of a liquid may be pictured as a layer of molecules, each of which is bound to the molecules below it by the attractive forces among them. One of the surface molecules may be removed only by overcoming the attractive forces holding it to the others. This is possible if the molecule is given sufficient translational kinetic energy to overcome the maximum potential energy of attraction and to enable it to move past the point of maximum attraction. Once it has passed this distance of maximum attraction, the molecule is free to move away from the surface under the effect of its translational energy and to become a gas molecule.

In the simple kinetic-theory mechanisms which have been discussed, it is frequently assumed that all molecules of a substance at a given temperature are endowed with the same kinetic energies and move at the same speeds. Actually it has been demonstrated that this is not true and that molecular speeds and energies vary over wide ranges above and below the average values. In every liquid and gas there are always highly energized molecules moving at speeds much higher than the average. When such a molecule comes to the surface of a liquid, with its velocity directed away from the main body, it may have sufficient energy to break away completely from the forces tending to hold it to the surface. This phenomenon of the breaking away of highly energized molecules takes place from every exposed liquid surface. As a result, molecules of the liquid continually tend to assume the gaseous or vapor state. This phenomenon is termed *vaporization* or *evaporation*.

When a liquid evaporates into a space of limited dimensions, the space will become filled with the vapor that is formed. As vaporization proceeds, the number of molecules in the vapor state will increase and cause an increase in the pressure exerted by the vapor. It will be recalled that the pressure exerted by a gas or vapor is due to the impacts of its component molecules against the confining surfaces. Since the original liquid surface forms one of the walls confining the vapor, there will be a continual series of impacts against it by the molecules in the vapor state. The number of such impacts will be dependent on or will determine the pressure exerted by the vapor.

However, when one of these gaseous molecules strikes the liquid surface, it comes under the influence of the attractive forces of the densely aggregated liquid molecules and will be held there, forming a part of the liquid once more. This phenomenon, the reverse of vaporization, is known as *condensation*. The rate of condensation is determined by the number of molecules striking the liquid surface per unit time, which in turn is determined by the pressure or density of the vapor.

It follows that when a liquid evaporates into a limited space, two opposing processes are in operation. The process of vaporization tends to change the liquid to the gaseous state. The process of condensation tends to change the gas which is formed by vaporization back into the liquid state. The rate of condensation is increased as vaporization proceeds and the pressure of the vapor increases. If sufficient liquid is present, the pressure of the vapor must ultimately reach such a value that the rate of condensation will equal the rate of vaporization. When this condition is reached, a dynamic equilibrium is established and the pressure of the vapor will remain unchanged, since the formation of new vapor is compensated by condensation. If the pressure of the vapor is changed in either direction from this equilibrium value, it will adjust itself and return to the equilibrium conditions, owing to the increase or decrease in the rate of condensation which results from the pressure change. The pressure exerted by the vapor at such equilibrium conditions is termed the *vapor pressure* of the liquid. All liquids and solids exhibit definite vapor pressures of greater or smaller degree at all temperatures.

The magnitude of the equilibrium vapor pressure is in no way dependent on the amounts of liquid and vapor, as long as any free liquid surface is present. This results from both the rate of loss and the rate of gain of molecules by the liquid being directly proportional to the area exposed to the vapor. At the equilibrium conditions when both rates are the same, a change in the area of the surface exposed will not affect the conditions in the vapor phase.

The nature of the liquid is the most important factor determining the magnitude of the equilibrium vapor pressure. Since all molecules are endowed with the same kinetic energies of translation at any specified temperature, the vapor pressure must be entirely dependent on the magnitudes of the maximum potential energies of attraction which must be overcome in vaporization. These potential energies are determined by the intermolecular attractive forces. Thus, if a substance has high intermolecular attractive forces, the rate of loss of molecules from its surface should be small and the corresponding equilibrium vapor pressure low. The magnitudes of the attractive forces are dependent on both the size and nature of the molecules, usually increasing with increased size and complexity. In general, among liquids of similar chemical natures, the vapor pressure at any specified temperature decreases with increasing molecular weight.

Superheat and Quality. A vapor that exists above its critical temperature is termed a gas. The distinction between a vapor and a gas is thus quite arbitrary, and the two terms are loosely interchanged. For example, carbon dioxide at room temperature is below its critical temperature and, strictly speaking, is a vapor. However, such a material is commonly referred to as a gas.

A vapor that exists under such conditions that its partial pressure is equal to its equilibrium vapor pressure is termed a *saturated vapor*, whether it exists alone or in the presence of other gases. The temperature at which a vapor is saturated is termed the *dew point* or *saturation temperature*. A vapor whose partial pressure is less than its equilibrium vapor pressure is termed a *superheated vapor*. The difference between its existing temperature and its saturation temperature is called its *degrees of superheat*.

If a saturated vapor is cooled or compressed, condensation will result, and what is termed a *wet vapor* is formed. If the vapor is in turbulent motion, considerable portions of the condensed liquid will remain in mechanical suspension as small drops in the vapor and be carried with it. The *quality* of a wet vapor is the fraction that the weight of vapor forms of the total weight of vapor and entrained liquid associated with it.

$$x = \frac{w_v}{w_v + w_L} \qquad (2)$$

where w_v = weight of vapor

     w_L = weight of liquid

     x = quality

     $100x$ = % quality

Thus, wet steam of 95% quality is a mixture of saturated water vapor and entrained drops of liquid water in which the weight of the vapor constitutes 95% of the total weight.

Boiling Point. When a liquid surface is exposed to a space in which the *total* gas pressure is less than the equilibrium vapor pressure of the liquid, rapid vaporization known as *boiling* takes place. Boiling results from the formation of tiny free spaces within the liquid itself. If the equilibrium vapor pressure is greater than the total pressure on the surface of the liquid, vaporization will take place in these free spaces that tend to form below the liquid surface. This vaporization will cause the formation of bubbles of vapor which crowd back the surrounding liquid and increase in size because of the greater pressure of the vapor. Such a bubble of vapor will rise to the surface of the liquid and join the main body of gas above it. Thus, when a liquid boils, vaporization takes place not only at the surface level but also at many interior surfaces of contact between the liquid and bubbles of vapor. The rising bubbles also break up the normal surface into more or less of a froth. The vapor once liberated from the liquid is at a higher pressure than the gas in which it finds itself and will immediately expand and flow away from the surface. These factors all contribute to make vaporization of a liquid relatively very rapid when boiling takes place. When the total pressure is such that boiling does not take place, vaporization will nevertheless continue, but at a slower rate, as long as the vapor pressure of the liquid exceeds the partial pressure of its vapor above the surface.

The temperature at which the equilibrium vapor pressure of a liquid equals the total pressure on the surface is known as the *boiling point*. The boiling point is dependent on the total pressure, increasing with an increase in pressure. Theoretically, any liquid may be made to boil at any desired temperature in the liquid range by sufficiently altering the total pressure on its surface. The temperature at which a liquid boils when under a total pressure of 1.0 atm is termed the *normal boiling point*. This is the temperature at which the equilibrium vapor pressure equals 760 mm of mercury or 1.0 atm.

Vapor Pressures of Solids. Solid substances possess a tendency to disperse directly into the vapor state and to exert a vapor pressure just as do liquids. The transition of a solid directly into the gaseous state is termed *sublimation*, a process entirely analogous to the vaporization of a liquid. A familiar example of sublimation is the disappearance of snow in subzero weather.

The vapor pressure of a solid is a function of the nature of the material and its temperature. Sublimation will take place whenever the partial pressure of the vapor in contact with a solid surface is less than the equilibrium vapor pressure of the solid. Conversely, if the equilibrium vapor pressure of the solid is exceeded by the partial pressure of its vapor, condensation directly from the gaseous to the solid state will result.

At the melting point the vapor pressures of a substance in the solid and liquid states are equal. At temperatures above the melting point the solid state cannot exist. However, by careful cooling a liquid can be caused to exist in an unstable, supercooled state at temperatures below its melting point. The vapor pressures of supercooled liquids are always greater than those of the solid state at the same temperature, and the liquid tends to change to the solid.

The vapor pressures of solids, even at their melting points, are generally small. However, in some cases these values become large and of considerable importance. For example, at its melting point of 114.5° C an iodine crystal exerts a vapor pressure of 90 mm of mercury.

Solid carbon dioxide at its melting point of $-56.7°$ C exerts a vapor pressure of 5.11 atm and a pressure of 1.0 atm at a temperature of $-78.5°$ C. It is therefore impossible for liquid carbon dioxide to exist in a stable form at pressures less than 5.11 atm.

Calculations dealing with the vapor pressures and sublimation of solids are analogous to those of the vaporization of liquids. The principles and methods outlined in the following sections are equally applicable to sublimation and to vaporization processes.

Effect of Temperature on Vapor Pressure

The forces causing the vaporization of a liquid are derived from the kinetic energy of translation of its molecules. An increase in kinetic energy of molecular translation should increase the rate of vaporization tnd therefore the vapor pressure. In Chapter 3 it was pointed out ahat the kinetic energy of translation is directly proportional to the absolute temperature. On the basis of this theory, an increase in temperature should cause an increased rate of vaporization and a higher equilibrium vapor pressure. This is found to be universally the case. It must be remembered that it is the temperature of the liquid surface that is effective in determining the rate of vaporization and the vapor pressure.

An exact thermodynamic relationship between vapor pressure and temperature is developed in Chapter 13 as

$$\frac{dp}{dT} = \frac{\Lambda}{T(V_G - V_L)} \tag{3}$$

where p = vapor pressure
T = absolute temperature
Λ = heat of vaporization at temperature T
V_G = volume of gas
V_L = volume of liquid

The above relationship is the Clapeyron equation. It is rigorous, and is applied to any vaporization equilibrium. Its use in this form is, however, greatly restricted because it presupposes a knowledge of Λ, V_G, and V_L and their variation with temperature.

The latent heat of vaporization Λ is the quantity of heat that must be added in order to transform a substance from the liquid to the vapor state at the same temperature. The heat of vaporization decreases as pressure increases, becoming zero at the critical point. This property is fully discussed in subsequent chapters. Values of the heats of vaporization at the normal boiling point of many compounds are listed in Table 26, page 274.

If the volume of liquid is neglected and the applicability of the ideal-gas law assumed, the above relation reduces to the Clausius-Clapeyron equation

$$\frac{dp}{p} = \frac{\lambda dT}{RT^2} \quad \text{or} \quad d \ln p = -\frac{\lambda d}{R}\left(\frac{1}{T}\right) \tag{4}$$

where R = gas-law constant
λ = molal heat of vaporization

The Clausius-Clapeyron equation in the form written above is accurate only when the vapor pressure is relatively low, where it may be assumed that the vapor obeys the ideal gas law and that the volume in the liquid state is negligible compared with that of the vapor state.

Where the temperature does not vary over wide limits, it may be assumed that the molal latent heat of vaporization is constant and equation 4 may be integrated, between the limits p_0, T_0, and p, T, to give

$$\ln \frac{p}{p_0} = \frac{\lambda}{R}\left(\frac{1}{T_0} - \frac{1}{T}\right) \tag{5}$$

or

$$\log \frac{p}{p_0} = \frac{\lambda}{2.303R}\left(\frac{1}{T_0} - \frac{1}{T}\right) \tag{6}$$

Equation 6 permits calculation of the vapor pressure of a substance at a temperature T if the vapor pressure p_0 at another temperature T_0 is known, together with the latent heat of vaporization λ. The results are accurate only over limited ranges of temperature in which it may be assumed that the latent heat of vaporization is constant, and at such conditions that the ideal-gas law is obeyed.

Illustration 1. The vapor pressure of ethyl ether is given in the International Critical Tables as 185 mm Hg at 0° C. The latent heat of vaporization is 92.5 cal per gram at 0° C. Calculate the vapor pressure at 20 and at 35° C.

Molecular weight	74	
λ	6850	cal per g-mole
R	1.99	cal per g-mole per K°
T_0	273° K	
p_0	185 mm Hg	

when $\qquad T = 293°$ K (20° C)

$$\log \frac{p}{185} = \frac{6850}{2.30 \times 1.99}\left(\frac{1}{273} - \frac{1}{293}\right) = 1495\,(0.003663 - 0.003413) = 0.374$$

$$\frac{p}{185} = 2.36, \qquad p = 437 \text{ at } 20° \text{ C}$$

when $T = 308° K (35° C)$

$$\log \frac{p}{185} = 1495 (0.003663 - 0.003247) = 0.621$$

$$\frac{p}{185} = 4.18, \qquad p = 773 \text{ mm Hg at } 35° \text{ C}$$

The values for the vapor pressure of ether which have been experimentally observed are 442 mm Hg at 20° C and 775.5 mm Hg at 35° C.

In the preceding illustration the Clausius-Clapeyron equation yields results that are satisfactory for many purposes. However, equation 5 is only an approximation which may lead to considerable error in some cases. It should be used only in the absence of experimental data.

Tables of physical data contain experimentally determined values of the vapor pressures of many substances at various temperatures. Because of the frequent requirement of accurate values of the vapor pressure of water, extensive data are presented in Table 5 expressed in English units.

Vapor-Pressure Plots

From experimental data various types of plots have been devised for relating vapor pressures to temperature. Use of an ordinary uniform scale of coordinates does not result in a satisfactory plot because of the extreme curvature encountered. A single chart cannot be used over a wide temperature range without sacrifice of accuracy at the lower temperatures, and the rapidly changing slope makes both interpolation and extrapolation uncertain.

A better method which has been extensively used is to plot the logarithm of the vapor pressure (log p) against the reciprocal of the absolute temperature $(1/T)$. The resulting curves, while not straight, show much less curvature than a rectangular plot and may be read accurately over wide ranges of temperature. Another method is to plot the logarithm of the pressure against temperature on a uniform scale. This method does not reduce the curvature of the vapor-pressure lines as much as the use of the reciprocal temperature scale but is more easy to construct and read.

As a means of deriving consistent vapor-pressure data for homologous series of closely related compounds Coates and Brown developed a special method of plotting which has proved particularly valuable for the hydrocarbons.[1] For this plot rectangular coordinate paper is

[1] J. Coates and G. G. Brown, A Vapor Pressure Chart for Hydrocarbons, *Univ. Mich. Dept. Eng. Research Cir. Ser. No.* **2**, Dec. 1928.

used with temperatures as abscissas and normal boiling points as ordinates. Curved lines of constant vapor pressure are then plotted from the experimental data available for the various members of the series. This method is particularly well adapted to extrapolating data obtained for the lower-boiling homologs of a series in order to estimate vapor pressures for the higher-boiling homologs.

Reference-Substance Plots. The methods of plotting described above all result in lines having some degree of curvature, which makes necessary a considerable number of experimental data for the complete definition of the vapor-pressure curve. Where only limited data are available there is great advantage to a method of plotting that will yield straight lines over a wide range of conditions. With such a method a complete curve can be established from only two experimental points and erratic data can be detected.

Where an accurate evaluation of a physical property has been developed over a wide range of conditions for one substance the resulting relationship frequently may be made the basis of empirical plots for other substances of not greatly different properties. This general method may be applied to vapor-pressure data by selecting a reference substance the temperature–vapor pressure relationship of which has been evaluated over a wide range. A function of the temperature at which some other substance exhibits a given vapor pressure may then be plotted against the same function of temperature at which the reference substance has the same vapor pressure. Or, conversely, a function of the vapor pressure of the substance at a given temperature may be plotted against the same function of vapor pressure of the reference substance at the same temperature.

By proper selection of the reference substance and the functions of the properties plotted, curves that approximate straight lines over wide ranges of conditions are obtained. The best results are obtained with reference substances as similar as possible in chemical structure and physical properties to the compounds of interest.

Equal-Pressure Reference-Substance Plots. The first reference-substance plot of vapor-pressure data was proposed by Dühring, who plotted the temperature at which the substance of interest has a given vapor pressure against the temperature at which the reference substance has the same vapor pressure. Dühring lines of sodium hydroxide solutions are plotted in Fig. 17, page 104, with water used as the reference substance. Each of these lines relates the temperature of the designated solution to the temperature at which water exerts the same vapor pressure. Vapor-pressure data for water appear in Table 5.

Tᴀʙʟᴇ 5. Vᴀᴘᴏʀ Pʀᴇssᴜʀᴇ ᴏғ Wᴀᴛᴇʀ

English units

Pressure of aqueous vapor over *ice* in 10^{-3} inches of Hg from $-144°$ to $32°$F

Temp °F	0.0	2.0	4.0	6.0	8.0
−140	0.00095	0.00075	0.00063		
−130	0.00276	0.00224	0.00181	0.00146	0.00118
−120	0.00728	0.00595	0.00492	0.00409	0.00339
−110	0.0190	0.0157	0.0132	0.0111	0.00906
−100	0.0463	0.0387	0.0325	0.0274	0.0228
−90	0.106	0.0902	0.0764	0.0646	0.0543
−80	0.236	0.202	0.171	0.146	0.125
−70	0.496	0.429	0.370	0.318	0.275
−60	1.02	0.882	0.764	0.663	0.575
−50	2.00	1.75	1.53	1.33	1.16
−40	3.80	3.37	2.91	2.59	2.29
−30	7.047	6.268	5.539	4.882	4.315
−20	12.64	11.26	10.06	8.902	7.906
−10	22.13	19.80	17.72	15.83	14.17
−0	37.72	33.94	30.55	27.48	24.65
0	37.72	41.85	46.42	51.46	56.93
10	62.95	69.65	76.77	84.65	93.35
20	102.8	113.1	124.4	136.6	150.0
30	164.6	180.3			

Pressure of aqueous vapor over *water* in inches of Hg from $+4°$ to $212°$F

Temp °F	0.0	2.0	4.0	6.0	8.0
0			0.05402	0.05929	0.06480
10	0.07091	0.07760	0.08461	0.09228	0.1007
20	0.1097	0.1193	0.1299	0.1411	0.1532
30	0.1664	0.1803	0.1955	0.2118	0.2292
40	0.2478	0.2677	0.2891	0.3120	0.3364
50	0.3626	0.3906	0.4203	0.4520	0.4858
60	0.5218	0.5601	0.6009	0.6442	0.6903
70	0.7392	0.7912	0.8462	0.9046	0.9666
80	1.0321	1.1016	1.1750	1.2527	1.3347
90	1.4215	1.5131	1.6097	1.7117	1.8192
100	1.9325	2.0519	2.1775	2.3099	2.4491
110	2.5955	2.7494	2.9111	3.0806	3.2589
120	3.4458	3.6420	3.8475	4.0629	4.2887
130	4.5251	4.7725	5.0314	5.3022	5.5852
140	5.8812	6.1903	6.5132	6.850	7.202
150	7.569	7.952	8.351	8.767	9.200
160	9.652	10.122	10.611	11.120	11.649
170	12.199	12.772	13.366	13.983	14.625
180	15.291	15.982	16.699	17.443	18.214
190	19.014	19.843	20.703	21.593	22.515
200	23.467	24.455	25.475	26.531	27.625
210	28.755	29.922			

TABLE 5.   (*Continued*)

Pressure of aqueous vapor over *water* in lb/sq in. for temperatures 210–705.5°F

Temp °F	0.0	2.0	4.0	6.0	8.0
210	14.123	14.696	15.289	15.901	16.533
220	17.186	17.861	18.557	19.275	20.016
230	20.780	21.567	22.379	23.217	24.080
240	24.969	25.884	26.827	27.798	28.797
250	29.825	30.884	31.973	33.093	34.245
260	35.429	36.646	37.897	39.182	40.502
270	41.858	43.252	44.682	46.150	47.657
280	49.203	50.790	52.418	54.088	55.800
290	57.556	59.356	61.201	63.091	65.028
300	67.013	69.046	71.127	73.259	75.442
310	77.68	79.96	82.30	84.70	87.15
320	89.66	92.22	94.84	97.52	100.26
330	103.06	105.92	108.85	111.84	114.89
340	118.01	121.20	124.45	127.77	131.17
350	134.63	138.16	141.77	145.45	149.21
360	153.04	156.95	160.93	165.00	169.15
370	173.37	177.68	182.07	186.55	191.12
380	195.77	200.50	205.33	210.25	215.26
390	220.37	225.56	230.85	236.24	241.73
400	247.31	252.9	258.8	264.7	270.6
410	276.75	282.9	289.2	295.7	302.2
420	308.83	315.5	322.3	329.4	336.6
430	343.72	351.1	358.5	366.1	374.0
440	381.59	389.7	397.7	405.8	414.2
450	422.6	431.2	439.8	448.7	457.7
460	466.9	476.2	485.6	495.2	504.8
470	514.7	524.6	534.7	544.9	555.4
480	566.1	576.9	587.8	589.9	610.1
490	621.4	632.9	644.6	656.6	668.7
500	680.8	693.2	705.8	718.6	731.4
510	744.3	757.6	770.9	784.5	798.1
520	812.4	826.6	840.8	855.2	870.0
530	885.0	900.1	915.3	930.9	946.6
540	962.5	978.7	995.0	1011.5	1028.2
550	1045.2	1062.3	1079.6	1097.2	1115.1
560	1133.1	1151.3	1169.7	1188.5	1207.4
570	1226.5	1245.8	1265.3	1285.1	1305.3
580	1325.8	1346.4	1367.2	1388.1	1409.5
590	1431.2	1453.0	1475.0	1497.4	1520.0
600	1542.9	1566.2	1589.4	1613.2	1637.1
610	1661.2	1686.0	1710.7	1735.6	1761.0
620	1786.6	1812.3	1838.6	1865.2	1892.1
630	1919.3	1947.0	1974.5	2002.7	2031.1
640	2059.7	2088.8	2118.0	2147.7	2178.0
650	2208.2	2239.2	2270.1	2301.4	2333.3
660	2365.4	2398.1	2431.0	2464.2	2498.1
670	2531.8	2566.0	2601.0	2636.4	2672.1
680	2708.1	2745.0	2782.0	2819.1	2857.0
690	2895.1	2934.0	2973.5	3013.2	3053.2
700	3093.7	3134.9	3176.7	3206.2*	

* At 705.5°F, the critical temperature.

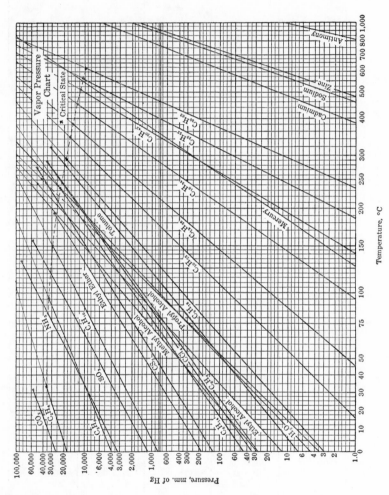

Fig. 15.   Vapor-pressure chart

Equal-Temperature Reference-Substance Plots.   Where the logarithm of the vapor pressure of a substance is plotted against the logarithm of the vapor pressure of a reference substance, both at the same temperature, a nearly straight line results.

The equal-temperature reference substance vapor pressure chart shown in Fig. 15 was constructed by plotting vapor pressures as ordinates against reference-substance vapor pressures on multicycle logarithmic paper.   From the vapor-pressure data of the reference substance an auxiliary abscissa scale of temperatures was established.   In extending the range of the chart to temperatures higher than the critical temperature of the first reference substance a second higher-boiling reference substance was selected and its vapor-pressure data were plotted over the temperature range of the first reference substance.   The vapor-pressure line of the second reference substance was then extended, and from it the extension of the auxiliary temperature abscissa scale was established.   In constructing Fig. 15 water was used as the primary reference substance and mercury was used as the reference substance for temperatures above the critical temperature of water.   In Fig. 15 the data for widely different types of materials yield lines with little curvature and the lines for a particular group of closely related compounds tend to converge at a single point which is characteristic of that group.   For example, single points of convergence were found for each of the following groups: the paraffin hydrocarbons, the benzene mono-halides, the alcohols, and the metals.   For a member of a group of materials having convergent lines only one experimental point and the point of convergence of the group are necessary to establish a complete relationship.

Figure 15 was constructed by Hougen and Watson[2a] giving credit to Cox[2] because of his original suggestion of the equal temperature method. The procedure was first published in detail by Othmer[3] and is referred to as the Othmer plot.   Extensive methods of correlating vapor pressures and other physical properties have been presented in a series of papers by Othmer and summarized in 1957.[3a]

When using water as the reference substance Calingaert and Davis[4] found that the vapor pressure relations of a substance as shown in Fig. 15 may be represented by the following equation:

$$\ln p = A - \frac{B}{T - 43} \tag{7}$$

[2] E. R. Cox, *Ind. Eng. Chem.*, **15**, 592 (1923).

[2a] O. A. Hougen and K. M. Watson, "Industrial Chemical Calculations", John Wiley & Sons (1931).

[3] D. F. Othmer, *Ind. Eng. Chem.*, **32**, 841–56 (1940).

[3a] D. F. Othmer, *Ind. Eng. Chem.*, **49**, 125–137 (1957).

[4] G. Calingaert and D. S. Davis, *Ind. Eng. Chem.*, **17**, 1287 (1925).

where p = vapor pressure

T = temperature, degrees Kelvin

A, B = empirical constants

Thus, by plotting log p against $1/(T - 43)$ a straight line should be obtained.

Illustration 2. The vapor pressure of chloroform is 61.0 mm Hg at 0° C and 526 mm Hg at 50° C. Estimate, from Fig. 15, the vapor pressure at 100° C.

Solution: The two experimental values of the vapor pressures at 0 and 50° C are represented by points on Fig. 15. A straight line is projected through these two points to the abscissa representing 100° C. The ordinate at this point is approximately 2450 mm Hg, the estimated vapor pressure at 100° C. The experimentally observed value is 2430 mm Hg.

Illustration 3. The vapor pressure of normal butyl alcohol at 40° C is 18.6 mm Hg. Estimate the temperature at which the vapor pressure is 760 mm Hg, the normal boiling point.

Solution: The experimental value of the vapor pressure at 40° C is represented by a point on Fig. 15. A straight line is drawn from this point to the point of convergence of the alcohol group. This point of convergence is located by extending the curves for methyl and propyl alcohols. The abscissa of the point at which this line crosses the 760-mm ordinate is about 117° C. The experimentally observed boiling point of normal butyl alcohol is 117.7° C.

Polarity of Molecules. The polarity of a molecule refers to the distribution of electrostatic charges caused by the unequal sharing of covalent atomic bonds. In a polar compound, these charges are unbalanced, giving the molecule a dipole moment and causing it to rotate when placed in an electrostatic field. Compounds that do not have a symmetrical arrangement such as water, ammonia, acids, and many organic compounds may be expected to be polar. Nonpolar compounds are, in general, chemically inactive, conduct electricity poorly, and do not ionize. Molecules having symmetrical arrangements such as methane and carbon tetrachloride are nonpolar. Nearly all hydrocarbons are nonpolar.

Critical Properties. Whether or not a pure substance can exist in the liquid state is dependent on its temperature. If the temperature is sufficiently high that the kinetic energies of translation of the molecules exceed the *maximum* potential energy of attraction between them, the liquid state of aggregation is impossible. The temperature at which the molecular kinetic energy of translation equals the maximum potential energy of attraction is termed the *critical temperature* T_c. Above the critical temperature the liquid state is impossible for a single component, and compression results only in a highly compressed gas, retaining all the properties of the gaseous state. Below the critical temperature a gas may be liquefied if sufficiently compressed.

The pressure required to liquefy a gas at its critical temperature is termed the *critical pressure* p_c. The critical pressure and temperature fix the *critical state* at which there is no distinction between the gaseous and liquid states. The volume at the critical state is termed the *critical volume* v_c. The density at the critical state is the *critical density* ρ_c. In Table 7A, page 92, are values of the critical data for the more common gases.

Reduced Conditions. At conditions equally removed from the critical state many properties of different substances are similarly related. This has given rise to the concept of reduced temperature, reduced pressure, and reduced volume. *Reduced temperature* is defined as the ratio of the existing temperature of a substance to its critical temperature, both being expressed on an absolute scale. Similarly, *reduced pressure* is the ratio of the existing pressure of a substance to its critical pressure, and the reduced volume the ratio of the existing molal volume to its critical molal volume. Thus,

$$\text{Reduced temperature } T_r = T/T_c$$

$$\text{Reduced pressure } \quad p_r = p/p_c$$

$$\text{Reduced volume } \quad v_r = v/v_c$$

Under conditions of equal reduced pressure and equal reduced temperature, substances are said to be in corresponding states. It will be later shown that many properties of gases and liquids, for example, the compressibilities of different gases, are nearly the same at corresponding states, that is, at equal reduced conditions.

Estimation of Critical Properties of Organic Substances

A knowledge of the critical properties of pure substances is of great importance in correlating or estimating related properties. Because of the difficulty of direct experimental measurements it is desirable to have reliable methods of estimating the values of the critical properties from molecular structure and easily determined properties. A comprehensive compilation of the critical constants of elements and compounds, together with an appraisal of various equations for predicting these constants, has been published by Kobe and Lynn.[5] This bulletin does not report the work of Riedel[6, 7] and of Michael and Thodos.[8]

[5] K. A. Kobe and R. E. Lynn, Jr. *Chem. Revs.*, **52**, Feb. 1953.

[6] L. Riedel, *Chem. Ing. Tech.*, **24**, 353–57 (1952).

[7] L. Riedel, *Z. Elektrochem.*, **53**, 222–28 (1949).

[8] A. V. Michael and G. Thodos, *Chem. Eng. Prog.*, Symposium No. 7, **49**, 131 (1953).

A further examination by Lydersen[9] of the original experimental data and methods of correlation shows that the group contribution method of Riedel is the most reliable over-all method as well as the simplest. This method was first proposed by Thomas.[10] Riedel's formulas for the atomic and group contributions to critical temperature and pressures have been slightly modified by Lydersen who also expanded the table to include the group contributions for critical volumes. The summarized group contributions for critical temperature, pressure, and volume are recorded in Table 6.

Critical temperatures may be estimated from the following equation expressed in terms of the normal boiling point and group contributions:[6,9]

$$\frac{T_b}{T_c} = 0.567 + \sum\Delta_T - (\sum\Delta_T)^2 \tag{8}$$

$\sum\Delta_T$ = summation of increments for the atoms and atomic groups as given in Table 6

The term $(\sum\Delta_T)^2$ is included to improve the correlation for large molecules.[9] The average deviation from the experimental data in using equation 8 is 1.0% based upon all available data (233 compounds) including old measurements dating back to 1882.

Critical pressures may be estimated from the following equation:[7]

$$\sqrt{\frac{M}{p_c}} = 0.34 + \sum\Delta_p \tag{9}$$

where M = molecular weight

$\sum\Delta_p$ = summation of increments for atoms and atomic groups as given in Table 6

p_c = critical pressure in atmospheres

The average deviation from experimental results in using equation 9 is 3.3% based upon all available data (159 compounds) including old measurements.

Critical volumes of organic compounds can be estimated from the following equation by Lydersen:

$$v_c = 40 + \sum\Delta_v \tag{10}$$

where v_c = critical volume in cubic centimeters per gram-mole

$\sum\Delta_v$ = summation of contributions for each atom or atomic group present as given in Table 6

[9] A. Lydersen, paper not yet published (1954).
[10] L. H. Thomas, *J. Am. Chem. Soc.*, **71**, 3411 (1949).

The average deviation of values calculated by this formula from experimental values is 2.4% based upon all available data (141 compounds) including old measurements.

Critical compressibility factors are calculated directly from experimental or estimated values of the three critical constants by the relations

$$z_c = \frac{p_c v_c}{RT_c} \tag{11}$$

The value of z_c obtained from estimated values of p_c, v_c, and T_c deviates from the experimental value by 3.4% based upon all available data (121 compounds).

Lydersen[9] has developed an independent correlation of the critical compressibility factor by relating it to the molal heat of vaporization, thus

$$z_c = \frac{1}{3.43 + 0.0067\lambda_b{}^2} \tag{12}$$

where λ_b = heat of vaporization at the normal boiling point in kilocalories per gram-mole.

This equation holds for both organic and inorganic compounds with an average deviation from experimental values of 3.8% based upon 156 compounds.

Critical Properties of Inorganic Substances

For *inorganic compounds* good correlations for critical constants are not available. In 1890 Guldberg[11] proposed the rule that the ratio of normal boiling point to the critical point was $\frac{2}{3}$.

For simple molecules containing one or two atoms (not counting the hydrogen atoms present)

$$T_b/T_c = 0.567 \tag{13}$$

and for other inorganic molecules

$$T_b/T_c = 0.635 \tag{14}$$

A tabulation of experimental values of critical constants for a selected list of inorganic and organic compounds is given in Tables 7A and 7B.

[11] C. M. Guldberg, *Z. physik. Chem.*, **5**, 374 (1890).

Illustration 3. Estimate the critical properties of chlorobenzene, $T_b = 405.2°$ K; $\lambda_b = 8735$ kcal per g-mole; $M = 112.50$.

		Δ_T	Δ_p	Δ_v
5 =CH	groups in ring	5(0.011)	5(0.154)	5(37)
		= 0.055	= 0.770	= 185
1 =C—	group in ring	0.011	0.154	36
1 =Cl		0.017	0.320	49
	$\Sigma\Delta_T =$	0.083	$\Sigma\Delta_p =$ 1.244	$\Sigma\Delta_v =$ 270
	$-(\Delta_T)^2 =$	−0.007		
		0.076		

$$\text{Constant} \quad 0.567 \qquad 0.34 \qquad 40$$

$$\frac{T_b}{T_c} = \quad 0.643 \qquad \sqrt{\frac{M}{p_c}} = \quad 1.584 \qquad v_c = \quad 310$$

Calculated values $T_c = 630.2°$ K $p_c = 44.9$ atm $v_c = 310$

Experimental values $T_c = 632.4°$ K $p_c = 44.6$ atm $v_c = 308$

$$z_c \text{ calculated from equation 12} \qquad = 0.254$$

$$z_c \text{ calculated from estimated constants} = 0.269$$

$$z_c \text{ experimental} \qquad = 0.265$$

Illustration 4. Estimate the critical properties of methyl propionate. $T_b = 352.°$ K; $\lambda_b = 7.718$ kcal per g-mole; $M = 88.10$.

		Δ_T	Δ_p	Δ_v
2 —CH₃ and 1 —CH₂	groups	3(0.020)	3(0.227)	3(55)
		= 0.060	= 0.681	= 165
1 —COO—	group	0.047	0.470	80
	$\Sigma\Delta_T =$	0.107	1.151	245
	$-(\Sigma\Delta_T)^2 =$	−.011		

$$\text{Constant} \quad 0.567 \qquad 0.34 \qquad 40$$

$$\frac{T_b}{T_c} = \quad 0.663 \quad \sqrt{\frac{M}{p_c}} = \quad 1.491 \quad v_c = \quad 285 \quad \text{ml/g-mole}$$

Estimated values $T_c = 531°$ K $p_c = 39.6$ atm $v_c = 285$ ml/g-mole

Experimental values $T_c = 531°$ K $p_c = 39.3$ atm $v_c = 282$ ml/g-mole

$$z_c \text{ calculated from equation 12} \qquad = 0.261$$

$$z_c \text{ calculated from estimated constants} \qquad = 0.259$$

$$z_c \text{ calculated from experimental constants} = 0.254$$

TABLE 6. INCREMENTS FOR USE IN CALCULATING CRITICAL CONSTANTS OF
ORGANIC COMPOUNDS

Use no increment for hydrogen.   All bonds indicated as free are connected with atoms other than hydrogen.

Atoms and atomic groups		Δ_T	Δ_p	Δ_v
—CH₃ and —CH₂—		0.020	0.227	55
	—CH₂— in ring	0.013	0.184	44.5
$-\overset{\mid}{\underset{\mid}{\text{CH}}}$		0.012	0.210	51
	when in ring	0.012	0.192	46.
=CH and =CH₂		0.018	0.198	45
	=CH in ring	0.011	0.154	37
$-\overset{\mid}{\underset{\mid}{\text{C}}}-$		0.0	0.210	41
	when in ring	(−0.007)	(0.154)	(31)
=C— and =C=		0.0	0.198	36
	when in ring	0.011	0.154	36
≡C— and ≡CH		0.005	0.153	(36)
—F		0.018	0.224	18
—Cl		0.017	0.320	49
—Br		0.010	(0.50)	(70)
—I		(0.012)	(0.83)	(95)
—O—		0.021	0.16	20
	when in ring	(0.014)	(0.12)	(8)
—OH (in alcohols)		0.082	0.06	(18)
—OH (in phenols)		(0.035)	(−0.02)	(3)
—CHO and >CO		0.048	0.33	73
>CO in ring		(0.041)	(0.2)	(63)
—COO—		0.047	0.47	80
—COOH		0.085	(0.4)	80
—NH₂		0.031	0.095	28
>NH		0.031	0.135	(37)
	when in ring	(0.024)	(0.09)	(27)
>N—		0.014	0.17	(42)
	when in ring	(0.007)	(0.13)	(32)
—CN		(0.060)	(0.36)	(80)
—SH and —S—		0.015	0.27	55
	—S— in ring	(0.008)	(0.24)	(45)

Numbers in parentheses are based upon too few experimental data to be reliable.

TABLE 7A. CRITICAL CONSTANTS OF INORGANIC COMPOUNDS

Compounds	Formula	Dipole Moment, (stat coulombs) (cm) 10^{18}	T_c, °K	p_c, atm	v_c, liters/ g-mole	z_c
Ammonia	NH_3	1.47	405.5	111.3	0.0725	0.243
Argon	A	<0.03	151.2	48.	0.075	0.290
Bromine	Br_2	0.0	584	102	0.144	0.306
Carbon dioxide	CO_2	0.0	304.2	72.9	0.094	0.275
Carbon disulfide	CS_2	0.0	552.	78.	0.170	0.293
Carbon monoxide	CO	0.11	133.	34.5	0.093	0.294
Chlorine	Cl_2	0.49	417.	76.1	0.124	0.276
Cyanogen	C_2N_2	—	400.	59.	—	—
Helium	He	<0.015	5.26	2.26	0.058	0.304
Hydrogen	H_2	<0.015	33.3	12.8	0.065	0.304
Hydrogen bromide	HBr	0.79	363.2	84.0	0.100	0.282
Hydrogen chloride	HCl	1.03	324.6	81.5	0.087	0.266
Hydrogen cyanide	HCN	2.93	456.7	53.2	0.139	0.197
Hydrogen sulfide	H_2S	0.93	373.6	88.9	0.098	0.284
Neon	Ne	<0.015	44.5	26.9	0.0417	0.307
Nitrogen	N_2	0.0	126.2	33.5	0.090	0.291
Nitric oxide	NO	0.16	179.2	65.0	0.058	0.256
Nitrogen peroxide	NO_2	0.29	431.	100.	0.082	0.232
Nitrous oxide	N_2O	0.17	309.7	71.7	0.0963	0.272
Oxygen	O_2	0.0	154.4	49.7	0.074	0.290
Ozone	O_3	0.0	268.	67.	0.0894	0.272
Phosgene	$COCl_2$	1.18	455.	56.	0.190	0.285
Phosphine	PH_3	0.55	324.5	64.5	0.113	0.274
Silicon tetrafluoride	SiF_4	—	259.1	36.7	—	—
Stannic chloride	$SnCl_4$	0	591.9	37.0	0.351	0.267
Sulfur	S	—	1313.	116.	—	—
Sulfur dioxide	SO_2	1.60	430.7	77.8	0.122	0.269
Sulfur trioxide	SO_3	0.0	491.4	83.8	0.126	0.262
Water	H_2O	1.84	647.3	218.2	0.056	0.230

Critical Pressures and Vapor Pressures of Organic Compounds.
It was found by Gamson and Watson[12] that the vapor-pressure data of all of over 40 substances investigated may be represented by the following equation:

$$\log p = \frac{-A}{T_r} + B - e^{-20(T_r - b)^2} \qquad (15)$$

where T_r = reduced temperature and A, B, and b are constants characteristic of the substance.

[12] B. W. Gamson and K. M. Watson, *Nat. Petrol. News, Tech. Section*, Sept. 6, 1944.

TABLE 7B. CRITICAL CONSTANTS OF ORGANIC COMPOUNDS

Name	Formula	T_c, °K	p_c, atm	v_c, liters/ g-mole	z_c
Acetic acid	$C_2H_4O_2$	594.8	57.1	0.171	0.200
Acetic anhydride	$C_4H_6O_3$	569.0	46.2	—	—
Acetone	C_3H_6O	508.7	46.6	0.213	0.238
Acetonitrile	C_2H_3N	547.9	47.7	0.173	0.184
Acetylene	C_2H_2	309.5	61.6	0.113	0.274
Aniline	C_6H_7N	698.8	52.3	0.274	0.250
Benzene	C_6H_6	562.1	48.6	0.260	0.274
Bromobenzene	C_6H_5Br	670.2	44.6	0.343	0.278
Butadiene (1,3)	C_4H_6	425.0	42.7	0.221	0.271
Butane (iso)	C_4H_{10}	408.1	36.0	0.263	0.262
Butane (n)	C_4H_{10}	425.2	37.47	0.255	0.274
Butene (cis-2)	C_4H_8	428.2	40.5	0.236	0.272
Butene (1)	C_4H_8	419.6	39.7	0.240	0.277
Butyl acetate (iso)	$C_6H_{12}O_2$	561.5	31.4	0.413	0.281
Butyric acid (n)	$C_4H_8O_2$	628.0	52.0	0.290	0.293
Carbon tetrachloride	CCl_4	556.4	45.0	0.276	0.272
Chlorobenzene	C_6H_5Cl	632.4	44.6	0.308	0.265
Chlorodifluoromethane	$CHClF_2$	369.6	48.5	0.165	0.264
Chloroform	$CHCl_3$	536.6	54.0	0.240	0.294
Cresol (m)	C_7H_8O	705.0	45.0	0.320	0.249
Cyclopentane	C_5H_{10}	511.76	44.55	0.260	0.276
Decane (n)	$C_{10}H_{22}$	619.0	20.8	0.602	0.247
Diethylamine	$C_4H_{11}N$	496.7	36.6	—	—
Diethyl ether	$C_4H_{10}O$	467.0	35.6	0.281	0.261
Diethyl sulfide	$C_4H_{10}S$	557.0	39.1	0.323	0.276
Dimethyl ether	C_2H_6O	400.1	52.6	0.187	0.300
Dimethyl sulfide	C_2H_6S	503.1	54.6	0.201	0.266
Dioxane	$C_4H_8O_2$	585.0	50.7	0.240	0.253
Diphenyl	$C_{12}H_{10}$	768.8	31.8	—	—
Eicosane (n)	$C_{20}H_{42}$	775.0	11.0	1.2	0.21
Ethane	C_2H_6	305.43	48.20	0.148	0.285
Ethylamine	C_2H_7N	456.4	55.5	0.181	0.268
Ethyl acetate	$C_4H_8O_2$	523.3	37.8	0.286	0.252
Ethyl alcohol	C_2H_6O	516.3	63.0	0.167	0.248
Ethyl bromide	C_2H_5Br	503.9	61.5	0.215	0.320
Ethyl chloride	C_2H_5Cl	460.4	52.0	—	—
Ethyl formate	$C_3H_6O_2$	508.5	46.8	0.229	0.257
Ethylene	C_2H_4	283.1	50.50	0.124	0.270
Ethyl benzene	C_8H_{10}	619.6	38.1	0.37	0.27
Ethyl mercaptan	C_2H_6S	499.0	54.2	0.207	0.274
Ethyl methyl ether	C_3H_8O	437.9	43.4	0.221	0.267
Ethyl methyl ketone	C_4H_8O	533.0	39.5	0.290	0.262
Ethylene oxide	C_2H_4O	468.0	71.0	0.138	0.255
Fluorobenzene	C_6H_5F	559.8	44.6	0.271	0.263
Freon 11	CCl_3F	471.2	43.2	0.248	0.277
Freon 12	CCl_2F_2	384.7	39.6	0.218	0.273

TABLE 7B.   (*Continued*)

CRITICAL CONSTANTS OF ORGANIC COMPOUNDS

Name	Formula	T_c, °K	p_c, atm	v_c, liters/ g-mole	z_c
Freon 13	$CClF_3$	302.0	38.2	0.180	0.278
Freon 21	$CHCl_2F$	451.7	51.0	0.197	0.271
Freon 113	$C_2Cl_3F_3$	487.3	33.7	0.325	0.274
Heptane (*n*)	C_7H_{16}	540.16	27.01	0.426	0.260
Hexane (*n*)	C_6H_{14}	507.9	29.92	0.368	0.264
Iodobenzene	C_6H_5I	721.0	44.6	0.351	0.265
Methane	CH_4	190.7	45.80	0.099	0.290
Methyl acetate	$C_3H_6O_2$	506.9	46.3	0.228	0.254
Methylamine	CH_3NH_2	430.2	73.1	—	—
Methyl chloride	CH_3Cl	416.3	65.9	0.143	0.276
Methyl alcohol	CH_4O	513.2	78.5	0.118	0.220
Methyl cyclopentane	C_6H_{12}	532.77	37.36	0.319	0.273
Methyl fluoride	CH_3F	317.8	58.0	0.113	0.251
Methyl formate	$C_2H_4O_2$	487.2	59.2	0.172	0.255
Methyl mercaptan	CH_4S	470.0	71.4	0.149	0.276
Methyl pentane (2)	C_6H_{14}	497.9	29.95	0.367	0.269
Methyl pentane (3)	C_6H_{14}	504.7	30.8	0.367	0.273
Naphthalene	$C_{10}H_8$	751.7	40.6	—	—
Nitromethane	CH_3NO_2	588.0	62.3	0.173	0.223
Nonane (*n*)	C_9H_{20}	595.0	22.5	0.543	0.250
Octane (*n*)	C_8H_{18}	569.4	24.64	0.486	0.256
Pentane (iso)	C_5H_{12}	461.0	32.9	0.308	0.268
Pentane (*n*)	C_5H_{12}	469.8	33.31	0.311	0.269
Pentane (neo)	C_5H_{12}	433.76	31.57	0.303	0.269
Propane	C_3H_8	369.9	42.01	0.200	0.277
Propionic acid	$C_3H_6O_2$	612.0	53.0	0.230	0.242
Propyl alcohol (*n*)	C_3H_8O	537.3	50.2	0.220	0.251
Propyl alcohol (iso)	C_3H_8O	508.8	53.0	0.219	0.278
Propyl amine (*n*)	C_3H_9N	497.0	46.8	—	—
Propyl benzene	C_9H_{12}	638.8	31.2	0.44	0.26
Propylene	C_3H_6	365.1	45.4	0.181	0.274
Triethylamine	$C_6H_{15}N$	535.4	30.0	0.403	0.275
Trimethylamine	C_3H_9N	433.3	40.2	0.254	0.287
Toluene	C_7H_8	594.0	41.6	0.320	0.273
Xylene (*m*)	C_8H_{10}	616.8	34.7	0.394	0.270
Xylene (*o*)	C_8H_{10}	631.6	36.9	0.390	0.278
Xylene (*p*)	C_8H_{10}	618.8	33.9	0.390	0.260

Data in Tables 7A and 7B were taken from:

Am. Petroleum Inst. Research Proj. **44**, edited by F. D. Rossini, Carnegie Institute of Technology (1952).

K. A. Kobe and R. E. Lynn, Jr., *Chem. Revs.*, **52**, no. 1 (1953).

Landolt-Börnstein, *Physikalisch-Chemische Tabellen*, 6th ed. (1951).

L. Riedel, *Chem.-Ing. Tech.*, 24, 353 (1952).

L. Riedel, *Z. Elektrochem.*, 53, 222 (1949).

In applying this equation over vapor-pressure ranges from a few tenths of a millimeter of mercury to the critical point, deviations were found to be generally less than 3%.

Because of the difficulty of evaluating the constants, equation 15 is not convenient for the extrapolation of fragmentary data. However, through generalized expressions for the constants of the equation, it may be used to predict a complete vapor-pressure curve from only a single measurement. The ranges of the constants are indicated in Table 8.

TABLE 8.   VAPOR-PRESSURE CONSTANTS

p in mm Hg

	A	B	b	p_c	T_c, ° K
Methane	2.3383	6.8800	0.000	34,810	190.7
Ethane	2.5728	7.1411	0.088	36,630	305.4
Ethylene	2.5463	7.1269	0.098	38,380	283.1
Propane	2.6606	7.1819	0.125	31,930	369.9
n-Octane	3.2316	7.5034	0.236	18,730	569.4
Water	3.1423	8.3610	0.163	165,800	647.3
Methyl alcohol	3.5876	8.3642	0.243	59,660	513.2
Diethyl ether	2.9726	7.4039	0.204	27,060	467.0
Acetone	3.0644	7.6173	0.180	35,420	508.7
Ammonia	2.9207	7.8519	0.163	84,590	405.5
Methylamine	2.9589	7.7066	0.239	55,560	430.2
Hydrogen cyanide	3.2044	7.7761	0.000	40,430	456.7
Methyl chloride	2.7195	7.4185	0.052	50,080	416.3
Carbon tetrachloride	2.7989	7.3329	0.158	34,200	556.4
Acetic acid	3.3908	8.0291	0.138	43,400	594.8

Applying equation 15 to the critical point gives

$$\log p_c = B - A - e^{-20(1-b)^2}$$

Since the exponential correction term is found to be negligible at reduced temperatures above 0.8, $B = \log p_c + A$, and equation 15 may be written

$$\log p_r = \frac{-A(1 - T_r)}{T_r} - e^{-20(T_r - b)^2} \qquad (16)$$

If the methods previously described are used to calculate the critical temperatures and pressures, it is possible to estimate a complete vapor-pressure relationship from a single point by use of equation 16 in conjunction with a generalized expression for the constant b.

It was found that, as the number of carbon atoms in a homologous series of organic compounds is increased above 2, the constant b increases. This relationship is shown graphically for the paraffin hydrocarbons in Fig. 16. For series of compounds other than the paraffin hydrocarbons, Fig. 16 may be used with the following equation for estimating the constant b

$$b = b' + \Delta b \qquad (17)$$

where b is the value for a compound containing n_c carbon atoms, and

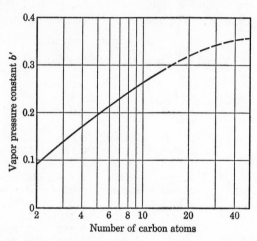

FIG. 16. Vapor-pressure constants of paraffin hydrocarbons

b' is the value read from Fig. 16 corresponding to n_c. Values of Δb are constants characteristic of the homologous series as given in Table 9.

TABLE 9. VAPOR-PRESSURE CONSTANTS

	Δb
Acids (organic)	0.05
Alcohols	0.22
Aldehydes	0.0(?)
Amines	0.12
Aromatic hydrocarbons (monocyclic)	−0.02
Esters	0.09
Ethers	0.04
Halogenated paraffins, mono-	0.08
Ketones	0.05
Naphthenes	−0.03
Nitriles	0.02
Phenols	0.0(?)
Olefins, mono-	0.01

Illustration 5. Estimate the vapor pressure of n-propylamine at 0° C. Its normal boiling point is 48.7° C.

From Fig. 16 at $n_c = 3$, $b' = 0.133$

From Table 9, $\Delta b = 0.12$

$$b = 0.133 + 0.12 = 0.253$$

At the normal boiling point,

$$T_r = \frac{322}{497} = 0.647; \qquad p_r = \frac{1}{46.8} = 0.0214$$

Substituting in equation 16 yields

$$-A\left(\frac{1 - 0.647}{0.647}\right) = \log 0.0216 + e^{-20(0.647-0.253)^2}, \qquad A = 2.9703$$

At 0° C,

$$T_r = \frac{273}{497} = 0.549, \qquad \log p_r = \frac{-2.9703(1 - 0.549)}{0.549} - e^{-20(0.549-0.253)^2} = -2.6139$$

$$p_r = 0.00243, \qquad p = 1.66 \text{ psi} \quad \text{or} \quad 85.8 \text{ mm Hg}$$

Vapor Pressure of Immiscible Liquids

In a system of immiscible liquids, where intimate mixing is maintained, an exposed liquid surface will consist of areas of each of the component liquids. Each of these components will vaporize at the surface and tend to establish an equilibrium value of the partial pressure of its vapor above the surface. As has been pointed out, the equilibrium vapor pressure of a liquid is independent of the relative proportions of liquid and vapor, but is determined by the temperature and the nature of the liquid. It follows from kinetic theory that the equilibrium vapor pressure of a liquid should be the same, whether it exists alone or as a part of a mixture, if a free surface of the pure liquid is exposed. In a nonhomogeneous mixture of immiscible liquids the vaporization and condensation of each component takes place at the respective surfaces of the pure liquids, independent of the natures or amounts of other components that may be present. Each component liquid actually exists in a pure state and as such exerts its normal equilibrium vapor pressure.

The total vapor pressure exerted by a mixture of immiscible liquids is the sum of the vapor pressures of the individual components at the existing temperature. When the vapor pressure of such a system equals the existing total pressure above its surface, the mixture will boil, giving off component vapors. Since each component of the mixture adds its own vapor pressure, depending only on the temperature, it follows that *the boiling point of a nonhomogeneous mixture must be lower than that of any one of its separate components.* This fact is made use of in the important industrial process of *steam distillation* of materials that are insoluble in water. If an immiscible material is mixed with water, it can be distilled at a temperature always below the boiling point of water corresponding to the existing total pressure. In this manner it is possible to distil waxes, fatty acids of high molecular weight, petroleum fractions, and the like, at relatively low temperatures and with less danger of decomposition than by other methods of distillation.

The composition of the vapors in equilibrium with or rising from a mixture of immiscible liquids is determined by the vapor pressures of the liquids. The partial pressure of each component in the vapor is equal to its vapor pressure in the liquid state. The ratio of the partial pressure to the total pressure gives the mole fraction or percentage by volume, from which the weight percentage of the component in the vapor may be calculated. The total vapor pressure exerted by a mixture of immiscible liquids is easily calculated as the sum of the vapor pressures of the component liquids. Conversely, the boiling point of the mixture under a specified total pressure is the temperature at which the sum of the individual vapor pressures equals the total pressure. This temperature is best determined by trial or by a graphical method in which a plot of total vapor pressure against temperature is prepared.

Illustration 6. It is proposed to purify benzene from small amounts of non-volatile solutes by subjecting it to distillation with saturated steam under atmospheric pressure of 745 mm Hg. Calculate (a) the temperature at which the distillation will proceed and (b) the weight of steam accompanying 1 lb of benzene vapor.

Solution: This problem may be solved by trial, using the data of Fig. 15.

Temperature, °C	Vapor Pressure C_6H_6, mm	Vapor Pressure H_2O, mm	Total Vapor Pressure, mm
60	390	150	540
70	550	235	785
65	460	190	650
68	510	215	725
69	520	225	745

The boiling point of the mixture is seen from the foregoing table to be 69° C.

Basis: 1 lb-mole of mixed vapor.

$$\text{Benzene} = \frac{520}{745} = 0.70 \text{ lb-mole or } 0.70 \times 78 = 55 \quad \text{lb}$$

$$\text{Water} = 0.30 \text{ lb-mole or} \qquad\qquad\qquad 5.4 \quad \text{lb}$$

$$\text{Steam per lb of benzene} = \frac{5.4}{55} \qquad\qquad = 0.099 \text{ lb}$$

In the preceding illustration it has been assumed that the steam and the liquid being distilled leave the still in the proportions determined by their vapor pressures. This will be so only when the liquids in the still are intimately mixed and when the steam that is introduced comes into intimate contact and equilibrium with the liquids. If these conditions are not realized, the proportion of steam in the vapors will be higher than that corresponding to the theoretical equilibrium.

Illustration 7. It is desired to purify myristic acid ($C_{13}H_{27}COOH$) by distilla-. tion with steam under atmospheric pressure of 740 mm. Calculate the temperature at which the distillation will proceed and the number of pounds of steam accompanying each pound of acid distilled.

$$\text{Vapor pressure of myristic acid at } 99° C = 0.032 \text{ mm Hg}$$

The vapor pressure of myristic acid is negligible in its effect on the boiling point of the mixture, which may be assumed to be that of water at 740 mm Hg, or 99° C.

Basis: 1 lb-mole of mixed vapors.

$$\text{Myristic acid} = \frac{0.032}{740} = 4.3 \times 10^{-5} \text{ lb-mole or } 4.3 \times 10^{-5} \times 228$$

$$= \quad 0.0098 \text{ lb}$$

Water = 1.0 lb-mole $\qquad\qquad\qquad\quad = \quad 18 \qquad \text{lb}$

Steam per lb of acid $= \dfrac{18}{0.0098} \qquad\qquad = 1840 \qquad \text{lb}$

Vaporization with Superheated Steam. The preceding illustration deals with an organic compound having a high boiling point which cannot be subjected to ordinary direct distillation at atmospheric pressure. By distillation with saturated steam the boiling point of the mixture is reduced below 100° C, but, as indicated by the results of the illustration, an enormous amount of steam must be used in order to obtain a small amount of product. An alternative method would be to conduct a direct distillation under a sufficiently reduced pressure to lower the boiling point to the desired temperature. However, the maintenance of high vacua in apparatus suitable for the vaporization of such materials is difficult and frequently impracticable.

These difficulties may be circumvented by maintaining the material to be vaporized at the highest permissible temperature and introducing superheated steam or some other inert gas. In this case there will be no liquid water in the system, and the superheated steam merely serves as a carrier which mixes with and removes the vapors of the material to be distilled. If the material being vaporized is allowed to reach equilibrium with its vapor, the partial pressure of the distillate vapor will be its equilibrium vapor pressure at the existing temperature. The partial pressure of the steam will be the difference between the existing total pressure and the partial pressure of the distillate vapor. The amount of steam required per unit quantity of distillate may, therefore, be diminished by either raising the temperature or lowering the total pressure. Distillation with superheated steam is frequently combined with reduced pressure in order to reduce the steam requirements for the distillation of high-boiling-point materials which will not withstand high temperatures.

Ordinarily the mixing of the steam with the material being vaporized will not be sufficiently intimate to result in equilibrium conditions. The steam will then leave the liquid without being completely saturated with distillate vapor.

Illustration 8. Myristic acid is to be distilled at a temperature of 200° C by use of superheated steam. It may be assumed that the relative saturation of the steam with acid vapors will be 80%.

(a) Calculate the weight of steam required per pound of acid vaporized if the distillation is conducted at an atmospheric pressure of 740 mm Hg.

(b) Calculate the weight of steam per pound of acid if a vacuum of 26 in. Hg is maintained in the apparatus.

Vapor pressure of myristic acid at 200° C = 14.5 mm Hg

Basis: 1 lb-mole of mixed vapors.

(a) Partial pressure of acid = 14.5 × 0.80 = 11.6 mm Hg

$$\text{Mass of acid} = \frac{11.6}{740} = 0.0157 \text{ lb-mole or } 0.0157 \times 228 = \quad 3.58 \text{ lb}$$

$$\text{Mass of water} = 0.9843 \text{ lb-mole or} \qquad\qquad\qquad 17.7 \quad \text{lb}$$

$$\text{Steam per lb of acid} = \frac{17.7}{3.58} \qquad\qquad\qquad = \quad 4.95 \text{ lb}$$

(b) Total pressure = 740 − (26 × 25.4) = 80 mm

$$\text{Mass of acid} = \frac{11.6}{80} = 0.145 \text{ lb-mole or } 33.1 \quad \text{lb}$$

$$\text{Mass of water} = 0.855 \text{ lb-mole or} \qquad 15.4 \quad \text{lb}$$

$$\text{Steam per lb of acid} = \frac{15.4}{33.1} \qquad\qquad 0.465 \text{ lb}$$

Solutions

The surface of a homogeneous solution contains molecules of all its components, each of which has an opportunity to enter the vapor state. However, the number of molecules of any one component per unit area of surface will be less than if that component exposed the same area of surface in the pure liquid state. For this reason the rate of vaporization of a substance will be less per unit area of surface when in solution than when present as a pure liquid. However, any molecule from a homogeneous solution which is in the vapor state may strike the surface of the solution at any point and will be absorbed by it, re-entering the liquid state. Thus, although the opportunity for vaporization of any one component is diminished by the presence of the others, the opportunity for the condensation of its vapor molecules is unaffected. For this reason, the equilibrium vapor pressure which is exerted by a component in a solution will be, in general, less than that of the pure substance.

This situation is entirely different from that of a nonhomogeneous mixture. In a nonhomogeneous mixture the rate of vaporization of either component, per unit area of total surface, is diminished because the effective surface exposed is reduced by the presence of the other component. However, condensation of a component can take place only at the restricted areas where the vapor molecules impinge upon its own molecules. Thus, both the rate of vaporization and the rate of condensation are reduced in the same proportion, and the equilibrium vapor pressure of each component is unaffected by the presence of the others.

Raoult's Law. The generalization known as *Raoult's law* states that the equilibrium vapor pressure that is exerted by a component in a solution is proportional to the mole fraction of that component. Thus,

$$p_A = P_A \left(\frac{n_A}{n_A + n_B + n_C + \cdots} \right) = N_A P_A \qquad (18)$$

where
$$p_A = \text{vapor pressure of component } A \text{ in solution with components } B, C, \ldots$$

$$P_A = \text{vapor pressure of } A \text{ in the pure state}$$

$$n_A, n_B, n_C \ldots = \text{moles of components } A, B, C, \ldots$$

$$N_A = \text{mole fraction of } A$$

From the kinetic theory of equilibrium vapor pressures it would be expected that this generalization would be correct when the following conditions exist:

1. No chemical combination or molecular association between unlike molecules takes place in the formation of the solution.

2. The sizes of the component molecules are approximately equal.

3. The attractive forces between like and unlike molecules are approximately equal.

4. The component molecules are nonpolar, and no component is concentrated at the surface of the solution.

Few combinations of liquids would be expected to fulfill all these conditions, and it is not surprising that Raoult's law represents only a more or less rough approximation to actual conditions.   Where the conditions are fulfilled, a solution will be formed from its components without thermal change, and without change in total volume.   A solution that exhibits these properties is termed an *ideal* or *perfect* solution.   Solutions that approximate the ideal are formed only by liquids of closely related natures such as the homologs of a series of nonpolar organic compounds.   For example, paraffin hydrocarbons of not too widely separated characteristics form almost ideal solutions in each other.   The behavior of the ideal solution is useful as a criterion by which to judge solutions and also as a means of approximately predicting quantitative data for solutions that would not be expected to deviate widely from ideal behavior.   For the accuracy required in the majority of industrial problems, a great many solutions of chemically similar materials may be included in this class.

Equilibrium Vapor Pressure and Composition.   If the validity of Raoult's law is assumed, it is necessary to have only the vapor-pressure data for the pure components in order to predict the pressure and composition of the vapor in equilibrium with a solution.   The total vapor pressure of the solution will be the sum of the partial pressures of the components, each of which may be calculated from equation (18).

Illustration 9.   Calculate the total pressure and the composition of the vapors in contact with a solution at 100° C containing 35% benzene (C_6H_6), 40% toluene ($C_6H_5CH_3$), and 25% orthoxylene ($C_6H_4(CH_3)_2$) by weight.

Vapor pressures at 100° C:

$$Benzene \ = 1340 \text{ mm Hg}$$
$$Toluene \ = \ \ 560 \text{ mm Hg}$$
$$o\text{-Xylene} = \ \ 210 \text{ mm Hg}$$

Basis: 100 lb of solution.

$$Benzene \ = \ \ 35 \text{ lb or } \frac{35}{78} \ = 0.449 \text{ lb-mole}$$

$$Toluene \ = \ \ 40 \text{ lb or } \frac{40}{92} \ = 0.435$$

$$o\text{-Xylene} = \ \ 25 \text{ lb or } \frac{25}{106} = 0.236$$

$$Total = \overline{100} \text{ lb or} \qquad \overline{1.120} \text{ lb-mole}$$

Vapor pressures:

$$Benzene \ = 1340 \times \frac{0.449}{1.120} = 1340 \times 0.401 = 536 \text{ mm Hg}$$

$$\text{Toluene} = 560 \times \frac{0.435}{1.120} = 560 \times 0.388 \quad = 217$$

$$o\text{-Xylene} = 210 \times \frac{0.236}{1.120} = 210 \times 0.211 \quad = \;\; 44$$

$$\text{Total} = \overline{797} \text{ mm Hg}$$

Molal percentage compositions:

	Liquid, %	Vapor, %
Benzene	40.1	536/797 = 67.3
Toluene	38.8	217/797 = 27.2
o-Xylene	21.1	44/797 = 5.5
	100.0	100.0

In a similar manner the vapor pressure of the solution at any other temperature might be calculated and a curve plotted relating total vapor pressure to temperature. From such a curve the boiling point of the solution at any specified pressure may be predicted. It will be noted that the composition of the vapor may differ widely from that of the solution, depending on the relative volatilities of the separate components. In the special case of a solution containing a nonvolatile component the vapor will contain none of this component, but its presence in the liquid will diminish the partial pressures of the other components in the same proportion that it reduces their mole fractions.

Nonvolatile Solutes. If one component of a binary solution has a negligible vapor pressure, its presence will have no effect on the composition of the vapor in equilibrium with the solution. The vapor will consist entirely of the volatile component, but its equilibrium pressure will be less than that of the pure liquid at the same temperature. Thus, a nonvolatile solute produces a *vapor-pressure lowering* or a *boiling-point elevation* in its solvent. If the components possess closely related characteristics, the system may approach ideal behavior. Then the total vapor pressure will be the product of the vapor pressure and the mole fraction of the solvent. With ionizing or associating solutes the effective mole fraction of the solute is dependent on the degree of ionization or association. For these reasons, the theories of ideal behavior are not valid in the estimation of vapor-pressure data for such solutions, which include those in which water is the solvent.

If the vapor pressure of a solution is known at two temperatures, these data will establish a straight line on a reference-substance chart prepared according to either the method of Othmer or that of Dühring, pages 85 and 81. In Fig. 17 are the Dühring lines corresponding to various concentrations of aqueous sodium hydroxide solutions. Where sufficient data are available, it is advisable to plot the temperature of the solution against that of the pure solvent, in this case water. When this method is used, the curve representing zero concentration of solute will be a straight line of unit slope. By interpolation between a set of Dühring lines the boiling point of a solution under any desired pressure or the vapor pressure at any temperature may be estimated.

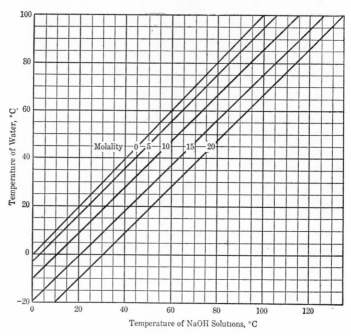

Fig. 17. Dühring lines of aqueous solutions of sodium hydroxide

An equal-temperature reference-substance plot may be applied to solutions using one of the components as the reference substance. Such a plot, developed by Othmer for sulfuric acid solutions, is shown in Fig. 18. This type of plot has the advantage over the Dühring plot of permitting estimation of thermal data from the slopes of the vapor-pressure lines.

The difference between the boiling point of a solution and that of the pure solvent is termed the *boiling-point elevation* of the solution. It will be noted that the lines of Fig. 17 diverge but slightly at the higher temperatures. It follows that the boiling-point elevation of a solution of sodium hydroxide is practically independent of temperature or pressure. Although several systems exhibit this behavior it can by no means be considered general, and the fact that the Dühring lines of sodium hydroxide solutions happen to be almost parallel is merely a characteristic of this system. It may be easily demonstrated that the boiling-point elevation of an ideal solution increases rapidly with an increase in temperature.

Relative Vapor Pressure. When it is desired to approximate the complete vapor-pressure data for a solution from only a single experi-

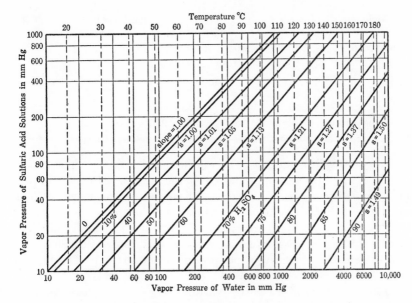

Fig. 18. Vapor pressure of sulfuric acid solutions (From Othmer, *Ind. Eng. Chem.* **32**, 847, 1940, with permission)

mental observation, a modified form of Raoult's law will frequently give good results:

$$p = kp_s \qquad (19)$$

where p = vapor pressure of solution

p_s = vapor pressure of pure solvent

k = a factor, dependent on concentration

For an ideal solution the factor k will equal the mole fraction of the solvent and will be independent of temperature or pressure. For nonideal solutions k may differ widely from the mole fraction, but in many cases it will be practically independent of temperature or pressure for a solution of a given composition. The factor k is sometimes termed the *relative vapor pressure* of a solution. The value of k for a solution may be obtained by a single determination of boiling point. Equation (19) may then be used to estimate the vapor pressures at other temperatures.

Illustration 10. An aqueous solution of sodium chloride contains 5 g-moles of NaCl per 1000 grams of water. The normal boiling point of this solution is 106° C. Estimate its vapor pressure at 25° C.

$$\text{Mole fraction of water} = \left(1 - \frac{5}{55.5 + 5}\right) = 1 - 0.0826 = 0.9174$$

From Table 5, p_s at 106° C = 940   mm Hg

$$p_s \text{ at } 25° \text{ C} = 23.5 \text{ mm Hg}$$

$$k = \frac{760}{940} = 0.81$$

Vapor pressure of solution at 25° C = 23.5 × 0.81 = 19.0   mm Hg

Experimentally observed value   = 18.97 mm Hg

In the preceding illustration it will be noted that the value of k is widely different from the mole fraction of the solvent. However, for this particular system it is, for all practical purposes, independent of temperature. In certain systems, notably aqueous solutions of strong bases, such constancy does not exist, and the relative vapor pressure may vary considerably. For example, a solution of caustic soda having a molality of 10 has a relative vapor pressure k of 0.584 at 100° C and only 0.479 at 25° C. When the type of behavior of a system is unknown, it is desirable to obtain at least two experimental points and to establish a Dühring line if it is desired to predict reliable values of vapor pressure.

Problems

1. (a) Obtaining the necessary data from a physical table, plot a curve relating the vapor pressure of acetic acid ($C_2H_4O_2$) in millimeters of mercury to temperature in degrees centigrade. Plot the curve for the temperature range from 20 to 140° C, using vapor pressures as ordinates and temperatures as abscissas, both on uniform scales.

(b) Ethylene glycol ($OHCH_2 \cdot CH_2OH$) has a normal boiling point of 197° C. At a temperature of 120° C it exerts a vapor pressure of 39 mm Hg. From these data construct a Dühring line for ethylene glycol, using water as the reference substance. From this line estimate the vapor pressure at 160° C and the boiling point under a pressure of 100 mm Hg. **Ans.** 216 mm Hg, 141° C.

(c) Ethyl bromide (C_2H_5Br) exerts a vapor pressure of 165 mm Hg at 0° C and has a normal boiling point of 38.4° C. From Fig. 15 estimate its vapor pressure at 60° C. **Ans.** 1550 mm Hg.

(d) Nonane (C_9H_{20}) has a normal boiling point of 150.6° C. From Fig. 15 estimate its boiling point under a pressure of 100 mm Hg. **Ans.** 87° C.

2. For one of the substances from the following list, prepare a table of data and graphs as indicated below:

Substance	Temperature Range, ° C	Pressure Units
Ammonia	−30 to 24	atm
Carbon dioxide	−38 to critical temp.	atm
Carbon disulfide	0 to 46.3	mm Hg
Chlorine	−34.6 to 30	atm
Sulfur dioxide	0 to 50	atm
Acetone	10 to 60	mm Hg
Carbon tetrachloride	25 to 80	mm Hg
Chloroform	10 to 70	mm Hg
Ethyl alcohol	35 to 80	mm Hg
Ethyl ether	34.6 to 120	atm
Methyl chloride	−24.0 to 40	atm

(a) From a handbook of physical-chemical data, tabulate the following values. Indicate source of data.

Column 1.   Temperature, ° C (t)
Column 2.   Absolute temperature, ° K (T)
Column 3. $1/T$
Column 4.   Vapor pressure (p)

(b) Using graph paper with uniform scales, plot vapor pressures as ordinates and temperature as abscissas. Draw a smooth curve through the plotted points.

(c) Using semilog paper, plot vapor pressures as ordinates on the logarithmic scale versus $1/T$ as abscissas. Choose the scale for $1/T$ so that the curve will have a slope as close to 45° as feasible. Draw a smooth curve through the plotted points.

(d) Using semilog paper, plot vapor pressure as ordinates and temperature, degrees centigrade, as abscissas.

3. Construct Dühring and equal-temperature reference-substance charts for ethyl alcohol, using water as the reference substance. Plot the curve using the following data:

Temperature, ° C	Vapor Pressure of Ethyl Alcohol, mm Hg
10	23.6
78.3	760.0

Using the Dühring line based on these figures, determine

(a) The vapor pressure at 60° C.   **Ans.**   355 mm Hg.
(b) The boiling point under 500 mm pressure.   **Ans.**   68° C.

4. From lines constructed on Fig. 15 determine the boiling point at 2000 mm Hg of one of the following substances. Use the vapor-pressure data that are given below to establish the lines on Fig. 15.

Ethyl Acetate	Ethyl Formate	Sulfur	
0° C 24.2 mm Hg	0° C 72.4 mm Hg	250° C	12 mm Hg
160° C 8.349 atm	200° C 28.0 atm	444.6° C	760.0 mm Hg
Ans. 112° C.	**Ans.** 85° C.	**Ans.** 515° C.	

5. The normal boiling point of benzene is 80° C. Estimate its critical properties by equations 8, 9, and 10.

6. In the following table, the normal boiling points are given for several hydrocarbons. Calculate the critical constants by the method of group contributions.

Compare these computed values with the experimental values to indicate the accuracy that may be expected by this method.

Hydrocarbon	Normal Boiling point, ° C
n-Butane	-0.3
n-Pentane	36.1
n-Hexane	69.0
n-Heptane	98.4
n-Octane	125.8
Cyclohexane	80.8

7. By the method of group contributions estimate the critical properties of acetone, acetic acid, chlorobenzene, and ethyl alcohol, and compare with the values in Table 7B.

8. Using the data of Fig. 16 and Tables 8 and 9, calculate the necessary data and plot vapor-pressure curves extending from 0.01 psi to the critical point for the following substances. Plot the curves on five-cycle semilogarithmic paper, using vapor pressures as ordinates on the logarithmic scale and temperatures in degrees Fahrenheit as abscissas.

    (*a*) Decane ($t_B = 174°$ C).
    (*b*) Toluene ($t_B = 110.8°$ C).
    (*c*) Propyl ethyl ether ($t_B = 64°$ C).

9. Using the data of Fig. 15 estimate the temperature required for the distillation of hexadecane ($C_{16}H_{34}$) at a pressure of 750 mm Hg in the presence of liquid water. Calculate the weight of steam evolved per pound of hexadecane distilled. **Ans.** 99.6° C, 43.6 lb steam.

10. A fuel gas has the following analysis by volume (in the third column are the normal boiling points of the pure components):

Components	%	Boiling Point, ° C
Ethane (C_2H_6)	2.0	-88
Propane (C_3H_8)	40.0	-44
Isobutane (C_4H_{10})	7.0	-10
Normal butane (C_4H_{10})	47.0	0
Pentanes (C_5H_{12})	4.0	$+30$ (average)
	100.0	

It is proposed to liquefy this gas for sale in cylinders and tankcars.

    (*a*) Calculate the vapor pressure of the liquid at 20° C and the composition of the vapor evolved. (The vapor pressures may be estimated from Fig. 15 and the normal boiling points.)

    (*b*) Calculate the vapor pressure of the liquid at 20° C if all the ethane were removed.

11. Assuming that benzene (C_6H_6) and chlorobenzene (C_6H_5Cl) form ideal solutions, plot curves relating total and partial vapor pressures to mole percentages of benzene in the solution at temperatures of 90, 100, 110, and 120° C. Also plot the curves relating total vapor pressure to the mole percentage of benzene in the vapor at 90 and 120° C. The normal boiling point of benzene is 79.6° C. Following are other vapor-pressure data:

Vapor Pressure, mm Hg

Temperature, ° C	Chlorobenzene	Benzene
90	208	1013
100	293	1340
110	403	1744
120	542	2235
132.1	760	2965

12. From the curves of problem 11, plot the isobaric boiling-point curves of benzene-chlorobenzene solutions under a pressure of 760 mm Hg. The points on both the liquid and vapor curves corresponding to temperatures of 90, 100, 110, and 120° C should be used in establishing the curves.

13. An aqueous solution of $NaNO_3$ containing 10 g-moles of solute per 1000 grams of water boils at a temperature of 108.7° C under a pressure of 760 mm Hg. Assuming that the relative vapor pressure of the solution is independent of temperature, calculate the vapor pressure of the solution at 30° C and the boiling-point elevation produced at this pressure. **Ans.** 23.50 mm Hg, 5.18° C.

14. The following table gives the vapor pressures of pure hexane and pure heptane.

Vapor Pressure, mm Hg

Temperature, ° C	Hexane	Heptane
69	760	295
70	780	302
75	915	348
80	1060	426
85	1225	498
90	1405	588
95	1577	675
99.2	1765	760

(a) Assuming that Raoult's law is valid, use the above data to calculate for each of the above temperatures the mole per cent x of hexane in the liquid and the mole per cent y of hexane in the vapor, at a total pressure of 760 mm Hg.

(b) Plot the results obtained in part a, using compositions of liquid and of vapor as abscissas and temperature as ordinates.

(c) Plot y as ordinates versus x as abscissas.

15. A solution of methanol in water containing 0.158 mole fraction alcohol boils at 84.1° C (760 mm pressure). The resulting vapor contains 0.553 mole fraction of methanol.

	At 80° C	At 100 ° C
Vapor pressure of methanol	1.764 atm	3.452 atm

How does the actual composition of the vapor compare with the composition calculated from Raoult's law?

Humidity and Saturation

When a gas or a gaseous mixture remains in contact with a liquid surface, it will acquire vapor from the liquid until the partial pressure of the vapor in the gas mixture equals the vapor pressure of the liquid at its existing temperature. When the vapor concentration reaches this equilibrium value the gas is said to be *saturated* with the vapor. It is not possible for the gas to contain a greater stable concentration of vapor, because, as soon as the vapor pressure of the liquid is exceeded by the partial pressure of the vapor, condensation takes place. The vapor content of a saturated gas is determined entirely by the vapor pressure of the liquid and may be predicted directly from vapor-pressure data.

The pure-component volume of the vapor in a saturated gas may be calculated from the relationships derived in Chapter 3. Thus, if the ideal-gas law is applicable,

$$V_v = V \frac{p_v}{p} \tag{1}$$

where V_v = pure-component volume of vapor
  p_v = partial pressure of vapor = vapor pressure of liquid at existing temperature
  V = total volume
  p = total pressure

From equation 1 the percentage composition by volume of a vapor-saturated gas may be calculated. When the ideal-gas law is applicable, the composition by volume of a vapor-saturated gas is independent of the nature of the gas but is dependent on the nature and temperature of the liquid and on the total pressure. The composition by weight varies with the natures of both the gas and the liquid, the temperature, and the total pressure.

For certain types of engineering problems it is convenient to use special methods of expression for the vapor content of a gas. The weight of vapor per unit volume of vapor-gas mixture, the weight of vapor per unit weight of vapor-free gas, and the moles of vapor per mole of vapor-free gas are three common and useful methods of expression. When a gas is saturated with vapor, the composition expressed by the first method is independent of both the nature of the gas and the total pressure but varies with the nature and temperature of the liquid. When the composition is expressed by the third method, it varies with the nature of the liquid, the temperature, and the pressure but is independent of the nature of the gas. From a knowledge of the equilibrium vapor pressure of the liquid the compositions of vapor-saturated gases may be readily calculated in any of these methods of expression, using the principles developed in Chapters 3 and 4.

Illustration 1. Ethyl ether at a temperature of 20° C exerts a vapor pressure of 442 mm Hg. Calculate the composition of a saturated mixture of nitrogen and ether vapor at a temperature of 20° C and a pressure of 745 mm Hg expressed in the following terms:

(a) Percentage composition by volume.
(b) Percentage composition by weight.
(c) Pounds of vapor per cubic foot of mixture.
(d) Pounds of vapor per pound of vapor-free gas.
(e) Pound-moles of vapor per pound-mole of vapor-free gas.

(a) *Basis:* 1.0 cu ft of mixture.

$$\text{Pure-component volume of vapor} = 1.0 \times \frac{442}{745} = 0.593 \text{ cu ft}$$

Composition by volume:

Ether vapor	59.3%
Nitrogen	40.7%

(b) *Basis:* 1.0 lb-mole of the mixture.

Vapor present = 0.593 lb-mole or	43.9 lb
Nitrogen present = 0.407 lb-mole or	11.4
Total mixture	55.3 lb

Composition by weight:

Ether vapor	79.4%
Nitrogen	20.6%

(c) *Basis:* Same as b.

$$\text{Volume} = 359 \times \frac{760}{745} \times \frac{293}{273} = 393 \text{ cu ft}$$

$$\text{Weight of ether per cu ft} = \frac{43.9}{393} = 0.112 \text{ lb}$$

This result is independent of the total pressure. For example, an increase in the total pressure would decrease the volume per mole of mixture but would correspondingly decrease the weight of vapor per mole of mixture.

(*d*) *Basis:* Same as *b*.

Weight of vapor per lb nitrogen $= \dfrac{43.9}{11.4} = 3.85$ lb

(*e*) *Basis:* Same as *b*.

Moles of vapor per mole of nitrogen $= \dfrac{0.593}{0.407} = 1.455$

Partial Saturation.   If a gas contains a vapor in such proportions that its partial pressure is less than the vapor pressure of the liquid at the existing temperature, the mixture is but partially saturated. The *relative saturation* of such a mixture may be defined as the percentage ratio of the partial pressure of the vapor to the vapor pressure of the liquid at the existing temperature.   The relative saturation is therefore a function of both the composition of the mixture and its temperature as well as of the nature of the vapor.

From its definition it follows that the relative saturation also represents the following ratios:

(*a*) The ratio of the percentage of vapor by volume to the percentage by volume that would be present were the gas saturated at the existing temperature and total pressure.

(*b*) The ratio of the weight of vapor per unit volume of mixture to the weight per unit volume present at saturation at the existing temperature and total pressure.

Another useful means for expressing the degree of saturation of a vapor-bearing gas may be termed the *percentage saturation*.   The percentage saturation is defined as the percentage ratio of the existing weight of vapor per unit weight of vapor-free gas to the weight of vapor that would exist per unit weight of vapor-free gas if the mixture were saturated at the existing temperature and pressure.   The percentage saturation also represents the ratio of the existing moles of vapor per mole of vapor-free gas to the moles of vapor that would be present per mole of vapor-free gas if the mixture were saturated at the existing temperature and pressure.

Care must be exercised that the *relative saturation* and the *percentage saturation* are not confused.   They approach equality when the vapor concentrations approach zero but are different at all other conditions. The quantitative relationship between the two terms is readily derived from their definition.   Thus,

$$\text{Relative saturation in } \% = \frac{p_v}{p_s}\,(100) = y_r \tag{2}$$

where p_v = partial pressure of vapor
p_s = vapor pressure of pure liquid

$$\% \text{ saturation} = \frac{n_v}{n_s}(100) = y_p \tag{3}$$

where n_v = moles of vapor per mole of vapor-free gas actually present
n_s = moles of vapor per mole of vapor-free gas at saturation

From Dalton's law,

$$\frac{n_v}{1} = \frac{p_v}{p - p_v} \quad \text{and} \quad \frac{n_s}{1} = \frac{p_s}{p - p_s} \tag{4}$$

or

$$\frac{n_v}{n_s} = \frac{p_v}{p_s}\left(\frac{p - p_s}{p - p_v}\right) \tag{5}$$

Hence

$$y_p = y_r\left(\frac{p - p_s}{p - p_v}\right) \tag{6}$$

where p = total pressure

Illustration 2. A mixture of acetone vapor and nitrogen contains 14.8% acetone by volume. Calculate the relative saturation and the percentage saturation of the mixture at a temperature of 20° C and a pressure of 745 mm Hg.

Vapor pressure of acetone at 20° C	= 184.8 mm Hg
Partial pressure of acetone = 0.148 × 745	= 110.0 mm Hg
Relative saturation = 110/184.8	= 59.7%

Basis: 1.0 lb-mole of mixture.

Acetone	= 0.148 lb-mole
Nitrogen	= 0.852 lb-mole
Moles of acetone per mole of nitrogen = 0.148/0.852 = 0.174	

Basis: 1.0 lb-mole of saturated mixture at 20° C and 745 mm Hg.

% by volume of acetone = 184.8/745	= 24.8%
Lb-moles of acetone	= 0.248
Lb-moles of nitrogen	= 0.752
Moles of acetone per mole of nitrogen = 0.248/0.752 =	0.329
% saturation = 0.174/0.329	= 52.9%

As indicated by this illustration, the percentage saturation is always smaller than the relative saturation.

The composition of a partially saturated gas-vapor mixture is fixed if the relative or percentage saturation and the temperature and pressure are specified. From this information and a knowledge of the equilibrium vapor pressure at this temperature the composition may be expressed in any other terms. Conversely, the relative or percentage saturation may be calculated if the composition, pressure, and temperature are specified. The temperature required to produce a specified degree of saturation may be calculated if the composition at a specified pressure is known.

Illustration 3. Moist air is found to contain 8.1 grains of water vapor per cubic foot at a temperature of 30° C. Calculate the temperature to which it must be heated in order that its relative saturation shall be 15%.

Basis: 1 cu ft of moist air.

$$\text{Water} = \frac{8.1}{7000} = 1.16 \times 10^{-3} \text{ lb or} \qquad\qquad 6.42 \times 10^{-5} \text{ lb-mole}$$

Pure-component volume of water vapor $= 6.42$
$$\times 10^{-5} \times 359 = 0.0230 \text{ cu ft at S.C.}$$

$$\text{Partial pressure of water vapor} = 760 \times \frac{0.0230}{1.0} \times \frac{303}{273} = 19.4 \text{ mm Hg}$$

Vapor pressure of water at temperature correspond-
$$\text{ing to 15\% relative saturation} = \frac{19.4}{0.15} = 130 \text{ mm Hg}$$

From the vapor-pressure data for water it is found that this pressure corresponds to a temperature of 57° C.

Humidity. Because of the widespread occurrence of water vapor in gases of all kinds, special attention has been given to this case and a special terminology has been developed. The humidity H of a gas is generally defined as the weight of water per unit weight of moisture-free gas. The molal humidity H_m is the number of moles of water per mole of moisture-free gas. When the vapor under consideration is water, the percentage saturation is termed the percentage humidity. The relative saturation becomes the relative humidity.

Considerable confusion exists in the literature in the use of these terms, and care must always be exercised to avoid misuse. The terminology recommended above is an extension of that proposed by Grosvenor.[1]

Dew Point. If an unsaturated mixture of vapor and gas is cooled, the relative amounts of the components and the percentage composition by volume will at first remain unchanged. It follows that, if the total pressure is constant, the partial pressure of the vapor will be unchanged by cooling. This will be the case until the temperature is lowered to such a value that the vapor pressure of the pure liquid at this temperature is equal to the existing partial pressure of the vapor in the mixture. The mixture will then be saturated, and any

---

[1] W. M. Grosvenor, *Trans. Am. Inst. Chem. Engrs.*, **1**, 184 (1908).

further cooling will result in condensation. The temperature at which the equilibrium vapor pressure of the liquid is equal to the existing partial pressure of the vapor is termed the *dew point* of the mixture.

The vapor content of a vapor-gas mixture may be calculated from dew-point data, or, conversely, the dew point may be predicted from the composition of the mixture.

Illustration 4. A mixture of benzene vapor and air contains 10.1% benzene by volume.

(*a*) Calculate the dew point of the mixture when at a temperature of 25° C and a pressure of 750 mm Hg.

(*b*) Calculate the dew point when the mixture is at a temperature of 30° C and a pressure of 750 mm Hg.

(*c*) Calculate the dew point when the mixture is at a temperature of 30° C and a pressure of 700 mm Hg.

Solution

(*a*) Partial pressure of benzene = 0.101 × 750 = 75.7 mm Hg
From the vapor-pressure data for benzene (Fig. 15) it is found that this pressure corresponds to a temperature of 20.0° C, the dew point.

(*b*) Partial pressure of benzene   = 75.7 mm
      Dew point   = 20.0° C
(*c*) Partial pressure of benzene = 0.101 × 700 = 70.7 mm Hg

The temperature corresponding to a vapor pressure of 70.7 mm Hg is found to be 18.7° C. From these results it is seen that the dew point does not depend on the temperature but does vary with the total pressure.

Vaporization Processes

The manufacturing operations of drying, air conditioning, and certain types of evaporation all involve the vaporization of a liquid into a stream of gases. In dealing with such operations it is of interest to calculate the relationships between the quantities and volumes of gases entering and leaving and the quantity of material evaporated. Such problems are of the general class that was discussed on page 60 under the heading of "Volume Changes with Change in Composition." The concentrations of vapor in these problems are generally expressed in terms of the dew points, the relative saturations, or the moles of vapor per mole of vapor-free gas. The first two methods of expression are convenient because they are directly determined from dew point or wet- and dry-bulb temperature measurements. From such data the partial pressures of vapor may be readily calculated and the partial-pressure method of solution might be used as described in Chapter 3.

The vaporization processes all require the introduction of energy in the form of heat. The effective utilization of this heat is frequently the most important factor governing the operation of the process, and

a knowledge of the relationships between the quantity of heat introduced and that dissipated in various ways is of great significance.
The calculation of such an energy balance is greatly simplified if the
quantities of all materials concerned are expressed in molal or weight
units rather than in volumes. These units have the advantage of
expressing quantity independent of change of temperature and pressure.
The same desirability of weight or molal units arises when relationships are derived for the design of vaporization equipment. For these
reasons it has become customary to express all data in either weight
or molal units where thermal calculations are to be made or where
design relationships are to be used. The molal units are preferable.
From data expressed in molal units, volumes at any desired conditions
may be readily obtained.

It will be noted that in any mixture, following the ideal-gas law,
the ratio of the number of moles of vapor to the number of moles of
vapor-free gas is equal to the ratio of the partial pressure of the vapor
to the partial pressure of the vapor-free gas.

Illustration 5. It is proposed to recover acetone, which is used as a solvent in an
extraction process, by evaporation into a stream of nitrogen. The nitrogen enters
the evaporator at a temperature of 30° C containing acetone such that its dew point
is 10° C. It leaves at a temperature of 25° C with a dew point of 20° C. The barometric pressure is constant at 750 mm Hg.

(a) Calculate the vapor concentrations of the gases entering and leaving the evaporator, expressed in moles of vapor per mole of vapor-free gas.

(b) Calculate the moles of acetone evaporated per mole of vapor-free gas passing
through the evaporator.

(c) Calculate the weight of acetone evaporated per 1000 cu ft of gases entering
the evaporator.

(d) Calculate the volume of gases leaving the evaporator per 1000 cu ft entering.

Vapor pressure of acetone:

    116 mm Hg at 10° C
    185 mm Hg at 20° C

Solution

(a) Entering gases:

Partial pressure of acetone	= 116 mm Hg
Partial pressure of nitrogen = 750 − 116	= 634 mm Hg
Moles of acetone per mole of nitrogen = 116/634 = 0.183	

Leaving gases:

Partial pressure of acetone	≈ 185 mm Hg
Partial pressure of nitrogen = 750 − 185	= 565 mm Hg
Moles of acetone per mole of nitrogen = 185/565 = 0.328	

(b) *Basis:* 1.0 lb-mole of nitrogen.

Acetone leaving the process	= 0.328 lb-mole
Acetone entering the process	= 0.183
Acetone evaporated	= 0.145 lb-mole

(c) *Basis:* 1.0 lb-mole of nitrogen.

Total gas entering the process = $1.0 + 0.183$ $\qquad$ = 1.183 lb-moles

Volume of gas entering = $1.183 \times 359 \times \dfrac{760}{750} \times \dfrac{303}{273}$ $\qquad$ = 477 cu ft

Molecular weight of acetone $\qquad$ = 58

Weight of acetone evaporated = 58×0.145 $\qquad$ = 8.4 lb

Acetone evaporated per 1000 cu ft of gas entering =

$\dfrac{8.4}{477} \times 1000$ $\qquad$ = 17.6 lb

(d) *Basis:* 1.0 lb-mole of nitrogen.

Total gas leaving the process = $1.0 + 0.328$ $\qquad$ = 1.328 lb-moles

Volume of gas leaving = $1.328 \times 359 \times \dfrac{760}{750} \times \dfrac{298}{273}$ $\qquad$ = 526 cu ft

Volume of gas leaving per 1000 cu ft entering the process =

$\dfrac{526}{477} \times 1000$ $\qquad$ = 1102 cu ft

Condensation

The relative saturation of a partially saturated mixture of vapor and gas may be increased in two ways without the introduction of additional vapor. If the temperature of the mixture is reduced, the vapor concentration corresponding to saturation is reduced, thereby increasing the relative saturation even though the existing partial pressure of the vapor is unchanged. If the total pressure is increased, the existing partial pressure of the vapor is increased, again increasing the relative saturation. Thus, by sufficiently increasing the pressure or reducing the temperature of a vapor-gas mixture it is possible to cause it to become saturated, the existing partial pressure of vapor equaling the vapor pressure of the liquid at the existing temperature. Further reduction of the temperature or increase of the pressure will result in condensation, since the partial pressure of the vapor cannot exceed the vapor pressure of the liquid in a stable system.

In problems dealing with condensation processes, four interdependent factors are to be considered: the initial composition, the final temperature, the final pressure, and the quantity of condensate. It may be desired to calculate any one of these factors when the others are known or specified. Such calculations are readily carried out by selecting as a basis a definite quantity of vapor-free gas and calculating the quantities of vapor that are associated with it at the various stages of the process. In a condensation process the final conditions will be those of saturation at the final temperature and pressure. Any one of the three methods of calculation demonstrated on page 60 under "Volume Changes with Change in Composition" may be used. However, for the reasons given in the preceding section it is generally desirable to express the vapor concentrations in moles of vapor per mole of vapor-free gas if any thermal or design calculations are to be carried out.

Illustration 6. Air at a temperature of 20° C and a pressure of 750 mm Hg has a relative humidity of 80%.

(a) Calculate the molal humidity of the air.

(b) Calculate the molal humidity of this air if its temperature is reduced to 10° C and its pressure increased to 35 psi, condensing out some of the water.

(c) Calculate the weight of water condensed from 1000 cu ft of the original wet air in cooling and compressing to the conditions of part b.

(d) Calculate the final volume of the wet air of part c.

Vapor pressure of water:

      17.5 mm Hg at 20° C
      9.2 mm Hg at 10° C

Solution

(a) Initial partial pressure of water = 0.80 × 17.5   = 14.0 mm Hg

      Initial molal humidity $= \dfrac{14.0}{750 - 14.0}$ = 0.0190

(b) Final partial pressure of water   = 9.2 mm Hg

      Final total pressure $= 35 \times \dfrac{760}{14.7}$ = 1810 mm Hg

      Final molal humidity $= \dfrac{9.2}{1810 - 9.2}$ = 0.0051

Basis: 1000 cu ft of original wet air.

(c) Partial pressure of dry air = 750 − 14   = 736 mm Hg

      Partial volume of dry air at S.C. $= 1000 \times \dfrac{736}{760} \times \dfrac{273}{293}$ = 903 cu ft

      Moles of dry air = 903/359   = 2.52 lb-moles
      Water originally present = 2.52 × 0.0190   = 0.0478 lb-mole
      Water finally present = 2.52 × 0.0051   = 0.0128
      Water condensed = 0.0350 lb-mole or   0.630 lb

(d) Total wet air finally present = 2.52 + 0.0128   = 2.53 lb-moles

      Final volume of wet air $= 2.53 \times 359 \times \dfrac{760}{1810} \times \dfrac{283}{273}$ = 396 cu ft

Illustration 7.   A mixture of dry flue gases and acetone at a pressure of 750 mm Hg and a temperature of 30° C has a dew point of 25° C.   It is proposed to condense 90% of the acetone in this mixture by cooling to 5° C and compressing.   Calculate the necessary pressure in pounds per square inch.

Vapor pressure of acetone:

At 25° C = 229.2 mm Hg
At 5° C = 89.1 mm Hg

Solution

Basis: 1.0 lb-mole of original mixture.

Partial pressure of acetone	=	229.2 mm Hg
Acetone present = 229.2/750	=	0.306 lb-mole
Flue gases present	=	0.694 lb-mole
Acetone present in final mixture = 0.10 × 0.306	=	0.0306 lb-mole
Final mixture of gas = 0.694 + 0.0306	=	0.725 lb-mole
Partial pressure of acetone in final mixture (vapor pressure at 5° C)	=	89.1 mm
Mole % of acetone in final mixture = 0.0306/0.725	=	4.22%

$$\text{Final pressure} = \frac{89.1}{0.0422} = 2110 \text{ mm Hg or } 40.8 \text{ psi}$$

Wet- and Dry-Bulb Thermometry

When a liquid evaporates into a large volume of an unsaturated gas which is initially at the same temperature as the liquid, if no heat is supplied from an external source, the liquid will spontaneously cool to supply part of the energy required for vaporization.   If the body of liquid has a large surface area in proportion to its mass, its temperature will quickly drop to an equilibrium value.   This equilibrium temperature attained is determined by a dynamic balance established between the rate of transfer of heat to the liquid from the warmer ambient gas and the rate of transfer of vapor from the liquid to the ambient gas with absorption by the liquid of the corresponding heat of vaporization.   The equilibrium temperature attained by a liquid which is vaporizing into a gas is termed the *wet-bulb* temperature and is always less than the actual *dry-bulb* temperature of the gas into which evaporation is taking place.   If the gas is initially saturated with the given vapor, then neither vaporization of the liquid nor depression of the wet-bulb temperature occurs.   It is therefore convenient to use the depression of the wet-bulb temperature as a measure of the degree of unsaturation of the mixture of gas and vapor.   This is a valid procedure provided the heat of vaporization comes only from cooling of the ambient air by the liquid.

It is possible to extend this principle of wet- and dry-bulb thermometry to the measurement of vapor content of gases in general; however, the only system for which the method has been extensively used is that involving mixtures of water vapor and air. This special application

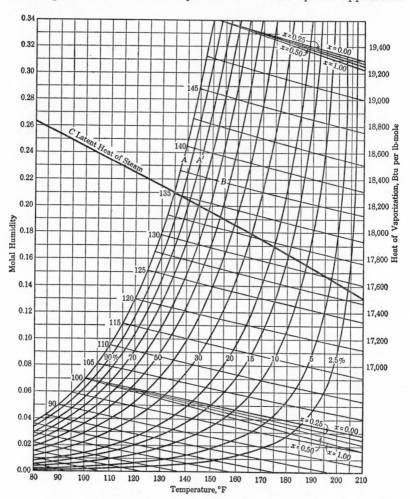

Fig. 19.   Molal humidity chart (high-temperature range). To obtain pounds of water per pound of dry air multiply molal humidity by 0.62

(Reproduced in *CPP Charts*)

of wet- and dry-bulb thermometry is termed *hygrometry* or *psychrometry*. The United States Weather Bureau has prepared extensive psychrometric tables and charts from which humidities of air may be obtained. For engineering purposes *humidity charts* have been developed which permit the determination of the humidities of air-water vapor mixtures from wet- and dry-bulb temperature measurements. These charts are ordinarily constructed for a total pressure of 1.0 atm. For use at other pressures, separate charts must be constructed or suitable corrections applied. Humidity charts are useful for the rapid solution of problems of vaporization, condensation, and air-conditioning where the processes occur at substantially constant atmospheric pressure.

Charts similar to humidity charts have been constructed for a few other systems of liquids and gases based on direct experimental data for each specific system. The construction of such charts is justified only where a considerable amount of attention is to be devoted to a single system.

The Humidity Chart. In Fig. 19 is plotted a molal humidity chart covering a range of molal humidities from 0.00 to 0.34, and a range of dry-bulb temperatures from 80 to 210° F. A chart of this type was described in the literature by Hatta.[2] A lower-range chart plotted on a semilog scale is shown in Fig. 20. Values of molal humidities are plotted as ordinates and dry-bulb temperatures as abscissas. Both charts are based on a total pressure of 29.92 in. Hg. The relationship between the molal heat of vaporization of water and temperature is shown in curve C of Fig. 19.

Curve A of Fig. 19 termed the *saturation curve*, is a plot of molal humidity H_m against temperature for a total pressure of 1.0 atm. Curve A is applicable not only for the water-air system but for any gas that is saturated with water vapor but not reacting therewith and for which Dalton's law applies. Curves A' express the relationships between molal humidities and temperatures corresponding to specified values of percentage humidity. These curves are also independent of the nature of the gas admixed with water vapor under the conditions stated previously.

Under these assumptions, the saturation curve A is calculated from the relation

$$H_{ms} = \frac{p_s}{p - p_s} \tag{7}$$

where H_{ms} = molal humidity at saturation
p_s = vapor pressure of water at temperature $t°$ F
p = total pressure = 29.92 in. Hg for Figs. 19 and 20

The percentage humidity lines A' are constructed from the saturation

[2] S. Hatta, *Chem. Met. Eng.*, **37**, 164 (1930).

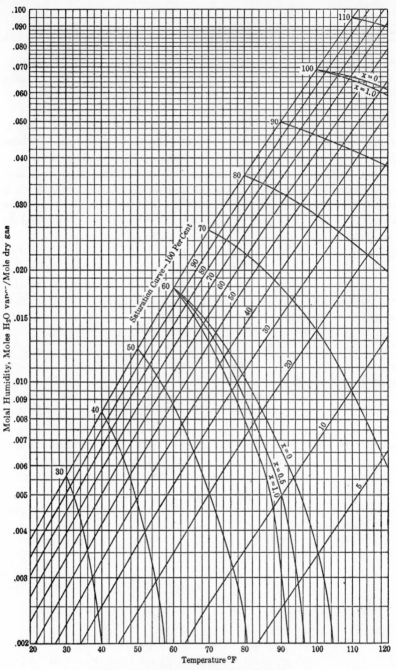

Pressure = 1 atmosphere = 29.92 in. of mercury
x = mole fraction of CO_2 in dry gas

FIG. 20.   Molal humidity chart (low-temperature range).   To use either Fig. 19 or 20 in obtaining the pounds of water per pound of dry air, multiply molal humidity, ordinates of the charts, by 18/29 or 0.62

(Reproduced in *CPP Charts*)

curve by direct proportionality,

$$H_m = \frac{y_p H_{ms}}{100} \tag{8}$$

where y_p = percentage humidity.

Curves B are lines of constant wet-bulb temperature for mixtures of water vapor with air. They are valid also for gases other than air, having the same molal heat capacity as air.

Any point on a given wet-bulb temperature line establishes the corresponding dry-bulb temperature, molal humidity, and percentage humidity of the mixture at atmospheric pressure. Any two of these properties fix the value of the other two. It will be noted that the wet-bulb and dry-bulb temperatures become identical at the saturation curve.

The construction of the wet-bulb temperature lines is described on page 285 for the special case of air–water vapor mixtures.

At wet-bulb temperatures of 100 and 150° F are groups of curves that indicate the effect on the wet-bulb temperature lines of the presence in air of carbon dioxide. The symbol x indicates the mole fraction of carbon dioxide in the dry gas mixture. All wet-bulb lines on the chart for which no value of x is designated correspond to a composition of $x = 0$. The wet-bulb temperature lines which correspond to various mixtures of air and carbon dioxide converge at the saturation curve where the wet-bulb temperature is equal to the dry-bulb temperature, regardless of the composition of the dry gas. These x lines permit application of the chart to mixtures of water vapor and carbon dioxide with air or with gases having the same molal heat capacity as air.

Any horizontal line on the humidity chart represents a change of temperature without change in molal humidity. At constant humidity and total pressure the dew point is also fixed. The dew point corresponding to any point on the chart can be obtained by projecting a line through this point, parallel to the temperature axis, to the saturation curve A. The abscissa of the intersection of this line with the saturation curve is the dew point of the mixture.

The use of the chart is demonstrated in the following illustrations:

Illustration 8. Air at a temperature of 100° F and atmospheric pressure has a wet-bulb temperature of 85° F.

(a) Estimate the molal humidity, the percentage saturation, and the dew point of this air.

(b) The air of part a is passed into an evaporator from which it emerges at a temperature of 120° F with a wet-bulb temperature of 115.3° F. Estimate the percentage saturation of the air leaving the evaporator, and calculate the weight of water evaporated per 1000 cu ft of entering air.

Solution: (a) The abscissa representing 100° F is located on Fig. 20 and followed vertically to its intersection with the 85° F wet-bulb temperature line. This point represents the initial conditions of the air. The percentage saturation is estimated from the position of this point with respect to the curves corresponding to various degrees of saturation. This value would be estimated to be about 52%. The molal humidity is read from the scale of ordinates as 0.037. The dew point is determined by following a horizontal line to its intersection with the saturation curve. The abscissa of this point of intersection is 80.5° F, the dew point.

(b) In the manner described above the percentage saturation of the air leaving the evaporator is estimated to be 84% and its molal humidity 0.110.

Basis: 1.0 lb-mole of moisture-free air.

Moles of wet air entering = $1.0 + 0.037$ = 1.037 lb-moles

Volume of wet air entering = $1.037 \times 359 \times \dfrac{560}{492}$ =   424 cu ft

Water evaporated = $0.110 - 0.037 = 0.073$ lb-mole or   1.31 lb

Water evaporated per 1000 cu ft of entering wet air $1.31 \times \dfrac{1000}{424} = 3.1$ lb

Illustration 9. A combustion gas has the following composition:

CO_2	12.1%
CO	0.1
O_2	7.6
N_2	80.2
	100.0%

(a) Estimate the wet-bulb temperature of this gas when moisture-free at a temperature of 200° F and atmospheric pressure.

(b) If the combustion gas has a dry-bulb temperature of 140° F and a wet-bulb temperature of 95° F, estimate its molal humidity.

Solution: (a) From the group of x curves corresponding to a wet-bulb temperature of 100° F, the angle between the curves for $x = 0$ and $x = 0.12$ is estimated. A line is established at this angle to the 85° wet-bulb line, which is closest to the point representing the conditions of the gas. A line parallel to this newly established 85° wet-bulb line, which corresponds to a composition $x = 0.12$, is projected from the point representing zero humidity and a dry-bulb temperature of 200° F. This line intersects the saturation curve at a temperature of 87° F, which is the wet-bulb temperature. These projections may be carried out with the aid of two draftsman's triangles.

(b) In the manner described above, the angle between the curve $x = 0$ and $x = 0.12$ at a wet-bulb temperature of 100° is estimated. A line is established at this angle to the 95° wet-bulb line, determining the 95° wet-bulb line corresponding to $x = 0.12$. The intersection of this line with the 140° dry-bulb temperature line establishes the point representing the conditions of the mixture. The ordinate of this point is 0.040, the molal humidity of the mixture.

The humidity charts of Figs. 19 and 20 are strictly applicable only to a barometric pressure of 29.92 in. of mercury. Where wet-bulb temperature readings are taken at pressures other than 29.92 in. of mercury, a correction must be applied to the chart readings to obtain the true humidity. In Fig. 21 the humidity correction to be added

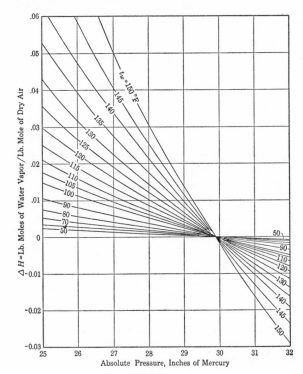

Fig. 21.   Pressure correction to humidity charts

(Reproduced in *CPP Charts*)

to the readings on the atmospheric pressure charts is plotted against total pressure for different wet-bulb readings. It will be observed that the correction increases with departure from atmospheric pressure and with increase in the wet-bulb temperature.

In other psychrometric charts, humidity is expressed as pounds of water vapor per pound of water-free air instead of on a molal basis. The same information may be obtained from Figs. 19 or 20 by multiplying molal humidity by the factor 18/29 or 0.62, which is the ratio of the molecular weight of water vapor to that of air.

In certain other charts, the unsaturation of the air is expressed in terms of relative humidity instead of percentage humidity. The relationship between the two is given by equation 6.

In Fig. 19, the curves for fixed values of percentage humidity all approach a common asymptote at 212° F; however, in psychrometric charts where the unsaturation of air is shown by lines of fixed relative humidity each curve approaches its own asymptote. The saturation curve approaches an asymptote at 212° F; the other curves approach asymptotes at temperatures above 212° F.

Illustration 10.   At 26.42 in. Hg the dry-bulb temperature of air is 150° F, and its wet-bulb temperature is 120° F.   Obtain the correct humidity from the charts.

From Fig. 21,	Molal humidity	= 0.116
From Fig. 23,	Molal humidity correction	= 0.020
Corrected molal humidity		= 0.136

The error in neglecting the pressure effect is 14.7%.

Adiabatic Vaporization.   An adiabatic system or process is one that neither gains heat from nor loses heat to the surroundings.   The significance of the wet-bulb temperature is based on adiabatic conditions.   If a wet-bulb thermometer receives radiant energy from its surroundings, its reading will not be related to humidity on standard psychrometric charts.   As previously explained, the temperature of a liquid vaporizing into an unsaturated gas will attain a wet-bulb temperature, dependent on the vapor content and temperature of the gas.   For water evaporating into unsaturated air under adiabatic conditions and at constant pressure, the wet-bulb temperature remains constant throughout the period of vaporization.   If evaporation continues until the air is saturated with water vapor, the final temperature of the gas will be the same as its initial wet-bulb temperature.   For air-water vapor mixtures, the wet-bulb temperature and adiabatic-cooling lines are practically the same under normal atmospheric pressure.   This identity is fortuitous only for air-water vapor mixtures and does not hold for other systems.

For example, from Fig. 19 it is seen that dry air at a temperature of 197° F has a wet-bulb temperature of 85° F.   Suppose that dry air at this temperature is brought in contact, in an adiabatic compartment, with water at a temperature of 85° F.   As vaporization takes place, the molal humidity of the air will increase, but, if its wet-bulb temperature is to remain constant, the dry-bulb temperature must correspondingly decrease along the 85° F wet-bulb temperature line.   Thus, if vaporization continues until the molal humidity of the air becomes 0.020, the dry-bulb temperature of the air will be reduced to 144° F. If the vaporization is continued until the air is saturated, the molal humidity will then be 0.042 and the dry-bulb temperature will be 85° F.   This molal humidity of 0.042 represents the maximum quantity of water that can be adiabatically evaporated into dry air at an initial temperature of 197° F.

Many types of industrial equipment for drying and evaporation are practically adiabatic in operation, the heat for the vaporization being almost entirely abstracted from the hot gases which enter the process. The humidity chart permits rapid calculation of the quantities of water which can be evaporated in such processes and of the temperatures and humidities throughout.

Illustration 11. Air enters a drier at atmospheric pressure, a dry-bulb temperature of 190° F, and a wet-bulb temperature of 90° F. It is found that 0.028 lb-mole of water is evaporated in the drier per pound-mole of dry air entering it. Assuming that the vaporization is adiabatic, estimate the dry-bulb temperature, the wet-bulb temperature, and the percentage saturation of the air leaving the drier.

Solution: On Fig. 19 it is found that a dry-bulb temperature of 190° and a wet-bulb temperature of 90° correspond to a molal humidity of 0.011.

Molal humidity of air leaving = 0.011 + 0.028 = 0.039.

If the process is adiabatic the wet-bulb temperature will remain constant at 90° F.

A wet-bulb temperature of 90° and a molal humidity of 0.039 correspond to a dry-bulb temperature of 116° F, the temperature of the air leaving the drier. The percentage saturation is estimated as 35%.

Illustration 12. Carbon dioxide is saturated with water vapor by passing it through a wetted chamber. The gas enters the chamber dry, at atmospheric pressure, and at a temperature of 120° F. It may be assumed that the vaporization in the saturator is adiabatic. Estimate the temperature and molal humidity of the saturated carbon dioxide leaving the chamber.

Solution: For pure carbon dioxide, wet-bulb curves corresponding to $x = 1.0$ must be used. In the group of 100° wet-bulb curves the angle is determined between the curves $x = 0$ and $x = 1.0$. A line is established at this angle to the 65° wet-bulb line, thus determining the 65° wet-bulb line for $x = 1.0$. A line parallel to this is projected from the point representing a dry-bulb temperature of 120° and zero humidity to the saturation curve. The intersection is at a temperature of 71° F, the wet-bulb temperature. The molal humidity at this point is 0.027.

Problems

Problems 1–21, inclusive, should be solved without use of the humidity charts.

1. (*a*) Calculate the composition, by volume and by weight, of air that is saturated with water vapor at a pressure of 750 mm Hg and a temperature of 70° F.

(*b*) Calculate the composition by volume and by weight of carbon dioxide that is saturated with water vapor at the conditions of part *a*. (The necessary data may be obtained from Table 5.)

Ans. (*a*) 2.50% by volume, 1.57% by weight.   (*b*) 2.50% by volume, 1.04% by weight, water content.

2. Nitrogen is saturated with benzene vapor at a temperature of 30° C and a pressure of 720 mm of Hg. Calculate the composition of the mixture, expressed in the following terms:

(*a*) Percentage by volume.

(b) Percentage by weight.

(c) Grains of benzene per cubic foot of mixture.

(d) Pounds of benzene per pound of nitrogen.

(e) Pound-moles of benzene per pound-mole of nitrogen.

3. Carbon dioxide contains 0.053 lb-mole of water vapor per pound-mole of dry CO_2 at a temperature of 35° C and a total pressure of 750 mm Hg.

(a) Calculate the relative saturation of the mixture. **Ans.** 89.4%.

(b) Calculate the percentage saturation of the mixture. **Ans.** 89.0%.

(c) Calculate the temperature to which the mixture must be heated in order that the relative saturation shall be 30%. **Ans.** 56.6° C.

4. A mixture of benzene and dry air at a temperature of 30° C and a pressure of 760 mm Hg is found to have a dew point of 15° C. Calculate:

(a) Percentage by volume of benzene.

(b) Moles of benzene per mole of air.

(c) Weight of benzene per unit weight of air.

5. An industrial heating gas is metered at a temperature of 69° F and a pressure of 752 mm Hg. The relative humidity in the meter is found to be 47%.

The gas has a heating value of 500 Btu per cu ft measured at 60° F under a pressure of 30 in. Hg and saturated with water vapor. Calculate the heating value per cubic foot measured in the above meter. **Ans.** 487 Btu per cu ft.

6. The heating value of an illuminating gas is best determined by means of the continuous-flow calorimeter. In such a determination a gas sample is burned that has a volume of 0.2 cu ft. This sample is measured, when saturated with water vapor, at a total pressure of 29.20 in. Hg and a temperature of 70° F. What volume would be occupied by the same quantity of moisture-free gas as contained in this sample, were it at the gas testers' standard conditions of 60° F, a pressure of 30.0 in. Hg, and saturated with water vapor?

7. It is desired to construct a drier for removing 100 lb of water per hour. Air is supplied to the drying chamber at a temperature of 66° C, a pressure of 760 mm Hg, and a dew point of 4.5° C. If the air leaves the drier at a temperature of 35° C, a pressure of 755 mm Hg, and a dew point of 24° C, calculate the volume of air, at the initial conditions, that must be supplied per hour. **Ans.** 112,800 cu ft.

8. Air, at a temperature of 60° C, a pressure of 745 mm Hg, and a percentage humidity of 10 is supplied to a drier at a rate of 50,000 cu ft per hr. Water is evaporated in the drier at a rate of 50 lb per hr. The air leaves at a temperature of 35° C, and a pressure of 742 mm Hg. Calculate:

(a) Percentage humidity of the air leaving the drier.

(b) Volume of wet air leaving the drier per hour.

9. For the best hygienic conditions, air in a living room should be at a temperature of 70° F and a relative humidity of 62%. It is also desirable that fresh air be admitted at such a rate that the air is completely renewed twice each hour. This requires that air be introduced at a volume rate, measured at the conditions of the room, equal to twice the volume of the room.

(a) Calculate the dew point of the air in the room. **Ans.** 56.4° F.

(b) If air is taken in from the outside at a temperature of 10° F and saturated, calculate the weight of water that must be evaporated, per hour, in order to maintain the above specified conditions in a room having a volume of 3000 cu ft. (Vapor pressure of water at 10° F, over ice = 1.6 mm Hg; barometric pressure = 743 mm Hg.) **Ans.** 3.70 lb.

10. Illuminating gas at a temperature of 100° F and a pressure of 760 mm Hg enters a gas holder carrying 20 grains of water vapor per cubic foot. If, in the holder, the gas is cooled to 35° F, calculate the weight of water condensed per 1000 cu ft of gas entering the holder. The pressure in the holder remains constant.

11. A gas mixture at a temperature of 27° C and a pressure of 750 mm Hg contains carbon disulfide vapor such that the percentage saturation is 70. Calculate the temperature to which the gas must be cooled, at constant pressure, in order to condense 40% of the CS_2 present. **Ans.** 14° C.

12. A compressed-air tank having a volume of 15 cu ft is filled with air at a gage pressure of 200 psi and a temperature of 80° F and saturated with water vapor. The tank was filled by compressing atmospheric air at a pressure of 14.5 psi, a temperature of 75° F, and a percentage humidity of 80%. Calculate the amount of water vapor condensed in compressing enough air to fill the tank, assuming that it originally contained air at atmospheric conditions.

13. Air at a temperature of 30° C and a pressure of 750 mm Hg has a percentage humidity of 60. Calculate the pressure to which this air must be compressed, at constant temperature, in order to remove 90% of the water present. **Ans.** 15.79 atm.

14. In a process in which benzene is used as a solvent it is evaporated into dry nitrogen. The resulting mixture at a temperature of 30° C and a pressure of 14.7 psi has a percentage saturation of 70. It is desired to condense 80% of the benzene present by a cooling and compressing process. If the temperature is reduced to 0° C, to what pressure must the gas be compressed?

15. Acetone is used as a solvent in a certain process. Recovery of the acetone is accomplished by evaporation in a stream of nitrogen, followed by cooling and compression of the gas-vapor mixture. In the solvent-recovery unit, 50 lb of acetone are to be removed per hour. The nitrogen is admitted at a temperature of 100° F and 750 mm Hg pressure, and the partial pressure of the acetone in the incoming nitrogen is 10.0 mm Hg. The nitrogen leaves at 85° F, 740 mm Hg, and a percentage saturation of 85.

(*a*) How many cubic feet of the incoming gas-vapor mixture must be admitted per hour to obtain the required rate of evaporation of the acetone? **Ans.** 734 cu ft per hr.

(*b*) How many cubic feet of the gas-vapor mixture leave the solvent recovery unit per hour? **Ans.** 1075 cu ft per hr.

16. The gas-vapor mixture in problem 15 leaves the evaporator at 85° F, 740 mm Hg, and a percentage saturation of 85. The mixture is compressed and cooled to 32° F, after which the condensate is removed. The nitrogen, with the residual acetone vapor, is expanded to 750 mm Hg pressure, heated to 100° F, and re-used in the solvent recovery system. The partial pressure of the acetone in the gas admitted to the solvent-recovery system is 10 mm Hg.

(*a*) To what pressure must the mixture be compressed at 32° F?

(*b*) On the basis of 1 cu ft of the gas-vapor mixture emerging from the solvent recovery unit, calculate:

(1) The volume of the gas-vapor mixture at 32° F after compression.

(2) The volume of the gas-vapor mixture entering the solvent recovery unit.

(3) The acetone condensed after compressing and cooling to 32° F.

17. A continuous drier is operated under such conditions that 250 lb of water are removed per hour from the stock being dried. The air enters the drier at 175° F and a pressure of 765 mm Hg. The dew point of the air is 40° F. The air emerges from the drier at 95° F, a pressure of 755 mm Hg, and at 90% relative humidity.

(a) How many cubic feet of the original air must be supplied per hour? **Ans.** 144,300 cu ft per hr.

(b) How many cubic feet of air emerge from the drier per hour? **Ans.** 133,-400 cu ft per hr.

18. An organic chemical must be dried in an atmosphere of hydrogen in a vacuum-type drier operating at 100 mm absolute pressure. The hydrogen–water vapor mixture leaves the drier at 80° F and 80% humidity. Calculate the number of pounds of hydrogen required per pound of water evaporated.

19. The combustion products of a hydrocarbon fuel burned with dry air leave a furnace at 500° F. The average volumetric analysis of the dry gas is as follows:

CO_2	13.30%
CO	0.04
O_2	2.66
N_2	84.00
	100.00%

The barometric pressure is 735 mm. Estimate the dew point of the stack gas.

20. Air at 80° F, 735 mm pressure, and 60% humidity is compressed and the total product is stored in a cylindrical tank 1 ft in diameter by 5 ft in height at 80° F and a pressure of 250 psi absolute. The air initially in the tank is the same as the intake to the compressor. Calculate the humidity of the air in the cylinder and the weight of water condensed.

21. Air at 25° C, 740 mm Hg, and 55% relative humidity is compressed to 10 atm.

(a) To what temperature must the gas-vapor mixture be cooled if 90% of the water is to be condensed? **Ans.** 16.0° C.

(b) On the basis of 1 cu ft of original air, what will be the volume of the gas-vapor mixture at 10 atm after cooling to the final temperature? **Ans.** 0.0930 cu ft.

In working out the following problems, the humidity charts in Figs. 19 and 20 are to be used as far as possible. The use of the large-scale chart in Fig. 20 is preferable within its range. Unless otherwise specified, the pressure is assumed to be one atmosphere.

22. Air at atmospheric pressure has a wet-bulb temperature of 72° F and a dry-bulb temperature of 88° F.

(a) Estimate the percentage saturation, molal humidity, and the dew point.

(b) Calculate the weight of water contained in 100 cu ft of the air.

23. Hydrogen is saturated with water vapor at atmospheric pressure and a temperature of 90° F. The wet gas is passed through a cooler in which its temperature is reduced to 45° F and a part of the water vapor condensed. The gas after leaving the cooler is heated to a temperature of 70° F.

(a) Estimate the weight of water condensed in the cooler per pound of moisture-free hydrogen. **Ans.** 0.355 lb.

(b) Estimate the percentage humidity and molal humidity of the gas in its final conditions. **Ans.** 40%, 0.0103 molal humidity.

24. It is desired to maintain the air entering a building at a constant temperature of 75° F and a percentage humidity of 40. This is accomplished by passing the air through a series of water sprays in which it is cooled and saturated with water. The air leaving the spray chamber is then heated to 75° F.

(a) Assume that the water and air leave the spray chamber at the same temperature. What is the temperature of the water leaving?

(b) Estimate the water content of the air in the building in pounds of water per pound of dry air.

(c) If the air entering the spray chamber has a temperature of 90° F and a percentage humidity of 65, how much water will be evaporated or condensed in the spray chamber per pound of moisture-free air?

25. In a direct-fired evaporator the hot combustion gases are passed over the surface of the evaporating liquid. The gases leaving the evaporator have a dry-bulb temperature of 190° F and a wet-bulb temperature of 145° F. The dry combustion gases contain 11% CO_2. Estimate the percentage saturation of the gases leaving the evaporator, their molal humidity, and their dew point. **Ans.** 15% saturation, 0.271 molal humidity, 142.5° F dew point.

26. An air-conditioning system takes warm summer air at 100° F, 90% humidity, and 760 mm Hg. This air is passed through a cold-water spray and emerges saturated. The air is then heated to 70° F.

(a) If it is required that the final percentage humidity be 40, what must be the temperature of the air emerging from the spray?

(b) On the basis of 10,000 cu ft of entering air, calculate the weight of water condensed by the spray.

27. Cold air at 20° F, 760 mm pressure, and 70% humidity is conditioned by passing through a bank of steam-heated coils, through a water spray, and finally through a second set of steam-heated coils. In passing through the first bank of steam-heated coils, the air is heated to 75° F. The water supplied to the spray chamber is adjusted to the wet-bulb temperature of the air admitted to the chamber, hence the humidifying unit may be assumed to operate adiabatically. It is required that the air emerging from the conditioning unit be at 70° F and 35% humidity.

(a) What should be the temperature of the water supplied to the spray chamber? **Ans.** 49° F.

(b) In order to secure air at the required final conditions, what must be the percentage humidity of the air emerging from the spray chamber? **Ans.** 57%.

(c) What is the dry-bulb temperature of the air emerging from the spray chamber? **Ans.** 56° F.

(d) On the basis of 1 cu ft of outside air, calculate the volume at each step of the process. **Ans.** First coil 1.115 cu ft, humidifier 1.082 cu ft, second coil 1.110 cu ft.

(e) Calculate the number of pounds of water evaporated per cubic foot of original air. **Ans.** 0.000305 lb.

28. At the top of a chimney the flue gas from a gas-fired furnace has a temperature of 180° F and a wet-bulb temperature of 133° F. The composition of the moisture-free flue gas is as follows:

CO_2	14.1%
O_2	6.0
N_2	79.9
	100.0%

(a) If the temperature of the gases in the stack were reduced to 100° F, calculate the weight of water in pounds that would be condensed per pound-mole of dry gas.

(b) The gas burned in the furnace is estimated to contain 7% CH_4, 27% CO, and 3% CO_2 by volume; it does not contain any carbon compounds other than these. The gas is burned at a rate of 10,000 cu ft per hr measured at a temperature of 65° F, saturated with water vapor. Estimate the weight of water condensed in the chimney per hour under the conditions of part a.

29. A fuel gas has the following analysis:

CO_2	3.2%	H_2	40.2%
C_2H_4	6.1	CO	33.1
C_6H_6	1.0	$C_{1.20}H_{4.40}$	10.2
O_2	0.8	N_2	5.4
			100.0%

Assume that this gas is metered at 65° F, 100% saturation, and is burned with air supplied at 70° F, 1 atm pressure, and 60% humidity. Assume that 20% excess air is supplied and that combustion is complete.

(a) What is the dew point of the stack gases? **Ans.** 131° F.

(b) If the gases leave the stack at 190° F, what is the wet-bulb temperature? **Ans.** 136° F.

(c) If the gases are cooled to 85° F before leaving the chimney, how many pounds of condensate drain down the chimney on the basis of 100 cu ft of fuel gas metered? **Ans.** 3.10 lb.

30. Air at 200° F and a dew point of 50° F is supplied to a drier which operates under adiabatic conditions.

(a) What is the minimum temperature to which the air will cool in the drier?

(b) What is the maximum evaporation in pounds that can be obtained on the basis of 800 cu ft of entering air?

(c) If maximum evaporation is obtained, what will be the volume of the emerging air on the basis of 800 cu ft of entering air?

31. Air enters a drier at a temperature of 240° F with a dew point of 55° F. The drier may be assumed to approach adiabatic vaporization in its operation, all the heat being supplied by the air. If the air leaves the drier saturated with water vapor, how much water can be evaporated per 1000 cu ft of entering wet air? **Ans.** 1.912 lb.

32. A drier for the drying of textiles may be assumed to operate adiabatically. The air enters the drier at a temperature of 140° F with a dew point of 68° F. It is found that 0.62 lb of water is evaporated per 1000 cu ft of wet air entering the drier. Estimate the percentage saturation and temperature of the air leaving the drier.

33. A humidifier is conditioning air to 120° F dry-bulb and 90° F wet-bulb temperature by heating outside air, passing it through an adiabatic spray chamber in which it reaches 90% humidity, and then reheating to the desired temperature. The outside air is at 40° F and is foggy, carrying 0.0004 lb of liquid water per cubic foot. To what temperature must the air be heated in the first heating operation? What is the temperature of the air as it emerges from the spray chamber before the final heating operation?

34. A leaf tobacco as received from the storage warehouse contains 12% water. Ten thousand pounds of this tobacco are stored in a "sweat room" for one week. The air in the room is maintained at 100° F and 85% relative humidity. The tobacco removed from "sweat room" at the end of the week contains 14% moisture. Assuming that only 5% of the water in the air is absorbed by the tobacco, calculate the volume of air handled per week for the 10,000-lb lot of tobacco leaf.

35. It is desired to supply 10,000 cu ft per min of air at 150° F, dew point 77° F, to a soap drier. The outdoor air is available at 60° F, wet bulb 50° F. Indicate three ways of carrying out the change, stating temperatures and humidities involved.

6

Solubility and Crystallization

Industrial problems involving phase equilibria among solids, liquids, and gases involve all the mutual equilibrium relations among these three phases in processes such as the dissolution of a solid by a liquid, the crystallization of a solid from a liquid, the absorption and desorption of gases by liquids, the distribution of a solute between immiscible liquids, and the adsorption and desorption of gases by solids.

Dissolution and Crystallization

A substance that is a solvent for a solid has a specific effect on the distribution of particles between the solid and its dispersed state. Thus a substance that is an excellent solvent for one solid may exert no appreciable influence on another. The solvent action of a liquid is presumed to result from a high affinity or attractive force between the liquid and the solid particles. When a solvent and a solid are brought into contact with each other, the attractive forces of the liquid aided by the thermal motion of the solid particles tend to break apart the structure of the solid and to disperse molecules or ions from its surface in a manner somewhat analogous to the vaporization of a liquid into a gas. As a result the ions or molecules enter the liquid as individually mobile units, thus forming what is termed a *solution* of the solid in the liquid. The dispersed solid which goes into a solution is usually termed the *solute*. The distinction between the solvent and solute in a solution is arbitrary since either component may be considered as solute or solvent. For example, a solution of naphthalene in benzene may equally well be considered as a solution of benzene in naphthalene.

The solute particles of a solution are free to move about as a result of their kinetic energies of translation. By bombarding the walls of the confining vessel these particles exert an osmotic pressure which is entirely analogous to the partial pressure exerted by each component of a gaseous mixture. Thus, when a solution is in contact with the solid from which it was formed, there will be a continual return of

134

dissolved particles to the solid surface. The dispersion of a solid into a liquid is termed *dissolution*. The reverse process is known as *crystallization*. As dissolution proceeds, the concentration of dispersed particles is increased, resulting in an increased osmotic pressure and a greater rate of crystallization. When the solute concentration becomes sufficiently high, the rate of crystallization will equal the rate of dissolution, and a dynamic equilibrium will be established, the concentration of the solution then remaining constant. This equilibrium is analogous to that between a liquid and its vapor. Under these conditions the solution is said to be saturated with the solute and is incapable of dissolving greater quantities of that particular solute under equilibrium conditions.

The concentration of solute in a saturated solution is termed the *solubility* of the solute in the solvent. Solubilities are dependent on the nature of the solute, the nature of the solvent, and the existing temperature. Pressure also has a small effect on solubility, which ordinarily may be neglected. This pressure effect is such that, in a solution whose volume is less than the sum of the volumes of its components in their pure states, solubility is increased by an increase in pressure. In the opposite case, in which the total volume is increased in the formation of a solution, an increase in pressure lowers the solubility.

The simple generalizations that apply to vapor-liquid equilibria are not applicable to liquid-solid systems. Little is known regarding the relationships between solubilities and the specific properties of the solute and solvent. The solubility relationships of each particular system must be individually determined by experimental means, and it is impossible to predict exactly the behavior of one system from that of another. This results because the *solution pressure* of a solid is dependent on the nature of the solvent with which it is in contact. In many cases it is impossible to predict, even qualitatively, the effect of temperature on solubility.

Solubilities of Solids That Do Not Form Compounds with the Solvent. In general, where no true chemical compounds are formed between a solute and solvent, the solubility increases with increasing temperature. In Fig. 22 are plotted the solubility data of a typical system, naphthalene in benzene.

Curve *EB* of Fig. 22 represents the conditions of temperature and concentration that correspond to saturation with *naphthalene* of solutions of naphthalene and benzene. A solution whose concentration and temperature fix a point on this curve will remain in dynamic equilibrium with solid crystals of naphthalene. If the temperature is lowered or the concentration increased by the removal of solvent, naphthalene crystals will be formed. The area to the left of curve *EB* represents conditions of homogeneous, partially saturated solutions. The area between curves *EB* and *ED* represents conditions of non-homogeneous mixtures of crystals of pure naphthalene in solutions of naphthalene and benzene. If a solution whose concentration and

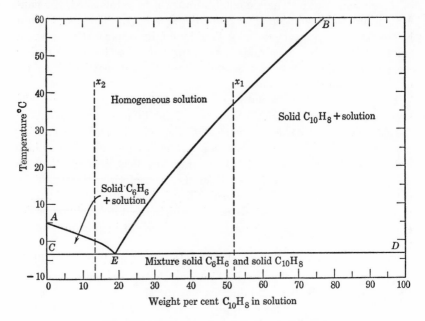

Fig. 22. Solubility of naphthalene in benzene

temperature are represented by point x_1 is cooled without change in composition, its conditions will vary along the dotted line parallel to the temperature axis. The temperature at which this dotted line intersects the solubility curve *EB* is the temperature at which pure naphthalene will begin to crystallize from the solution. As the temperature is reduced further, more naphthalene will crystallize, and the remaining saturated solution will diminish in concentration, its condition always represented by a point on curve *EB*. When the temperature corresponding to line *CED* is reached, the system will consist of pure naphthalene crystals and a saturated solution whose concentration corresponds to point *E*. Further removal of heat at constant temperature will cause this remaining solution to solidify to a mixture of crystals of pure benzene and crystals of pure naphthalene. The line *CED* represents the completion of solidification, and the area below it corresponds to completely solid systems of napththalene and benzene crystals.

The point E is termed the *eutectic* point, and the corresponding temperature and composition are, respectively, the *eutectic temperature* and the *eutectic composition*. If a solution of eutectic composition is cooled, it will undergo no change until reaching the eutectic temperature, when it will completely solidify without change in temperature. A solution of eutectic composition solidifies completely at one definite temperature which is also the lowest solidification point possible for the system.

The curve AE represents the conditions of temperature and composition that correspond to saturation with benzene of solutions of naphthalene and benzene. Whereas curve EB represents the solubility of naphthalene in benzene, curve AE represents the solubility of benzene in naphthalene. The area AEC represents conditions of nonhomogeneous mixtures of benzene crystals in saturated solutions of naphthalene and benzene. If a solution whose composition and temperature are represented by point x_2 is cooled, its conditions will vary along a line parallel with the temperature axis. At the temperature of the intersection of this line with curve AE, crystals of pure benzene will be formed. As the temperature is further reduced, more pure benzene will crystallize, and the remaining saturated solution will vary in composition along curve AE towards E. When the eutectic temperature is reached, the remaining saturated solution will be of eutectic composition, represented by point E.

Further removal of heat at constant temperature will cause further solidification until complete solidification results.

The curve AE is termed the *freezing-point curve* of the solution and represents the temperatures at which solvent begins to freeze out. The point A represents the freezing point of the pure solvent, benzene. From the same viewpoint the curve EB might be considered as a freezing-point curve along which solute begins to freeze out. Thus, either curve may be considered as a solubility curve or a freezing-point curve.

The *percentage saturation* of a solution may be defined as the percentage ratio of the existing weight of solute per unit weight of solvent to the weight of solute that would exist per unit weight of solvent if the solution were saturated at the existing temperature. The percentage saturation of a solution may be varied by changing either its temperature or composition. The effects of such changes on the percentage saturation of a solution may be predicted by locating the point representing the conditions of the solution on a solubility chart such as Fig. 22. A change in temperature will move this point along a line parallel to the temperature axis. A change in composition will move it along a line parallel to the concentration axis. A process in which both composition and temperature are changed simultaneously is best considered as taking place in two steps: a change in composition at constant temperature and a change in temperature with constant composition.

Solubilities of Solids That Form Solvates with Congruent Points. Many solutes possess the property of forming definite chemical compounds with their solvents. Such compounds of definite proportions between solutes and solvents are termed *solvates*, or, if the solvent is water, *hydrates*. Several solvates of different compositions may be formed by a single system, each a stable form under certain conditions of temperature and composition. The presence of such solvates greatly complicates the solubility relationships of the system.

The system ferric chloride and water is an excellent illustration of the effects of the formation of many hydrates. In Fig. 23 are plotted the solubility curves of this system. Point A represents the freezing point of the pure solvent, water, and curve AB the conditions of equilibrium in a solution that is saturated with the solid solvent, ice. This curve is analogous to curve AE of Fig. 22 and represents the solubility of water in ferric chloride. The area $AB1$ represents non-homogeneous mixtures of pure ice in equilibrium with saturated solutions.

Point B of Fig. 23 is a eutectic point analogous to point E of Fig. 22. Curve BC represents the conditions of saturation of a solution of ferric chloride in water with the solid hydrate, $FeCl_3 \cdot 6H_2O$. If a solution, whose conditions are represented by point x_1, is cooled, it will become saturated with $FeCl_3 \cdot 6H_2O$ when conditions corresponding to this concentration on curve BC are reached. Further cooling will result in separation of crystals of $FeCl_3 \cdot 6H_2O$ which will be in equilibrium with saturated solutions whose compositions correspond to the abscissas of curve BC at the existing temperatures. If cooling is continued, more $FeCl_3 \cdot 6H_2O$ crystals will be formed and the concentration of the remaining solution diminished along curve BC until it reaches a value of B, corresponding to the eutectic temperature. On further removal of heat at constant temperature, the remaining solution will solidify into a mixture of crystals of pure ice and crystals of pure $FeCl_3 \cdot 6H_2O$. The area $BC2$ represents nonhomogeneous mixtures of pure $FeCl_3 \cdot 6H_2O$ crystals and saturated solutions. The area below line $1B2$ represents mixtures of crystals of pure ice and crystals of pure $FeCl_3 \cdot 6H_2O$, which are not soluble in each other in the solid state.

As the concentration and temperature of a saturated solution are increased along curve BC, a concentration corresponding to point C is reached, which is the composition of the pure hydrate, $FeCl_3 \cdot 6H_2O$. At this point the entire system is solid hydrate in equilibrium with solution of the same composition. If a solution of this composition is

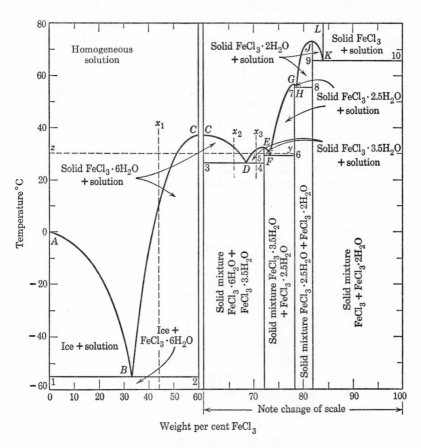

Fig. 23.   Solubility of ferric chloride in water

cooled, when its temperature reaches that of point C it will completely solidify into $FeCl_3 \cdot 6H_2O$ without change of temperature. This behavior is analogous to the freezing of a pure compound, and the temperature of point C represents the melting point of $FeCl_3 \cdot 6H_2O$. A point such as C, at which a hydrate is in equilibrium with a saturated solution of the same composition, is termed a *congruent point*. The curves $ABC21$ may be considered as representing the complete solubility–

freezing-point relationship of a system in which water is the solvent and $FeCl_3 \cdot 6H_2O$ the solute. Point A is the melting point of the pure solvent and C that of the pure solute; B represents the eutectic point of this system.

From the negative slope of curve CD it follows that at concentrations higher than that of point C the concentration of $FeCl_3$ in a saturated aqueous solution may be increased by *lowering* the temperature. This behavior may be readily understood by considering that the curves $3CDE4$ represent the solubility–freezing-point data of a new system in which the solvent is $FeCl_3 \cdot 6H_2O$ and the solute is $FeCl_3 \cdot 3.5H_2O$. In this sytem curve CD is analogous to AB and represents the conditions of equilibrium between pure, solid solvent, $FeCl_3 \cdot 6H_2O$, and a solution of $FeCl_3 \cdot 3.5H_2O$ in $FeCl_3 \cdot 6H_2O$. If a solution at conditions represented by point x_2 is cooled, pure $FeCl_3 \cdot 6H_2O$ will crystallize out when a temperature corresponding to this concentration on curve CD is reached. Further removal of heat at constant temperature will result in the formation of more pure solvent crystals, $FeCl_3 \cdot 6H_2O$, and the remaining saturated solution will increase in concentration along curve CD. When the temperature of curve $3D4$ is reached, this remaining solution will have the composition of point D. Further cooling will result in complete solidification, without further change in temperature, into a mixture of crystals of pure $FeCl_3 \cdot 6H_2O$ and crystals of pure $FeCl_3 \cdot 3.5H_2O$. Point D represents the eutectic point of this system and is analogous to point B. The area $CD3$ represents conditions of nonhomogeneous mixtures of pure crystals of $FeCl_3 \cdot 6H_2O$ and saturated solutions of $FeCl_3 \cdot 3.5H_2O$ in $FeCl_3 \cdot 6H_2O$.

Curve DE represents the conditions of equilibrium between crystals of pure solute, $FeCl_3 \cdot 3.5H_2O$, and a saturated solution of $FeCl_3 \cdot 3.5H_2O$ in $FeCl_3 \cdot 6H_2O$. As the concentration and temperature of a saturated solution are increased along curve DE, a concentration corresponding to point E is reached which is the composition of the hydrate $FeCl_3 \cdot 3.5H_2O$. At this point the entire system must be solid $FeCl_3 \cdot 3.5H_2O$ in equilibrium with solution of the same concentration. Point E is, therefore, a second congruent point, representing the melting point of $FeCl_3 \cdot 3.5H_2O$. Area $DE4$ represents conditions of nonhomogeneous mixtures of crystals of pure $FeCl_3 \cdot 3.5H_2O$ and saturated solutions. The area below line $3D4$ represents entirely solid systems composed of mixtures of crystals of pure $FeCl_3 \cdot 6H_2O$ and $FeCl_3 \cdot 3.5H_2O$.

By similar analysis, the curves $5EFG6$ may be considered as representing the solubility–freezing-point data of a system in which the solute is $FeCl_3 \cdot 2.5H_2O$ and the solvent $FeCl_3 \cdot 3.5H_2O$. Curves $7GHJ8$ represent the system of $FeCl_3 \cdot 2H_2O$, solute, and $FeCl_3 \cdot 2.5H_2O$, solvent.

Curves $9JKL$ represent the system of $FeCl_3$, solute, and $FeCl_3 \cdot 2H_2O$, solvent. Points F, H, and K are the eutectic points of these systems. Points G and J are the congruent melting points of $FeCl_3 \cdot 2.5H_2O$ and $FeCl_3 \cdot 2H_2O$, respectively.

By means of a chart such as Fig. 23 the changes taking place in even very complicated systems may be readily predicted or explained. An illustration of the peculiar phenomena that may take place in complex systems is furnished by the isothermal concentration of a dilute aqueous solution of ferric chloride along the line zy. Such concentration may be carried out by evaporation at constant temperature. At the intersection of zy with BC, crystals of $FeCl_3 \cdot 6H_2O$ will begin to form. At line $C2$ the system will be entirely solid $FeCl_6 \cdot 6H_2O$. Further concentration results in the appearance of liquid solution, and at curve CD the system will be once more entirely liquid. As concentration is continued, solidification again begins at curve DE, becoming complete at line $E5$. Further concentration causes the reappearance of liquid, and liquefaction is completed at curve EF. At curve FG solidification begins and becomes complete at $G6$. This surprising alternation of the liquid and solid states while concentration is progressing may be easily demonstrated experimentally.

Many other systems, both organic and inorganic, behave in a manner similar to that of aqueous ferric chloride and form one or more solvates and congruent points.

Solubilities of Solids That Form Solvates without Congruent Points. In certain systems solvates are formed that are not stable and decompose before the temperature of a congruent point is reached. Such solvates undergo direct transition from the solid state into other chemical compounds. These transitions take place at sharply defined temperatures which are termed *transition points*. The system of sodium sulfate and water illustrates this type of behavior. In Fig. 24 are plotted the solubility and freezing-point data of this system.

Curve AB of Fig. 24 represents conditions of equilibrium between ice and aqueous solutions of sodium sulfate. Point B is the eutectic point of the system of $Na_2SO_4 \cdot 10H_2O$, the solute, and water, the solvent. Curve BC represents the equilibrium between crystals of $Na_2SO_4 \cdot 10H_2O$ and saturated solution. As the temperature of the system is increased the concentration of the saturated solution increases along curve BC. Normally this increase would continue until a congruent point was reached. However, at a temperature of $32.384°$ C, $Na_2SO_4 \cdot 10H_2O$ becomes unstable and is decomposed into Na_2SO_4 and water. The solubility of the anhydrous Na_2SO_4 as indicated by curve CD diminishes with increasing temperature. Points on this curve represent conditions of equilibrium between crystals of anhydrous Na_2SO_4 and saturated solutions. A solution whose conditions are represented by point y_1 will become saturated if either heated or cooled sufficiently. If it is cooled, crystals of $Na_2SO_4 \cdot 10H_2O$ will form when the conditions of curve BC are reached. If it is heated, crystals of anhydrous Na_2SO_4

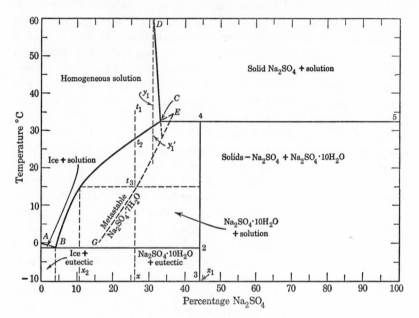

Fig. 24. Solubility of sodium sulfate in water

will form at the temperature corresponding to composition y_1 on curve CD.

The significance of the areas of Fig. 24 is similar to that of the other diagrams that have been discussed. The area of the small triangle to the left of line AB represents a region of equilibrium between crystals of pure ice and saturated solution. Area $BC42$ is a region of equilibrium between crystals of pure $Na_2SO_4 \cdot 10H_2O$ and saturated solution. The area below line $B2$ represents a solid mixture of crystals of ice and crystals of $Na_2SO_4 \cdot 10H_2O$. Line $C45$ indicates the transition temperature at which the decahydrate decomposes to form the anhydrous salt. Line 324 indicates the composition of the pure decahydrate, $Na_2SO_4 \cdot 10H_2O$. The area above curve $54CD$ represents conditions of equilibria between crystals of pure Na_2SO_4 and saturated solutions. The area below and to the right of curves 3245 is a region of entirely solid mixtures of Na_2SO_4 and $Na_2SO_4 \cdot 10H_2O$. Line 45 therefore represents a temperature of complete solidification.

Many systems which form several solvates show solubility relationships both with and without congruent points and transition points in the same system. For example, zinc chloride forms five hydrates. Four of these hydrates decompose, as does $Na_2SO_4 \cdot 10H_2O$, exhibiting transition points before congruent concentrations are reached. The fifth hydrate, $ZnCl_2 \cdot 2.5H_2O$, exhibits a true congruent point, as do the hydrates of ferric chloride.

Effect of Particle Size on Solubility. The solution pressure and solubility of a solid are affected by its particle size in a manner analogous to the effect of particle size on vapor pressure. The solubility of a substance is increased with increase in its degree of subdivision. This increasing solubility with diminishing particle size is demonstrated by the behavior of crystals that are in equilibrium with their saturated solutions. Where such an equilibrium exists, the total amounts of solid and liquid must remain unchanged. However, the equilibrium is dynamic, resulting from equality between the rates of dissolution and crystallization. As a result of the effect of particle size on solubility, the small crystals in such a solution will possess higher solution pressures than the large ones and will tend to disappear with corresponding increase in size of the large crystals. This growth of large crystals at the expense of the small ones in a saturated solution is a familiar phenomenon of considerable industrial importance. For the same reasons, an irregular crystalline mass will change its shape in a saturated solution. Like vapor pressures, solubilities are not noticeably affected by particle size until submicroscopic dimensions are approached.

Supersaturation. Just as spontaneous condensation of a vapor is made difficult because of the high vapor pressure of small drops, spontaneous crystallization is interfered with by the high solubility of small crystals. In order to produce spontaneous crystallization the concentration of a solution must be sufficiently high so that the small crystalline nuclei which are formed by simultaneous molecular or ionic impacts do not immediately redissolve. Such a concentration will be much greater than that which is in equilibrium with large crystals of the same solid. Once crystallization is started and nuclei are formed, it will continue and the crystals will grow until the normal equilibrium conditions are reached. For these reasons it is relatively easy to obtain solutions whose concentrations are higher than the values normally corresponding to saturation. Such solutions are *supersaturated* with respect to large crystals but are only partially saturated with respect to the tiny nuclei. Supersaturated solutions may be formed by careful exclusion of all crystalline particles of solute and by slow changes in temperature or concentration without agitation.

Because of the phenomenon of supersaturation, it is possible to extend the curves of solubility diagrams, such as Fig. 24, into regions where the equilibria that they represent would not normally be stable. The dotted curves of Fig. 24 represent such equilibria that have been experimentally observed in supersaturated solutions of this system. Such equilibria are termed *metastable* and possess the tendency to revert to the normal, stable state corresponding to their conditions of temperature and concentration.

The dotted curve GE of Fig. 24 represents the metastable equilibrium between crystals of $Na_2SO_4 \cdot 7H_2O$ and its saturated solutions. If a solution at conditions y_1 is carefully cooled, normal crystallization of $Na_2SO_4 \cdot 10H_2O$ may be prevented and a supersaturated solution produced at conditions y'_1. If cooling is continued, the crystallization of $Na_2SO_4 \cdot 7H_2O$ may be induced at a temperature corresponding to composition y_1 on curve GE. A supersaturated solution at conditions y'_1 is capable of dissolving $Na_2SO_4 \cdot 7H_2O$ and is unsaturated with respect to this compound. On the other hand, its conditions are unstable, and any disturbance such as agitation, a sudden temperature change, or the introduction of a crystal of $Na_2SO_4 \cdot 10H_2O$ will cause it to assume its normal equilibrium conditions with the crystallization of $Na_2SO_4 \cdot 10H_2O$.

Metastable equilibria are of little industrial significance, but supersaturation is commonly encountered. It may be produced in two ways. By the exclusion of all particles of solid solute the formation of crystalline nuclei may be entirely prevented as described above. Another type of supersaturation may result from sudden changes of the conditions of a saturated solution, even though crystallization has started and crystalline solute is present. The reason is that crystallization, especially in certain types of viscous solutions, is a slow process. Sudden cooling of such a solution will produce temporary conditions of supersaturation simply because the system is slow in adjusting itself to its equilibrium conditions. Agitation of the solution will hasten this adjustment.

In the majority of crystallization processes it is desirable to avoid supersaturation. Supersaturation of the type resulting from the absence of crystalline nuclei is prevented by *seeding* saturated solutions with crystals of solute. Spontaneous nuclei formation is also favored by the presence of rough, adsorbent surfaces. The crystallization of sugar on pieces of string to form rock candy and the scratching of the wall of a beaker to cause crystallization of an analytical precipitate are familiar illustrations of this principle. Supersaturation due to slow crystallization rates is avoided by using correspondingly slow rates of change of the conditions that promote supersaturation and in some cases by agitation.

Dissolution

Problems arise in which it is required to calculate the amount of solute that can be dissolved in a specified quantity of solvent or solution, or, conversely, the quantity of solvent required to dissolve a given amount of solute to produce a solution of specified degree of saturation. Where solvates are not formed in the system, such calculations introduce no new difficulties. From the solubility data is determined the quantity of solute that may be dissolved in a unit quantity of solvent to form a saturated solution at the existing temperature. The amount of solute that may be dissolved in a solution is then the difference between the amount already present and the amount that may be present if the solution is saturated at the specified conditions, both quantities being based on the same quantity of solvent.

Illustration 1. A solution of sodium chloride in water is saturated at a temperature of 15° C. Calculate the weight of NaCl that can be dissolved by 100 lb of this solution if it is heated to a temperature of 65° C.

Solubility of NaCl at 15° C = 6.12 lb-moles per 1000 lb of H_2O
Solubility of NaCl at 65° C = 6.37 lb-moles per 1000 lb of H_2O

Basis: 1000 lb of water.

NaCl in saturated solution at 15° C = 6.12 × 58.5	= 358 lb
% NaCl by weight = 358/1358	= 26.4%
NaCl in saturated solution at 65° C = 6.37 × 58.5	= 373 lb
NaCl that may be dissolved per 1000 lb of H_2O = 373 − 358	= 15 lb
Water present in 100 lb of original solution = 100 × (1.0 − 0.264)	= 73.6 lb

$$\text{NaCl dissolved per 100 lb of original solution} = \frac{15 \times 73.6}{1000} \qquad = \quad 1.1 \text{ lb}$$

It will be noted that the solubility of NaCl changes but little with change in temperature.

Where the substance to be dissolved is a solvated compound, the problem is complicated by the fact that both solute and solvent are added to the solution. Such calculations are carried out by equating the total quantities of solute entering and leaving the process. Algebraic expressions are formed for the sum of the quantity of solute to be dissolved plus that originally present in the solution, and for the quantity of solute in the final solution. Since the total quantity of solute must be constant, these two expressions are equal and may be equated and solved. This method is demonstrated in the following illustration:

Illustration 2. After a crystallization process a solution of calcium chloride in water contains 62 lb of $CaCl_2$ per 100 lb of water. Calculate the weight of this solution necessary to dissolve 250 lb of $CaCl_2 \cdot 6H_2O$ at a temperature of 25° C. Solubility at 25° C = 7.38 lb-moles of $CaCl_2$ per 1000 lb of H_2O.

Basis: x = weight of water in the required quantity of solution.

$$\text{CaCl}_2 \cdot 6\text{H}_2\text{O to be dissolved} \quad = \frac{250}{219} \qquad = 1.14 \text{ lb-moles}$$

$$\text{Total CaCl}_2 \text{ entering process} = 1.14 + \frac{62x}{111 \times 100} \quad = 1.14 + \frac{0.559x}{100} \text{ lb-moles}$$

$$\text{Total water entering process} = x + (1.14 \times 6 \times 18) = x + 123 \text{ lb}$$

$$\text{Total CaCl}_2 \text{ leaving process} \qquad\qquad = 7.38 \frac{x + 123}{1000} \text{ lb-moles}$$

Equating gives

$$1.14 + \frac{0.559x}{100} = 7.38 \frac{x + 123}{1000}, \qquad 1140 + 5.59x = 7.38x + 908$$

$$1.79x = 232, \qquad x = 130 \text{ lb}$$

Total weight of solution required = $130 + (130 \times 0.62) = 211$ lb

Crystallization

The crystallization of a solute from a solution may be brought about in three different ways. The composition of the solution may be changed by the removal of pure solvent, as by evaporation, until the remaining solution becomes supersaturated and crystallization takes place. The second method involves a change of temperature to produce conditions of lower solubility and consequent supersaturation and crystallization. A third method by which crystallization may be produced is through a change in the nature of the system. For example, inorganic salts may be caused to crystallize from aqueous solutions by the addition of alcohol. Other industrial processes involve the *salting out* of a solute by the addition of a more soluble material which possesses an ion in common with the original solute. The calculations involved in this third type of crystallization processes are frequently very complicated and require a large number of data regarding the particular systems involved. Such systems involve more than two components and require application of the principles of complex equilibria which are discussed in a later section.

Where Solvates Are Not Formed. The most important crystallization processes of industry are those that combine the effect of increasing the concentration by the removal of solvent with the effect of change of temperature. Where crystallization is brought about only through change in temperature, the yields of crystals and the necessary conditions may be calculated on the basis of the quantity of solvent, which remains constant throughout the process. From the solubility data may be obtained the quantity of solute that will be dissolved in this quantity of solvent in the saturated solution which will remain after crystallization. The difference between the quantity of solute originally present and that remaining in solution will be the quantity of crystals formed. Such problems are of two types: one in which it is desired to calculate the yield of crystals produced by a specified temperature change, and the converse, in which the amount of temperature change necessary to produce a specified yield is desired. The percentage yield of a crystallization process is the percentage that the yield of crystallized solute forms of the total quantity of solute originally present.

Illustration 3. A solution of sodium nitrate in water at a temperature of 40° C contains 49% NaNO$_3$ by weight.

(a) Calculate the percentage saturation of this solution.

(b) Calculate the weight of NaNO$_3$ that may be crystallized from 1000 lb of this solution by reducing the temperature to 10° C.

(c) Calculate the percentage yield of the process.

$$\text{Solubility of NaNO}_3 \text{ at } 40° \text{ C} = 51.4\% \text{ by weight}$$
$$\text{Solubility of NaNO}_3 \text{ at } 10° \text{ C} = 44.5\% \text{ by weight}$$

Basis: 1000 lb of original solution.

(a) $\%$ saturation $= \dfrac{49}{51} \times \dfrac{48.6}{51.4} = 91.0\%.$

(b)   Yield of NaNO$_3$ crystals $= x$ lb.

From a NaNO$_3$ balance,

$$1000(0.49) = (1000 - x)(0.445) + x$$

Hence $x = 81$ lb

(c) $\%$ yield $= \dfrac{81}{490} = 16.5\%.$

Illustration 4. A solution of sodium bicarbonate in water is saturated at 60° C. Calculate the temperature to which this solution must be cooled in order to crystallize 40% of the NaHCO$_3$.

Solubility of NaHCO$_3$ at 60° C = 1.96 lb-moles per 1000 lb H$_2$O

Basis: 1000 lb of H$_2$O.

NaHCO$_3$ in original solution   = 1.96 lb-moles
NaHCO$_3$ in final solution = 1.96 × 0.60   = 1.18 lb-moles

From the solubility data of NaHCO$_3$ it is found that a saturated solution containing 1.18 lb-moles per 1000 lb of H$_2$O has a temperature of 23° C. The solution must be cooled to this temperature to produce the specified percentage yield.

Calculations of the yields and necessary conditions of crystallization by concentration may be carried out by consideration of the quantity of solvent remaining after concentration has taken place.   The quantity of solute that will be dissolved in this quantity of solvent in the saturated solution remaining after crystallization may be calculated from solubility data.   The quantity of crystals formed in the process will be the difference between the quantity of solute originally present and that finally remaining in solution.   If the concentration is accompanied or followed by a temperature change, the problem is unchanged.   It is only necessary to consider the final temperature in order to determine the quantity of solute remaining in solution.   In such processes three variable factors are present: the yield, the temperature change, and the degree of concentration.   Problems arise in which it is necessary to evaluate any one of these factors if the other two are specified.

Illustration 5.   A solution of potassium dichromate in water contains 13% $K_2Cr_2O_7$ by weight.   From 1000 lb of this solution are evaporated 640 lb of water. The remaining solution is cooled to 20° C.   Calculate the amount and the percentage yield of $K_2Cr_2O_7$ crystals produced.

Solubility of $K_2Cr_2O_7$ at 20° C = 0.390 lb-mole per 1000 lb H_2O

Basis: 1000 lb of original solution.

Water	= 870 lb
$K_2Cr_2O_7$	= 130 lb
Water remaining after concentration = 870 − 640	= 230 lb
$K_2Cr_2O_7$ in solution after crystallization at 20° C	

$$= \frac{230}{1000} \times 0.390 = 0.090 \text{ lb-mole or } 0.090 \times 294 \qquad = 26.4 \text{ lb}$$

Yield of $K_2Cr_2O_7$ crystals = 130 − 26.4   = 103.6 lb

$$\% \text{ yield} = \frac{103.6}{130} \qquad\qquad = 79.7\%$$

Care must always be taken that the true solvent in a crystallizing system is recognized.   For example, in Fig. 22, curve EB represents conditions under which naphthalene is the solute and benzene the solvent.   However, curve EA represents the solubility of benzene as solute in naphthalene as solvent.   If a solution having a concentration of naphthalene less than that of point E is cooled to produce crystallization, pure benzene will crystallize as the solute and the naphthalene will be the solvent, remaining constant in quantity throughout the process.   Similarly, in aqueous solutions of salts, if the concentration is less than the eutectic value, cooling will produce the crystallization of water as pure ice and the system may be treated as a solution of water in salt.

Illustration 6.   A solution of sodium nitrate in water contains 100 grams of $NaNO_3$ per 1000 grams of water.   Calculate the amount of ice formed in cooling 1000 grams of this solution to a temperature of −15° C.

Concentration of saturated, water-in-$NaNO_3$ solution at $-15°$ C $= 6.2$
g-moles of $NaNO_3$ per 1000 grams of H_2O

Basis: 1000 grams H_2O.

$NaNO_3$ in original solution	$= 100$ grams

$$\% \ NaNO_3 \text{ by weight} = \frac{100}{1100} \qquad = \quad 9.1\%$$

Basis: 1000 grams of original solution.

$NaNO_3 = 91$ grams or $91/85$	$= \quad 1.07$ g-moles
Water	$= 909$ grams

Water dissolved in $NaNO_3$ in residual solution $= \dfrac{1000}{6.2} \times 1.07 = 173$ grams

Weight of ice formed $= 909 - 173$ $\qquad = 736$ grams

Where Solvates Are Present. Where solvates are involved, it becomes necessary to consider the solvent chemically combined with the solute which is removed from solution when solvated crystals are precipitated or which is added to the solution when solvated crystals are dissolved. The calculations involved are most easily performed by establishing a material balance for either component. A binary system of weight W containing $x\%$ of component A and $(100 - x)\%$ of component B will be considered. It is assumed that this solution separates under a given temperature change into two phases, phase 1 having a weight w_1 and a composition of $x_1\%$ of component A, and phase 2 having a weight $W - w_1$, and a composition of $x_2\%$ of component A. A material balance of component A gives

$$xW = x_1w_1 + x_2(W - w_1) \tag{1}$$

or

$$\frac{w_1}{W} = \frac{x - x_2}{x_1 - x_2} \tag{2}$$

If separation results from evaporation of a weight w_2 of pure component B, the material balance of component A becomes

$$xW = x_1w_1 + x_2(W - w_1 - w_2) \tag{3}$$

Illustration 7. An aqueous solution of sodium sulfate is saturated at $32.5°$ C. Calculate the temperature to which this solution must be cooled in order to crystallize 60% of the solute as $Na_2SO_4 \cdot 10H_2O$.

From Fig. 24 the solubility at $32.5°$ is seen to be $32.5\% \ Na_2SO_4$.

Basis: 1000 lb of initial solution.

Na_2SO_4 crystallized $= 325 \times 0.6$	$= 195$ lb
$Na_2SO_4 \cdot 10H_2O$ crystallized $= 195/0.441$	$= 442$ lb
Water in these crystals $= 442 - 195$	$= 247$ lb
Water left in solution $= 675 - 247$	$= 428$ lb

Na_2SO_4 left in solution $= 325 \times 0.4 \quad = 130$ lb
Composition of final solution $= 130/(130 + 428) = 23.3\%$ Na_2SO_4

From Fig. 24 it is found that this concentration corresponds to a temperature of 27° C, the required crystallizing temperature.

Illustration 8. A solution of ferric chloride in water contains 64.1% $FeCl_3$ by weight. Calculate the composition and yield of the material crystallized from 1000 lb of this solution if it is so cooled as to produce the maximum amount of crystallization from a residual liquid.

From Fig. 23 it is seen that, if a solution of this composition is cooled, the hydrate $FeCl_3 \cdot 6H_2O$ will crystallize. The maximum crystallization from a liquid residue will result from cooling to the eutectic temperature, 27° C. Further cooling would cause complete solidification of the system. From Fig. 23 the solubility of $FeCl_3$ at the eutectic temperature is 68.3% by weight.

Basis: 1000 lb of original solution.

$$\% \; FeCl_3 \text{ in } FeCl_3 \cdot 6H_2O = \frac{162.2}{162.2 + 108} = 60.0\%$$

Let $y =$ pounds of $FeCl_3 \cdot 6H_2O$ crystallized.
Material Balance of $FeCl_3$

Original solution	Final solution	Crystals

$$(1000)(0.641) = (1000 - y)(0.683) + 0.600y$$

or $\qquad\qquad\qquad y = 511$ lb $FeCl_3 \cdot 6H_2O$ crystals

Illustration 9. A solution of sodium sulfate in water is saturated at a temperature of 40° C. Calculate the weight of crystals and the percentage yield obtained by cooling 100 lb of this solution to a temperature of 5° C.

From Fig. 24 it is seen that at a temperature of 5° C the decahydrate will be the stable crystalline form. The solubilities read from Fig. 24 are as follows:

At 40° C: 32.6% Na_2SO_4
At 5° C: 5.75% Na_2SO_4

Basis: 100 lb of original solution, saturated at 40° C.

$$\% \; Na_2SO_4 \text{ in } Na_2SO_4 \cdot 10H_2O \text{ crystals} = \frac{142}{142 + 180} = 44.1\%$$

Let $y =$ pounds of $Na_2SO_4 \cdot 10H_2O$ crystals formed.
Material Balance of Na_2SO_4

Original solution	Final solution	Crystals

$$0.326(100) = 0.057(100 - y) + 0.441y$$

or $\qquad\qquad\qquad y = 69.5$ lb $Na_2SO_4 \cdot 10H_2O$ formed

$$\text{Weight of } Na_2SO_4 \cdot 10H_2O \text{ in original solution} = 32.6\,\frac{322}{142} = 74 \text{ lb}$$

$$\% \text{ yield} = \frac{69.5}{74} = 94\%$$

Calculations from Line Segments of Equilibrium Diagrams.

In the separation of crystals from a solution the ratio of the weight of crystals to the weight of the original solution is given by equation 2. The ratios of equation 2 can be obtained directly from the line segments on a binary equilibrium diagram when compositions are plotted in weight percentage. This method is illustrated by Fig. 24 for the system Na_2SO_4 and H_2O. Starting with a homogeneous solution at a temperature t_1 containing $x\%$ Na_2SO_4 (component A), and cooling, crystals of $Na_2SO_4 \cdot 10H_2O$ (phase 1) will start to separate at a temperature t_2. As the temperature drops, crystallization will continue, the crystals having a uniform composition of 44.1% Na_2SO_4 corresponding to the decahydrate as represented by the right-hand boundary line of this two-phase field. The percentage of Na_2SO_4 in the remaining liquid phase becomes progressively less as crystallization proceeds, and its decreasing composition will be represented by points on the solubility curve corresponding to the existing temperature. When the final temperature t_3 is reached, the composition of the residual liquid will be represented by abscissa x_2, where x_2 represents the percentage of component A (Na_2SO_4) in the liquid phase. The weight of the $Na_2SO_4 \cdot 10H_2O$ crystals relative to the entire weight of the system as given by equation 2 will be seen to be equal to the ratio of the line segments $x - x_2$ to $x_1 - x_2$.

In general, the percentage by weight of any phase in a two-phase equilibrium mixture at a given temperature may be obtained from the equilibrium diagram by extending the concentration line across the field showing the phases present. The weight ratio of phase 1 to the total weight will then be the ratio of the line segment extending between the composition of the entire system and the composition of phase 2 to the line segment extending between the compositions of phases 1 and 2. Phase 1, represented by segment $x - x_2$, always corresponds to that phase which is richest in component A. For example, in the above system phase 1 represents the $Na_2SO_4 \cdot 10H_2O$ crystal phase for compositions above the eutectic and represents the liquid phase below the eutectic composition where ice separates as the solid phase.

Illustration 9 will be solved from line segments on the equilibrium diagram. By referring to Fig. 24 the value of x corresponding to saturation at 40° C is 32.6%. At 5° C, the composition x_1 of the solid phase is 44.1%, and x_2 of the liquid phase is 5.75%. If the above rule of segments is applied, the percentage weight of crystals separated is $\dfrac{32.6 - 5.75}{44.1 - 5.75} = 69.5\%$. By this ratio of segment lines the composition of any equilibrium mixture of two phases can be estimated from the equilibrium diagram.

It must be emphasized that the line-segment rule can be directly applied only when compositions are expressed in weight percentages and never when they are expressed in molality or mole fraction.

Solubilities in Complex Systems
Fractional Crystallization

All the systems thus far discussed have contained only two primary components. Where three or more components are involved, the solubility relationships become very complex because of the effect that the presence of each solute has on the solubility of the others. These mutual effects cannot be predicted without experimental data on the particular case under consideration. For salts that are ionized in solution the following qualitative generalization applies: Where the same ion is formed from each of two ionized solute compounds, the solubility of each compound is diminished by the presence of the other. This principle is called the common ion effect. For example, the solubility of NaCl in water may be diminished by the addition of another solute which forms the chloride ion. This principle is applied in the crystallization of pure NaCl by bubbling HCl into concentrated brine. Similarly, the solubility of NaCl in water is reduced by the addition of sodium hydroxide. This accounts for the recovery of nearly pure caustic soda by evaporation of cathode liquor in the electrolysis of sodium chloride.

The mutual solubility relationships in complex systems are of great importance in many industrial processes. Soluble substances frequently may be purified or separated from other substances by properly conducted *fractional crystallization* processes. In such processes conditions are so adjusted that only certain of the total group of solutes will crystallize, the others remaining in solution. For the production of very pure materials, it is frequently necessary to employ several successive fractional crystallizations. In such a scheme of recrystallization the crystal yield from one step is redissolved in a pure solvent and again fractionally crystallized to produce further purification.

Complete data have been developed for solubilities in many complex systems. The presentation and use of such data for more complicated cases are beyond the scope of this text. In an excellent monograph by Blasdale[1] the scientific principles involved in such problems are thoroughly discussed.

In Fig. 25 are the solubility data of Caspari[2] for the system of sodium

---

[1] W. G. Blasdale, *Equilibria in Saturated Salt Solutions*, Chemical Catalog Co., New York (1927).

[2] W. A. Caspari, *J. Chem. Soc.*, **125**, 2381 (1924).

sulfate and sodium carbonate in water, presented in an isometric, three-coordinate diagram. A diagram of this type permits visualization of the relationships existing in such systems of only three components. The solubility data determine a series of surfaces which, with the axial planes, form an irregular-shaped enclosure in space. Any conditions of concentration and temperature that fix a point lying within

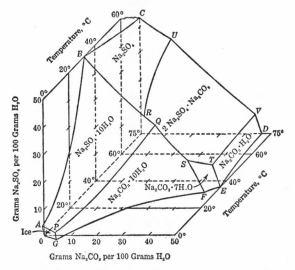

Fig. 25. Solubility of sodium carbonate–sodium sulfate in water

this enclosure correspond to a homogeneous solution. Points lying outside the enclosure represent conditions under which at least one solid substance is present. The surfaces themselves represent conditions of equilibria between the indicated solids and saturated solutions. The line formed by the intersection of two of these surfaces represents conditions of equilibria between a mixture of two solid substances and saturated solution.

For example, curves $OABC$ of Fig. 25 represent the solubility in water of Na_2SO_4 alone, corresponding to Fig. 24. The addition of Na_2CO_3 to such a system lowers the solubility of the Na_2SO_4, as indicated by the surfaces $ABRQP$ and $BCUR$. The former surface corresponds to equilibria between crystals of pure $Na_2SO_4 \cdot 10H_2O$ and solutions containing both Na_2SO_4 and Na_2CO_3. Surface $BCUR$ represents similar equilibria with anhydrous Na_2SO_4 as the solid. Similarly, surface $GFSQP$ represents equilibria between crystals of pure $Na_2CO_3 \cdot 10H_2O$ and solutions containing both Na_2CO_3 and Na_2SO_4. These solutions are, therefore, saturated with respect to Na_2CO_3 but unsaturated with respect to Na_2SO_4. If Na_2SO_4 were added to a solution whose conditions correspond to a point on surface $GFSQP$, the salt would dissolve and $Na_2CO_3 \cdot 10H_2O$ would crystallize.

The line PQ represents solutions that are saturated with both Na_2CO_3 and Na_2SO_4. Such solutions are incapable of dissolving greater quantities of either salt and are in equilibrium with crystals of each. It so happens that when crystals of $Na_2CO_3 \cdot 10H_2O$ and $Na_2SO_4 \cdot 10H_2O$ are formed together in a solution, each solid compound is slightly soluble in the other, forming what are termed *solid solutions*. Therefore, line PQ represents equilibria between crystals of $Na_2CO_3 \cdot 10H_2O$ containing small amounts of $Na_2SO_4 \cdot 10H_2O$ in solid solution, crystals of $Na_2SO_4 \cdot 10H_2O$ containing small amounts of $Na_2CO_3 \cdot 10H_2O$ in solid solution, and liquid solution containing both Na_2SO_4 and Na_2CO_3.

At the higher temperatures the system is complicated by the decomposition of the hydrates and by the formation of a stable double salt of definite composition, $2Na_2SO_4 \cdot Na_2CO_3$. The surface $RQSTVU$ represents equilibria between pure crystals of this double salt and liquid solutions. The other surfaces and lines of the diagram may be interpreted in a similar manner from the composition of the equilibrium solid which is marked on each surface.

An isometric diagram such as Fig. 25, though valuable as an aid to visualization, is not suitable as a basis for quantitative calculations. Data for calculation purposes are better presented as a family of isothermal solubility curves on a triangular diagram as shown in Fig. 26. The coordinate scales represent weight percentages of Na_2SO_4, Na_2CO_3, and H_2O, respectively. Each line of Fig. 26 represents the solubility relationships at one indicated temperature. By interpolation between these lines, solubilities at any desired temperature may be obtained. Points along curve AB represent conditions of saturation with both solutes. The curves running upward to the left from this curve represent solutions that are saturated with sodium carbonate but only partially saturated with sodium sulfate. Similarly, the curves running upward to the right represent saturation with sodium sulfate. The hydrates, $Na_2SO_4 \cdot 10H_2O$ (44.1% Na_2SO_4) and $Na_2CO_3 \cdot 10H_2O$ (37.1% Na_2CO_3), are represented by points C and D, respectively.

Point x_1 on Fig. 26 represents a composition that at a temperature of 25° C corresponds to a solution saturated with sodium carbonate but unsaturated with sodium sulfate. If such a solution is cooled, pure $Na_2CO_3 \cdot 10H_2O$ will crystallize, and the composition of the residual solution will change along the dotted line Dx_1. At a temperature of 22.5° C, corresponding to the intersection of this dotted line with curve AB, the remaining solution will become saturated with both sodium carbonate and sodium sulfate. Further cooling will result in crystallization of both decahydrates. The concentration of the residual solution will then diminish along curve AB. If cooling were stopped at 22.5° C, a yield of pure $Na_2CO_3 \cdot 10H_2O$ crystals would be obtained, and a separation of one solute from the other would result.

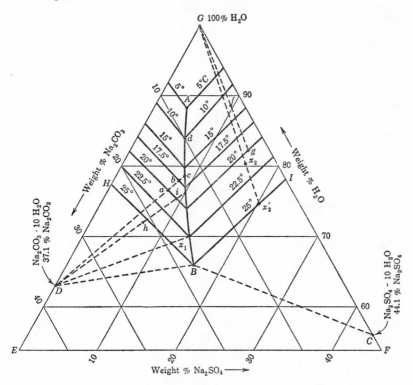

Fig. 26.   Solubility of the system Na_2CO_3—Na_2SO_4—H_2O at low temperatures

Point x_2 on Fig. 26 represents a composition that at a temperature of 25° C corresponds to a solution unsaturated with both solutes. If water is evaporated from such a solution at a temperature of 25° C, the concentration of each solute will be increased. Since the relative proportions of the two solutes remain unchanged, the composition of the solution will vary along a straight line passing through point G of the diagram. If evaporation is continued, a composition x'_2 will be reached at which the solution is saturated with sodium sulfate. Further evaporation will result in crystallization of pure $Na_2SO_4 \cdot 10H_2O$, and the composition of the residual solution will vary along the 25° C isothermal line from x'_2 to B. When the residual solution reaches a composition B, it will be saturated with both solutes.   Further evaporation will produce crystallization of both solutes, and the composition of the residual solution will remain unchanged at B.   By such isothermal evaporation processes it is possible to separate the original solution into pure $Na_2SO_4 \cdot 10H_2O$ and a solution of composition B if the evaporation is not carried beyond the point of initial saturation with both solutes.

Points lying below a solubility isotherm on Fig. 26 represent the compositions of systems which at that temperature are heterogeneous mixtures.   For example, at 25° C, any point lying in the field IBC will represent a mixture of solid $Na_2SO_4 \cdot 10H_2O$ and a solution saturated with sodium sulfate but not with sodium carbonate.   Similarly, a point in the field HBD represents a system of solid $Na_2CO_3 \cdot 10H_2O$ and solution.   A point in the field $EDBCF$ corresponds to a system saturated with both decahydrates.   It will consist of a solution of composition B and a mixture of crystals of the two decahydrates.

Illustration 10.   An aqueous solution at a temperature of 22.5° C contains 21 grams of Na_2CO_3 and 10 grams of Na_2SO_4 per 100 grams of water.

(a)   Calculate the composition and weight of the crystals formed by cooling 1000 grams of this solution to a temperature of 17.5° C.

(b)   Repeat the calculation of part a to correspond to a final temperature of 10° C.

Basis: 1000 grams of original solution.

Initial composition, point a on Fig. 26:

$$Na_2SO_4 = \frac{10}{131} = \quad 7.64\%$$

$$Na_2CO_3 = \frac{21}{131} = \quad 16.02$$

$$H_2O \qquad = \underline{\quad 76.34 \quad}$$
$$\qquad\qquad\qquad 100 \quad \%$$

(a)   Upon cooling, $Na_2CO_3 \cdot 10H_2O$ will crystallize out along line aD, giving a composition at 17.5° C corresponding to point b:

$$Na_2SO_4 \qquad 8.5\%$$

$$Na_2CO_3 \qquad 13.8\%$$

Let x = weight of $Na_2CO_3 \cdot 10H_2O$ crystallized.   Then a material balance for Na_2CO_3 gives

$$(0.1602)(1000) = 0.138(1000 - x) + 0.371(x)$$

$$x = 98 \text{ g } Na_2CO_3 \cdot 10H_2O$$

(b)   When the original solution cools to 10° C, crystallization of $Na_2CO_3 \cdot 10H_2O$ will proceed along Da until it strikes line AB, after which crystallization of both decahydrates will follow line AB.

Composition of final solution at d is

$$
\begin{array}{ll}
\text{Na}_2\text{SO}_4 & 6.2\% \\
\text{Na}_2\text{CO}_3 & 10.0 \\
\text{H}_2\text{O} & \underline{83.8} \\
& \overline{100\ \ \%}
\end{array}
$$

Let x = weight of $\text{Na}_2\text{CO}_3 \cdot 10\text{H}_2\text{O}$ crystallized

y = weight of $\text{Na}_2\text{SO}_4 \cdot 10\text{H}_2\text{O}$ crystallized

From a Na_2CO_3 balance

$$(0.1602)(1000) = 0.371(x) + (0.10)(1000 - x - y)$$

From a Na_2SO_4 balance

$$(0.0764)(1000) = 0.441(y) + (0.062)(1000 - x - y)$$

Hence $x = 251$ grams, $y = 79$ grams

Where evaporation of solvent is involved in a crystallization process, two types of problems may arise. It may be desired to calculate the quantity of solvent that must be removed in order to produce a specified yield of crystals, or it may be desired to calculate the yield of crystals resulting from the removal of a specified quantity of solvent. In the first type of problem it is necessary to determine the composition of the solution remaining after the process. The quantity of solvent to be evaporated will then be the difference between that in the initial and the final solutions. If both solutes crystallize in the process, the residual solution will have a composition corresponding to saturation with both solutes at the final temperature. If only one solute crystallizes, the composition of the final solution may be determined graphically.

Illustration 11. An aqueous solution contains 9.8% Na_2SO_4 and 1.8% Na_2CO_3. How much water must be evaporated from 100 lb of this solution to saturate the solution with one solute at 20° C without crystallization?

Upon evaporation, the composition of the solution will follow along a line fG, starting from the original composition corresponding to f on Fig. 26.

The line fG crosses the 20° C line at a composition of 15.1% Na_2SO_4 and 2.8% Na_2CO_3.

Let x = water evaporated. A water balance gives

$$100(0.884) = x + (100 - x)(0.821)$$

or $x = 35.2$ lb

If a specified quantity of solvent is to be evaporated from a solution and it is desired to calculate the resultant yield of crystals, the composition of the entire final mixture of crystals and solution is readily determined by subtraction of the quantity of evaporated solvent. If, in the final mixture, the concentration of only one of the solutes is greater than that corresponding to saturation, only this one, A, will be crystallized. The entire quantity of the other solute, B, must then be in the residual solution, fixing its composition with respect to this solute. The complete composition of the residual solution will be that corresponding to this concentration of solute B and saturation with solute A at the existing temperature. The quantity of solute A that will be crystallized will be the difference between the total quantity and that remaining in solution.

Illustration 12. A solution contains 19.8% Na_2CO_3 and 5.9% Na_2SO_4 by weight. Calculate the weight and composition of crystals formed from 100 lb of this solution when

(a) 10 lb of water is evaporated and the solution cooled to 20° C.
(b) 20 lb of water is evaporated and the solution cooled to 20° C.

Basis: 100 lb of solution.

(a) The residue after evaporation consists of

$$\frac{19.8}{90} = 22\% \ Na_2CO_3, \qquad \frac{5.9}{90} = 6.55\% \ Na_2SO_4$$

This composition corresponds to h on Fig. 26. Upon cooling, $Na_2CO_3 \cdot 10H_2O$ will crystallize out along line Dh reaching at 20° C a composition corresponding to point i, 9.4% Na_2SO_4 and 15.3% Na_2CO_3.

Let x = lb $Na_2CO_3 \cdot 10H_2O$ crystallized. Then

$$90(0.22) = x(0.371) + (90 - x)(0.153)$$

or $$x = 27.5 \text{ lb}$$

(b) After evaporation of 20 lb water the residue consists of

$$\frac{19.8}{80} = 24.8\% \ Na_2CO_3, \qquad \frac{5.9}{80} = 7.4\% \ Na_2SO_4$$

At 20° C the solution becomes saturated with respect to both solutes, corresponding to a composition of 14.8% Na_2CO_3 and 11.4% Na_2SO_4.

Let x = lb $Na_2CO_3 \cdot 10H_2O$ crystallized
   y = lb $Na_2SO_4 \cdot 10H_2O$ crystallized

From a Na_2CO_3 balance

$$0.248(80) = 0.371x + 0.148(80 - x - y)$$

From a Na_2SO_4 balance

$$0.074(80) = 0.44y + 0.114(80 - x - y)$$

or $$x = 38.1 \text{ lb } Na_2CO_3 \cdot 10H_2O \text{ crystallized}$$

$$y = 3.45 \text{ lb } Na_2SO_4 \cdot 10H_2O \text{ crystallized}$$

The system of sodium sulfate, sodium carbonate, and water has been selected as an illustration, not because of its industrial importance but because it exhibits most of the phenomena to be found in such ternary systems. Many other systems involve similar formations and decompositions of hydrates, double salts, and solid solutions and may be dealt with by means of similar diagrams and methods. Several systems of industrial importance show peculiarities of individual behavior which form the bases of important processes. For example, in the system $NaNO_3-NaCl-H_2O$ lowering the temperature decreases the solubility of $NaNO_3$ but *increases* that of NaCl in solutions that are saturated with both salts. This peculiarity makes it possible to crystallize pure $NaNO_3$ by cooling a solution saturated with both salts. This principle is used on a large scale for the commercial production of Chile saltpeter. In the system $NaOH-NaCl-H_2O$ the solubility of NaCl in solutions containing high concentrations of NaOH becomes very small. This fact is taken advantage of in the separation of NaCl from NaOH solutions in electrolytic production. The solution is merely concentrated by evaporation at a relatively high temperature where the solubility of NaOH is great. As the concentration increases, NaCl crystallizes, and solutions may be produced containing only traces of NaCl.

Vapor Pressure and Relative Humidity above Solutions

Certain aqueous solutions are used for drying gases, dehumidification of air, and control of humidity in air conditioning. Solutions of lithium chloride, calcium chloride, glycerol, and triethylene glycol are used for this purpose in large industrial applications. Sulfuric acid, phosphoric acid, and caustic soda solutions are used as laboratory desiccants. It is convenient to show in a single diagram the relation of vapor pressure, relative humidity, temperature, and composition of the water-desiccant system.

Such a diagram is shown for the calcium chloride–water system in Fig. 27. The area to the left of curve *abcde* represents unsaturated solutions of calcium chloride in equilibrium with water vapor. The curved lines are isotherms showing a decrease in vapor pressure with increase in concentration. The nearly vertical dotted lines represent conditions at constant relative humidity. Curve *abc* represents the solubility of the hexahydrate in water with corresponding changes in temperature, vapor pressure, and relative humidity. Line *cf* represents pure hexahydrate crystals containing 50.7% $CaCl_2$. In area *abcf*, crystals of the hexahydrate and its saturated solution are present. The horizontal lines inside this area represent the vapor pressure of the saturated solution at various temperatures, with corresponding relative humidities.

The vapor pressure of the hexahydrate is lower than that of the saturated solution, but any tendency for condensation on the surface of the hexahydrate is immediately reversed by the formation of a saturated solution having a vapor pressure equal to that of the system. At point c, 86° F, crystals of hexahydrate are transformed to the tetrahydrate, and crystals of the hexahydrate do not exist above this temper-

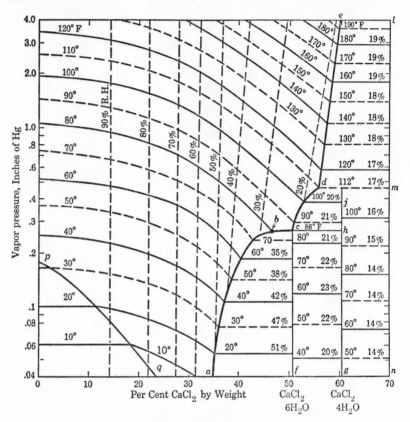

Fig. 27. Vapor pressure and relative humidity over calcium chloride solutions

ature. At point d, 112° F, crystals of tetrahydrate are converted to the dihydrate, and crystals of the tetrahydrate do not exist above this temperature. In area fchg, crystals of both hexa- and tetrahydrates are present. The horizontal lines represent vapor pressures of the hexahydrate. Three phases exist in equilibrium and in mutual contact in this area. The vapor pressure of the hexahydrate is higher than that of the tetrahydrate, but any tendency to condense water on the tetrahydrate is immediately followed by hexahydrate formation with a vapor pressure equal to that of the system. In the area cdjh, crystals of tetrahydrate are in equilibrium with a saturated solution of tetrahydrate. The horizontal lines represent vapor pressures of the saturated solutions of tetrahydrate. In area *delm*, crystals of dihydrate are in equilibrium with a saturated solution of dihydrate. The horizontal vapor-pressure lines represent the vapor pressure of the saturated solution of the dihydrate at various temperatures. In the area *gjmn* two solid phases of tetrahydrate and dihydrate are present. The horizontal lines represent the vapor pressure of the tetrahydrate at various temperatures. The vapor pressure of the dihydrate is lower, but the tendency for water to condense on the dihydrate is reversed by formation of the tetrahydrate with a vapor pressure equal to that of the system.

Area	Phases Present Besides Water Vapor	Vapor-Pressure Lines Represent
abcf	Saturated solution and crystals of hexahydrate	Saturated solution of hexahydrate
fchg	Crystals of hexahydrate and of tetrahydrate	Crystals of hexahydrate
cdjh	Saturated solution and crystals of tetrahydrate	Saturated solution of tetrahydrate
delm	Saturated solution and crystals of dihydrate	Saturated solution of dihydrate
gjmn	Crystals of tetrahydrate and dihydrate	Crystals of tetrahydrate
*pq*0	Ice and saturated solution	Saturated solution of ice

When the calcium chloride–water system is used in a drying and regenerating cycle, during the drying stage the vapor pressure of water in the gas to be dried must exceed the equilibrium value of the system, and in regeneration the vapor pressure of water in the hot air blown through must be less than the equilibrium vapor pressure of the system. Regeneration of calcium chloride from solution is usually not considered feasible because of the relative cheapness of the salt.

It will be observed that in using a homogeneous solution of $CaCl_2$ the lowest attainable relative humidity is 17%. This occurs at 112° F.

Illustration 13. Ninety pounds of anhydrous $CaCl_2$ are placed in a room of 90,000 cu-ft capacity. The air is initially at 80° F and 90% relative humidity. Estimate the conditions of air and desiccant when equilibrium has been attained at 80° F, assuming a pressure of 1.0 atm.

As the drying of the air proceeds, fresh air leaks in, thus maintaining constant pressure in the room. It is assumed that the temperature and humidity of this entering air are the same as of the air initially in the room.

In order to solve the problem, it is necessary to make an assumption, subject to later verification, regarding the desiccant phases that are present when the system reaches equilibrium. It will be assumed that crystals of $CaCl_2 \cdot 6H_2O$ and a saturated solution of calcium chloride and water will be present when equilibrium is reached.

Referring to Fig. 27, it is seen that the air at equilibrium with hexahydrate crystals and saturated solution has a relative humidity of 27%.

Original partial pressure of water vapor = (0.90)(1.0321) = 0.929 in. Hg
Final partial pressure of water vapor   = (0.27)(1.0321) = 0.2787 in. Hg

$$\text{Original humidity} = \frac{0.929}{29.92 - 0.929} \times \frac{18.02}{29} = 0.01988 \text{ lb } H_2O \text{ per lb dry air}$$

$$\text{Final humidity} = \frac{0.2787}{29.92 - 0.2787} \times \frac{18.02}{29} = \underline{0.00583} \text{ lb } H_2O \text{ per lb dry air}$$

Water removed per lb of dry air   = 0.01405 lb

$$\text{Final dry air} = 90,000 \times \frac{29.92 - 0.28}{29.92} \times \frac{492}{540} \times \frac{1}{359} \times 29 = 6570 \text{ lb}$$

Water removed = 6570 × 0.01405 = 92.5 lb

The weights of hexahydrate crystals and saturated solution are determined by a calcium chloride balance. Let x equal the weight of hexahydrate crystals. The total weight of the desiccant system is 182.5 lb, and the weight of the saturated solution is $182.5 - x$. The hexahydrate crystals contain 50.7% $CaCl_2$, and the saturated solution contains 46.0% $CaCl_2$.

$$90 = 0.507x + 0.46(182.5 - x)$$

$$x = 128.5 \text{ lb hexahydrate crystals}$$

$$(182.5 - x) = 53.9 \text{ lb saturated solution}$$

The positive values that are obtained verify the initial assumption regarding the phases present in the desiccant system when equilibrium is reached.

Distribution of a Solute between Immiscible Liquids

When a solute is added to a system of two immiscible liquids, the solute is distributed between the liquids in such proportions that a definite equilibrium ratio exists between its concentrations in the two phases.

In solvent extraction the component A to be removed from a given solution is arbitrarily designated as the *solute* and the initial solvent as the *raffinate solvent B*. Upon extraction with another solvent, the phase rich in the extraction solvent is designated as the *extract phase E* and the phase rich in the raffinate solvent as the *raffinate phase R*.

In dilute solutions the equilibrium distribution of a solute between immiscible solvents may be expressed by the *distribution coefficient K*, which is the ratio of the concentrations in the two phases. Thus,

$$K = \frac{C_E}{C_R} \tag{4}$$

where C_E, C_R = concentrations of solute in phases E and R, respectively. If sufficient solute is present to saturate the system, each phase must

contain solute in the concentration corresponding to its normal saturation conditions.   Therefore, the distribution coefficient at saturation
is the ratio of the solubilities of the solute in the two liquids.

TABLE 10. DISTRIBUTION OF PICRIC ACID BETWEEN BENZENE AND WATER

C_R = concentration of picric acid in water, g-moles per liter of solution
C_E = concentration of picric acid in benzene, g-moles per liter of solution
K = distribution coefficient at 15° C = C_E/C_R

C_E	K
0.000932	2.23
0.00225	1.45
0.01	0.705
0.02	0.505
0.05	0.320
0.10	0.240
0.18	0.187

In ideal systems in which dissociation and association are absent, the
distribution coefficient is independent of concentration.   Otherwise, it
shows marked variation with concentration, as indicated by the values
in Table 10 for the distribution of picric acid, $HOC_6H_2(NO_2)_3$, between
water and benzene, C_6H_6.

The effect of temperature on the distribution coefficient is small if
the temperature coefficients of solubility are approximately equal in
the two phases.   Specific data are necessary for predicting the effects
of a temperature change.

Distribution Calculations.   The distribution of a solute between
two immiscible liquids is of industrial importance in the separation and
purification of organic compounds.   Often one liquid will be an aqueous solution and the other an immiscible organic solvent.   Equilibrium concentrations of a solute in such systems may be varied by the
addition of a second solute which is soluble in only one of the liquids.
The addition of such a solute is, in effect, a change in the nature of one
of the liquids.

Illustration 14.   Picric acid exists in aqueous solution at 17° C in the presence
of small amounts of organic impurities whose effects on its solubility may be
neglected.   The picric acid is to be extracted with benzene in which the inorganic
materials are insoluble.

(a) If the aqueous solution contains 0.20 g-mole of picric acid per liter, calculate
the volume of benzene with which 1 liter of the solution must be extracted in order
to form a benzene solution containing 0.02 g-mole of picric acid per liter.   (Neglect
the difference between the volume of a solution and that of the pure solvent.)

(b) Calculate the percentage recovery of picric acid from the aqueous solution.

Basis: 1 liter of original aqueous solution.

Picric acid   = 0.20 g-mole

From Table 10,

$$K = 0.505 \text{ in final system} = C_E/C_R$$

Final concentration of picric acid in
　aqueous solution $= 0.02/0.505$　　　$= 0.0396$ g-mole per liter

Picric acid in final benzene solution
　$= 0.20 - 0.0396$　　　　　　　　　$= 0.16$ g-mole

Benzene required $= \dfrac{0.16}{0.02}$　　　　　$= 8.0$ liters per liter of aqueous solution

% extraction of picric acid $= \dfrac{0.16}{0.20}$　$= 80\%$

If definite quantities of two immiscible solvents and a solute are mixed together, the final concentration of solute in either solution will be unknown. If the distribution coefficient varies considerably with concentration, it will also be unknown.

Illustration 15. One liter of a benzene solution containing 0.10 g-mole of picric acid per liter is agitated with 2.0 liters of water. Estimate the final concentration of picric acid in each solvent.

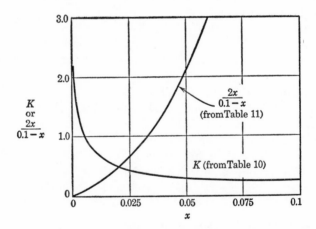

Fɪɢ. 28.　Evaluation of distribution solubilities

Solution: Let x = gram-moles of picric acid in the final benzene solution. Then $0.1 - x$ = gram-moles of picric acid in the final aqueous solution.

$$C_E = x \quad \text{and} \quad C_R = \frac{0.1 - x}{2}$$

The ratio of $\dfrac{2x}{0.1 - x}$ is equal to $\dfrac{C_E}{C_R}$, and at equilibrium $K = \dfrac{C_E}{C_R}$.

For various values of x the values of C_R and C_E given in Table 11 result:

TABLE 11

x	$\dfrac{0.1 - x}{2}$	$\dfrac{C_E}{C_R} = \dfrac{2x}{0.1 - x}$
C_E	C_R	
0	0.05	0
0.02	0.04	0.5
0.04	0.03	1.33
0.06	0.02	3.0
0.08	0.01	8.0
0.10	0	∞

Values of $C_E/C_R = K$ from Table 10 and values of C_E/C_R from Table 11 are plotted against C_E in Figure 28. The intersection of the two curves gives the required values.

$$x = C_E = 0.020$$
$$K = 0.50$$
$$C_R = \frac{0.020}{0.50} = 0.040$$

Picric acid in final benzene solution = 0.020 g-mole: $C_E = 0.020$
Picric acid in final aqueous solution = 0.080 g-mole: $C_R = 0.040$

Calculations Where Volume Changes are Not Negligible. In calculations involving concentrated solutions the differences between the volume of a solution and that of the pure solvent cannot be neglected as was done in illustration 14. In such cases it is convenient to express the concentrations and distribution coefficients in terms of the weight of solute per unit weight of solvent. The units in which distribution data are ordinarily expressed, as in Table 10, may be readily converted into these terms if density-concentration data are available for both solutions.

Stagewise Extraction, Fresh Solvent in Each Batch. The solute in a given solution may be extracted batchwise in multiple stages, fresh solvent being used in each stage. This procedure is common in the laboratory but is uneconomical of solvent for industrial use.

According to the flow chart of Fig. 29, let

R_o = initial raffinate solution to be extracted
E_1, E_2 = extract solution leaving stages 1, 2, respectively
R_1, R_2 = raffinate solution leaving stages 1, 2, respectively
Y = mass of solute A per unit mass of extract solvent C
X = mass of solute A per unit mass of raffinate solvent B
X_o = mass of solute A per unit mass of raffinate solvent B in original feed, R_o
b = mass of raffinate solvent B
c = mass of extract solvent C, assumed the same for each stage

Where the two solvents are completely immiscible even in the presence of solute A and where b and c are the same in each stage, the following material balances result:

	Total Balance	Solute Balance	
Stage 1	$c + R_o = E_1 + R_1$	$bX_o = bX_1 + cY_1$	(5)
Stage 2	$c + R_1 = E_2 + R_2$	$bX_1 = bX_2 + cY_2$	(6)
Stage n	$c + R_{n-1} = E_n + R_n$	$bX_{n-1} = bX_n + cY_n$	(7)

Hence
$$\frac{Y_1}{X_1 - X_o} = \frac{Y_2}{X_2 - X_1} = \frac{Y_n}{X_n - X_{n-1}} = -\frac{b}{c} \qquad (8)$$

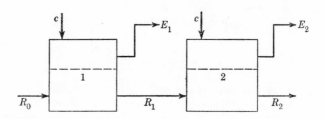

FIG. 29.   Multistage extraction with fresh solvent in each stage

The relationship $Y_n = f(X_n)$ is obtained from solubility data for solution A in the two phases in equilibrium.

For the special case where $Y_n = kX_n$ and where k is constant over the entire composition range,

$$X_n = \left(\frac{b}{b + ck}\right)^n X_o \qquad (9)$$

Stagewise Extraction, Countercurrent.   For countercurrent extraction in stages the flow chart is designated in Fig. 30.

Where the two solvents are completely immiscible even in the presence of solute, and where c is defined as in the preceding section, the following material balances result:

	Total Balance	Solute Balance	
First stage	$R_o + E_2 = R_1 + E_1$	$bX_o + cY_2 = bX_1 + cY_1$	(10)
Second stage	$R_1 + E_3 = R_2 + E_2$	$bX_1 + cY_3 = bX_2 + cY_2$	(11)
nth stage	$R_{n-1} + E_o = R_n + E_n$	$bX_{n-1} + cY_o = bX_n + cY_n$	(12)
Over-all	$R_o + E_o = R_n + E_1$	$bX_o + cY_o = bX_n + cY_1$	(13)

$$Y_o = \frac{b}{c}(X_n - X_o) + Y_1 = Y_{n+1} \qquad (14)$$

Where the two phases leaving each stage are in equilibrium, the above equations can be solved from the solubility relations where $Y_n = f(X_n)$.

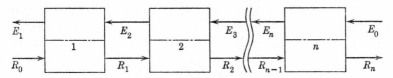

FIG. 30.   Countercurrent stagewise extraction

Partially Miscible Liquids

When two partially miscible liquids are brought together, a range of compositions occurs in which two liquid phases exist in equilibrium with each other.   This behavior is typified by the water-phenol system

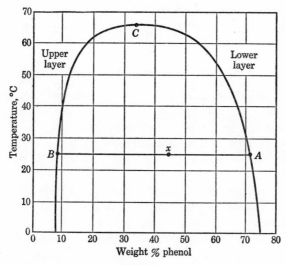

FIG. 31.   Solubility of phenol in water

illustrated in Fig. 31.   At a temperature of 50° C mixtures containing more than 11.8% and less than 62.6% phenol by weight will separate into two layers, the upper containing 11.8% phenol and the lower 62.6%.   Addition of either component to such a two-phase system will not change the composition of either phase but will shift the proportions in which they are present.   Two liquid phases existing in equilibrium with each other in this manner with their compositions independent of the composition of the total two-phase mixture are termed *conjugate solutions*.

If the temperature of the phenol-water mixture is increased above 50° C, the compositions of the conjugate solutions are changed, the phenol content of the upper layer increasing and that of the lower layer decreasing.   When a temperature of 66° C is reached, the compositions of the conjugate phases become equal, and the mixture becomes homogeneous.   The composition at which equality of composition of the two phases is reached is termed the *critical solution composition*, and the temperature at which miscibility becomes complete is termed the *critical solution temperature*.   In Fig. 31, point C corresponds to a critical solution composition of 34% phenol and a critical solution temperature of 66° C.

The weights of the two phases in a heterogeneous binary system are readily calculated by component material balances if the composition of the entire mixture is known, together with compositions of the conjugate solutions as shown in Fig. 31.   It may be demonstrated that, if point x represents the composition of the entire mixture, composed of phases of compositions A and B, the weight of phase A is proportional to line segment Bx, and the weight of phase B is proportional to segment Ax.

Ternary Liquid Mixtures

The compositions of ternary systems are conveniently represented by points on a triangular diagram covering the entire composition range as described in Chapter 1.   Compositions may be expressed on either a mass-fraction or mole-fraction basis.   The use of such diagrams will now be extended to show the phase relations in ternary liquid systems.

Where two components B and C are partly miscible with each other and complete miscibility exists between B and A and between A and C, the solubility relationships are shown in Fig. 32.

The lower area FPG represents the region of immiscibility and the border line curve FPG represents the solubility limits (or saturation curve) of the system.   A mixture of composition K will separate into two phases of composition E and R.   Phases in equilibrium with each other are termed *conjugate phases*, and the line ER connecting the conjugate phases E and R is termed a *tie line*.   Any number of tie lines connecting conjugate phases may be constructed in the two-phase region.   The mass ratio of the two phases resulting from the separation of mixture K is given by the ratio of opposite line segments, thus

$$\frac{\text{Mass of phase } E}{\text{Mass of phase } R} = \frac{KR}{KE} \tag{15}$$

The tie line disappears at point P, designated as the *plait point*. This point represents the composition where the two conjugate phases become mutually soluble. The solubility curve FPG is obtained ex-

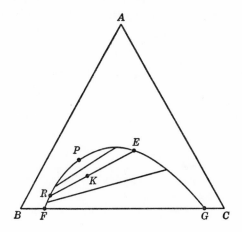

Fɪɢ. 32.   Type 1 system

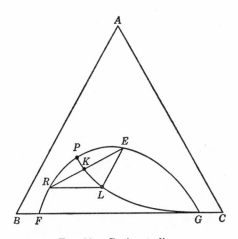

Fɪɢ. 33.   Conjugate line

perimentally by measuring the compositions of several pairs of conjugate solutions. The tie lines for any particular system are usually not parallel, nor is the plait point necessarily at the peak of the borderline curve. Where A is the solute to be recovered and C is the extraction solvent, separation is increased as the plait point approaches B and the tie lines increase in slope.

To avoid the confusion of showing many tie lines any single desired tie line can be conveniently constructed from a single so-called *conjugate line*. In Fig. 33 *GLP* is the conjugate line. The point *L* on the conjugate line is used for constructing the tie line *ER* by drawing line *EL* parallel to the base line *AB* and *LR* parallel to the base line *BC*.

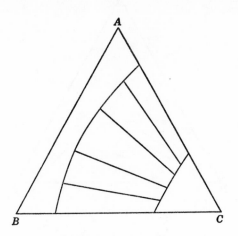

FIG. 34.   Type 2 system

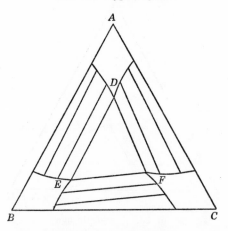

FIG. 35.   Type 3 system

Types of Ternary Liquid Systems. With respect to the partial miscibility of three components three types of ternary liquid systems exist. Type 1 represents a system where only one pair of components is partly or wholly immiscible; the other two pairs are completely miscible in all proportions. Type 1 is shown in Figs. 32 and 33 and is represented by such systems as chloroform–water–acetic acid, toluene–water–acetic acid, isopropyl ether–water–acetic acid and benzene–ethyl alcohol–water.   In type 2, two pairs of components are partly or wholly immiscible, and the regions of immiscibility unite as shown in Fig. 34.   In type 3, all three pairs of components are partly or wholly

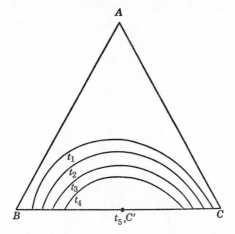

FIG. 36.   Effect of temperature on type 1 system

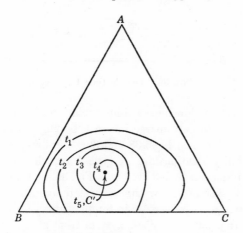

FIG. 37.   Effect of temperature on type 1 system

immiscible, and the three regions of immiscibility unite as shown in Fig. 35.

In type 2 (Fig. 34) component pairs CA and CB are partly miscible as shown in the ternary diagram whereas pair AB is soluble in all proportions.   In type 3 (Fig. 35) all three pairs are partly miscible as shown in the ternary diagram. The shaded areas represent compositions of immiscibility. The tie lines connecting conjugate phases are shown in the shaded areas. Type 2 is represented by the system *n*-heptane–methyl cyclohexane–water. Type 3 is of rare occurrence.

Effect of Temperature on Phase Equilibria in Ternary Liquid Systems. Usually the area of immiscibility of a ternary system is diminished by an increase in temperature. In Figs. 36 and 37, the area of immiscibility for a type-1 system disappears at temperature t_5 called the *critical-solution temperature*, corresponding to a *critical-solution composition* of C'. At temperatures above t_5 immiscibility disappears for all ranges of composition.

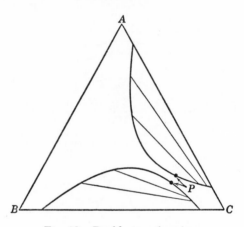

Fig. 38.   Double type 1 system

A ternary system showing solubility relationships of type 1 for two pairs of components may merge into a type-2 system by lowering the temperature as shown in Figs. 38 and 39. Similarly, a ternary system showing solubility relationships of type 1 for all three pairs of components may become a type-3 system of Fig. 35, by lowering the temperature. A few systems show the reverse temperature effect, that is, solubility increases with decrease in temperature.

If in the formation of a type-2 system from a double type-1 system or in the formation of a type-3 system from a triple type-1 system, the plait points of the different two-phase regions do not merge, then a three-phase region results as shown in Fig. 35 for type 3 where compositions D, E, and F represent the three conjugate phases. When this condition occurs, the composition of all three phases is fixed by the pressure and temperature of the system.

Selectivity of Solvent. In the operation of solvent extraction two components miscible in all proportions can be partly separated by the addition of a third liquid component which produces a region of immiscibility and hence a separation into two conjugate phases. For example, if solvent C is added to a binary solution of composition R_o as shown in Fig. 40, a composition F is reached that separates into phase E richer in component A and phase R leaner in component A than the original solution. In solvent extraction, phase R, which is lean in the solute A and rich in solvent B, is designated as the *raffinate phase*, and phase E, which is rich in solute A and in extract solvent C, is termed the *extract phase*.

From inspection of Fig. 40, it will be observed that separation of binary system AB by solvent extraction with solvent C can occur

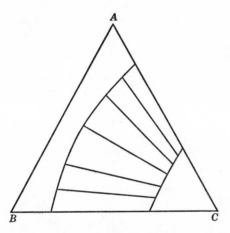

FIG. 39.   Development of type 2 system from the double type 1 system of Fig. 38 by lowering temperature

only in the composition range from B to T' where the line CT' is tangent at point T to the solubility curve.

A solvent is selective for a given solute when the mass fraction of the solute in the extract layer is greater than in the raffinate layer where both mass fractions are expressed on an extract solvent-free basis.

In the system of Fig. 40, A is selectively absorbed by the extract layer. By projection of CE to E' and of CR to R' the ratio of BE'/BR' represents the selectivity of the solvent in the separation of A and B. The removal of solvent C from E produces a binary solution E' and the removal of solvent C from R in the binary solution R'. The selectivity of a solvent for different systems increases as the plait point approaches point B.

In Fig. 41, the selectivity diagram for the system given in illustration

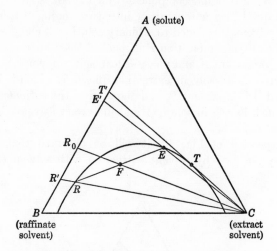

Fig. 40.   Selectivity in solvent extraction

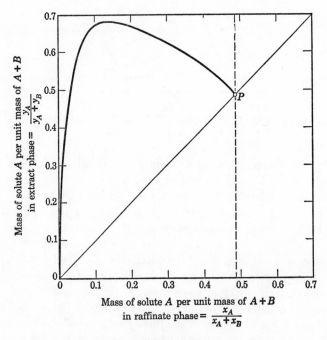

Fig. 41.   Selectivity diagram

16 is constructed by plotting values of BE', the mass fraction of A in the extract layer, against values of BR', the mass fraction of A in the raffinate layer, both on a solvent C-free basis.   Maximum selectivity is obtained at the greatest divergence from the 45° line.   Zero selectivity occurs at the plait point P.

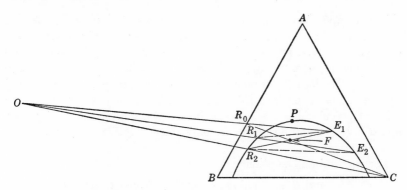

Fig. 42.   Two-stage countercurrent extraction

Partially Miscible Solvents, Countercurrent Staging (Type 1). A stepwise countercurrent multicontact extraction where the two solvents B and C are only partially miscible and where the solute A is completely miscible in binary solutions with B and C can be solved by the direct use of the ternary equilibrium diagrams by the method of Hunter and Nash.[3]   The flow chart is represented by Fig. 30 and the ternary diagram in Fig. 42.

For the purpose of material balances the letters in Fig. 42 indicate the mass and composition of each stream.   Thus R_0 represents the mass and composition of the initial feed consisting of solute A and raffinate solvent B, and C the mass and composition of the pure extract solvent C. F represents the mass and composition of the mixture resulting from extracting feed R_0 with solvent C in the ratio of R_0F/FC = solvent C/feed R_0.

Since F lies in the two-phase region, it separates into solutions, extract E_1 and raffinate R_2.

An over-all material balance over extractor 1 gives

$$R_0 + E_2 = R_1 + E_1 \tag{16}$$

and over extractor 2, where $C = E_3$ in Fig. 42,

$$R_1 + C = R_2 + E_2 \tag{17}$$

[3] T. G. Hunter and A. W. Nash, *J. Soc. Chem. Ind.*, **51**, 957 (1943).

An over-all material balance for two stages gives

$$C + R_0 = E_1 + R_2 \tag{18}$$

From these three relations the following results:

$$R_2 - C = R_1 - E_2 = R_0 - E_1 \tag{19}$$

The composition of all six mixtures corresponding to R_0, R_1, R_2, E_1, E_2, and C are shown on the triangular diagram (Fig. 42). In accordance with the properties of triangular diagrams (Chapter 1), the three lines represented by equation 19 must intersect at a common point O. The location of point O thus lies on the three lines represented by the extension of lines R_2C, R_1E_2 and R_0E_1. The point O may lie off the triangular diagram corresponding to a negative value and in such a case is of geometric significance only since negative compositions are meaningless.

For equilibrium conditions,

E_1 and R_1 are conjugate solutions from extractor 1

E_2 and R_2 are conjugate solutions from extractor 2.

The net effect of adding solvent C to feed R_0 in the ratio of R_0F/FC in two stages is to produce a final extract of composition E_1 and a final raffinate of composition R_2.

The intermediate raffinate of composition R_1 is in equilibrium with the final extract E_1 and the intermediate extract of composition E_2 is in equilibrium with final raffinate R_2, with the ratio

$$\frac{\text{Final raffinate}}{\text{Final extract}} = \frac{FE_1}{FR_2} \tag{20}$$

The construction of a chart of this type requires an assumption of the composition E_1 of the final extract. Line E_1F is extended to R_2. Then locate O as the intersection of CR_2 and E_1R_0. R_1 and E_2 are located on the tie lines conjugate with E_1 and R_2, respectively. The extension of line E_2R_1 should intersect at O; if not, the location of E_1 should be adjusted and the process repeated.

Illustration 16. A raffinate solution weighing 100 lb has an initial composition by weight fraction of solute $x_A = 0.25$, raffinate solvent $x_B = 0.75$, and extract solvent C, $x_C = 0$. The solubility data of conjugate phases are given in Table 12.

It is desired to extract solute A from this feed, using pure extract solvent C in two countercurrent stages to produce a final extract $E_1 = 0.40$ weight fraction of component A. Evaluate compositions and weights of the various streams.

Solution: The conjugate and solubility data of Table 12 are plotted on a right-angle triangular diagram (Fig. 43), with point R_0 located at $x_A = 0.25$, $x_B = 0.75$, and point E_1 at $y_A = 0.4$. A line is drawn through R_0 and E_1 and extended to

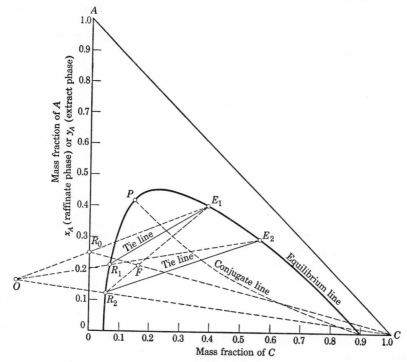

Fig. 43. Countercurrent two-stage extraction. Illustration 16

TABLE 12. COMPOSITION OF CONJUGATE PHASES

Raffinate Phase			Extract Phase		
x_A	x_B	x_C	y_A	y_B	y_C
0	0.950	0.05	0	0.10	0.90
0.025	0.925	0.0505	0.079	0.101	0.82
0.050	0.899	0.051	0.150	0.108	0.742
0.075	0.873	0.052	0.210	0.115	0.675
0.100	0.846	0.054	0.262	0.127	0.611
0.125	0.819	0.056	0.300	0.142	0.558
0.150	0.791	0.059	0.338	0.159	0.503
0.175	0.763	0.062	0.365	0.178	0.457
0.200	0.734	0.066	0.390	0.196	0.414
0.250	0.675	0.075	0.425	0.246	0.329
0.300	0.611	0.089	0.445	0.280	0.275
0.350	0.544	0.106	0.450	0.333	0.217
0.400	0.466	0.134	0.430	0.405	0.165
0.416	0.434	0.150	0.416	0.434	0.150

the left. The tie line E_1R_1 is drawn through E_1 by aid of the conjugate line. Line CR_0 is drawn. The final raffinate is first estimated to be at point R_2. This locates the tie line R_2E_2 and the point E_2. Lines E_2R_1 amd CR_2 are drawn and extended to the left. If all three lines E_1R_o, E_2R_1 and CR_2 intersect at a common point, the assumed value of R_2 is correct; if not, a new value must be assumed. In this instance the correct value is obtained when the compositions of E_2 and R_2 are as follows:

Composition R_2	Composition E_2
$x_A = 0.121$	$y_A = 0.295$
$x_B = 0.824$	$y_B = 0.140$
$x_C = 0.055$	$y_C = 0.565$

The intersection of lines CR_o and E_1R_2 locates point F. The ratio of solvent C to initial raffinate R_o is given by the ratio of line segments:

$$\frac{\text{Solvent } C}{\text{Raffinate } R_o} = \frac{R_oF}{FC}$$

or $\dfrac{x}{100} = \dfrac{43.5}{220}$ or $x = 19.8$ lb solvent C per 100 lb feed R_o

From the construction of Fig. 43, the following values are obtained:

Initial feed R_o,	$x_A = 0.25$
Final extract E_1,	$y_A = 0.40$
Intermediate extract E_2,	$y_A = 0.295$
Final raffinate R_2,	$x_A = 0.121$
Intermediate raffinate R_1,	$y_A = 0.211$

The amounts of raffinate and extract at any point are obtained from the following material balances:

Stage 1

Over-all balance, $E_1 + R_1 = E_2 + R_o$ (a)

Component A, $0.40E_1 + 0.211R_1 = 0.295E_2 + 0.25R_o$ (b)

Stage 2

Over-all balance, $E_2 + R_2 = C + R_1$ (c)

Component A, $0.295E_2 + 0.121R_2 = 0.0C + 0.211R_1$ (d)

From equations a through d, there results, where $R_o = 100$ and $C = 19.8$,

$$E_1 = 37.63 \text{ lb}$$
$$E_2 = 38.33 \text{ lb}$$
$$R_1 = 100.68 \text{ lb}$$
$$R_2 = 82.15 \text{ lb}$$

Correlation of Liquid-Solubility Data in Ternary Systems. Equations expressing solubility data in ternary systems of type 1 have been developed by Othmer and Tobias[4] applicable to a great number of systems; thus

$$\left(\frac{1 - x_B}{x_B}\right) = u\left(\frac{1 - y_C}{y_C}\right)^v \tag{21}$$

[4] D. F. Othmer and P. E. Tobias, *Ind. Eng. Chem.*, **34**, 690, 696 (1942).

where u and v are constants dependent on the system and temperature. Knowledge of the composition of two sets of conjugate solutions permits evaluation of the constants u and v in equation 21. Once u and v are established, values of x_B corresponding to assumed values of y_C are readily evaluated from equation 21 and a system of tie lines and conjugate lines is established. This method does not give the concentration of component A.

Hand[5] showed that the following relationships ordinarily hold for solutions of type 1.

$$\left(\frac{y_A}{y_C}\right) = u' \left(\frac{x_A}{x_B}\right)^{v'} \qquad (22)$$

where u' and v' are constants dependent on the system and the temperature. Both methods of correlation show good results with a great number of systems of type 1.

For the type-2 ternary system represented by aniline A, methylcyclohexane B, and n-heptane C, the solubility relation according to Varteressian and Fenske[6] is represented by the relation

$$\frac{y_B}{y_C} = 1.90 \left(\frac{x_B}{x_C}\right) \qquad (23)$$

Treybal[7] gives a comprehensive discussion of the phase relationships in solvent extraction.

Ponchon Diagram for Solvent Extraction. A special scheme of expressing solubility data is the so-called Ponchon diagram. This diagram is useful for calculations involving continuous countercurrent extraction with reflux. In this method S_E is plotted against Y, and S_R is plotted against X where

For the extract phase:

S_E = mass of solvent C per unit mass of $A + B$

$$S_E = \frac{y_C}{y_A + y_B} \qquad (24)$$

Y = mass of solute A per unit mass of $A + B$

$$Y = \frac{y_A}{y_A + y_B} \qquad (25)$$

[5] D. B. Hand, *J. Phys. Chem.*, 34, 1961 (1930).
[6] K. A. Varteressian, and M. R. Fenske, *Ind. Eng. Chem.*, **29**, 270 (1937).
[7] R. E. Treybal, *Liquid Extraction*, McGraw-Hill Book Co. (1951).

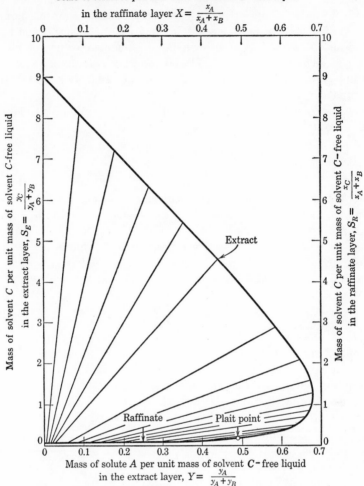

Fig. 44. Ponchon diagram for solvent extraction

For the raffinate phase:

S_R = mass of solvent C per unit mass of $A + B$

$$S_R = \frac{x_C}{x_A + x_B}$$

X = mass of solute A per unit mass of $A + B$

$$X = \frac{x_A}{x_A + x_B}$$

These two plots are combined into the same diagram as shown in Fig. 44 for the data of illustration 16. The selectivity of this same system is shown in Fig. 41.

Solubility of Gases

When a gas is brought into contact with the surface of a liquid, some of the molecules of the gas striking the liquid surface will dissolve. These dissolved molecules of gas will continue in motion in the dissolved state, some returning to the surface and re-entering the gaseous state. The dissolution of the gas in the liquid will continue until the rate at which gas molecules leave the liquid is equal to the rate at which they enter. Thus, a state of dynamic equilibrium is established, and no further change will take place in the concentration of gas molecules in either the gaseous or liquid phases. The concentration of gas that is dissolved in a liquid is determined by the partial pressure of the gas above the surface.

Henry's Law. For many gases the relationship between the concentration of gas dissolved in a liquid and the equilibrium partial pressure of the gas above the liquid surface may be expressed by Henry's law. The ordinary statement of this law is that *the equilibrium value of the mole fraction of gas dissolved in a liquid is directly proportional to the partial pressure of that gas above the liquid surface,* or

$$N_1 = \frac{1}{H} p_1 \qquad (26)$$

where p_1 = equilibrium partial pressure of gas in contact with liquid
N_1 = mole fraction of gas dissolved in liquid
H = Henry's constant, characteristic of the system

This relationship is satisfactory at low concentrations, corresponding to low partial pressures of gas and high values of H. The factor H is a function of the specific nature of the gas and liquid and in general increases with increase in temperature.

When pressures and concentrations are low, the solubility data of a gas-liquid system are expressed by data relating values of Henry's constant H to temperature. In Fig. 45 are curves expressing this relationship for several common gases in water, the numerical values of $1/H$ corresponding to pressures in atmospheres. The data for hydrogen sulfide and carbon dioxide lead to considerable error if used for pressures above 1 atmosphere.

Illustration 17. Calculate the volume of oxygen, in cubic inches, that may be dissolved in 10 lb of water at a temperature of 20° C and under an oxygen pressure of 1 atm.

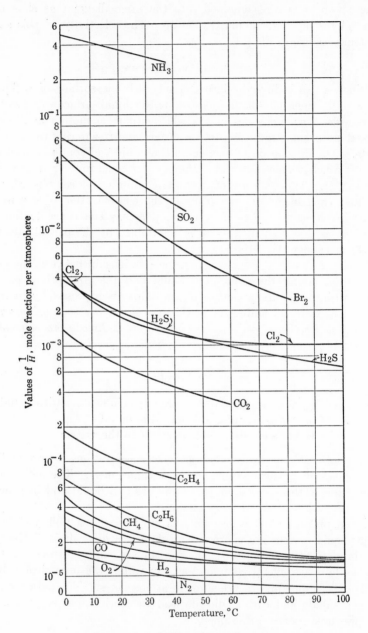

Fɪɢ. 45.   Solubility of gases in water

Plotted from data reported in Lange's *Handbook of Chemistry* (1949)

Solution:

For oxygen and water at 20° C, $1/H$, from Fig. 45 $\quad = 25 \times 10^{-6}$
Mole fraction of $O_2 = 1 \times 25 \times 10^{-6}$ $\qquad = 25 \times 10^{-6}$
Mole fraction of water $\qquad\qquad\qquad\qquad\qquad = 1.00$

Lb-moles of dissolved $O_2 = \dfrac{10}{18} \times 25.1 \times 10^{-6}$ $\quad = 13.9 \times 10^{-6}$

Volume of dissolved $O_2 = 13.9 \times 10^{-6} \times 359 \times \dfrac{293}{273} = 5.37 \times 10^{-3}$ cu ft

or 9.3 cu in. measured at 20° C and a pressure of 1 atm.

Raoult's Law. According to Raoult's law (page 101) the solubility of a gas is expressed as

$$N = p/p_s \qquad (27)$$

where N = mole fraction of solute in liquid phase
$\quad p$ = partial pressure of solute in gas phase
$\quad p_s$ = vapor pressure of pure solute at temperature of system

Hildebrand and Scott[8] have shown that Raoult's law is applicable at temperatures beyond the critical temperature of the gas where p_s is obtained by extrapolation of the vapor-pressure curve of the pure liquid. This relationship predicts the variation of the solubility of a nonpolar gas with temperature through the variation of p_s with temperature, whereas with Henry's law the constant H is of empirical significance only and permits no such prediction. Raoult's law holds fairly well with nonpolar gases in nonpolar liquids. For the solubility of nonpolar gases in polar liquids an empirical correction factor α is applied, thus,

$$N = \alpha \frac{p}{p_s} \qquad (28)$$

This factor is influenced primarily by the nature of the solvent and is less than unity except where chemical reactions are involved. In Table 13 correction factors for the solubility of nonpolar gases in various solvents are given at 25° C and one atmosphere partial pressure, calculated from a compilation of solubility data by Hildebrand and Scott.[8] In this table μ is the dipole moment of the solvent and δ is the square of the heat of vaporization of the solvent on the basis of unit volume. Solubility of nonpolar gases in related solvents can be estimated from this table except where chemical reactions occur. It will be observed that the solubility of nonpolar gases in water is only

[8] J. H. Hildebrand and R. L. Scott, *Solubility of Nonelectrolytes*, Reinhold Publishing Corp. (1950).

TABLE 13. CORRECTION FACTOR α FOR SOLUBILITIES OF NONPOLAR GASES IN SOLVENTS OF VARIOUS POLARITIES

At 25° C and 1 atm Partial Pressure

Solvents	δ	$\mu(10^{18})$	H_2	N_2	CO	O_2	A	CH_4	C_2H_4	C_2H_2	C_2H_6
Ideal			1.0	1.0	1.0	1.0	1.0	1.0	1.0	1.0	1.0
n-hexane	7.29	0	0.81	1.25		1.46		1.06	1.06		0.709
Ethyl ether	7.45	1.2	0.69	1.25	1.32	1.50		1.30			
Cyclohexane	8.20	0	0.475	0.722			0.925	0.81			
Carbon tetrachloride	8.64	0	0.408	0.642	0.691	0.91		0.816	0.965	0.534	0.851
m-Xylene	8.82	0.5	0.517	0.614	0.712			0.736			
Methyl acetate	9.44	1.7	0.384	0.597	0.675	0.688		0.571	0.77	3.40	0.432
Acetone	9.86	2.8	0.288	0.592	0.59	0.70	0.576	0.636	0.494	3.54	
Benzene	9.16	0	0.327	0.440	0.50	0.618	0.552	0.591	0.76	0.845	0.605
Chloroform	9.25	1.1	0.275	0.445	0.506	0.560					
Chlorobenzene		1.6	0.333	0.431	0.493	0.60		0.595	0.796	0.721	0.692
Nitrobenzene		4.1	0.195	0.263	0.308						
Methyl alcohol	14.4	1.7	0.196	0.235	0.254	0.241	0.277	0.203			
Carbon bisulfide	10.0	0	0.116	0.145	0.161						
Aniline		1.5	0.134	0.115	0.156						
Water	23	1.8	0.019	0.012	0.014	0.017	0.016	0.0068	0.0058	0.0362	0.0132
p/p_s at 25° C			8	10	12.8	13.2	16	35	152	208	250

Source: J. H. Hildebrand and R. L. Scott, *Solubility of Nonelectrolytes*, Reinhold Publishing Corp. (1950), with permission.

about 1% of their solubilities in ideal solutions.   For this same system it will be observed that Henry's constant can be calculated by the relation $H = p_s/\alpha$.

Deviations from Laws of Henry and Raoult.   At high pressures or for gases of relatively high solubilities the direct proportionality of the gas-solubility laws breaks down.   Aqueous solutions of ammonia, carbon dioxide, and hydrochloric acid are examples of systems whose

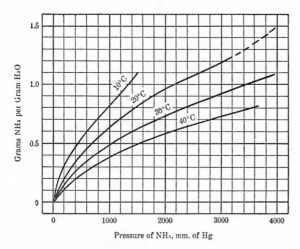

Fig. 46.   Solubility of ammonia in water

behaviors deviate widely from those predicted by the solubility laws except at low pressures.   These deviations result from chemical reaction of the gas with the liquid and subsequent ionization of the dissolved molecules.

In general, experimentally determined data relating temperatures, pressures, and solubilities over the entire desired range are necessary to predict solubilities in such systems.   These data may be expressed by *solubility isotherms* relating the concentration of a dissolved gas to its partial pressure at a constant temperature.   In Fig. 46 are solubility isotherms of ammonia in water at various temperatures. It will be noted that all the lines have considerable curvatures contrary to the ideal solubility laws.   The form of the 20° C isotherm is typical of the behavior to be expected of a gas that is below its critical temperature and dissolved in a solvent with which it is miscible when in the liquid state.

If measurements were continued at higher pressures, the slope of the curve would be expected to increase, becoming asymptotic to the line corresponding to a pressure of 6420 mm, the vapor pressure of liquid ammonia at 20° C. Gaseous ammonia at 20° C could not exist at higher partial pressures, and at this pressure liquid ammonia and water are soluble in all proportions. The isotherms corresponding to higher temperatures exhibit similar points of inflection if extended to higher pressures.

From the data of Fig. 46 it is apparent that the gas-solubility laws should not be used in dealing at high pressures with a gas having considerable affinity for the solvent. At high pressures the solubility calculated from the simple solubility laws will, in general, be higher than the correct value. In general, the solubility of a gas in a liquid is diminished by the addition of a nonvolatile solute with which it does not react chemically.

Problems

1. From the *International Critical Tables*, Vol. IV, plot a curve relating the solubility of sodium carbonate in water to temperature. Plot solubilities as ordinates, expressed in percentage by weights of Na_2CO_3, and temperature as abscissas, expressed in degrees centigrade, up to 60° C. On the same axes plot the freezing-point curve of the solution, or the solubility curve of ice in sodium carbonate, locating the eutectic point of the system.

2. From the data of Figs. 22, 23, and 24 and problem 1, tabulate in order the successive effects produced by the following changes:

(a) A solution of naphthalene in benzene is cooled from 20 to −10° C. The solution contains 0.6 g-mole of naphthalene per 1000 grams of benzene.

(b) A solution of sodium carbonate in water is cooled from 40 to −5° C. The solution contains 4.5 g-moles of Na_2CO_3 per 1000 grams of water.

(c) A mixture of aqueous sodium sulfate solution and crystals is heated from 10 to 60° C. The original mixture contains 3.3 g-moles of Na_2SO_4 per 1000 grams of water.

(d) Pure crystals of $Na_2SO_4 \cdot 10H_2O$ are heated from 20 to 40° C.

(e) A solution of $FeCl_3$ in water is cooled from 20 to −60° C. The solution contains 2.5 g-moles of $FeCl_3$ per 1000 grams of water.

(f) A solution of $FeCl_3$ is evaporated at a temperature of 34° C. The original solution contains 5 g-moles of $FeCl_3$ per 1000 grams of water, and it is evaporated to a concentration of 25 g-moles of $FeCl_3$ per 1000 grams of water.

(g) An aqueous solution of $FeCl_3$ is cooled from 45 to 10° C. The solution contains 18 g-moles of $FeCl_3$ per 1000 grams of water.

3. In a solution of napthalene in benzene the mole fraction of naphthalene is 0.12. Calculate the weight of this solution necessary to dissolve 100 lb of naphthalene at a temperature of 40° C. **Ans.** 123 lb.

4. An aqueous solution of sodium carbonate contains 5 grams of Na_2CO_3 per 100 cc of solution at 15° C. If 10 gal of this solution are measured at 15° C and then heated to 60° C, how many pounds of Na_2CO_3 can be dissolved at 60° C?

5. A solution of sodium carbonate in water is saturated at a temperature of 10° C. Calculate the weight of $Na_2CO_3 \cdot 10H_2O$ crystals which can be dissolved in 100 lb of this solution at 30° C. **Ans.** 192 lb.

6. A solution of naphthalene in benzene contains 9.5 lb-moles of naphthalene per 1000 lb of benzene.

(*a*) Calculate the temperature to which this solution must be cooled in order to crystallize 70% of the naphthalene.

(*b*) Calculate the composition of the solid product if 90% of the naphthalene is crystallized.

7. The concentration of naphthalene in a solution in benzene is 1.4 lb-moles per 1000 lb of benzene. This solution is cooled to $-3°$ C. Calculate the weight and composition of the material crystallized from 100 lb of the original solution. **Ans.** 18.7 lb. benzene.

8. A solution of naphthalene in benzene contains 20% naphthalene by weight. Calculate the weight of benzene that must be evaporated from 100 lb of this solution in order that 85% of the naphthalene may be crystallized by cooling to 20° C.

9. A batch of saturated Na_2CO_3 solution, weighing 1000 lb, is to be prepared at 50° C.

(*a*) If the monohydrate ($Na_2CO_3 \cdot H_2O$) is available as the source of Na_2CO_3, how many pounds of this material and how many pounds of water would be needed to form the required quantity of solution? **Ans.** 375 lb $Na_2CO_3 \cdot H_2O$.

(*b*) If the decahydrate ($Na_2CO_3 \cdot 10H_2O$) is available as the source of Na_2CO_3, how many pounds of this material and how many pounds of water would be required? By means of a sketch, show how the solubility chart was used in solving the problem. **Ans.** 865 lb $Na_2CO_3 \cdot 10H_2O$.

10. A system containing 40% Na_2CO_3 and 60% H_2O has a total weight of 500 lb. At 50, 35, and 20°C, report on:

(*a*) The nature and composition of the phases constituting the system.

(*b*) The weight of each phase.

By means of a sketch, indicate how the solubility chart was used in solving the problem.

11. A solution containing 35% Na_2CO_3 weighs 5000 lb.

(*a*) To what temperature must the system be cooled in order to recover 98% of the Na_2CO_3? **Ans.** 7.5° C.

(*b*) What will be the weight of the crystals recovered and of the residual mother liquor? **Ans.** 4627 lb $Na_2CO_3 \cdot 10H_2O$; 373 lb residual liquor.

By means of a sketch, indicate how the solubility chart was used in solving the problem.

12. In Fig. 47 the temperature-composition diagram of lithium chloride is presented, showing the composition of the system as pounds of water per pound of lithium chloride. Explain the significance of each separate area, boundary line, and point of intersection.

13. A dilute solution of Na_2CO_3 and water weighs 2000 lb, and contains 5% Na_2CO_3. It is required that 95% of the Na_2CO_3 be recovered as the decahydrate. The lowest temperature that can be obtained is 5° C. Specify the process to be employed for securing the degree of recovery required. Present calculations with respect to the quantitative relations involved in the process specified.

By means of a sketch, indicate how the solubility chart was used in solving the problem. **Ans.** Evaporate 1683 lb water, cool to 5° C, recover 256.5 lb $Na_2CO_3 \cdot 10H_2O$.

14. An aqueous solution of ferric chloride contains 12 lb-moles of $FeCl_3$ per 1000 lb of water. Calculate the yield of crystals formed by cooling 1000 lb of this solution to 28° C.

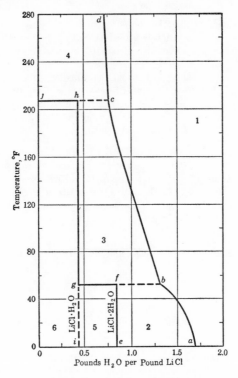

Fɪɢ. 47.   Solubility of lithium chloride

15. A solution of ferric chloride in water contains 15 g-moles of $FeCl_3$ per 1000 grams of water.

(a) Calculate the composition of the resulting crystals in percentage of each hydrate formed when this solution is cooled to 0° C. **Ans.** 89.3% $FeCl_3 \cdot 3.5H_2O$.

(b) Calculate the percentage of eutectic crystals present in the total crystal mass. **Ans.** 42.0%.

16. An aqueous solution contains 3.44% Na_2SO_4 and 21.0% Na_2CO_3 by weight. It is desired to cool this solution to the temperature that will produce a maximum yield of crystals of pure sodium carbonate decahydrate. Calculate the proper final temperature and the yield of crystals per 100 lb of original solution, using the data of Fig. 26.

17. It is desired to crystallize a maximum amount of pure $Na_2CO_3 \cdot 10H_2O$ from the solution of problem 11 by evaporating water at a temperature of 25° C. Calculate the quantity of water that must be evaporated and the yield of crystals produced per 100 lb of original solution. **Ans.** 32.4 lb water evaporated; 46.4 lb $Na_2CO_3 \cdot 10H_2O$ crystallized.

18. A solution contains 25 grams of Na_2SO_4 and 4.0 grams of Na_2CO_3 per 100 grams of water. From 100 lb of this solution 20 lb of water are evaporated, and the residual solution is cooled to 20° C. Calculate the weight and composition of the crystals formed in the process.

19. Calculate the weight of water that must be evaporated from 100 lb of the solution of problem 18 in order to crystallize 70% of the Na_2SO_4 as the pure decahydrate at a temperature of 15° C. **Ans.** 11.3 lb.

20. A solution has the following initial composition:

$$Na_2SO_4 \cdot 10H_2O = 80 \text{ parts by weight}$$
$$Na_2CO_3 \cdot 10H_2O = 60 \text{ parts by weight}$$
$$\text{Solvent water} = 100 \text{ parts by weight}$$

The solution is at 50° C and weighs 4200 lb. If this solution is cooled to 5° C:

(a) At what temperature does crystallization start? What phase crystallizes out first?

(b) At what temperature will a second crystalline phase start to separate? What is its composition?

(c) Calculate the maximum weight of pure crystals that can be obtained in the first stage of crystallization when but one solid phase is separating.

(d) Calculate the respective weights of the two solid phases that separate during the second stage of crystallization.

(e) With respect to the residual mother liquid at 5° C, report its total weight, and the weight of each of the three components that were present in the original solution.

By means of a sketch, show how Fig. 26 was used in solving the problem.

21. A solution has the following initial composition:

$$Na_2SO_4 \cdot 10H_2O = 40 \text{ parts by weight}$$
$$Na_2CO_3 \cdot 10H_2O = 40 \text{ parts by weight}$$
$$\text{Solvent water} = 100 \text{ parts by weight}$$

The solution originally weighs 1500 lb. If evaporation is carried out at 25° C:

(a) How much water must be evaporated before crystallization starts? **Ans.** 658 lb.

(b) What is the first solid to crystallize from the solution? How much of this pure crystalline material can be removed before a second solid starts to separate? **Ans.** 85.1 lb $Na_2SO_4 \cdot 10H_2O$.

(c) How much water must be removed by evaporation before a second crystalline phase starts to form? What is the nature of this second crystalline phase? **Ans.** 71 lb H_2O; $Na_2CO_3 \cdot 10H_2O$.

By means of a sketch, show how Fig. 26 was used in solving the problem.

22. A solution has the following initial composition:

$$Na_2SO_4 \cdot 10H_2O = 10 \text{ parts by weight}$$
$$Na_2CO_3 \cdot 10H_2O = 200 \text{ parts by weight}$$
$$\text{Solvent water} = 100 \text{ parts by weight}$$

The solution weighs 3600 lb. It is required that 80% of the $Na_2CO_3 \cdot 10H_2O$ be recovered in pure form. The lowest temperature that can be obtained is 15° C. How many pounds of water must be removed by evaporation if 80% recovery is obtained on cooling to 15° C?

By means of a sketch, show how Fig. 26 was used in solving the problem.

23. A solution has the following initial composition:

$$Na_2SO_4 \cdot 10H_2O = 80 \text{ parts by weight}$$
$$Na_2CO_3 \cdot 10H_2O = 10 \text{ parts by weight}$$
$$\text{Solvent water} = 100 \text{ parts by weight}$$

The solution weighs 1600 lb. Compare the maximum yield of $Na_2SO_4 \cdot 10H_2O$ obtainable by each of the following three processes:

    (a) Cooling to 10° C. **Ans. 481 lb.**

    (b) Evaporation at a constant temperature of 25° C. **Ans. 610 lb.**

    (c) Evaporation of a limited amount of water at 25° C, followed by cooling to 10° C. **Ans. 630 lb.**

By means of a sketch, indicate how Fig. 26 was used in solving the problem.

24. The residual liquor from a crystallizing operation has the following composition:

$$Na_2SO_4 \cdot 10H_2O = 10 \text{ parts by weight}$$
$$Na_2CO_3 \cdot 10H_2O = 60 \text{ parts by weight}$$
$$\text{Solvent water} = 100 \text{ parts by weight}$$

This liquor is to be used to extract the soluble material from a powdered mass that has the following composition:

Na_2SO_4	1.25%
Na_2CO_3	10.50
Insoluble matter	88.25
	100.00%

    (a) What is the minimum weight of the residual liquor required to dissolve the soluble matter present in 600 lb of the solid mass, assuming that the leaching operation is carried out at 25° C?

    (b) If the liquor obtained from the leaching operation is subsequently cooled, what is the maximum possible percentage recovery of pure Na_2CO_3?

By means of a sketch, show how Fig. 26 was used in connection with the solution of this problem.

25. From the data of *International Critical Tables*, plot a solubility chart similar to Fig. 26 for the system $NaNO_3$—$NaCl$—H_2O. Include the solubility isotherms of 15.5, 50, and 100° C.

26. A mixture of $NaNO_3$ and $NaCl$ is leached with water at 100° C to form a solution which is saturated with both salts. From the chart of problem 25, calculate the weight and composition of the crystals formed by cooling 50 lb of this solution to 15.5° C.

27. A solution of picric acid in benzene contains 30 grams of picric acid per liter. Calculate the quantity of water with which 1 gal of this solution at 18° C must be shaken in order to reduce the picric acid concentration to 4.0 grams per liter in the benzene phase. **Ans. 3.49 gal.**

28. One gallon of an aqueous solution of picric acid containing 0.15 lb of picric acid is shaken with 2.5 gal of benzene at 18° C. Calculate the number of pounds of picric acid in each phase after the treatment.

29. A mixture of phenol and water contains 45% phenol by weight. Calculate the weight fractions of upper and lower layers formed by this mixture at a temperature of 45° C. **Ans.** 63.3% lower layer.

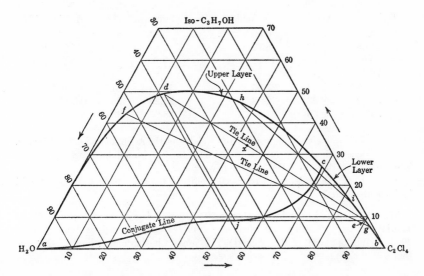

FIG. 48. Solubility curve and conjugate line for tetrachloroethylene–isopropyl alcohol–water system at 77° F in weight per cent. [From Bergelin, Lockhart, and Brown, *Trans. Am. Inst. Chem. Eng.*, **39**, 173 (1943)]

30. From Fig. 48 calculate the weight fractions and compositions of the upper and lower layers of a mixture of 30% water, 20% isopropyl alcohol, and 50% tetra-chloroethylene at a temperature of 77° F.

31. Assuming the applicability of Henry's law, calculate the percentage CO_2 by weight that may be dissolved in water at a temperature of 20° C in contact with gas in which the partial pressure of CO_2 is 450 mm Hg. **Ans.** 0.102%.

32. Assuming the applicability of Henry's law, calculate the partial pressure of H_2S above an aqueous solution at 30° C that contains 3.0 grams of H_2S per 1000 grams of water.

33. Calculate the volume in cubic feet of NH_3 gas under a pressure of 1 atm and at a temperature of 20° C that can be dissolved in 1 gal of water at the same temperature. **Ans.** 100 cu ft.

34. An aqueous solution of ammonia at 10° C is in equilibrium with ammonia gas having a partial pressure of 500 mm Hg.

    (*a*) Calculate the percentage ammonia, by weight, in the solution.

    (*b*) Calculate the partial pressure of the ammonia in this solution if it were warmed to a temperature of 40° C.

35. Correlate the equilibrium data of illustration 16 by the Othmer and Tobias and Hand equations by determining the constants u and v in equation 21 and the constants u' and v' in equation 22.

36. A mixture of 35% chloroform and 65% acetic acid by weight weighing 100 lb is extracted with 50 lb of water to remove the acid.

The conjugate layers of the system have the following values at 18° C:

Raffinate Phase			Extract Phase		
Chloroform	Water	Acetic Acid	Chloroform	Water	Acetic Acid
99.01	0.99	0.00	0.84	99.16	0.00
91.85	1.38	6.77	1.21	73.69	25.10
80.00	2.28	17.72	7.30	48.58	44.12
70.13	4.12	25.75	15.11	34.71	50.18
67.15	5.20	27.65	18.33	31.11	50.56
59.99	7.93	32.08	25.20	25.39	49.41
55.81	9.58	34.61	28.85	23.28	47.87

(a) Determine the composition and weights of the raffinate and extract layers.

(b) The raffinate of part a is treated with one-half its weight of water. Calculate the composition and weight of the new layers.

(c) If all the water were removed from the raffinate of part b what would be its composition?

37. One thousand pounds of water at 20° C containing 3.5 lb of benzoic acid in solution are treated with two successive portions of benzene, each portion weighing 200 lb. What per cent of the original acid is present in the final raffinate? Assume that equilibrium is attained in each stage.

EQUILIBRIUM DATA

| Grams acid per 1000 grams H_2O | 0.915 | 1.135 | 1.525 | 2.04 | 2.56 | 3.99 | 5.23 |
| Grams acid per 1000 grams C_6H_6 | 1.025 | 1.60 | 2.91 | 5.33 | 7.94 | 20.1 | 36.0 |

38. Assuming that equilibrium is attained without changing the volume of the gas phase, what will be the concentration of the solution formed when 50 cu ft of a gaseous mixture of 10% NH_3 and 90% air are brought into contact with 2 gal of water at 20° C and 760 mm pressure, the water already containing 5 grams of ammonia? Solve by using data on the ammonia-water system from Perry[9].

39. The equilibrium relationship for the solubility of acetone in water is $p = 330c$ at 20° C where p is in millimeters of mercury and c is in pounds of acetone per pound of water. Assuming that equilibrium is attained without changing the volume and temperature of the gas phase, what will be the concentration of the solution formed (in pounds of acetone per pound of water) when 10 cu ft of a gaseous mixture of 1% acetone vapor and 99% air at 20° C and 760 mm pressure are brought into contact with 2 cu ft of water at 20° C and 760 mm pressure?

40. Ten pounds of water at 20° C are placed in a cylinder with a movable piston. One thousand cubic feet (measured at 20° C and 1 atm) of a mixture of 20% NH_3 and 80% air (each by volume) are brought into contact with the water. If equilibrium is attained at 20° C and a total pressure of 1 atm, what is the partial pressure of the NH_3, in millimeters, above the solution, and what is the concentration of

[9] J. H. Perry, *Chemical Engineers' Handbook*, McGraw Hill Book Co., Inc. (1950).

the solution, in grams of NH_3 per 1000 grams of water? At 20° C the vapor pressure of water is 17.535 mm. Assume no vapor-pressure lowering of water due to the presence of solute. Solve by using data on the ammonia-water system from Perry.[10]

41. Twenty-five pounds of dry gaseous ammonia and 434 cu ft of dry air measured at 20° C and 760 mm pressure are brought into contact with 100 lb of water in a vessel in which the gas occupies a volume of 1000 cu ft. When the system reaches equilibrium at 20° C, the volume remaining constant, what will be the concentration of ammonia in the liquid phase, expressed as pounds per 100 lb of water? What will be the pressure of the ammonia in the gas space, in atmospheres, and what will be the total pressure in the gas space, in atmospheres?

At 20° C the partial pressure of ammonia, in millimeters of mercury, over aqueous solutions of ammonia is as follows:

Lb NH_3 per 100 lb H_2O	7.5	10	15	20	25
Pressure of NH_3, mm Hg	50.0	69.6	114	166	227

42. Following are data on the solubility of carbon dioxide in water at 12.4° C:

Pressure, atm	1	5	10	15	20	25
ml CO_2 (measured at 12.4° C and 1 atm) per ml water	1.086	5.15	9.65	13.63	17.11	20.31

In the process of preparing hydrogen from water gas the following gas mixture is available: 62.8% H_2, 33.4% CO_2, 3.2% N_2, 0.6% CH_4. It is desired to reduce the CO_2 content by scrubbing the gases with water. A comparison is wanted for operation at 25 atm total pressure and at 70 atm. It is assumed that the absorption tower can operate at 12.4° C.

Plot equilibrium curves for total pressures of 25 and of 70 atm, showing moles of CO_2 per mole of inert gas against moles of CO_2 per mole of water. The solubility of H_2, CH_4, and N_2 may be considered negligible.

For each of the two pressures specified, determine the minimum number of pounds of water required per cubic foot of dry entering gas to reduce the CO_2 content to 1% (dry basis).

43. As applied to chlorine dissolved in water, the values for H in the equation for Henry's law, $p = HN$ (where p is the partial pressure of solute in atmospheres, N is the mole fraction of solute in the liquid phase, and H is the proportionality constant) are given as follows:

Temperature, ° C	10	15	20	25	30	35
$H \times 10^{-2}$	3.91	4.65	5.29	5.97	6.60	7.28

A mixture of 100 cu ft of Cl_2 and 100 cu ft of N_2, each measured saturated with water vapor at 15° C and 760 mm is brought into contact with 50 lb of water, the entire system being maintained at 15° C and 760 mm pressure. Neglecting the solubility of nitrogen and assuming that the chlorine does not react appreciably with the water under the given conditions, estimate the values of p and N at equilibrium.

44. A pharmaceutical manufacturer has an order for a few hundred pounds of $CuSO_4 \cdot 3H_2O$. A tray drier that is available has an automatic temperature control set at 122° F. Air for the drying is preheated before entering the drier proper, and no heat is furnished within the drier except that from the air. The drier is so small that drying conditions may be assumed substantially constant. What are the

[10] *Ibid.*

limits for the wet-bulb temperatures of the air in the drier? At 122° F the dissociating pressures are:

$$CuSO_4 \cdot 5H_2O = CuSO_4 \cdot 3H_2O + 2H_2O \text{ (vapor)} \qquad 45.4 \text{ mm}$$
$$CuSO_4 \cdot 3H_2O = CuSO_4 \cdot H_2O + 2H_2O \text{ (vapor)} \qquad 30.9 \text{ mm}$$
$$CuSO_4 \cdot H_2O = CuSO_4 + H_2O \text{ (vapor)} \qquad 4.5 \text{ mm}$$

The crystals of $CuSO_4 \cdot 3H_2O$ are to be packed in bottles at 77° F. Would the air used in the drying operation be suitable for the packing room if it were cooled at constant humidity from 122 to 77° F? Give a quantitative answer. At 77° F the dissociation pressures corresponding to the above reactions are 7.8, 5.6, and 0.8 mm, respectively.

45. It is desired to remove methanol from naphthalene crystals by passing the crystals on a belt through a chamber in which warm air is blown countercurrent to the crystals. To recover the methanol, the air, on leaving the drying chamber, passes through a condenser, and then is recirculated back into the drier through a bank of steam coils which returns it to its original temperature. The air is to enter the drying chamber at 110° F and at a humidity of 10%. It leaves at 5° above its adiabatic saturation temperature. The crystals may be assumed to remain at the adiabatic saturation temperature throughout the process and leave absolutely dry.

(a) Tabulate the dew point, absolute humidity, adiabatic saturation temperature, humid heat, and humid volume of the air both as it enters and leaves the drying chamber.

(b) To what temperature must the air be cooled in the condenser in order to maintain the desired terminal conditions in the drier? How many Btu per 10,000 cu ft (standard conditions) of alcohol-free air recirculated are transferred to the cooling water of the condenser?

(c) If the wet crystals enter the drier at the rate of 1000 lb per hr, carrying 10% methanol on the dry basis, what volume of air (under entering conditions) must be recirculated per minute?

46. Construct a plot of per cent humidity versus temperature for $LiCl \cdot H_2O$. The transition point at which $LiCl \cdot H_2O$ changes to $LiCl$ is 98° C. The decomposition pressures of the hydrate are as follows:

Temperature, °K	Decomposition pressure, atmospheres
303	0.0005
308	0.001
313	0.0016
318	0.0021
323	0.0037
328	0.005
333	0.008
338	0.0116
343	0.015
373.6	0.118
375.1	0.132

At 100° C and 760 mm pressure, the vapor pressure of the water above the LiCl solution is related to the concentration in the following way:

$$100R = \frac{100(p_o - p)}{Mp_o}$$

where M = gram-formula weight of LiCl (anhydrous) per 1000 grams of water
  p_o = vapor pressure of pure water
  p = vapor pressure of water above LiCl solution
  R = a quantity determined by experiment

M	1.00	4.00	6.00	8.00	9.00	10.00	12.00	15.00	23.00
$100R$	3.39	4.35	4.82	5.14	5.24	5.21	5.02	4.69	3.60

Assume that the relative vapor-pressure lowering is independent of the temperature.

From these data and the solubility of LiCl, calculate points for the two curves at 30, 40, 50, 60, 70, and 80° C.

Material Balances

A material balance of an industrial process is an exact accounting of all the materials that enter, leave, accumulate, or are depleted in the course of a given time interval of operation. The material balance is thus an expression of the law of conservation of mass in accounting terms. If direct measurements were made of the weight and composition of each stream entering or leaving a process during a given time interval and of the change in material inventory in the system during that time interval, no calculations would be required. Seldom is this feasible, and hence calculation of the unknowns becomes indispensable.

The general principle of material-balance calculations is to establish a number of independent equations equal to the number of unknowns of composition and mass. For example, if two streams enter a process and one stream leaves, with no change in inventory in the system during the time interval, the mass and composition of each stream establishes the complete material balance. For calculating the complete material balance the greatest number of unknowns permissible is three, selected from among six possible items. Variations in solving the problem will depend on the particular items that are unknown, whether they be of composition or mass, or of streams entering or leaving. The following guides serve to direct the course of calculations:

1. If no chemical reaction is involved, nothing is gained by establishing material balances for the chemical elements present. In such processes, material balances should be based on the chemical compounds rather than on the elements, or on components of fixed composition even if not pure chemical compounds.

2. If chemical reactions occur, it becomes necessary to develop material balances based on chemical elements, or on radicals, compounds, or substances that are not altered, decomposed, or formed in the process.

3. For processes wherein no chemical reactions occur, use of weight units such as grams or pounds is preferable. For processes in which chemical reactions occur, it is desirable to utilize the gram-mole or pound-mole, or the gram-atom or pound-atom.

4. The number of unknown quantities to be calculated cannot exceed the number of independent material balances available; otherwise, the problem is indeterminate.

5. If the number of independent material-balance equations exceeds the number of unknown weights that are to be computed, it becomes a matter of judgment to determine which of the equations should be selected to solve the problem. If all the analytical data used in setting up the equations were perfect, it would be immaterial which equations were selected for use. However, analytical data are never free from error, and a certain amount of discretion is necessary in order to select the most nearly accurate equations for solving the problem. In general, equations based on components forming the largest percentage of the total mass are most dependable.

6. Recognition of the maximum number of truly independent equations is important. Any material-balance equation that can be derived from other equations written for the process cannot be regarded as an additional independent equation. For example, in the following illustration 1, it would be possible to write material-balance equations based on water, HNO_3, H_2SO_4, nitrogen, sulfur, hydrogen, oxygen, chlorine, and total weights. Of these nine equations only three are independent. If the equations based on H_2SO_4, HNO_3, and total weights are selected, the other seven equations can be deduced from computations alone.

7. If any two or more substances exist in fixed ratio with respect to one another in each stream where they appear, only one independent material-balance equation may be written with respect to these substances. Although a balance may be written for any one substance in question, it is generally best to combine the substances appearing in constant ratio into a single group and develop a single equation for this combined group.

8. A substance that appears in but one incoming stream and one outgoing stream serves as a reference for computations and is termed a *tie substance*. Knowledge of the percentage of a tie substance in two streams establishes the relationship between the weights of the streams so that, if one is known, the other can be calculated.

Illustration 1. The waste acid from a nitrating process contains 23% HNO_3, 57% H_2SO_4, and 20% H_2O by weight. This acid is to be concentrated to contain 27% HNO_3 and 60% H_2SO_4 by the addition of concentrated sulfuric acid containing 93% H_2SO_4 and concentrated nitric acid containing 90% HNO_3. Calculate the weights of waste and concentrated acids that must be combined to obtain 1000 lb of the desired mixture.

Basis: 1000 lb of final mixture.

Let x = weight of waste acid

y = weight of concentrated H_2SO_4

z = weight of concentrated HNO_3

Over-all Balance
$$x + y + z = 1000 \qquad (a)$$

H_2SO_4 Balance
$$0.57x + 0.93y = 1000 \times 0.60 = 600 \qquad (b)$$

HNO_3 Balance
$$0.23x + 0.90z = 1000 \times 0.27 = 270 \qquad (c)$$

Solving the simultaneous equations a, b, and c gives

$$x = 418 \text{ lb waste acid}$$
$$y = 390 \text{ lb concentrated } H_2SO_4$$
$$z = 192 \text{ lb concentrated } HNO_3$$

These results may be verified by a material balance of the water in the process:

Water entering $= (418 \times 0.20) + (390 \times 0.07) + (192 \times 0.10) = 130$ lb

Since the final solution contains 13% H_2O, this result verifies the calculations.

Processes Involving Chemical Reactions. Material balances of processes involving chemical reactions fall into two general classes:

(*a*) The compositions and weights of the various streams entering the process are known. It is required to calculate the compositions and weights of the streams leaving the process for a specified degree of completion of the reaction.

(*b*) The compositions and weights of the entering streams are partially known. It is required to calculate the compositions and weights of all entering and leaving streams and to determine the degree of completion of the reaction.

In these calculations it is desirable to work with molal rather than ordinary weight units, particularly for components undergoing chemical transformation. The limiting reactant should be selected. The quantity of each reacting material may then be specified in terms of the percentage excess it forms of that theoretically required. The calculation is then completed on the basis of the limiting reactant that is present in a unit quantity of the reactants. The amounts of the new products formed in the reaction are determined from the degree of completion. The unconsumed reactants and inert materials pass into the product unchanged.

Illustration 2. A producer gas made from coke has the following composition by volume:

CO	28.0%
CO_2	3.5
O_2	0.5
N_2	68.0
	100.0%

This gas is burned with such a quantity of air that the oxygen from the air is 20% in excess of the *net* oxygen required for complete combustion. If the combustion is 98% complete, calculate the weight and composition in volumetric per cent of the gaseous products formed per 100 lb of gas burned.

Discussion. The carbon monoxide is the limiting reactant, while the oxygen is the excess reactant. The amount of oxygen supplied by the air is expressed as the percentage in excess of the *net* oxygen demand, this latter term referring to the total oxygen required for complete combustion, minus that present in the fuel. Since the composition of the fuel is known on a molal basis, it is most convenient to choose 100 lb-moles of the fuel gas as the basis of calculation, and at the close of the solution to convert the results over to the basis of 100 lb of gas burned.

Basis of Calculation: 100 lb-moles of producer gas.

Constituent	Molecular Weight	Mole %	Weight, lb
CO	28.0	28.0	$28.0 \times 28.0 =$ 784
CO_2	44.0	3.5	$3.5 \times 44.0 =$ 154
O_2	32.0	0.5	$0.5 \times 32.0 =$ 16
N_2	28.2	68.0	$68.0 \times 28.2 =$ 1917
		100.0	2872

Oxygen Balance

O_2 required to combine with all CO present $1/2 \times 28.0 =$		14.0 lb-moles
O_2 in the producer gas	$=$	0.5
Net O_2 demand (by difference)	$=$	13.5 lb-moles
O_2 supplied by air $= 13.5 \times 1.20$	$=$	16.2 lb-moles
O_2 consumed $= 0.98 \times 28.0 \times 1/2$	$=$	13.7 lb-moles
O_2 in products $= 16.2 + 0.5 - 13.7$	$=$	3.0 lb-moles
or 3.0×32	$=$	96.0 lb

Carbon Balance

C in fuel gas $= 28.0 + 3.5$	$=$	31.5 lb-atoms
C in CO of products of combustion $= 0.02 \times 28.0$	$=$	0.56 lb-atom
C in CO_2 of products of combustion	$=$	30.94 lb-atoms
Carbon monoxide in products	$=$	0.56 lb-mole
or 0.56×28	$=$	15.7 lb
Carbon dioxide in products	$=$	30.94 lb-moles
or 30.94×44	$=$	1359 lb

Nitrogen Balance

N_2 in producer gas	$=$	68.0 lb-moles
N_2 from air $= 79/21 \times 16.2$	$=$	60.9 lb-moles
N_2 in products (by addition)	$=$	128.9 lb-moles
or 128.9×28.2	$=$	3637 lb

PRODUCTS OF COMBUSTION PER 100 LB-MOLES OF PRODUCER GAS

	Lb	Molecular Weight	Lb-moles	Mole or volume %
CO_2	1359	44.0	30.9	18.92
O_2	96	32.0	3.0	1.84
CO	16	28.0	0.56	0.34
N_2	3637	28.2	128.9	78.9
Total	5108		163.4	100.0

$$\text{Total weight of products per 100 lb producer gas} = \frac{(5108)(100)}{2872} = 178 \text{ lb}$$

Illustration 3. A solution of sodium carbonate is causticized by the addition of partly slaked commercial lime. The lime contains only calcium carbonate as an impurity. A small amount of free caustic soda is in the original solution. The mass obtained from the causticization has the following analysis:

$$
\begin{array}{ll}
CaCO_3 & 13.48\% \\
Ca(OH)_2 & 0.28 \\
Na_2CO_3 & 0.61 \\
NaOH & 10.36 \\
H_2O & 75.27 \\
\hline
& 100.00\%
\end{array}
$$

The following items are desired:

(a) The weight of lime charged per 100 lb of the causticized mass, and the composition of the lime.

(b) The weight of the alkaline liquor charged per 100 lb of the causticized mass, and the composition of the alkaline liquor.

(c) The reacting material that is present in excess and its percentage excess.

(d) The degree of completion of the reaction.

Discussion. The problem as stated cannot be solved before additional data are obtained. The needed additional information is either the analysis of the lime or the analysis of the alkaline liquor.

If the analysis of the lime were determined, the problem could be solved by the following steps: (1) By using calcium as a tie substance, the weight of lime would be determined. (2) An over-all material balance would establish the weight of alkaline liquor. (3) By a carbon balance, the weight of Na_2CO_3 in the alkaline liquor would be computed. (4) By using sodium as a tie substance, the weight of NaOH in the alkaline liquor would be calculated. (5) The weight of water in the alkaline liquor would be determined by difference.

If the analysis of the alkaline liquor were determined instead, the problem would be solved according to the following procedure: (1) By using sodium as a tie substance, the total weight of the alkaline liquor would be calculated. (2) An over-all material balance would establish the weight of lime. (3) By means of a carbon balance, the weight of $CaCO_3$ in the lime would be determined. (4) A calcium balance would give the weight of active CaO [free CaO plus the CaO in the $Ca(OH)_2$] in the lime. (5) The weight of free CaO and the weight of $Ca(OH)_2$ in the lime could then be computed from the available values for the weight of lime, the weight of $CaCO_3$, and the weight of active CaO.

From the foregoing, it is apparent that it is only a matter of convenience whether the lime or the alkaline liquor is analyzed. The normal choice would be to analyze the alkaline liquor, because the analysis is rapid and accurate.

An analysis of the alkaline liquor used in the process gave the following results:

$$\begin{array}{ll} \text{NaOH} & 0.594\% \\ \text{Na}_2\text{CO}_3 & 14.88 \\ \text{H}_2\text{O} & \underline{84.53} \\ & 100.00\% \end{array}$$

Basis of Calculation: 100 lb of the causticized mass.

Reactions for the process:

$$\text{CaO} + \text{H}_2\text{O} \qquad \to \text{Ca(OH)}_2$$
$$\text{Na}_2\text{CO}_3 + \text{Ca(OH)}_2 \to 2\text{NaOH} + \text{CaCO}_3$$

Molecular weights:

$$\begin{array}{ll} \text{CaO} = 56.1 & \text{Na}_2\text{CO}_3 = 106.0 \\ \text{NaOH} = 40.0 & \text{CaCO}_3 = 100.1 \\ \text{Ca(OH)}_2 = 74.1 & \text{H}_2\text{O} = 18.02 \end{array}$$

Conversion into molal quantities:

Alkaline Liquor.   Basis: 1 lb.

	Lb	Lb-moles	Lb-atoms Na	Lb-atoms C
NaOH	0.00594	0.000149	0.000149	
Na$_2$CO$_3$	0.1488	0.001404	0.002808	0.001404
H$_2$O	0.8453			
	1.0000		0.002957	0.001404

Causticized Mass.   Basis: 100 lb.

	% or lb	Lb-moles	Lb-atoms Ca	Lb-atoms Na	Lb-atoms C
CaCO$_3$	13.48	0.1347	0.1347		0.1347
Ca(OH)$_2$	0.28	0.00377	0.0038		
Na$_2$CO$_3$	0.61	0.00575		0.0115	0.00575
NaOH	10.36	0.2590		0.2590	
H$_2$O	75.27				
	100.00		0.1385	0.2705	0.1405

Sodium Balance

Object: To evaluate weight of alkaline liquor.

Na in causticized mass	= 0.2705 lb-atom
Na in 1 lb alkaline liquor	= 0.002957 lb-atom
Weight of alkaline liquor 0.2705/0.002957	= 91.50 lb

Over-all Material Balance

Object: To determine the total weight of lime.

Weight of causticized mass	= 100.00 lb
Weight of alkaline liquor	= 91.50 lb
Weight of lime = 100.00 − 91.50	= 8.50 lb

Carbon Balance

Object: To evaluate the weight of $CaCO_3$ in the lime.

C in causticized mass	= 0.1405 lb-atom
C in Na_2CO_3 = 91.50 × 0.001404	= 0.1285 lb-atom
C in $CaCO_3$ = 0.1405 − 0.1285	= 0.0120 lb-atom
Weight of $CaCO_3$ = 0.0120 × 100.1	= 1.20 lb

Calcium Balance

Object: To determine the active CaO [that present in the $Ca(OH)_2$ plus the free CaO] in the lime.

Ca in the causticized mass = Ca in the lime	= 0.1385 lb-atom
Ca present as $CaCO_3$ in the lime (see carbon balance)	= 0.0120 lb-atom
Ca present in $Ca(OH)_2$ and in free CaO =	
0.1385 − 0.0120	= 0.1265 lb-atom

Over-all Balance of Constituents in the Lime

Object: To determine free CaO and $Ca(OH)_2$ in the lime

Total weight of lime charged	= 8.50 lb
Weight of $CaCO_3$ (see carbon balance)	= 1.20 lb
Weight of $CaO + Ca(OH)_2$ = 8.50 − 1.20	= 7.30 lb
Weight of total active CaO = 0.1265 × 56.1	= 7.10 lb
H_2O present in the $Ca(OH)_2$ = 7.30 − 7.10	= 0.20 lb
$Ca(OH)_2$ in lime = 0.20 × (74.1/18.02)	= 0.82 lb
Weight of free CaO = 7.30 − 0.82	= 6.48 lb

Results:

(a) Weight of lime = 8.50 lb

Analysis of lime

	Lb	%
$CaCO_3$	1.20	14.1
$Ca(OH)_2$	0.82	9.6
CaO	6.48	76.3
	8.50	100.0

(b) Weight of alkaline liquor   = 91.50 lb

Analysis of alkaline liquor was determined experimentally.

(c) Determination of excess reactant.

Total active CaO	= 0.1265 lb-mole
Na_2CO_3 in liquor = 91.50 × 0.001404	= 0.1285 lb-mole

Since, according to the reaction equation, 1 mole of Na_2CO_3 requires 1 mole of active CaO, it is concluded that the Na_2CO_3 is present in excess, and that the active CaO is the limiting reactant.

Excess of Na_2CO_3 = 0.1285 − 0.1265	= 0.0020 lb-atom
% excess = (0.0020/0.1265) × 100	= 1.6%

(d) Degree of completion

$Ca(OH)_2$ in causticized mass	= 0.00377 lb-mole
$CaO + Ca(OH)_2$ in lime charged	= 0.1265 lb-mole
Degree of completion of the reaction =	
100 − (0.00377/0.1265) × 100	= 97.0%

In each of the preceding three illustrations, there was but one stream of material emerging from the process. In many processes, there are two or more emergent streams. For example, there may be an evolution of gas or vapor from the reactants, as in the calcination of limestone, or there may be a separable residue which is removed from the major product, as when a precipitate is separated from a liquid solution. While the complexity of the problem tends to increase as the number of streams of material involved in the process increases, the general methods of solution are as indicated for the illustrations already presented. Unknown weights are evaluated through the use of material balances. Frequent use is made of tie substances for determining unknown stream weights. The following illustrative problem is typical of the treatment applied to a process wherein there is more than one stream of emergent material.

Illustration 4. The successive reactions in the manufacture of HCl from salt and sulfuric acid may be represented by the following equations:

$$NaCl + H_2SO_4 \rightarrow NaHSO_4 + HCl$$
$$NaCl + NaHSO_4 \rightarrow Na_2SO_4 + HCl$$

In practice the salt is treated with aqueous sulfuric acid, containing 75% H_2SO_4, in slight excess of the quantity required to combine with all the salt to form Na_2SO_4. Although the first reaction proceeds readily, strong heating is required for the second. In both steps of the process HCl and water vapor are evolved from the reaction mass.

"Salt cake" prepared by such a process was found to have the following composition:

Na_2SO_4	91.48%
$NaHSO_4$	4.79
NaCl	1.98
H_2O	1.35
HCl	0.40
	100.00%

The salt used in the process is dry, and may be assumed to be 100% NaCl.

(a) Calculate the degree of completion of the first reaction and the degree of completion of the conversion to Na_2SO_4 of the salt charged.

(b) On the basis of 1000 lb of salt charged, calculate the weight of acid added, the weight of salt cake formed, and the weight and composition of the gases driven off.

Discussion. The logical choice of a basis of calculation is 1000 lb of salt, since the results are wanted on that basis.

The solution of part *b* is accomplished through a series of material balances. It will be noted that sodium serves as a tie substance between the salt cake and the salt charged, and that sulfur serves as a tie substance between the salt cake and the aqueous acid. Accordingly, the problem may be solved by the following successive steps: (1) A sodium balance establishes the weight of salt cake. (2) A sulfur balance serves to determine the weight of aqueous acid used. (3) A chlorine balance establishes the weight of HCl driven off in the gases. (4) A water balance determines the weight of H_2O in the gases driven off.

Basis of Calculation: 1000 lb salt charged.

Conversion into molal quantities:

　　Salt. Basis: 1000 lb.

　　　　$NaCl = 1000/58.5 = 17.08$ lb-moles

　　Sulfuric Acid. Basis: 1 lb.

　　　　$H_2SO_4 = 0.75$ lb, or $0.75/98.1 = 0.00764$ lb-mole

　　Salt Cake. Basis: 1 lb.

	Lb.	Molecular Weight	Lb-moles	Lb-atoms Na	Lb-atoms S	Lb-atoms Cl
Na_2SO_4	0.9148	142.1	0.00644	0.01288	0.00644	
$NaHSO_4$	0.0479	120.1	0.000398	0.000398	0.000398	
NaCl	0.0198	58.5	0.000338	0.000338		0.000338
H_2O	0.0135	18				
HCl	0.0040	36.46	0.000110			0.000110
Total	1.000			0.01362	0.00684	0.000448

Part *a* can be worked out by focusing attention on the salt cake. Since no H_2SO_4 is present in the product, the first reaction went to completion.

Total Na present	$= 0.01362$ lb-atom
Na in Na_2SO_4	$= 0.01288$ lb-atom
Conversion of NaCl to $Na_2SO_4 =$	
$(0.01288/0.01362) \times 100$	$= 94.5\%$

Sodium Balance

Object: To determine the weight of salt cake.

Na in 1000 lb salt charged	$= 17.08$ lb-atoms
Na in 1 lb salt cake	$= 0.01362$ lb-atom
Weight of salt cake $= 17.08/0.01362$	$= 1253$ lb

Sulfur Balance

Object: To determine the weight of aqueous acid used.

S in salt cake $= 1253 \times 0.00684$	$= 8.58$ lb-atoms
S in 1 lb of aqueous acid	$= 0.00764$ lb-atom
Weight of aqueous acid $= 8.58/0.00764$	$= 1123$ lb

Water Balance

Object: To determine the weight of H_2O in gas evolved.

H_2O in aqueous acid $= 1123 \times 0.25$	$= 281$ lb
H_2O in salt cake $= 1253 \times 0.0135$	$= 17$ lb
H_2O driven off $= 281 - 17$	$= 264$ lb

Chlorine Balance

Object: To determine the weight of HCl in gas evolved.

Cl in salt charged	$= 17.08$ lb-atoms
Cl in salt cake $= 1253 \times 0.000448$	$= 0.56$ lb-atom
Cl in HCl driven off $= 17.08 - 0.56$	$= 16.52$ lb-atoms
Weight HCl in gases $= 16.52 \times 36.46$	$= 603$ lb

Composition of leaving gases:

	Lb	%
HCl	603	69.6
H₂O	264	30.4
	867	100.0

Over-all Balance

Object: To verify the accuracy of the calculations.

Total weight of reactants = 1000 + 1123 = 2123 lb
Total weight of products = 1253 + 867 = 2120 lb

Stepwise Countercurrent Processing

The leaching of solids, the washing of precipitates, the extraction of oil from crushed seeds, the softening of water, and the sorption of gases are processes that are often carried out in stages by continuous or discontinuous methods.

Countercurrent extraction offers the most economical use of solvent, permitting high concentrations in the final extract and high recovery from the initial solid with the least amount of solvent. Since the continuous countercurrent flow of the solid is difficult to achieve, the same results are approached by arranging several tanks in series, entering the supply of fresh solvent into the tank containing the most nearly exhausted solid, and then advancing the solvent progressively from one tank to the next until the most concentrated solution discharges from the tank where the fresh feed enters. Ordinarily the operation is discontinuous in that the solvent is pumped from one tank to the next intermittently and allowed to remain in each tank until equilibrium extraction is approached.

Leaching of fine solids where the particles become suspended in the agitated liquid is accomplished by a method of suspension, settling, and decantation. Countercurrent results are achieved by flowing the supernatant liquid from one tank to the next and pumping the sludge from one tank to the next in the opposite direction. The fresh solvent encounters the most nearly exhausted sludge, and the spent solvent discharges from the tank where the fresh sludge is introduced. This method of leaching is also designated as countercurrent decantation.

In a continuous countercurrent extraction or washing process the precipitate to be washed or solid to be leached is fed to a tank as a sludge, as shown in Fig. 49. Here the sludge is mixed with a clear solution from tank 2. Upon slow agitation and settling, the resultant clear solution is discharged as a product from tank 1 and the settled sludge from tank 1 is pumped to tank 2. In tank 2 the sludge is mixed with clear solution from tank 3, and, upon slow agitation and settling, the resultant clear solution from tank 2 passes to tank 1 and the sludge from tank 2 is pumped to tank 3. In the last tank L the sludge from the previous tank $L - 1$ is combined with fresh solvent. Upon slow agitation and settling, the clear solution from the last tank is discharged as a residual product.

As an approximation it is assumed that a uniform concentration of solution is attained in each tank, and that the solution leaving each tank as a clear liquid has the same concentration as that retained in

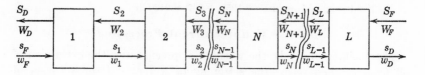

Fig. 49.   Countercurrent leaching

the outgoing sludge. It is also assumed that all the soluble constituents in the solid are dissolved and that the resultant solutions retained by the solid and surrounding it have the same concentration.

Two cases will be considered, first, the general case, where the mass of solution retained by the solid is a known function of concentration, and, second, a special case where the mass of solution retained by unit mass of solid remains constant from tank to tank, and is independent of the solution concentration. With reference to Fig. 49 the following symbols are used.

Capital letters for properties of the clear solution.

Lower-case letters for properties of the solution retained by the solid as a sludge.

$$S, s = \text{mass of solute per unit mass of inert solid}$$
$$W, w = \text{mass of solution per unit mass of inert solid}$$
$$X, x = \text{concentration of solute in solution}$$

Subscript F = feed stream

Subscript D = discharge stream

N = any tank

L = last tank (where final sludge discharges and fresh solution enters)

$$x_N = \frac{s_N}{w_N}, \quad X_N = \frac{S_N}{W_N}, \quad \text{and} \quad x_N = X_N \tag{1}$$

$$x_F = \frac{s_F}{w_F} \tag{2}$$

$$x_D = \frac{s_D}{w_D} = x_L = X_L \tag{3}$$

$$X_F = \frac{S_F}{W_F} \tag{4}$$

$$X_D = \frac{S_D}{W_D} = X_1 = x_1 \tag{5}$$

Material balances over *all* tanks give

$$s_F + S_F = s_D + S_D \tag{6}$$

$$w_F + W_F = w_D + W_D \tag{7}$$

From equations 5, 6, and 7,

$$X_D = \frac{S_D}{W_D} = \frac{s_F + S_F - s_D}{w_F + W_F - w_D} \tag{8}$$

Material balances from tank 1 to tank N, inclusive, give

$$s_F + S_{N+1} = s_N + S_D \tag{9}$$

$$w_F + W_{N+1} = w_N + W_D \tag{10}$$

From equations 9 and 10,

$$x_{N+1} = \frac{s_{N+1}}{w_{N+1}} = \frac{S_{N+1}}{W_{N+1}} = \frac{S_D - s_F + s_N}{W_D - w_F + w_N} \tag{11}$$

From equations 6 and 7, equation 11 may also be written as

$$x_{N+1} = \frac{S_F - s_D + s_N}{W_F - w_D + w_N} \tag{12}$$

In the general case the mass of solution retained per unit mass of solid is a function of the concentration of the solute in the solution as represented in Fig. 50.

The solute retained by unit mass of solid is given by the relation $s = wx$; hence s may also be represented as a function of x as shown in Fig. 50. The s curve of Fig. 50 is derived from the w curve. The method of calculations involved in this type of separation as developed by Ruth[1] follows.

In any extraction process the mass and composition of the entering solution and entering sludge should be known. From the desired fraction recovery the solute in the final discharge sludge is calculated. Hence, S_F, W_F, X_F, s_F, w_F, x_F are known and s_D is calculated.

[1] B. F. Ruth, *Chem. Eng. Prog.* **44**, 71 (1948).

Since s_D is known, x_D and w_D can be obtained from Fig. 50; then S_D can be obtained from equation 6 and W_D from equation 7, and $X_D = S_D/W_D$. All terminal conditions are hence known or calculated.

The calculations, tank to tank, follow:

Material balance for tank 1,

$$s_F + S_2 = S_D + s_1 \tag{13}$$

$$w_F + W_2 = W_D + w_1 \tag{14}$$

Material balance for tank 2,

$$s_1 + S_3 = S_2 + s_2 \tag{15}$$

$$w_1 + W_3 = W_2 + w_2 \tag{16}$$

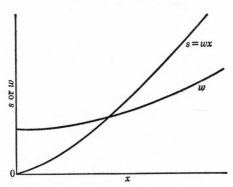

FIG. 50.　Masses of solute and solution per unit mass of solid as a function of concentration of solution

Since $X_D = x_1$ is known, s_1 and w_1 can be obtained at x_1 from Fig. 50. Hence S_2 is obtained from equation 13 and W_2 from equation 14. In this manner the condition at the terminals of each tank can be obtained.

If the value of s_D is greater than s_D previously calculated, then a fractional number of tanks is required for the desired recovery. When it is desired to keep all tanks the same size, the desired fractional recovery may be adjusted to result in an integral number of tanks of equal size.

In the *special case* where the mass of solution retained by the unit solid in the sludge is constant, the values of both w and W are the same for each tank and $w/W = c$.

Equation 12 may then be written as

$$x_{N+1} = \frac{s_{N+1}}{w} = \frac{S_F - s_D + s_N}{W} \tag{17}$$

$$s_{N+1} = (S_F - s_D + s_N)c \tag{18}$$

Let $S_F - s_D = b$.

Then

$$s_{N+1} = c(b + s_N) \quad \text{or} \quad s_N = c(b + s_{N-1}) \tag{19}$$

This equation relates the solute leaving the sludge of any tank $N + 1$ with the solute from the preceding tank N. Hence a series may be written beginning with tank 1 as

$$s_1 = c(b + s_0) = c(b + s_F) \tag{20}$$

$$s_2 = c(b + s_1) = cb + c^2b + c^2s_F \tag{21}$$

$$s_3 = c(b + s_2) = cb + c^2b + c^3b + c^3s_F \tag{22}$$

or $\quad s_N = c(b + s_{N-1}) = cb + c^2b + c^3b + \cdots + c^Nb + c^Ns_F \tag{23}$

For the case where no solute is present in the fresh feed, $b = -s_D = -s_N$. Substitution in equation 23 gives

$$-b = cb + c^2b + c^3b + \cdots + c^Nb + c^Ns_F \tag{24}$$

or $\quad b(1 + c + c^2 + c^3 + \cdots + c^N) = -c^Ns_F \tag{25}$

or, since $\quad -b = s_D,$

$$\frac{s_D}{s_F} = \frac{c^N}{1 + c + c^2 + \cdots c^N} = U \tag{26}$$

The ratio s_D/s_F gives the fraction U of solute unextracted from the solid by N tank in series.

It should be emphasized that Equation 26 is valid only when: (1) The fresh liquid feed contains no solute. (2) The ratio, w/W, is constant throughout the system, including the left side of the first unit (Fig. 49). This requirement is met only if, in the feed sludge, the ratio of mass of solvent plus solute to insoluble matter is identical with the ratio of mass of retained solution to insoluble matter in the sludge lines from all units of the system.

Illustration 5. A sludge contains 1000 lb of sodium carbonate, 1000 lb of insoluble matter, and 2000 lb of water. The sodium carbonate is to be extracted from this sludge with 10,000 lb of water, three thickeners being used in series with countercurrent flow of sludge and water. The fresh water enters the third thickener, overflows to the second, and is then passed to an agitator, where it is mixed with the feed sludge. The resultant sludge from the agitator is passed to the first thickener. The sludge is pumped from one thickener to the next and discharged as waste from the third thickener. The sludge holds 3 lb of solution for each pound of insoluble matter as it leaves each thickener. The concentrated sodium carbonate is drawn off and recovered from the first thickener. Calculate the weight of sodium carbonate recovered, assuming that all sodium carbonate is entirely dissolved to form a uniform solution in each agitator.

On the basis of 1 lb of inert solid,

$$w_D = 3.0$$
$$s_F = 1.0 \qquad c = \frac{w}{W_F} = \frac{3}{10} = 0.30$$
$$w_F = 3.0 \qquad N = 3$$
$$W_F = 10$$

From equation 26,

$$\frac{s_D}{s_F} = \frac{(0.30)^3}{1 + 0.3 + 0.3^2 + 0.3^3} = \frac{0.027}{1.417} = 0.01905$$

Na_2CO_3 lost in outgoing sludge = 0.01905 (1000) = 19.05 lb
Na_2CO_3 recovered in clear solution leaving   =   981 lb

Inventory Changes

The inventory of a chemical process comprises the mass and composition of materials and their distribution in various vessels of the process at any one time. In continuous-flow processes under steady-state conditions the inventory remains constant in each vessel. In actual continuous-flow processes fluctuations in inventory may take place even though steady-state conditions are attempted. Over a long duration of time slight variations in inventory will be negligible compared to the total throughput, and an average material balance may be taken, neglecting fluctuations in inventory. However, these fluctuations become significant over short periods of operation and must be considered in the interpretation of experimental data and in arriving at a reliable material balance.

Inventory changes may result from variations in liquid or solid levels in the various vessels of the plant, from changes in the quantities of materials held in intermediate storage between successive units, and from changes in compositions at various points in the process. Less commonly inventory changes might be encountered as a result of changes in the operating pressure. Particular attention must be given to the separation of solids from solutions, causing a change in liquid level with or without change in the inventory of the liquid.

Where a change in inventory occurs, the material input to the plant is equal to the output plus the net accumulation of inventory or minus the net depletion in all parts of the process. A material balance may be obtained by either adding accumulation items to the output or subtracting them from the input.

If the accumulation is an unchanged charge material, it is subtracted from the input; if the accumulation is a finished product, it is added to the output.

If accumulation or depletion of an intermediate product takes place, the problem requires conversion of the quantity of accumulation into the corresponding quantities of either charge or final product materials. In order to make this conversion, a set of yield figures must be assumed, relating the intermediate product to the charge materials from which it originated or to the products to be derived from it, to whichever it is more similar. Where calculation of yields is the object of the material balance, it is necessary to assume approximate yields on which to base these corrections. If the inventory changes are small, little error is introduced by these assumptions. However, the calculation may be corrected by means of a second approximation based on the yields calculated from the first assumptions.

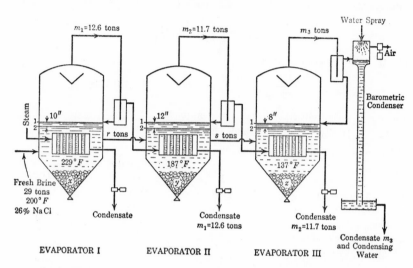

Fig. 51. Inventory changes in triple-effect evaporation

Illustration 6. A triple-effect, vertical-tube, submerged-type evaporator (Fig. 51) is fed with 29 tons of salt-brine solution during a given test period. The brine is fed to the first evaporator at 200° F and contains 26% NaCl by weight. The water evaporated from the first effect is condensed in the steam chest of the second effect and weighed as 12.6 tons; the water evaporated in the second effect is 11.7 tons. The salt crystallized during evaporation is allowed to collect in the conical hopper of each evaporator. During the test period the liquid level was allowed to drop in each evaporator. A drop of one inch corresponds to 12.1 cu ft of brine.

The following data were obtained:

Feed, 29 tons during period, 26% NaCl, and 200° F.

	Evaporator		
	I	II	III
Steam temperature in steam chest	250° F	214° F	173° F
Brine temperature	229° F	187° F	137° F
Level change, in.	10 in. drop	12 in. drop	8 in. drop
Water evaporated	12.6 tons	11.7 tons	. . .
Solubility of NaCl, % by weight	29.0	27.9	27.0
Specific gravity of NaCl crystals	2.15		

It is desired to calculate the tons of NaCl crystallized in each evaporator and the tons of water evaporated from evaporator III.

Solution: In each effect the brine solution is saturated. Its density may be obtained from Fig. 13, page 41. The depletion of brine in each evaporator is not only that corresponding to the drop in liquid level but also that displaced by the salt crystallizing out and accumulating in the hopper. For each tone of salt crystallizing, $G/2.15$ tons of brine are displaced, where 2.15 is the specific gravity of the salt crystals and G is that of the brine solution. For example, in the first evaporator,

$$\frac{G}{2.15} = \frac{1.169}{2.150} = 0.544$$

and the total depletion of brine in tons is

$$\frac{(10)(12.1)(62.4)(1.169)}{2000} + 0.544x = 4.413 + 0.544x$$

where x = tons of salt crystallized in evaporator I

62.4 = density of water in pounds per cu ft

The results are tabulated as follows:

| | Evaporator | | |
	I	II	III
% NaCl by weight	29.0	27.9	27.0
Specific gravity of brine (Fig. 13)	1.169	1.173	1.183
Salt crystallized, tons	x	y	z
Drop in level, in.	10	12	8
Drop in level, tons of brine	4.413	5.314	3.573
Brine displaced by salt	$0.544x$	$0.546y$	$0.550z$
Salt in total brine depletion	$1.280 + 0.1577x$	$1.483 + 0.1522y$	$0.9647 + 0.1486z$
Water in total brine depletion	$3.133 + 0.3860x$	$3.831 + 0.3934y$	$2.608 + 0.4017z$
Water evaporated, tons	12.6	11.7	m_3

From an over-all water balance,

$$29(0.74) + 3.133 + 0.3860x + 3.831 + 0.3934y + 2.608 + 0.4017z = 12.6 + 11.7 + m_3 \quad (a)$$

or $$m_3 = 6.73 + 0.3860x + 0.3934y + 0.4017z \quad (b)$$

From an over-all salt balance,

$$29(0.26) + 1.280 + 0.1577x + 1.483 + 0.1522y + 0.9647 + 0.1486z = x + y + z \quad (c)$$

$$0.8423x + 0.8478y + 0.8514z = 11.27 \quad (d)$$

Let r = tons of brine going from evaporator I to evaporator II

     s = tons of brine going from evaporator II to evaporator III

From salt and water balances for each evaporator the following are obtained:

Evaporator I,

Salt $$1.280 + 0.1577x + 0.26(29) = x + 0.29r \quad (e)$$

Water $$3.133 + 0.3860x + 0.74(29) = 12.6 + 0.71r \quad (f)$$

$$x = 3.92$$

$$r = 19.02$$

Evaporator II,

$$\text{Salt} \qquad 1.483 + 0.1522y + 0.29(19.02) = y + 0.279s \qquad (g)$$

$$\text{Water} \qquad 3.831 + 0.3934y + 0.71(19.02) = 11.7 + 0.721s \qquad (h)$$

$$y = 4.82$$
$$s = 10.45$$

Evaporator III,

$$\text{Salt} \qquad 0.9647 + 0.1486z + 0.279(10.45) = z \qquad (i)$$

$$z = 4.56$$

From equation b,

$$m_3 = 6.73 + 0.3860x + 0.3934y + 0.4017z = 11.97 \qquad (j)$$

Material Balance

	Input	Tons			Output	Tons
(a)	Brine	29.00		(a)	Water from I	12.60
	Depletion Items			(b)	Water from II	11.70
(a)	Brine in I			(c)	Water from III	11.97
	$4.413 + 0.544x =$				Accumulation Items	
	$4.413 + (0.544)(3.92) =$	6.55		(a)	Salt in I, x	3.92
(b)	Brine in II			(b)	Salt in II, y	4.82
	$5.314 + 0.546y =$			(c)	Salt in III, z	4.56
	$5.314 + (0.546)(4.82) =$	7.94				
(c)	Brine in III					
	$3.573 + 0.550z =$					
	$3.573 + (0.550)(4.56) =$	6.08				
		49.57				49.57

While equation (d) is not required to solve the problem, it may be used to check the accuracy of the calculations.

Recycling Operations

The recycling of fluid streams in chemical processing is a common practice to increase yields, to enrich a product, to conserve heat, or to improve operations. In fractionating columns part of the distillate is refluxed through the column to enrich the product. In ammonia synthesis the gas mixture leaving the converter after recovery of ammonia is recycled through the converter. In the operation of driers part of the exit air stream may be recirculated to conserve heat.

Better wetting of a tower packing is provided in scrubbing towers by recirculating part of the exit liquid. Recycling occurs in nearly every stage of petroleum processing.

In a recycling operation the *total* or *combined feed* is made up of a mixture of the *fresh* or *net feed* with the *recycle stock*. The *gross products* of the operation are a mixture of the *net products*, which are withdrawn from the system, and the recycle stock. In such an operation two types of material balances are of interest. In the *over-all material balance* the net feed is equated against the net products. In a *once-through material balance* the total feed is equated against the gross products.

The performance terms commonly used in chemical processing are defined in Table 14. The definitions of these terms are somewhat arbitrary, depending upon the specific process involved. Evaluations of these terms are given for a process involving dehydrogenation of propane in illustration 7 and in problem 26.

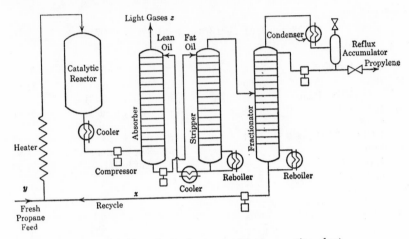

FIG. 52. Flow chart of propane dehydrogenation plant

Illustration 7. Propane is catalytically dehydrogenated to produce propylene by being quickly heated to a temperature of 1000 to 1200° F and passed over a granular solid catalyst. As the reaction proceeds, carbon is deposited on the catalyst, necessitating its periodic reactivation by burning off the carbon with oxygen-bearing gases.

In a laboratory experiment in which pure propane is fed to the reactor, gases of the following composition leave the reactor:

Gas	Mole %
Propane	44.5
Propylene	21.3
Hydrogen	25.4
Ethylene	0.3
Ethane	5.3
Methane	3.2
	100.0

Based on these data it is desired to design a plant to produce 100,000 lb per day of propylene in a mixture of 98.8% purity. The flow diagram of the proposed process is shown in Fig. 52.

Fresh propane feed, mixed with propane recycle stock, is fed to a heater from which it is discharged to a catalytic reactor, operating under such conditions as

TABLE 14. DEFINITIONS OF TERMS USED IN CHEMICAL PROCESSING

Limiting Reactant: The reactant present in such proportions that its consumption will limit the extent to which the reaction can proceed.

Reactant Ratio: The number of moles of an excess reactant per mole of limiting reactant in the reactor feed.

Conversion per Pass: The percentage of the limiting reactant in the combined reactor feed that is converted and disappears.

Fresh-Feed Conversion: The rate of conversion of the limiting reactant to other products expressed as a percentage of the rate at which the limiting reactant enters the operation as fresh feed.

Selectivity: The quantity of the limiting reactant that goes to form the desired product in the reactor effluent, expressed as a percentage of the total quantity of the limiting reactant that is converted.

Yield per Pass: The net yield of any product in the reactor effluent expressed as a percentage of the limiting reactant in the combined reactor feed. Yields may be expressed on a molal, weight, or volume basis, and the basis must always be designated.

Ultimate Yield: The yield of any product in the reactor effluent expressed as a percentage of that portion of the limiting reactant in the reactor feed which is converted and disappears during the course of the reaction.

Over-all Yield: The recovered yield of any product expressed as a percentage of the limiting reactant constituting fresh feed to the operation. The over-all yield differs from the ultimate yield only as a result of inefficiencies in separation and recycling.

Recycle Ratio: The ratio of the quantity of a reactant recycled to the quantity of that same reactant entering a recycling operation as fresh feed.

Combined Feed Ratio: The ratio of the total quantity of a reactant present in the reactor feed of a recycling operation to the quantity of that same reactant entering the operation as fresh feed.

Liquid-Volume Hourly Space Velocity: The liquid volume at 60° F of limiting reactant fed per hour per unit volume of effective reactor or catalyst bed.

Gaseous Hourly Space Velocity: The gaseous volume at 60° F and 14.7 psi of limiting reactant fed per hour per unit volume of effective reactor or catalyst bed.

Weight Hourly Space Velocity: The weight of limiting reactant fed per hour per unit weight of catalyst in the reactor.

Catalyst per Combined Feed: The weight ratio of catalyst circulation to reactor feed in a continuous moving-bed operation. In a fixed-bed operation it is the reciprocal of the product of the weight space velocity times the process period length in hours.

Catalyst Residence Time: The process period length in a fixed-bed operation. In a continuous moving-bed operation it is equal to the weight of catalyst in the reactor divided by the catalyst circulation rate or the reciprocal of the product of the weight space velocity times the catalyst per feed ratio.

Space-Time Yield: The net yield of a product from the reactor per hour per unit of effective reactor volume or catalyst-bed volume or weight. The units must be completely designated.

to produce the same conversion of propane as was obtained in the laboratory experiment. The reactor effluent gases are cooled and compressed to a suitable pressure for separation of the light gases. This separation is accomplished by absorption of the propane and propylene, together with small amounts of the lighter gases, in a cooled absorption oil which is circulated through an absorption tower. The "fat oil" from the bottom of the absorber is pumped to a stripping tower where, by application of heat at the bottom of the tower, the dissolved gases are distilled away from the oil which is then cooled and recirculated to the absorber.

The gases from the stripping tower are passed to a high-pressure fractionating tower which separates them into propane recycle stock as a bottoms product and propylene and lighter gases as overhead. The following compositions are established as preliminary design bases estimated from the vaporization characteristics of the gases:

1. The light gases from the absorber are to contain 1.1% propane and 0.7% propylene by volume. Substantially all hydrogen, ethylene, and methane leaving the cooler will appear in the light gases. The ethane, however, will appear in both the light gases and the product.

2. The propane recycle stock is to contain 98% propane and 2% propylene by weight.

3. The propylene product is to contain 98.8% propylene, 0.7% ethane, and 0.5% propane by weight.

The total feed is passed over the catalyst at the rate of 4.1 lb-moles per cu ft of catalyst per hour. The density of the catalyst is 54 lb per cu ft.

It may be assumed that the small amount of propylene in the feed to the reactor passes through unchanged.

Required

(a) The amount of carbon formed on the catalyst, expressed as weight per cent of the propane fed to the catalyst chamber.

(b) The process period in minutes required in building a carbon deposit equal to 2% by weight of the catalyst.

(c) An over-all material balance.

(d) A once-through material balance.

(e) Ultimate yield of recovered propylene expressed as mole percentage of propane in fresh feed.

(f) Yield per pass of propylene made in reactor expressed as mole percentage of total propane entering reactor.

(g) Recycle ratio, weight units.

(h) Combined feed ratio, weight units.

Preliminary Calculations

Gas formed by decomposition of propane

Basis: 1 lb-mole of gas.

	Lb-moles	Molecular Weight	Lb	Lb-atoms C	Lb-atoms H
Propane C_3H_8	0.445	44.09	19.62	1.335	3.560
Propylene C_3H_6	0.213	42.08	8.96	0.639	1.278
Hydrogen H_2	0.254	2.016	0.512		0.508
Ethylene C_2H_4	0.003	28.05	0.084	0.006	0.012
Ethane C_2H_6	0.053	30.07	1.594	0.106	0.318
Methane CH_4	0.032	16.04	0.513	0.032	0.128
	1.000		31.28	2.118	5.804

Gas formed from 1 mole of propane (by hydrogen balance) $= \dfrac{8.000}{5.804} = 1.378$ lb-moles.

Process Design Calculations

Basis of Design: 1 hour of operation.

Production rate of pure propylene $= \dfrac{100,000}{24} = 4167$ lb per hr

or $\qquad\qquad\qquad\qquad\qquad \dfrac{4167}{42.08} = 99.03$ lb-moles per hr

Evaluation of unknown stream weights:

Let x = pound-moles recycle

y = pound-moles fresh feed

z = pound-moles light gases

Combined feed:

Propane $= y + 0.9791x$ lb-moles

Propylene $= 0.02095x$ lb-moles

Gases leaving reactor:

Gases formed from the propane entering	$= 1.378(y + 0.9791x)$
	$= 1.378y + 1.349x$ lb-moles
Propylene passing through unchanged	$= 0.02095x$ lb-moles
Total gas leaving	$= 1.378y + 1.370x$
Propane leaving	$= (1.378y + 1.349x)(0.445)$
	$= 0.613y + 0.600x$ lb-moles
Propylene leaving	$= 0.213(1.378y + 1.349x) + 0.02095x$
	$= 0.2935y + 0.3083x$ lb-moles

Material Balances

Lb-moles per hr

	Leaving Reactor	Recycle	Product	Light Gases
Propane balance:	$0.613y + 0.600x$	$= 0.9791x$	$+ \dfrac{99.03}{0.9855}(0.00476)$	$+ 0.011z$
Propylene balance:	$0.2935y + 0.3083x$	$= 0.02095x$	$+ 99.03$	$+ 0.007z$
Total:	$1.378y + 1.370x$	$= x$	$+ \dfrac{99.03}{0.9855}$	$+ z$

Simplifying the above three equations gives

$$-0.379x + 0.613y - 0.011z = 0.478$$
$$0.2873x + 0.2935y - 0.007z = 99.03$$
$$0.370x + 1.378y - z = 100.5$$

Solving:

$x = 210.8$ lb-moles recycle

$y = 134.2$ lb-moles fresh feed

$z = 162.9$ lb-moles light gases

Gases Leaving Reactor

Formed from propane $= (1.378)(134.2) + (1.349)(210.8) = 469.3$

Propylene from recycle $= (0.02095)(210.8)$ $= 4.42$

	Lb-moles	Molecular Weight	Lb
$C_3H_8 = (0.445)(469.3)$	$= 208.8$	44.09	9,206
$C_3H_6 = (0.213)(469.3) + 4.4$	$= 104.4$	42.08	4,393
$H_2 = (0.254)(469.3)$	$= 119.3$	2.016	240
$C_2H_4 = (0.003)(469.3)$	$= 1.4$	28.05	39.2
$C_2H_6 = (0.053)(469.3)$	$= 24.9$	30.07	749
$CH_4 = (0.032)(469.3)$	$= \underline{15.0}$	16.04	$\underline{241}$
	473.8		14,868

$$\text{Molecular weight} = \frac{14,868}{473.8} = 31.38$$

Light gases:

	Lb-moles	Molecular Weight	Lb
$C_3H_8 = (0.011)(162.9) =$	1.792	44.09	79.0
$C_3H_6 = (0.007)(162.9) =$	1.140	42.08	48.0
H_2 (from reactor)	119.3	2.016	240.5
C_2H_4 (from reactor)	1.4	28.05	39.2
CH_4 (from reactor)	15.0	16.04	240.6
C_2H_6: Leaving reactor 24.9			
In product 0.98			
In light gases 23.9	$\underline{23.9}$	30.07	$\underline{718.7}$
	162.5		1,366.0

$$\text{Molecular weight} = \frac{1366}{162.5} = 8.41$$

Recycle:

	Lb-moles	Molecular Weight	Lb
$C_3H_8 (0.9791)(210.8)$	206.4	44.09	9,100
$C_3H_6 (0.02095)(210.8)$	$\underline{4.42}$	42.08	$\underline{186}$
	210.8		9,286

$$\text{Molecular weight} = \frac{9,286}{210.8} = 44.05$$

Product:

Total weight $= \dfrac{100,000}{(0.988)(24)} = 4{,}217 \text{ lb}$

C_3H_6 $\dfrac{100,000}{24} = 4{,}167 \text{ lb}$

C_3H_8 $(0.005)(4{,}217) = 21.1 \text{ lb}$

C_2H_6 $(0.007)(4{,}217) = 29.5 \text{ lb}$

Fresh feed:

    (134.2) (44.09)   = 5,917 lb

Carbon deposit:

    Propane to reactor   = 134.2 + 206.4 = 340.6 lb-moles

    C in propane   = (3)(340.6) (12)   = 12,262 lb

    C in gases formed from propane = (1.378)(340.6) (2.118)(12) = 11,929

    C deposited on catalyst   =   333 lb

Required results:

(a) Amount of carbon deposited on catalyst, as weight per cent of the propane fed to the reactor:

$$\text{Carbon deposit} = \frac{333}{5917 + 9100} = 2.22\%$$

(b) Process period:

    Total feed   = 210.8 + 134.2 =   345 lb-moles per hr

    Volume of catalyst $= \dfrac{345.0}{4.1}$ =   84.2 cu ft

    Weight of catalyst   = (84.2)(54)   =   4,547 lb

    Carbon tolerance   = (0.02)(4,547) =   90.9 lb

    Rate of carbon deposition   =   333 lb per hr

    Process period $= \dfrac{90.9}{333}$ (60)   =   16.4 min

(c) Over-all material balance:

Input:

Fresh feed	5,917 lb

Output:

Light gases	1,366 lb	
Product	4,217	
Carbon deposit	333	
Total	5,916	5,916 lb

(d) Once-through material balance:

Input:

Fresh feed	5,917 lb	
Recycle	9,286	
Total	15,203	15,203 lb

Output:

Light gases	1,366 lb	
Product	4,217	
Carbon deposit	333	
Recycle	9,286	
Total	15,202	15,202 lb

(e) Ultimate yield of propylene:

Propylene formed in converter   =   100.2 lb-moles

Propane consumed in converter   =   131.8 lb-moles

$$\% \text{ yield} = \frac{100.2}{131.8} (100) \qquad = \qquad 76.07\%$$

(f) Yield per pass:

Propylene produced in reactor $= 99.03 + 1.14 =$ 100.2 lb-moles

Propane entering reactor $= 134.2 + 206.4 =$ 340.6 lb-moles

$$\% \text{ yield} = \frac{100.2}{340.6} (100) \qquad = \qquad 29.4\%$$

(g) Recycle ratio:

Weight of propane in recycle   =   9100 lb

Weight of propane in fresh feed   =   5917 lb

$$\text{Recycle ratio} = \frac{9100}{5917} \qquad = 1.54$$

(h) Combined feed ratio:

Weight of propane in recycle   =   9,100 lb

Weight of propane in fresh feed   =   5,917

Total weight   =   15,017 lb

$$\text{Total feed ratio} = \frac{15,017}{5917} \qquad = 2.54$$

In dealing with operations in which all streams are subject to exact analysis, design problems are generally approached from data obtained in laboratory or pilot-plant experiments of the once-through type, in which the conditions resulting from recycling are simulated by merely varying the composition of the total feed. However, in dealing with extremely complex mixtures such as encountered in petroleum refining, data from experiments in which recycling is actually carried out are necessary as a design basis for the establishment of ultimate yields and the characteristics of the recycle stock. In the former case the design problem is approached by first setting up a once-through material balance from which the recycle quantities and ultimate yields are derived. In the latter case both the over-all and the once-through material balances must be based directly on experimental data. Illustration 7 is typical of the first class of problem while an example of the second class is shown in Chapter 12, page 486.

A common example of recycling is the recirculation of air in the drying of solids, shown diagrammatically in Fig. 53. On the basis of 1 lb of stock (dry basis) passing through the drier, r lb of air (dry basis) are passed through countercurrently. If y is the fraction of air (dry basis) recirculated, then $(1 - y)$ is the fraction of fresh air supplied (dry basis) and also the fraction of spent air rejected (dry basis).

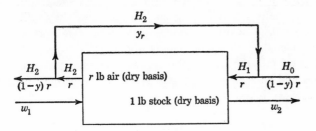

Fig. 53.   Recirculation of air in drying

Accordingly, the following material balances result:

Over-all Balance

$$w_1 - w_2 = r(1 - y)(H_2 - H_0) \tag{27}$$

Once-Through Balance

$$w_1 - w_2 = r(H_2 - H_1) \tag{28}$$

where

w_1 = moisture content of stock entering, pound per pound dry stock
w_2 = moisture content of stock leaving, pound per pound dry stock
H_2 = humidity of air leaving, pound per pound dry air
H_1 = humidity of air entering drier, pound per pound dry air
H_0 = humidity of fresh air supply, pound per pound dry air

From equations 27 and 28,

$$(H_2 - H_0)y = H_1 - H_0 \tag{29}$$

The fraction of air to be recirculated depends on the relative costs of drier, power, and heat, and is hence established by minimizing the total costs.   With increased recirculation for the same drying capacity, a larger drier is required with increased costs of equipment, power and radiation offset by a reduced cost of heat requirements.

Equations 27, 28, and 29 are valid under conditions of steady flow and steady state.   If the recycling were suddenly stopped or started, these equations would not apply during the interval required to reach steady-state conditions.   The time interval in attaining any fractional approach to steady-state conditions after sudden stopping or starting or recycling or sudden changes in composition of feed is discussed on page 228.

Accumulation of Inerts in Recycling. One limitation sometimes encountered in the recycling of fluid streams is the gradual accumulation of inerts or impurities in the recycled stock. Unless some provision is made for removing such impurities, they will gradually accumulate until the process comes to a stop. This problem can be solved by bleeding off a fraction of the recycle stock. The recycle stock bled off may be passed through some special process to remove impurities and to recover the useful components, or may be discarded if such recovery is too costly.

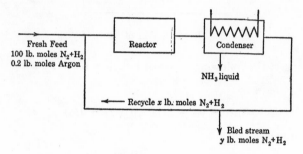

Fresh Feed
100 lb. moles N_2+H_2
0.2 lb. moles Argon

NH_3 liquid

Recycle x lb. moles N_2+H_2

Bled stream
y lb. moles N_2+H_2

Fig. 54. Purging of inerts in recycle stock

Where processes take place in the medium of a special solvent, the solvent may be circulated throughout the plant in a closed system. As the product is withdrawn from the solvent, some impurities remain in solution and gradually accumulate. The accumulation is fixed at a certain limiting concentration by continually withdrawing a definite fraction of the recycled solvent. This type of control is maintained in the electrolytic refining of copper wherein the electrolyte is continually recirculated while a portion is bled off and replaced by fresh electrolyte. The spent electrolyte is treated separately by a special process for the recovery of residual copper, for the removal of accumulated impurities such as nickel sulfate, and the purified sulfuric acid is returned to the process. In the synthesis of ammonia from atmospheric nitrogen and hydrogen the percentage conversion of a 1:3 mixture is 25% in a single pass through the reactor. The ammonia formed is removed by cooling and condensation under high pressure, and the unconverted nitrogen and hydrogen are recirculated to the reactor. The argon from the atmospheric nitrogen is allowed to accumulate to a fixed upper limit by bleeding off a fraction of the recycled gas or by leakage from the system.

Illustration 8. In the operation of a synthetic ammonia plant, shown diagrammatically in Fig. 54, a 1:3 nitrogen-hydrogen mixture is fed to the converter resulting in a 25% conversion to ammonia. The ammonia formed is separated by condensation, and the unconverted gases are recycled to the reactor. The initial nitrogen-hydrogen mixture contains 0.20 part of argon to 100 parts of N_2-H_2 mixture. The toleration limit of argon entering the reactor is assumed to be 5 parts to 100 parts of N_2 and H_2 by volume. Estimate the fraction of recycle that must be continually purged.

Basis of Calculations: 100 lb-moles N_2-H_2 in fresh feed.

Let
$$x = \text{moles } N_2 \text{ and } H_2 \text{ recycled to reactor}$$
$$y = \text{moles } N_2 \text{ and } H_2 \text{ bled off or purged}$$

Moles N_2 and H_2 entering reactor	$= 100 + x$
Moles N_2 and H_2 leaving reactor	$= 0.75(100 + x)$
Moles of NH_3 formed	$= \dfrac{0.25(100 + x)}{2}$
Moles of argon in fresh feed	$= 0.20$
Moles of argon in total feed	$= 0.05(100 + x)$
Moles of argon per mole of N_2-H_2 mixture leaving condenser	$= \dfrac{0.05}{0.75} = 0.0667$
Moles of argon bled off	$= 0.0667y$

When a steady state of operation is attained, the argon purged is equal to the argon in the fresh gas supply. Hence

$$0.0667y = 0.20 \quad \text{or} \quad y = 3.00$$

From an H_2-N_2 balance around bleed point

$$0.75(100 + x) = x + y$$

Since $y = 3.00$, $\quad x = 288$ lb-moles.

Summary

$$
\begin{aligned}
\text{Fresh } N_2 \text{ and } H_2 &= 100 \text{ lb-moles} \\
\text{Recycle } N_2 \text{ and } H_2 &= 288 \text{ lb-moles} \\
\text{Purged } N_2 \text{ and } H_2 &= 3.0 \text{ lb-moles} \\
\text{Ammonia formed} &= 48.5 \text{ lb-moles} \\
\text{Argon} &= 0.20 \text{ lb-mole}
\end{aligned}
$$

$$\text{Recycle ratio} = \frac{288}{100} = 2.88$$

$$\text{Purge ratio} = \frac{3}{288} = 0.0104$$

In the actual operation of a high-pressure plant the unavoidable leakage is generally sufficient to keep argon in the reactor below the toleration limit so that no special provision need be applied for venting part of the recycle.

By-passing Streams. *By-passing* of a fluid stream by splitting it into two parallel streams is often practiced where accurate control in concentration is desired. For example, in conditioning air with a single stationary bed of silica gel to a lower humidity a more nearly uniform humidity is obtained in the final air mixture if part of the air is

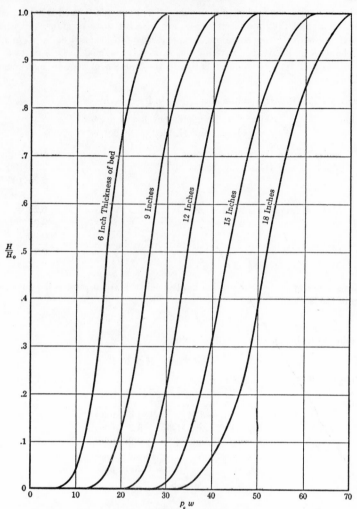

p_s = vapor pressure of water in atmosphere at temperature of bed
w = pounds of dry air passed through bed per square foot

Fig. 55. Fraction of moisture removed from air by a bed of silica gel

dehumidified to a low moisture content and then mixed with the un-
conditioned air to produce the desired mixture. The fraction of the
entering humidity remaining in the air after drying in beds of silica
gel of various thicknesses is shown in Fig. 55 plotted against the product
of $p_s w$ where

w = the total pounds of dry air passed through gel per square
foot of cross section

p_s = the vapor pressure of water in atmospheres at the temperature of the bed

$w = G\tau = u\rho\tau$

where G = pounds of dry air per (square foot)(minute)

τ = time in minutes

ρ = density of entering air, pounds of dry air per cubic foot of mixture

u = velocity of entering air, feet per minute

It will be observed that the humidity of the air leaving the gel varies greatly with time. For an extended period of time, as indicated by the abscissas, the air leaves with nearly no humidity; then it rises abruptly as more air is passed through. The bed behaves as though a wedge of water of the shape of the curves shown in Fig. 55 were advancing through the bed; at first the advancing wedge is nearly imperceptible, then suddenly the full wedge appears. The mathematics underlying Fig. 55 is presented in Part III.

Illustration 9. One thousand cubic feet of air per minute at 85° F, 70% relative humidity is to be conditioned to a relative humidity of 30% and kept at 85° F, using a stationary bed of silica gel 12 in. thick and 4 ft in diameter. To maintain the relative humidity constant at 30%, part of the air is by-passed around the drier and mixed with the part that is dried. Constant humidity in the final air mixture is controlled automatically by a damper regulating the fraction of air to be by-passed. This is accomplished by a constant wet-bulb controller. Assuming isothermal operation:

(a) Construct a plot showing the fraction of air to be by-passed against time.

(b) Calculate the time of operation before regeneration becomes necessary.

$$\text{Area of bed} = \frac{\pi D^2}{4} = 12.52 \text{ sq ft}$$

$$\text{Total rate of air flow} = \frac{1000}{359} \times \frac{492}{545} \times 0.9716 \times 29 = 70.9 \text{ lb per min (dry basis)}$$

$$\text{Rate of air flow per sq ft of bed} = \frac{70.9}{12.52} = 5.66 \text{ lb per min (dry basis)}$$

H_0 = initial humidity = 0.0182, p_s = vapor pressure at 85° F = 0.0406 atm

H_1 = desired humidity = 0.0077,   partial pressure at 70% r.h. = 0.0284 atm

H = humidity leaving drier,   partial pressure at 30% r.h. = 0.0122 atm

Let x = fraction of air by-passed. From a material balance for water,

$$0.0182x + H(1 - x) = 0.0077$$

$$x = \frac{0.0077 - H}{0.0182 - H} \qquad (a)$$

Let Δw = dry air passed through drier in time interval $\Delta\tau$. Then, $\dfrac{\Delta w}{1-x} = 5.66\Delta\tau$
= total dry air delivered at 30% relative humidity.

The calculated results are tabulated at stated intervals of $p_s w$, the abscissas of Fig. 55. The corresponding values of H/H_0 are obtained from Fig. 55 and tabulated in column 2. From the value of $p_s = 0.0406$ atm at 85° F, the values of w are tabulated in column 3. The dry air delivered for each interval of $p_s w$ is obtained as the difference of succeeding items in column 3 and tabulated in column 4 as Δw. The value of the humidity of air H leaving the gel is obtained from column 2 with $H_0 = 0.0182$ and tabulated in column 5. The average humidity H_a during each interval is obtained from Fig. 55 as the mean value over the interval of $p_s w$ indicated in column 1. The average fraction x of air by-passed during the interval Δw is obtained from the equation $x = \dfrac{0.0077 - H_a}{0.0182 - H_a}$ and tabulated in column 7. The

1	2	3	4	5	6	7	8	9	10
$p_s w$	H/H_0 (Fig. 55)	w	Δw Interval	H	H_a for Interval Δw	x	$\dfrac{\Delta w}{1-x}$	$\Delta\tau$	τ min
0	0	0	.....	0	....	.....	...	.....	0
....	.....	...	554.0	....	0	0.423	960	169.6	
22.5	0	554		0	....	.....			170
....	.....	...	61.5	....	0.00032	0.413	105	18.6	
25.0	0.035	616		0.000637	....	.....			188
....	.....	...	61.5	....	0.00128	0.379	99	17.5	
27.5	0.105	677		0.00191	....	.....			206
....	.....	...	61.5	....	0.00291	0.313	90	15.9	
30.0	0.215	739		0.00392	....	.....			222
....	.....	...	61.5	....	0.00519	0.193	76	13.4	
32.5	0.355	800		0.00646	....	.....			235
....	.....	...	24.0	....	0.00708	0.055	25	4.4	
33.5	0.423	824		0.0077	....	.....			239

$\Sigma\Delta\tau = 239$ min

total air delivered per interval Δw is equal to $\dfrac{\Delta w}{1-x}$ and is tabulated in column 8.

The time interval $\Delta\tau$ corresponding to the air interval Δw is obtained by dividing items in column 8 by 5.66; these are tabulated in column 9. The total time elapsed is obtained from a summation of items in column 9, $\tau = \Sigma\Delta\tau$, and tabulated in column 10.

A plot of x against τ gives the desired schedule for by-passing air. For the first 170 min, 42% of the air is by-passed, after which time the by-pass is gradually decreased to no by-pass, according to schedule, until a total of 239 min have elapsed. The silica gel must then be regenerated. The results of this illustration are plotted in Fig. 56.

Time Lag in Stirred Vessels. When the composition of a stream flowing through a series of vessels is suddenly altered, a time lag will occur before new steady-state conditions are established. This time lag progressively increases from vessel to vessel, increases with the size of each vessel, and decreases with increase in flow rate. Under conditions of no mixing of the altered stream with the preceding stream, the time lag will be a minimum and can be calculated directly from the ratio of the volumetric content of each vessel to the volumetric rate

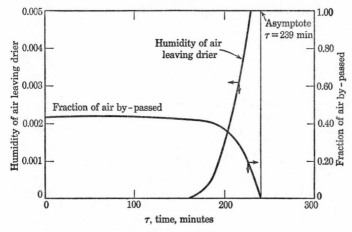

Fig. 56.   By-passing air for humidity control

of flow. With complete and instantaneous mixing in each vessel, the time lag will be a maximum for any fractional attainment of new steady-state conditions; for complete attainment of altered steady-state concentrations throughout the series of vessels the time lag is theoretically infinite. The calculation of time lag is of importance in the taking of samples and in establishing the time of purging required for new conditions to be established. For the special case of complete and instantaneous mixing in a series of vessels of equal size, Kandiner[2] has developed the following equation for the time lag expressed as a function of the number of tanks in series, the flow rate of the stream, the size of each tank, and the fractional approach to new steady-state conditions.

$$Y_n = 1 - e^{-F\tau/V} \sum_1^n \frac{1}{(n-1)!} \left(\frac{F\tau}{V}\right)^{n-1} \qquad (30)$$

where Y_n = fractional change of composition leaving reactor n corresponding to the altered feed composition entering the first vessel

[2] H. J. Kandiner, *Chem. Eng. Prog.*, **44**, 383 (1948), with permission.

n = number of reactors in series
F = feed rate, volume per unit time
V = volume of solution in each tank
τ = time lag

It is assumed that when $\tau = 0$, the composition of solution fed to the first tank is suddenly altered. Under these conditions at $\tau = 0$, $Y_n = 0$; at $\tau = \infty$, $Y_n = 1.0$, and $Y_o = 1.0$ at all values of τ.

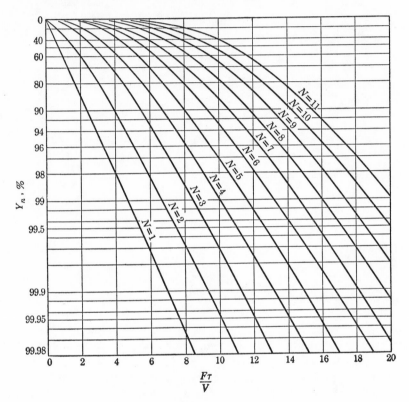

Fig. 57. Time lag in stirred vessels

Equation 30 is dimensionless when F, τ, and V are expressed in consistent units. The evaluation of equation 30 is presented graphically in Fig. 57 where Y_n is plotted against $F\tau/V$ for parameters of n up to 11. This relationship is based on the assumption that mixing and chemical equilibrium have been attained in each tank at all times. Kandiner also gives corrections for conditions of partial mixing and unequal-size vessels.

Illustration 10. A solution is flowing through a series of five vessels at a rate of 50 gpm. Each tank holds 400 gal of liquid, and complete mixing is assumed in each vessel at all times. If the composition of the feed entering the first tank is suddenly changed, calculate the time elapsed for at least 95% of the ultimate change to be realized in the effluent from the fifth tank.

From Fig. 57 at $Y_n = 0.95$ and $n = 5$, the corresponding value of $F\tau/V$ is 9.2. Hence $\tau = 9.2\ (V/F) = 9.2\ (\frac{400}{50}) = 73.6$ min. With no mixing, the time lag required at the effluent of the fifth tank would be

$$\tau = \frac{(400)\,(5)}{50} = 40 \text{ min}$$

The actual time lag with incomplete mixing will be between 40 and 73.6 min for attainment of 95% new conditions. The calculated time lag of 73.6 min is a conservative value for actual conditions and 95% completion.

Problems

1. In the manufacture of soda-ash by the LeBlanc process, sodium sulfate is heated with charcoal and calcium carbonate. The resulting "black ash" has the following composition:

Na_2CO_3	42%
Other water-soluble material	6
Insoluble material (charcoal, CaS, etc.)	52
	100%

The black ash is treated with water to extract the sodium carbonate. The solid residue from this treatment has the following composition:

Na_2CO_3	4 %
Other water-soluble salts	0.5
Insoluble matter	85
Water	10.5
	100 %

(a) Calculate the weight of residue remaining from the treatment of 1.0 ton of black ash. **Ans.** 1224 lb.

(b) Calculate the weight of sodium carbonate extracted per ton of black ash. **Ans.** 792 lb.

2. A contract is drawn up for the purchase of paper containing 5% moisture at a price of 7 cents per pound. It is provided that, if the moisture content varies from 5%, the price per pound shall be proportionately adjusted in order to keep the price of the bone-dry paper constant. In addition, if the moisture content exceeds 5%, the purchaser shall deduct from the price paid to the manufacturer the freight charges incurred as a result of the excess moisture. If the freight rate is 90 cents per 100 lb, calculate the price to be paid for 3 tons of paper containing 8% moisture.

3. A laundry can purchase soap containing 30% of water at a price of $7 per 100 lb f. o. b. the factory. The same manufacturer offers a soap containing 5% of water. If the freight rate is 60 cents per 100 lb, what is the maximum price that the laundry should pay the manufacturer for the soap containing 5% water? **Ans.** $9.71.

4. The spent acid from a nitrating process contains 33% H_2SO_4, 36% HNO_3, and 31% H_2O by weight. This acid is to be strengthened by the addition of concentrated sulfuric acid containing 95% H_2SO_4, and concentrated nitric acid containing 78% HNO_3. The strengthened mixed acid is to contain 40% H_2SO_4 and 43% HNO_3. Calculate the quantities of spent and concentrated acids that should be mixed together to yield 1500 lb of the desired mixed acid.

5. The waste acid from a nitrating process contains 21% HNO_3, 55% H_2SO_4, and 24% H_2O by weight. This acid is to be concentrated to contain 28% HNO_3 and 62% H_2SO_4 by the addition of concentrated sulfuric acid containing 93% H_2SO_4 and concentrated nitric acid containing 90% HNO_3. Calculate the weights of waste and concentrated acids that must be combined to obtain 1000 lb of the desired mixture. **Ans.** Spent acid **127** lb, conc. H_2SO_4 **592** lb, conc. HNO_3 **281** lb.

6. In the manufacture of straw pulp for the production of a cheap straw-board paper, a certain amount of lime is carried into the beaters with the cooked pulp. It is proposed to neutralize this lime with commercial sulfuric acid, containing 67% H_2SO_4 by weight.

In a beater containing 5000 gal of pulp it is found that there is a lime concentration equivalent to 0.50 gram of CaO per liter.

(a) Calculate the number of pound-moles of lime present in the beater.

(b) Calculate the number of pound-moles and pounds of H_2SO_4 that must be added to the beater in order to provide an excess of 1.0% above that required to neutralize the lime.

(c) Calculate the weight of commercial acid that must be added to the beater for the conditions of b.

(d) Calculate the weight of calcium sulfate formed in the beater.

7. Phosphorus is prepared by heating in the electric furnace a thoroughly mixed mass of calcium phosphate, sand, and charcoal. It may be assumed that in a certain charge the following conditions exist: The amount of silica used is 10% in excess of that theoretically required to combine with the calcium to form the silicate; the charcoal is present in 40% excess of that required to combine, as carbon monoxide, with the oxygen that would accompany all the phosphorus as the pentoxide.

(a) Calculate the percentage composition of the original charge. **Ans.** $Ca_3(PO_4)_2$ 52.3%, sand 33.5%, charcoal 14.2%.

(b) Calculate the number of pounds of phosphorus obtained per 100 lb of charge, assuming that the decomposition of the phosphate by the silica is 90% complete and that the reduction of the liberated oxide of phosphorus, by the carbon, is 70% complete. **Ans.** 6.60 lb.

8. A coal containing 87% total carbon and 7% unoxidized hydrogen is burned in air.

(a) If air is used 40% in excess of that theoretically required, calculate the number of pounds of air used per pound of coal burned.

(b) Calculate the composition, by weight, of the gases leaving the furnace, assuming complete combustion.

9. In the Deacon process for the manufacture of chlorine, a dry mixture of hydrochloric acid gas and air is passed over a heated catalyst which promotes oxidation of the acid. Air is used in 30% excess of that theoretically required.

(a) Calculate the weight of air supplied per pound of acid. **Ans.** 1.230 lb.

(b) Calculate the composition, by weight, of the gas entering the reaction chamber. **Ans.** HCl 44.8%, O_2 12.8%, N_2 42.4%.

(c) Assuming that 60% of the acid is oxidized in the process, calculate the composition, by weight, of the gases leaving the chamber. **Ans.** HCl 17.92%, O_2 6.90%, N_2 42.45%, Cl_2 26.10%, H_2O 6.63%.

10. In order to obtain barium in a form that may be put into solution, the natural sulfate, barytes, is fused with sodium carbonate. A quantity of barytes, containing only pure barium sulfate and infusible matter, is fused with an excess of pure, anhydrous soda ash. Upon analysis of the fusion mass it is found to contain 11.3% barium sulfate, 27.7% sodium sulfate, and 20.35% sodium carbonate. The remainder is barium carbonate and infusible matter.

(a) Calculate the percentage completion of the conversion of the barium sulfate to the carbonate and the complete anlysis of the fusion mass.

(b) Calculate the composition of the original barytes.

(c) Calculate the percentage excess in which the sodium carbonate was used above the amount theoretically required for reaction with all the barium sulfate.

11. In the manufacture of sulfuric acid by the contact process, iron pyrites, FeS_2, are burned in dry air, the iron being oxidized to Fe_2O_3. The sulfur dioxide thus formed is further oxidized to the trioxide by conducting the gases mixed with air over a catalytic mass of platinum black at a suitable temperature. It will be assumed that in the operation sufficient air is supplied to the pyrites burner so that the oxygen shall be 40% in excess of that required if all the sulfur actually burned were oxidized to the trioxide. Of the pyrites charged, 15% is lost by falling through the grate with the "cinder" and not burned.

(a) Calculate the weight of air to be used per 100 lb of pyrites charged. **Ans.** 513 lb.

(b) In the burner and a "contact shaft" connected with it, 40% of the sulfur burned is converted to the trioxide. Calculate the composition, by weight, of the gases leaving the contact shaft. **Ans.** SO_2 10.05%, SO_3 8.37%, O_2 8.76%, N_2 72.82%.

(c) By means of the platinum catalytic mass, 96% of the sulfur dioxide *remaining* in the gases leaving the contact shaft is converted to the trioxide. Calculate the total weight of SO_3 formed per 100 lb of pyrites charged. **Ans.** 110.6 lb.

(d) Assuming that all gases from the contact shaft are passed through the catalyzer, calculate the composition by weight of the resulting gaseous products. **Ans.** SO_2 0.40%, SO_3 20.43%, O_2 6.34%, N_2 72.83%.

(e) Calculate the over-all degree of completion of the conversion of the sulfur in the pyrites charged to SO_3 in the final products. **Ans.** 82.9%.

12. In the LeBlanc soda process the first step is carried out according to the following reaction:

$$2NaCl + H_2SO_4 = NaCl + NaHSO_4 + HCl$$

The acid contains 80.0% H_2SO_4. It is supplied in 5% excess of that theoretically required for the above reaction.

(a) Calculate the weight of acid supplied per 1000 lb of salt charged.

(b) Assume that the reaction goes to completion, all the acid forming bisulfate, and that in the process 90% of the HCl formed and 25% of the water present are removed. Calculate the weights of HCl and water removed per 1000 lb of salt charged.

(c) Assuming the conditions of part b, calculate the percentage composition of the remaining salt mixture.

13. In the common process for the manufature of nitric acid, sodium nitrate is treated with aqueous sulfuric acid containing 95% H_2SO_4 by weight. In order that the resulting "niter cake" may be fluid, it is desirable to use sufficient acid so that there will be 34% H_2SO_4 by weight in the final cake. This excess H_2SO_4 will actually be in combination with the Na_2SO_4 in the cake, forming $NaHSO_4$, although for purposes of calculation it may be considered as free acid. It may be assumed that the cake will contain 1.5% water, by weight, and that the reaction will go to completion, but that 2% of the HNO_3 formed will remain in the cake. Assume that the sodium nitrate used is dry and pure.

(a) Calculate the weight and percentage composition of the niter cake formed per 100 lb of sodium nitrate charged. **Ans.** H_2SO_4 34.0%, H_2O 1.5%, HNO_3 1.12%, Na_2SO_4 63.38%; 131.8 lb.

(b) Calculate the weight of aqueous acid to be used per 100 lb of sodium nitrate. **Ans.** 107.9 lb.

(c) Calculate the weights of nitric acid and water vapor distilled from the niter cake, per 100 lb of $NaNO_3$ charged. **Ans.** HNO_3 72.6 lb, H_2O 3.5 lb.

14. Pure carbon dioxide may be prepared by treating limestone with aqueous sulfuric acid. The limestone used in such a process contained calcium carbonate and magnesium carbonate, the remainder being inert insoluble materials. The acid used contained 12% H_2SO_4 by weight. The residue from the process had the following composition:

$CaSO_4$	8.56%
$MgSO_4$	5.23
H_2SO_4	1.05
Inerts	0.53
CO_2	0.12
Water	84.51
	100.00 %

During the process the mass was warmed and carbon dioxide and water vapor were removed.

(a) Calculate the analysis of the limestone used.

(b) Calculate the percentage of excess acid used.

(c) Calculate the weight and analysis of the material distilled from the reaction mass per 1000 lb of limestone treated.

15. Barium carbonate is commercially important as a basis for the manufacture of other barium compounds. In its manufacture, barium sulfide is first prepared by heating the natural sulfate, barytes, with carbon. The barium sulfide is extracted from this mass with water and the solution treated with sodium carbonate to precipitate the carbonate of barium.

In the operation of such a process it is found that the solution of barium sulfide formed contains also some calcium sulfide, originating from impurities in the barytes.

The solution is treated with sodium carbonate, and the precipitated mass of calcium and barium carbonates is filtered off. It is found that 16.45 lb of dry precipitate are removed from each 100 lb of filtrate collected. The analysis of the precipitate is:

$$CaCO_3 \quad 9.9\ \%$$
$$BaCO_3 \quad 90.1$$

The analysis of the filtrate is found to be:

$$Na_2S \quad\quad 6.85\%$$
$$Na_2CO_3 \quad 2.25$$
$$H_2O \quad\quad\ 90.90$$
$$\overline{\quad\quad\quad\quad 100.00\%}$$

The sodium carbonate for the precipitation was added in the form of anhydrous soda ash which contained calcium carbonate as an impurity.

(a) Determine the percentage excess sodium carbonate used above that required to precipitate the BaS and CaS. **Ans.** 24.2%.

(b) Calculate the composition of the original solution of barium and calcium sulfides. (*Note:* Barium sulfide is actually decomposed in solution, existing as the compound $OHBaSH \cdot 5H_2O$. However, in this reaction the entire calculation may be carried out and the compositions expressed as though the compound in solution were BaS.) **Ans.** BaS 12.16%, CaS 0.89%, H_2O 86.95%.

(c) Calculate the composition of the dry soda ash used in the precipitation. **Ans.** Na_2CO_3 97.07%, $CaCO_3$ 2.93%.

16. A mixture of 55% CaO, 10% MgO, and 35% NaOH by weight is used for treating ammonium carbonate liquor in order to liberate the ammonia.

(a) Calculate the number of pounds of CaO to which 1 lb of this mixture is equivalent.

(b) Calculate the composition of the mixture, expressing the quantity of each component as the percentage it forms of the total reacting value of the whole.

(c) Calculate the number of pounds of ammonia theoretically liberated by 1 lb of this mixture.

17. A water is found to contain the following metals, expressed in milligrams per liter: Ca 32; Mg 8.4; Fe (ferrous) 0.5.

(a) Calculate the "total hardness" of the water, expressed in milligrams of equivalent $CaCO_3$ per liter, the calcium of which would have the same reacting value as the total reacting value of the metals actually present. **Ans.** 115.5.

(b) Assuming that these metals are all combined as bicarbonates, calculate the cost of the lime required to soften 1000 gal of the water. Commercial lime, containing 95% CaO, costs $8.50 per ton. **Ans.** 0.242 cent.

18. A glass for the manufacture of chemical ware is composed of the silicates and borates of several basic metals. Its composition is as follows:

SiO_2	66.2%		ZnO	10.3%
B_2O_3	8.2		MgO	6.0
Al_2O_3	1.1		Na_2O	8.2
				100.0%

Determine whether the acid or the basic constituents are in excess in this glass, and the percentage excess reacting value above that theoretically required for a neutral glass.   (Assume that Al_2O_3 acts as a base and B_2O_3 as an acid, HBO_2.)

19. In illustration 6, assume that the level is kept constant in each effect with the same evaporations as noted, namely, 12.6 and 11.70 tons from the first two evaporators.   How much salt would be crystallized in each effect?   **Ans.**   (I) 3.92 tons, (II) 4.75 tons, (III) 0.25 tons.

20. In illustration 6 assume that the salt was removed from the hoppers, that the drops in liquid levels were the same, and that the water evaporated was 11.1 tons in the first effect and 10.2 tons in the second effect.   Calculate the salt crystallized in each evaporator and the water evaporated from the third effect.

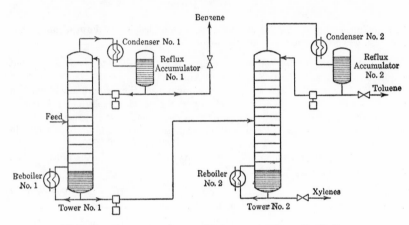

Fig. 58.   Inventory changes in three-component distillation

21. A mixture of benzene, toluene, and xylenes is separated by continuous fractional distillation in two towers, the first of which produces benzene as an overhead product and a bottoms product of toluene and xylenes, which is charged to the second tower.   This tower produces toluene as overhead and xylenes as bottoms. A flow diagram of the operation is shown in Fig. 58.   During a 12-hr period of operation the following products are removed from the towers:

	No. 1 Tower, Net overhead	No. 2 Tower, Net overhead	No. 2 Tower, Bottoms
Gallons at 60° F	9,240	6,390	10,800
Composition, % by volume			
Benzene	98.5	1.0	....
Toluene	1.5	98.5	0.7
Xylenes	....	0.5	99.3

During the period the liquid levels in the accumulator sections of the plant varied, as shown by level indicators measuring the heights of the levels above arbitrary fixed points.   The liquid volume, corrected to 60° F of the accumulator sections per inch of height and the initial and final levels were as follows:

	Corrected Liquid Vol., gal per in.	Initial Level, in.	Final Level in.
No. 1 Tower reflux accumulator	24	36	58
No. 1 Tower bottom accumulator	46	60	45
No. 2 Tower reflux accumulator	35	80	65
No. 2 Tower bottom accumulator	18	52	54

Calculate:

(a) The volume and composition of the feed during the 12-hr operating period.

		Gal	Volumetric %
Ans.	Benzene	9678	37.54
	Toluene	5757	22.34
	Xylene	10344	40.12

(b) The rate of flow from the bottom of tower No. 1. **Ans.** 16,700 gal per 12 hr.

(c) The rate of production of distillate and bottoms from each tower, assuming that the liquid levels and compositions remained constant for the same rate of feed.

Ans. No. 1 overhead 9,768 gal per 12 hr
No. 1 bottoms 16,010 gal per 12 hr
No. 2 overhead 5,623 gal per 12 hr
No. 2 bottoms 10,388 gal per 12 hr

22. Stock containing 1.562 lb of water per pound of dry stock is to be dried to 0.099 lb/lb. For each pound of stock (dry basis) 52.5 lb of dry air pass through the drier, leaving at a humidity of 0.0525 lb/lb. The fresh air is supplied at a humidity of 0.0152.

(a) Calculate the fraction of air recirculated.

(b) If the size of drier required is inversely proportional to the average wet-bulb depression of the air in the drier, calculate the relative size of driers required with and without recirculation when operated at a constant temperature of 140° F.

23. It is required to condition 1000 cu ft of air per minute from 70° F and 80% relative humidity to a constant value of 10% relative humidity by means of a stationary bed of silica gel. Part of the stream may be by-passed and a damper regulator automatically controlled to maintain a constant wet-bulb temperature in the final air mixture. The air velocity through the bed shall not exceed 100 ft per min (total cross-section basis), and the time cycle, before regeneration is necessary, shall be 3 hr. Assume isothermal operation. Calculate the diameter and thickness of bed required. **Ans.** Diameter = 3.568 ft, thickness = 12.6 in.

24. Ten tons per hour of dry seashore sand containing 1% NaCl by weight is to be washed by 10 tons per hour of salt-free water running countercurrent to the sand through two classifiers in series. This operation is carried out 24 hr per day. Assume that perfect mixing of the sand and wash water occurs in each classifier. The sand is discharged from the classifiers with 50% liquid (dry sand basis).

(a) If the exit sand is dried in a rotary kiln drier, what percentage of NaCl will it retain?

(b) What hourly wash water rate would be needed in a single classifier to wash the sand equally as well as in a?

25. The causticizing of soda ash follows the reaction:

$$Na_2CO_3 + Ca(OH)_2 = 2NaOH + CaCO_3$$

After most of the liquor containing the NaOH and some residual Na_2CO_3 or $Ca(OH)_2$ is decanted, an appreciable amount of NaOH still remains with the $CaCO_3$ sludge.

In a certain plant, the $CaCO_3$ is to be recovered and marketed as "precipitated chalk." The sludge from the precipitation tanks contains 5% $CaCO_3$ on a total weight basis, 0.1% dissolved alkalinity (figured as NaOH), and the remainder water. One thousand tons per day of this sludge is fed continuously to two thickeners in series in which it is washed countercurrently with 200 tons per day of neutral water. The pulp removed from the bottom of each of the thickeners contains 20% solids (total weight basis). A filter takes the pulp from the last thickener and concentrates the solids to 50% (total weight basis), returning the filtrate to the system as wash water.

Calculate the per cent by weight alkalinity (as NaOH) remaining in the final $CaCO_3$ after it has been completely dried. Sketch the equipment and show flow paths, rates, and compositions.

26. In illustration 7 evaluate all the terms given in Table 14.

Ans.	Limiting reactant	= propane
	Reactant ratio	= none
	Conversion per pass	= 38.7%
	Fresh feed conversion	= 98.2%
	Selectivity	= 75.8%
	Yield per pass	= 29.4% (molal basis)
	Ultimate yield	= 75.8% (molal basis)
	Over-all yield	= 73.9%
	Recycle ratio	= 1.54
	Combined feed ratio	= 2.54
	Liquid hourly space velocity	= 0
	Gaseous hourly space velocity	= 1535 cu ft at 60° F, 1 atm/(hr)(ft³ catalyst)
	Weight hourly space velocity	= 3.30 lb C_3H_8/(hr)(lb catalyst)
	Catalyst per combined-feed	= 1.11 lb per lb
	Catalyst residence time	= 16.4 min.
	Space time yield	= 1.19 lb-moles/(hr)(cu ft)

8

Thermophysics

In Chapter 3, the general concepts of energy, temperature, and heat were introduced under the broad classification of potential and kinetic energies. Both these forms were subclassified into external forms determined by the position and motion of a mass of matter relative to the earth or other masses of matter and into internal forms determined by the inherent composition, structure, and state of matter itself, independent of its external position or motion as a whole.

Internal Energy. The internal energy of a substance is defined as the total quantity of energy that it possesses by virtue of the presence, relative positions, and movements of its component molecules, atoms, and subatomic units. A part of this energy is contributed by the translational motion of the separate molecules and is particularly significant in gases where translational motion is nearly unrestricted in contrast to the situation in liquids and solids. Internal energy also includes the rotational motion of molecules and of groups of atoms which are free to rotate within the molecules. It includes the energy of vibration between the atoms of a molecule and the motion of electrons within the atoms. These kinetic portions of the total internal energy are determined by the temperature of the substance and by its molecular structure. The remainder of the internal energy is present as potential energy resulting from the attractive and repulsive forces acting between molecules, atoms, electrons, and nuclei. This portion of the internal energy is determined by molecular and atomic structures and by the proximity of the molecules and atoms to one another. At the absolute zero of temperature all translational energy disappears, but a great reservoir of potential energy and a small amount of vibrational energy remain.

The total internal energy of a substance is unknown, but the amount relative to some selected temperature and state can be accurately determined. The crystalline state and hypothetical gaseous state at absolute zero temperature are commonly used as references for scientific studies, whereas engineering calculations are based on a variety of reference conditions arbitrarily selected.

External Energy. The external energy of a body is dependent on its position and motion relative to the earth. This includes its external potential energy due to its position expressed relative to an arbitrary datum plane and is equal to Z per unit mass, where Z is the height of its center of gravity above the datum point. It includes its external kinetic energy which is equal to $\frac{1}{2}mu^2$, where u is its linear velocity. When expressed in foot-pounds the kinetic energy of a fluid stream per pound-mass is $\bar{u}^2/2g_c$ where $\bar{u}$ is the average velocity of the stream in feet per second and g_c is the constant of the fundamental force equation, equal to 32.17 ft per sec per sec. It includes the surface energy of the body, designated as E_σ per unit mass. Surface energy is negligible in magnitude but is of important consideration where large surfaces are involved as in the formation of sprays and emulsions. In fluid streams flowing under the restraint of pressure the external energy includes the flow energy of the stream and is equal to pV per unit mass, where p is the pressure opposing the fluid and V its specific volume.

Energy in Transition; Heat and Work. In reviewing the several forms of energy previously referred to, it will be noted that some are capable of storage, unchanged in form. Thus, the potential energy of an elevated weight or the kinetic energy of a rotating flywheel is stored as such until by some transformation they are converted, in part at least, to other forms.

Heat represents energy in transition under the influence of a temperature difference. When heat flows from a hot metal bar to a cold one the internal energy stored in the cold bar is increased at the expense of that of the hot bar, and the amount of heat energy in transition may be expressed in terms of the change in internal energy of the source or of the receiver. Under the influence of a temperature gradient heat flows also by the bodily convection and mixing of hot and cold fluids and by the emission of radiant energy from a hotter to a colder body without the aid of any tangible intermediary.

It is inexact to speak of the storage of heat. The energy stored within a body is internal energy, and, when heat flows into the body, it becomes internal energy and is stored as such. Zemansky[1] writes as follows:

> The phrase, "the heat in a body" has absolutely no meaning. Perhaps an analogy will clinch the matter. Consider a fresh water lake. During a shower, a certain amount of rain enters the lake. After the rain has stopped, there is no rain in the lake. There is water in the lake. Rain is a word used to denote water that is entering the lake from the air above. Once it is in the lake, it is no longer rain.

Another form of energy in transition of paramount interest is *work*,

[1] M. W. Zemansky, *Heat and Thermodynamics*, McGraw-Hill Book Co. (1937), with permission.

which is defined as the energy that is transfered by the action of a mechanical force moving under restraint through a tangible distance. It is evident that work cannot be stored as such but is a manifestation of the transformation of one form of energy to another. Thus, when a winch driven by a gasoline engine is used to lift a weight, the internal energy of the gasoline is transformed in part to the potential energy of the elevated weight, and the work done is the energy transferred from one state to the other.

Energy Units. The basic concept that mechanical work is equal to force times the distance through which the force acts leads to definitions of units of mechanical energy. The common units are as follows: (1) *The erg* is the amount of work done (energy expended) when a force of one dyne acts through a distance of one centimeter. (2) *The joule.* Since one erg is an inconveniently small unit, the joule, equal to 10^7 ergs, is more commonly used. (3) *The newton-meter* is the work done where a force of one newton acts over a distance of one meter. 1 newton-meter = 1 joule. (4) *The foot-pound* of mechanical energy is expended when a force of one pound acts through a distance of 1 foot. (5) *The foot-poundal.* This energy unit is based on the poundal as the unit of force, which is defined as the force required to impart an acceleration of one foot per second per second to a body having a mass of one pound. Accordingly, one foot-poundal is the work done when a force of one poundal acts through a distance of one foot. The foot-poundal is 1/32.174 as great as the foot-pound. In most technical calculations, it is more convenient to express mechanical energy in foot-pounds than in foot-poundals.

The common units of energy in the field of electrical engineering are the watt-second and the kilowatt-hour. These energy units are usually thought of as being inherently electrical in nature, yet they are, in reality, mechanical energy units. The watt-second, for example, is defined as being equal to 1 (joule) = 10^7 (erg) = 10^7 (dyne)(cm).

In problems dealing with the production, generation, and transfer of heat, it is customary to use special units of energy called heat units. For many years, these units of thermal energy were defined in terms of the heat capacity of water. A variety of units developed because of the fact that various masses of water and various temperature scales were selected to define the units. Furthermore, it was soon recognized that the heat capacity of water varies with temperature, and, accordingly, a temperature specification was included in the definitions of the units of thermal energy. The units of thermal energy bear no derivable relation to mechanical energy units, and it was therefore necessary to determine, by experiment, the "mechanical equivalent of heat," which related the two independent sets of units.

The common units of thermal energy as formerly defined were as follows: (1) *The gram-calorie:* the energy required to heat one gram of water through a temperature range of one degree centigrade. Because of the variable heat capacity of water it was customary either to specify the temperature of the water or to take a mean value over a specified temperature range. *The 15-degree gram-calorie* was defined as the energy required to heat one gram of water from 14.50 to 15.50 degrees centigrade, at a pressure of one atmosphere. *The mean gram-calorie* was defined as 1/100 of the energy required to heat one gram of water from 0 to 100 degrees centigrade at a pressure of one atmosphere. (2) *The kilogram-calorie.* Because the gram-calorie is a rather small unit, it frequently is more convenient to use a unit 1000 times as great, the kilogram-calorie. The 15-degree kilogram-calorie and the mean kilogram-calorie were formerly defined in a manner similar to the way in which the corresponding gram-calories were defined, except that one kilogram of water was involved in the definition. (3) *The British thermal unit (Btu):* the energy required to heat one pound of water through a temperature range of one degree Fahrenheit. Because of the variable heat capacity of water, it was necessary with this unit, just as with the gram-calorie, either to specify the temperature of the water or to use a mean value. The 60-degree Btu and the mean Btu between 32 and 212 degrees Fahrenheit were in common use. In both instances, a constant pressure of one atmosphere was included in the definition. (4) *The centigrade heat unit (Chu)*, also known as the pound-calorie: the energy required to heat one pound of water through a temperature range of one degree centigrade. Just as with the gram-calorie, the kilogram-calorie, and the Btu, the variable heat capacity of water made it necessary to use either the 15-degree centigrade heat unit or the mean (between 0 and 100° C) centigrade heat unit. As with the other units, a constant pressure of one atmosphere was included in the definition of the unit.

While thermal energy units were defined for many years as indicated above, it is now customary to define them arbitrarily in terms of mechanical units, with no reference to the heating of water. At the present time, there are two "defined" gram-calories in wide use, the United States National Bureau of Standards thermochemical gram-calorie and the so-called steam gram-calorie (also called the I.T. gram-calorie because it was defined in 1929 by the International Steam Table Conference). These two gram-calories with their arbitrary defining equations and equivalents are as follows:[2]

[2] *Selected Values of Chemical Thermodynamic Properties*, as of July 1, 1953, edited by D. D. Wagman, National Bureau of Standards.

1 National Bureau of Standards thermochemical gram-calorie
= 4.1840   absolute   joule   (defining equation)

= 4.1833 international joule

= 0.001162222 absolute watt-hour

= 0.001162030   international   watt-hour

1 I.T. gram-calorie
= (1/860)   international   watt-hour (defining equation)

= 0.001162791   international   watt-hour

= 0.001162983 absolute watt-hour

= 4.18605 international joule

= 4.18674 absolute joule

The I.T. gram-calorie was used in calculating the data for Keenan and Keyes[3] steam tables, while the thermochemical gram-calorie is used in the publications of the National Bureau of Standards. Actually, the difference between the two "defined" gram-calories is slight, as shown by the following conversion factor, and may be ignored in all but the most precise calculations.

1 I.T. gram-calorie = 1.000654 thermochemical gram-calorie

The modern definition of the British thermal unit no longer is based on the energy required to heat a pound of water one degree Fahrenheit. Instead, it is defined through the relation

$$1 \text{ (I.T. g-cal)}/\text{(gram)} = \tfrac{9}{5} \text{ (Btu)}/\text{(lb)}$$

Rearranging gives

$$1 \text{ Btu} = \tfrac{5}{9} \text{ (I.T. g-cal)}\text{(lb)}/\text{(gram)}$$

In the United States the pound-mass is legally defined as equal to 453.5924277 gram. Substituting yields

$$1 \text{ Btu} = \tfrac{5}{9} \times 453.5924277 \text{ I.T. g-cal}$$
$$= 251.996 \text{ I.T. g-cal}$$
$$= 252.161 \text{ thermochemical g-cal}$$

An accurate table for the conversion of energy units is given in the appendix.

Although the basic definition of the gram-calorie is no longer as-

[3] J. H. Keenan and F. G. Keyes, *Thermodynamic Properties of Steam*, John Wiley & Sons (1936).

sociated with the heat capacity of water, it is still true that as a very close approximation the National Bureau of Standards thermochemical calorie represents the energy required to heat one gram of water from 14.50 to 15.50° C (the old 15° calorie). Also, the I.T. gram-calorie still represents, as a very close approximation, 1/100 of the energy required to raise one gram of water from 0 to 100° C (the old mean calorie). From a practical engineering standpoint, it is therefore legitimate still to think of the gram-calorie as being the energy required to heat one gram of water one centigrade degree, the Btu as being the energy required to heat one pound of water one Fahrenheit degree, and the centigrade heat unit as being the energy required to heat one pound of water one centigrade degree. If these concepts of the gram-calorie, the Btu, and the Chu are retained, it is easy to deduce the interrelationships among the three thermal units by memorizing a few conversion factors. The essential ones are:

$$1 \text{ lb} = 453.6 \text{ grams}$$

$$1 \text{ centigrade degree} = \tfrac{9}{5} \text{ Fahrenheit degrees}$$

Assume that 1 lb of water is heated 1 F°. The heat absorption is 1 Btu. We may also say that 453.6 grams of water have been heated $\tfrac{5}{9}$ C °; hence $\tfrac{5}{9} \times 453.6 = 252$ g-cal have been absorbed. Furthermore, we may say that 1 lb of water has been heated $\tfrac{5}{9}$ C °, with the absorption of $1 \times \tfrac{5}{9} = \tfrac{5}{9}$ Chu. The foregoing considerations give the first line of equivalents in Table 15. The other two lines of equivalents in Table 15 can be calculated from the values given in the first; they could also be deduced directly by considering the heating of 1 gram of water 1 C °, or the heating of 1 lb of water 1 C °.

TABLE 15. EQUIVALENT VALUES FOR UNITS OF HEAT ENERGY

$$1 \text{ Btu} \quad = \quad 252 \text{ g-cal} \quad = \quad \tfrac{5}{9} \text{ Chu}$$

$$\tfrac{1}{252} \text{ Btu} \quad = \quad 1 \text{ g-cal} \quad = \quad \frac{5}{9 \times 252} \text{ Chu}$$

$$\tfrac{9}{5} \text{ Btu} \quad = \quad \tfrac{9}{5} \times 252 \text{ g-cal} \quad = \quad 1 \text{ Chu}$$

Heats of reaction, heats of vaporization, and heats of fusion may be expressed as gram-calories per gram-mole, Btu per pound-mole, or Chu per pound-mole. The relation between these sets of units may be deduced from the following considerations.

Assume that in a given instance a heat of reaction is z Btu per lb-mole. The z Btu of heat energy is equivalent of $\tfrac{5}{9} \times 453.6 \times z$ g-cal, and the pound-mole is equivalent to 453.6 g-mole. Accordingly, the heat of reaction is $z \times \dfrac{\frac{5}{9} \times 453.6}{453.6}$, or $\frac{5}{9} \times z$ g-cal per g-mole. Further-more, the z Btu of heat energy is equivalent to $\frac{5}{9} \times z$ Chu. Thus the heat of reaction is $z \times \frac{5}{9}$ Chu per lb-mole. These considerations lead to the first line of equivalents below. The second line is obtained from the first by multiplication.

1 Btu per lb-mole = $\frac{5}{9}$ g-cal per g-mole = $\frac{5}{9}$ Chu per lb-mole

$\frac{9}{5}$ Btu per lb-mole = 1 g-cal per g-mole = 1 Chu per lb-mole

The same conversion factors apply when heats of reaction, vaporization, or fusion are given on an ordinary weight basis (pounds or grams) rather than on a molal basis.

Heat capacity may be expressed as $(\mathrm{Btu})/(\mathrm{lb\text{-}mole})(\mathrm{R}°)$, or as $(\mathrm{g\text{-}cal})/(\mathrm{g\text{-}mole})(\mathrm{K}°)$, or as $(\mathrm{Chu})/(\mathrm{lb\text{-}mole})(\mathrm{K}°)$. It may be shown that a numerical value of heat capacity is the same, regardless of which set of units is employed.

For example, suppose that a substance has a heat capacity of $z(\mathrm{Btu})/(\mathrm{lb\text{-}mole})(\mathrm{R}°)$. The z Btu are equivalent to $\frac{5}{9} \times 453.6 \times z$ g-cal. The pound-mole is equivalent to 453.6 g-mole, and the degree Rankine equals $\frac{5}{9}$ K°. Accordingly,

$$z \frac{(\mathrm{Btu})}{(\mathrm{lb\text{-}mole})(\mathrm{R}°)} = z \frac{\frac{5}{9} \times 453.6}{453.6 \times \frac{5}{9}} \frac{(\mathrm{g\text{-}cal})}{(\mathrm{g\text{-}mole})(\mathrm{K}°)} = z \frac{(\mathrm{g\text{-}cal})}{(\mathrm{g\text{-}mole})(\mathrm{K}°)}$$

Furthermore, the z Btu are equivalent to $\frac{5}{9}$ Chu, and the degree Rankine equals $\frac{5}{9}$ K°. Accordingly,

$$z \frac{(\mathrm{Btu})}{(\mathrm{lb\text{-}mole})(\mathrm{R}°)} = z \frac{\frac{5}{9}}{\frac{5}{9}} \frac{(\mathrm{Chu})}{(\mathrm{lb\text{-}mole})(\mathrm{K}°)} = z \frac{(\mathrm{Chu})}{(\mathrm{lb\text{-}mole})(\mathrm{K}°)}$$

The final conclusion is that the numerical value of a heat capacity is the same in these three sets of units. That is,

$1(\mathrm{Btu})/(\mathrm{lb\text{-}mole})(\mathrm{R}°) = 1(\mathrm{g\text{-}cal})/(\mathrm{g\text{-}mole})(\mathrm{K}°) =$
$$1(\mathrm{Chu})/(\mathrm{lb\text{-}mole})(\mathrm{K}°)$$

Heat capacities may, of course, be expressed on a regular weight (pound, gram) basis, rather than on a molal basis. The heat capacity on a regular weight basis is obtained by dividing the molal heat capacity by the molecular weight. Just as in the case of molal heat capacities, the numerical value of a heat capacity on a weight basis is the same in the three different sets of units. That is,

$$1(\mathrm{Btu})/(\mathrm{lb})(\mathrm{R}°) = 1(\mathrm{g\text{-}cal})/(\mathrm{g})(\mathrm{K}°) \qquad 1(\mathrm{Chu})/(\mathrm{lb})(\mathrm{K}°)$$

In the foregoing discussion, the temperature intervals have been indicated as Rankine degrees and Kelvin degrees. Since temperature *intervals* rather than specific temperatures on a temperature scale are involved, it would be equally correct to indicate Fahrenheit degrees instead of Rankine degrees, and Centigrade degrees instead of Kelvin degrees.

Energy Balances. In accordance with the principle of conservation of energy, also called the first law of thermodynamics, and referred to in Chapter 3, page 47, energy is indestructible, and the total amount of energy entering any system must be exactly equal to that leaving plus any accumulation within the system. A mathematical or numerical expression of this principle is termed an *energy balance*, which in conjunction with a material balance is of primary importance in problems of process design and operation.

In establishing a general energy balance for any process, it is convenient to use as a basis a unit time of operation, for example, one hour for a continuous operation and one cycle for a batch or intermittent operation. It is necessary to distinguish between a *flow process*, which is one in which streams of materials continually enter and leave the system, and a *nonflow process*, which is intermittent in character and in which no continuous streams of material enter or leave the system during the course of operation. A steady-flow process is also characterized by a steady state of flow, and by a constancy of temperatures and compositions at any given location in the process, in contrast to changing conditions of temperature and composition in a batch or nonflow process.

In an energy balance the inputs are equated to the outputs plus the accumulation of energy inventory within the system over the unit period of time in a flow process, or for a given cycle of operation for the nonflow process. The separate forms of energy are conveniently classified as follows, neglecting electrostatic and magnetic forms, which are ordinarily small:

(a) Internal energy, designated by the symbol U per unit mass or mU for mass m.

(b) The energy added in forcing a stream of materials into the system under the restraint of pressure. This flow work is equal to mpV, where p is pressure of the system and V is the volume per unit mass. A similar flow-work term is involved in forcing a stream of materials from the system. These terms appear only in the energy balance of a flow process.

(c) The external potential energies, in foot-pounds, of all materials entering and leaving the system. External potential energy, expressed relative to an arbitrarily selected datum plane, is equal to $mZ \dfrac{g_L}{a_n}$ where

Z is the height, in feet, of the center of gravity of the mass of material above the datum plane, g_L is the local value for acceleration of gravity, expressed as feet per second per second, and g_c is the constant of the fundamental force equation, numerically equal to 32.174 and having the units

$$\frac{(\text{lb-mass})(\text{ft})}{(\text{lb-force})(\text{sec})^2}.$$

(d) The kinetic energies of all streams entering or leaving the system. The kinetic energy of a single stream is equal to $\frac{1}{2}m\bar{u}^2$, where $\bar{u}$ is its average velocity, and energy is expressed in ergs, joules, or foot-poundals, depending on the units of m and $\bar{u}$. Expressed in foot-pounds the kinetic energy of a stream is $m\bar{u}^2/2g_c$, where g_c again is the constant of the force equation, with a numerical value of 32.174.

(e) The surface energies of all materials entering and leaving the system. Surface energy per unit mass is designated as E_σ and is generally negligible except where large surface areas are involved, as in the formation of sprays or emulsions.

(f) The net energy *added* to the system as heat, and designated as q. This net heat input represents the difference between the sum of all heat flowing into the system from all sources and the sum of all heat flowing out of the system from all sources.

(g) The net energy removed as *work done by the system*, designated as w. This net work includes all forms of work *done by* the system, such as mechanical and electrical work minus all such work added to the system from all sources.

(h) The net change in energy content within the system during the course of the operation and designated by the symbol ΔE. In a steady-state flow process with constant inventory, the energy-change term ΔE is zero. In a general process ΔE represents the change in energy content of the system as a result of any change in inventory of the system, of change in temperature, of change in composition, of change in potential energy of elevation, or of change in kinetic energy from stirring, for example. In a nonflow process, the energy terms of the entering and leaving streams disappear, and the term ΔE becomes of sole consideration.

The general energy equation may be set up, designating input items by the subscript 1 and output items by the subscript 2. All input-energy items are balanced against the output and change items, the summation symbol $\sum$ indicating the sum of each form of energy in all streams entering or leaving. Thus,

$$\sum m_1 U_1 + \sum m_1 p_1 V_1 + \sum \frac{m_1 \bar{u}_1{}^2}{2g_c} + \sum m_1 Z_1 \frac{g_L}{g_c} + \sum m_1 E_{\sigma 1} + q =$$

$$= \sum m_2 U_2 + \sum m_2 p_2 V_2 + \sum \frac{m_2 \bar{u}_2{}^2}{2g_c} + \sum m_2 Z_2 \frac{g_L}{g_c}$$

$$+ \sum m_2 E_{\sigma 2} + w + \Delta E \tag{1}$$

Simplifications of this general equation result in most specific cases. In a steady-flow process, without fluctuations in temperature, composition, and inventory, the term ΔE becomes zero. In a nonflow process, where no stream of materials enters or leaves the system during the course of operation, the flow-work, kinetic-energy, and potential-energy terms of the streams do not appear. For such a nonflow process where surface energy is negligible the general equation reduces to

$$q = w + \Delta E \tag{2}$$

where

$$\Delta E = \sum \left(m'_2 U'_2 + m'_2 Z'_2 \frac{g_L}{g_c} + m'_2 \frac{\bar{u}'_2{}^2}{2g_c} - m'_1 U'_1 - m'_1 Z'_1 \frac{g_L}{g_c} - \frac{m'_1 \bar{u}'_1{}^2}{2g_c} \right) \tag{3}$$

where the prime values refer to the properties of the inventory, with subscript 1 referring to initial conditions and subscript 2 to final conditions. The kinetic-energy term in a nonflow process results from agitation or stirring.

For a *nonflow process at constant volume*, where no work and no potential energy changes are involved, and if the kinetic- and surface-energy terms are absent or negligible, equation 1 combined with 3 becomes

$$q = \sum m'_2 U'_2 - \sum m'_1 U'_1 \tag{4}$$

For a *nonflow process at constant pressure*, where changes in kinetic-, potential-, and surface-energy terms are negligible,

$$q = \sum m'_2 U'_2 - \sum m'_1 U'_1 + w \tag{5}$$

Where work is performed only as a result of expansion against the constant pressure p,

$$w = \sum m'_2 p V'_2 - \sum m'_1 p V'_1 \tag{6}$$

Combining equations 5 and 6 yields

$$q = \sum m'_2 (U'_2 + p V'_2) - \sum m'_1 (U'_1 + p V'_1) \tag{7}$$

Where only a single phase is present, equation 7 becomes

$$q = m'(U'_2 + p V'_2) - m'(U'_1 + p V'_1) = m' \Delta (U' + p V') \tag{8}$$

In many flow processes of the type encountered in chemical-engineering practice where large internal-energy changes are involved and where the energy associated with work and changes in potential, kinetic, and surface energies are relatively small, the general energy equation 1 reduces to

$$\sum m_1(U_1 + p_1V_1) + q = \sum m_2(U_2 + p_2V_2) \qquad (9)$$

Equation 9 may be satisfactorily applied to the great majority of transformation processes where the primary objective is manufacture and not the production of power. However, this equation should always be recognized as an approximation, and the significance of the neglected terms should be verified when they are in doubt.

Enthalpy. In the energy equations for both flow and nonflow processes it will be seen that the term $(U + pV)$ repeatedly occurs. It is convenient to designate this term by the name *enthalpy*[4] and by the symbol H; thus

$$H = U + pV \qquad (10)$$

In a flow system the term pV represents flow energy, but in a nonflow system it merely represents the product of pressure and volume, having the units of energy but not representing energy.

In a *nonflow* process proceeding at *constant pressure* and without generation of electrical energy, the heat added is seen, from equation 7, to be equal to the increase in enthalpy of the system, or

$$q = \sum m'_2 H'_2 - \sum m'_1 H'_1$$

In a flow system where the kinetic-energy and potential-energy terms are negligible and where ΔE is zero, the heat added is equal to the gain in enthalpy plus the work done, including both electrical and mechanical work, or

$$q = \sum m_2 H_2 - \sum m_1 H_1 + w \qquad (11)$$

Where the work done is negligible in relative magnitude, the heat added is equal to the gain in enthalpy, or

$$q = \sum m_2 H_2 - \sum m_1 H_1 \qquad (12)$$

To summarize: For most industrial flow processes, such as the operation of boilers, blast furnaces, chemical reactors, or distillation equipment, the kinetic-energy, potential-energy, and work terms are negligible or cancel out, and the heat added is equal to the increase in enthalpy. Similarly, in nonflow processes at constant pressure where work other than that of expansion is negligible, the heat added is equal to the increase in enthalpy. However, where work, kinetic, or

---

[4] This word was coined by Kamerlingh Onnes (1909), who purposely placed the accent on the second syllable to distinguish it in speech from the commonly associated term, entropy.

potential energies are not negligible, or in nonflow processes at constant volume, the increase in enthalpy is *not* equal to the heat added. The distinction between increases of enthalpy and additions of heat must be kept constantly in mind.

Like internal energy, absolute values of enthalpy for any substance are unknown, but accurate values can be determined relative to some arbitrarily selected reference state. The temperature, form of aggregation, and pressure of the reference state must be definitely specified. The values so reported are *relative enthalpies*. For a gas the reference state of zero enthalpy is usually taken at 32° F and one atmosphere pressure, and for steam as liquid water at 32° F under its own vapor pressure at 32° F. For example, the enthalpy of steam at 200° F and an absolute pressure of 10 psi represents the increase in enthalpy when liquid water at 32° F and its own vapor pressure is vaporized and heated to a temperature of 200° F while its pressure is increased to 10 psi.

Heat Balance. "Heat balance" is a loose term referring to a special form of energy balance which has come into general use in all thermal processes where changes in kinetic energy, potential energy, and work done are negligible. For such processes, so-called heat balances are applicable to flow processes at any pressure and to nonflow processes at constant pressure.

From consideration of the general energy equation it is evident that neither heat nor enthalpy input and output items balance in the general case. Although the term "heat balance" has become entrenched in engineering literature, its use is undesirable because of the misleading implications of the name. Accordingly, all such balances will be referred to by the proper term of "energy balance," even where the kinetic, potential, and work items are neglected.

Illustration 1. Application of the General Energy Balance. The principle of the general energy balance of equation 1 may be illustrated by application to the recovery of hydrogen and carbon dioxide from water gas (Fig. 59). Dust-free water gas is compressed, mixed with steam, heated, and passed to a reactor where in contact with a catalyst the CO is converted to CO_2. The products from the reactor are cooled, with resultant condensation of water vapor, compressed further, and passed into an absorber where CO_2 is dissolved in water at high pressure. Hydrogen gas is delivered at high pressure in a nearly pure state. The high pressure carbon dioxide solution generates power in a turbine and the CO_2 gas and water are thereby released at atmospheric pressure, and separated.

The following symbols are used to designate the various streams. All mass and energy units correspond to the period of time selected.

m_{1a} = mass of water gas entering reactor
m_{1b} = mass of steam entering reactor

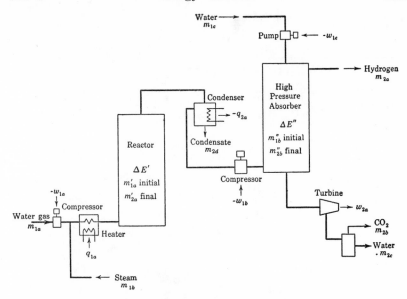

FIG. 59. Energy balance of a process for the production of hydrogen and carbon dioxide from water gas

m_{1c} = mass of water entering absorber
m_{2a} = mass of hydrogen stream leaving absorber
m_{2b} = mass of CO_2 stream leaving turbine
m_{2c} = mass of water leaving turbine
m_{2d} = mass of water leaving condenser
m'_{1a} = initial inventory in reactor
m'_{2a} = final inventory in reactor
m''_{1b} = initial inventory in absorber
m''_{2b} = final inventory in absorber
$-w_{1a}$ = net work delivered to water gas in compression
$-w_{1b}$ = net work delivered to absorber gases in compression
$-w_{1c}$ = net work delivered to water in pumping to absorber
$+w_{2a}$ = work delivered by turbine
q_{1a} = heat added in heating gas mixture entering reactor
$-q_{2a}$ = heat removed from gas stream leaving reactor by condenser
$\sum -q_r$ = heat lost by radiation from all parts of plant
$\Delta E'$ = change in energy content of mass in reactor over period of run
$\Delta E''$ = change in energy content of mass in absorber over period of run

The markings to U, p, V, u, Z all correspond to those used above.
Applying these conditions to the general equation 1 gives

$$\sum m_1 U_1 = m_{1a} U_{1a} + m_{1b} U_{1b} + m_{1c} U_{1c} \tag{a}$$

$$\sum m_2 U_2 = m_{2a} U_{2a} + m_{2b} U_{2b} + m_{2c} U_{2c} + m_{2d} U_{2d} \tag{b}$$

$$\sum m_1 p_1 V_1 = m_{1a} p_{1a} V_{1a} + m_{1b} p_{1b} V_{1b} + m_{1c} p_{1c} V_{1c} \tag{c}$$

$$\sum m_2 p_2 V_2 = m_{2a} p_{2a} V_{2a} + m_{2b} p_{2b} V_{2b} + m_{2c} p_{2c} V_{2c} + m_{2d} p_{2d} V_{2d} \tag{d}$$

$$\sum \frac{m_1 u_1{}^2}{2g_c} = \frac{m_{1a} u_{1a}{}^2}{2g_c} + \frac{m_{1b} u_{1b}{}^2}{2g_c} + \frac{m_{1c} u_{1c}{}^2}{2g_c} \tag{e}$$

$$\sum \frac{m_2 u_2{}^2}{2g_c} = \frac{m_{2a} u_{2a}{}^2}{2g_c} + \frac{m_{2b} u_{2b}{}^2}{2g_c} + \frac{m_{2c} u_{2c}{}^2}{2g_c} + \frac{m_{2d} u_{2d}{}^2}{2g_c} \tag{f}$$

$$\sum m_1 Z_1 \frac{g_L}{g_c} = m_{1a} Z_{1a} \frac{g_L}{g_c} + m_{1b} Z_{1b} \frac{g_L}{g_c} + m_{1c} Z_{1c} \frac{g_L}{g_c} \tag{g}$$

$$\sum m_2 Z_2 \frac{g_L}{g_c} = m_{2a} Z_{2a} \frac{g_L}{g_c} + m_{2b} Z_{2b} \frac{g_L}{g_c} + m_{2c} Z_{2c} \frac{g_L}{g_c} + m_{2d} Z_{2d} \frac{g_L}{g_c} \tag{h}$$

$$q = q_{1a} + q_{2a} + q_r \tag{i}$$
$$w = w_{2a} + w_{1a} + w_{1b} + w_{1c} \tag{j}$$
$$\Delta E = \Delta E' + \Delta E'' \tag{k}$$

$$\Delta E' = m'_{2a} \left(U'_{2a} + p'_{2a} V'_{2a} + \frac{u'_{2a}{}^2}{2g_c} + Z'_{2a} \frac{g_L}{g_c} \right)$$
$$- m'_{1a} \left(U'_{1a} + p'_{1a} V'_{1a} + \frac{u'_{1a}{}^2}{2g_c} + Z'_{1a} \frac{g_L}{g_c} \right) \tag{l}$$

$$\Delta E'' = m''_{2b} \left(U''_{2b} + p''_{2b} V''_{2b} + \frac{u''_{2b}{}^2}{2g_c} + Z''_{2b} \frac{g_L}{g_c} \right)$$
$$- m''_{1b} \left(U''_{1a} + p''_{1b} V''_{1b} + \frac{u''_{1b}{}^2}{2g_c} + Z''_{1b} \frac{g_L}{g_c} \right) \tag{m}$$

In general, the inventory of both mass and energy remains constant, the kinetic-energy terms are negligible, the potential-energy terms cancel, and the equation reduces to

$$\sum m_1 H_1 + q = \sum m_2 H_2 + w$$

It should be noted that the work terms refer to the net work energy added to the system or supplied to the turbine and as a result of mechanical inefficiencies do not correspond to the work required to drive the pumps and compressors or to that generated by the turbine. The heat developed as a result of these inefficiencies has been neglected. In chemical processing the work terms are usually negligible in the total energy balance although they may be of major importance in cost.

Heat Capacity of Gases

In a general sense *heat capacity* is defined as the amount of heat required to increase the temperature of a body by one degree. *Specific heat* is the ratio of the heat capacity of a body to the heat capacity of an equal mass of water. Specific heat is a property, characteristic of a substance and independent of any system of units, but dependent on the temperatures of both the substance and the reference water. Water at 15° C is usually chosen as the reference.

The heat capacity of any quantity of a substance is expressed mathematically as

$$C = \frac{dq}{dT} \tag{13}$$

where C = heat capacity

dq = heat added to produce a temperature change dT

If a substance is heated at constant volume under nonflow conditions and all the heat added goes to increasing the internal energy, then, from equation 4, $dq = dU$. The heat capacity at constant volume C_V is thus equal to the change of internal energy with temperature,

$$C_V = \left(\frac{\partial U}{\partial T}\right)_V \tag{14}$$

If a substance is heated at constant pressure, the heat added goes to increase the internal energy of the substance and to supply the energy equivalent of the mechanical work of expansion; thus $dq = dU + p\, dV$, and equation 13 becomes

$$C_p = \left(\frac{\partial q}{\partial T}\right)_p = \left(\frac{\partial U}{\partial T}\right)_p + p\left(\frac{\partial V}{\partial T}\right)_p = \left(\frac{\partial H}{\partial T}\right)_p \tag{15}$$

where C_p = heat capacity at constant pressure.

For an ideal gas $pV = nRT$ and the internal energy is independent of volume or pressure $\left(\frac{\partial U}{\partial T}\right)_V^* = \left(\frac{\partial U}{\partial T}\right)_p^*$; hence equation 14 becomes on the basis of one mole,

$$c_p = c_V + R \tag{16}*$$

where c_p, c_V = molal heat capacities at constant volume and constant pressure, respectively.

The heat capacity of a substance that expands with rise in temperature is greater when heated at constant pressure than when heated at constant volume by the heat equivalent of the external work done in expansion. For an ideal gas the molal heat capacities under the two different conditions differ by the magnitude of the constant R. The numerical value of R is 1.987 cal per g-mole per K°, or 1.987 Btu per lb-mole per R°.

For an ideal monatomic gas, such as helium, at a low pressure, it may be assumed that, as a result of the simple molecular structure, the only form of internal energy is the translational kinetic energy of the molecules. This energy per mole of gas from equation 3, Chapter 3, page 48, is equal to the internal energy u in this particular case. Thus,

$$u = \tfrac{1}{2}mnu^2 = \tfrac{3}{2}RT \tag{17}$$

Then from equation 13, for a monatomic gas,

$$c_V = \tfrac{3}{2}R \tag{18}$$

Since R is approximately 2.0 cal per g-mole per K°, the molal heat capacity of a monatomic gas at constant volume is equal to 3.0 cal per g-mole per K°. The molal heat capacity at constant pressure, from equation 16, will be 3.0 + 2.0 or 5.0 cal. For monatomic gases the ratio of heat capacities, κ is

$$\kappa = \frac{c_p}{c_V} = \frac{5}{3} = 1.67 \qquad (19)$$

For all gases, other than monatomic gases, the molal heat capacity at constant volume is greater than 3.0. For a multiatomic gas an increase in internal energy is used not only to impart additional translational kinetic energy, as evidenced by an increase in temperature and an increasing velocity of translation, but also to impart increased energies of rotation and vibration of the molecular and atomic units.

In Table 16 are given values of molal heat capacities of common gases at zero pressure taken from the Bureau of Standards publication, *Selected Values of Chemical Thermodynamic Properties* (1953), at temperature intervals from 300 to 5000° K for H_2, N_2, CO, air, O_2, NO, H_2O, and CO_2 and from 300 to 1500° K for HCl, Cl_2, CH_4, SO_2, C_2H_4, SO_3 and C_2H_6. These heat capacities are arranged approximately in order of increasing values at 300° K. Figure 60 shows molal heat capacities as a function of temperature in degrees Fahrenheit.

Empirical Equations for Heat Capacities.[*] For temperature intervals above 300° K, the molal heat capacities of common gases may be well represented by quadratic equations over the temperature interval from 300 to 1500° K; thus,

$$c_p = a + bT + cT^2 \qquad (20)$$

where T is in degrees Kelvin. At low temperatures approaching 0° K the heat capacity of gases is not given accurately by equation 20 but is a more complex function of temperature requiring an exponential equation for accurate expression such as the Debye-Einstein equation.

The constants a, b, and c of equation 20 have been established for common gases from the data of Table 16 for the temperature interval 300 to 1500° K at zero pressure and are presented in Table 17, together with percentage maximum deviations of the calculated values from the experimental values.

Specific Heats of Gaseous Hydrocarbons. For the relationship between molal heat capacity and temperature of hydrocarbon gases Fallon and Watson[5] proposed the use of two separate equations for each gas to cover two corresponding temperature ranges, thus:

[*] See Appendix for Kobe's table of equations for heat capacities.
[5] J. F. Fallon and K. M. Watson, *Nat. Petrol. News, Tech. Section*, June 7, 1944.

For temperatures from 50 to 1400° F:

$$c_p = a + bT + cT^2 \tag{21}$$

For temperatures from -300 to $+200°$ F (not above 200° F):

$$c_p = 7.95 + uT^v \tag{22}$$

where T is in degrees Rankine and c_p is the molal heat capacity.

TABLE 16.　Molal Heat Capacities of Gases at Constant Pressure ($p = 0$)

Units: g-cal/(g-mole)(K°)

$T°$ K	H_2	N_2	CO	Air	O_2	NO	H_2O	CO_2	HCl	Cl_2	CH_4	SO_2	C_2H_4	SO_3	C_2H_6
50	9.072		6.955			7.562									
100	6.729		6.955		6.958	7.714									
150	6.348		6.955		6.958	7.451									
200	6.561	6.957	6.956	6.958	6.961	7.278									
250	6.769	6.959	6.958	6.964	6.979	7.183									
300	6.895	6.961	6.965	6.973	7.019	7.134	8.026	8.894	6.96	8.12	8.552	9.54	10.45	12.13	12.648
400	6.974	6.991	7.013	7.034	7.194	7.162	8.185	9.871	6.97	8.44	9.736	10.39	12.90	14.06	15.68
500	6.993	7.070	7.120	7.145	7.429	7.289	8.415	10.662	7.00	8.62	11.133	11.12	15.16	15.66	18.66
600	7.008	7.197	7.276	7.282	7.670	7.468	8.677	11.311	7.07	8.74	12.546	11.71	17.10	16.90	21.35
700	7.035	7.351	7.451	7.463	7.885	7.657	8.959	11.849	7.17	8.82	13.88	12.17	18.76	17.86	23.72
800	7.078	7.512	7.624	7.627	8.064	7.833	9.254	12.300	7.29	8.88	15.10	12.53	20.20	18.61	25.83
900	7.139	7.671	7.787	7.785	8.212	7.990	9.559	12.678	7.42	8.92	16.21	12.82	21.46	19.23	27.69
1000	7.217	7.816	7.932	7.928	8.335	8.126	9.861	12.995	7.56	8.96	17.21	13.03	22.57	19.76	29.33
1100	7.308	7.947	8.058	8.050	8.440	8.243	10.145	13.26	7.69	8.99	18.09	13.20	23.54	20.21	30.77
1200	7.404	8.063	8.168	8.161	8.530	8.342	10.413	13.49	7.81	9.02	18.88	13.35	24.39	20.61	32.02
1300	7.505	8.165	8.265	8.258	8.608	8.426	10.668	13.68	7.93	9.04	19.57	13.47	25.14	20.96	33.11
1400	7.610	8.253	8.349	8.342	8.676	8.498	10.909	13.85	8.04	9.06	20.18	13.57	25.79	21.28	34.07
1500	7.713	8.330	8.419	8.416	8.739	8.560	11.134		8.14	9.08	20.71	13.65	26.36	21.58	34.90
1600	7.814	8.399	8.481	8.483	8.801	8.614	11.34	14.10							
1700	7.911	8.459	8.536	8.543	8.859	8.660	11.53	14.2							
1800	8.004	8.512	8.585	8.597	8.917	8.702	11.71	14.3							
1900	8.092	8.560	8.627	8.647	8.974	8.738	11.87	14.4							
2000	8.175	8.602	8.665	8.692	9.030	8.771	12.01	14.5							
2100	8.254	8.640	8.699	8.734	9.085	8.801	12.14	14.6							
2200	8.328	8.674	8.730	8.771	9.140	8.828	12.26	14.6							
2300	8.398	8.705	8.758	8.808	9.195	8.852	12.37	14.7							
2400	8.464	8.733	8.784	8.841	9.249	8.874	12.47	14.8							
2500	8.526	8.759	8.806	8.873	9.302	8.895	12.56	14.8							
2750	8.667	8.815	8.856	8.945	9.431	8.941	12.8	14.9							
3000	8.791	8.861	8.898	9.006	9.552	8.981	12.9	15.0							
3250	8.899	8.900	8.933	9.060	9.663	9.017	13.1	15.1							
3500	8.993	8.934	8.963	9.108	9.763	9.049	13.2	15.2							
3750	9.076	8.963	8.990	9.150	9.853	9.079	13.2	15.3							
4000	9.151	8.989	9.015	9.187	9.933	9.107	13.3	15.3							
4250	9.220	9.013	9.038	9.221	10.003	9.133	13.4	15.4							
4500	9.282	9.035	9.059	9.251	10.063	9.158	13.4	15.5							
4750	9.338	9.056	9.078	9.278	10.115	9.183	13.5	15.5							
5000	9.389	9.076	9.096	9.308	10.157	9.208	13.5	15.6							

Source: *Selected Values of Chemical Thermodynamic Properties*, as of July 1, 1953, edited by D. D. Wagman, National Bureau of Standards.

In Table 18 are values of the constants of these equations for the light paraffins and olefins at low pressures. The last line in the table gives the differences between the molal heat capacities of paraffin

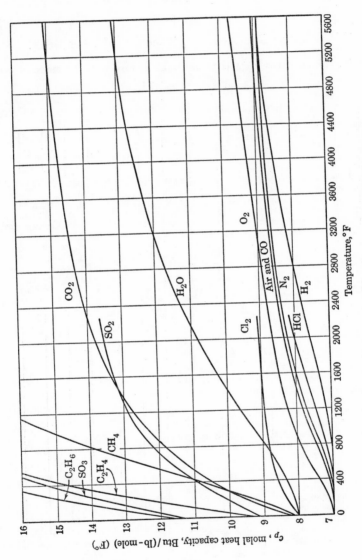

Fig. 60. Molal heat capacities of gases at constant pressure

TABLE 17. EMPIRICAL CONSTANTS FOR MOLAL HEAT CAPACITIES OF GASES AT
CONSTANT PRESSURE ($p = 0$)

$c_p = a + bT + cT^2$, where T is in degrees Kelvin; g-cal/(g-mole) (K°)

Temperature range 300 to 1500° K

Gas	a	$b(10^3)$	$c(10^6)$	% Maximum Deviation, 300 to 1500° K
H_2	6.946	−0.196	0.4757	0.5
N_2	6.457	1.389	−0.069	1.4
O_2	6.117	3.167	−1.005	1.5
CO	6.350	1.811	−0.2675	1.5
NO	6.440	2.069	−0.4206	1.6
H_2O	7.136	2.640	0.0459	1.3
CO_2	6.339	10.14	−3.415	2.0
SO_2	6.945	10.01	−3.794	1.5
SO_3	7.454	19.13	−6.628	3.7
HCl	6.734	0.431	+0.3613	1.0
C_2H_6	2.322	38.04	−10.97	1.4
CH_4	3.204	18.41	−4.48	3.5
C_2H_4	3.019	28.21	−8.537	1.9
Cl_2	7.653	2.221	−0.8733	1.5
Air	6.386	1.762	−0.2656	1.0
NH_3*	5.92	8.963	−1.764	1.0

* W. M. D. Bryant, *Ind. Eng. Chem.*, 25, 820 (1933).

hydrocarbon gases containing more than three carbon atoms and those of the corresponding olefins.

For specific heats of vaporized petroleum fractions the following equation is recommended[5] for the temperature range 0 to 1400° F:

$$C_p = (0.0450K - 0.233) + (0.440 + 0.0177K) \times 10^{-3}t \\ - 0.1530 \times 10^{-6}t^2 \qquad (23)$$

TABLE 18. MOLAL HEAT CAPACITIES OF HYDROCARBON GASES

For 50 to 1400° F, $c_p = a + bT + cT^2$

For −300 to 200° F, $c_p = 7.95 + uT^v$

T = degrees Rankine

Compound	a	$b \times 10^3$	$-c \times 10^6$	u	v
Methane	3.42	9.91	1.28	6.4×10^{-12}	4.00
Ethylene	2.71	16.20	2.80	8.13×10^{-11}	3.85
Ethane	1.38	23.25	4.27	6.20×10^{-5}	1.79
Propylene	1.97	27.69	5.25	2.57×10^{-3}	1.26
Propane	0.41	35.95	6.97	3.97×10^{-3}	1.25
n-Butane	2.25	45.40	8.83	0.93×10^{-2}	1.19
i-Butane	2.30	45.78	8.89	0.93×10^{-2}	1.19
Pentane	3.14	55.85	10.98	3.9×10^{-2}	1.0
Paraffin minus olefin	−1.56	8.26	1.72	......	...

where t is in degrees Fahrenheit, C_p is specific heat, and K is the characterization factor as defined on page 404. Equation 23 was found to be independent of specific gravity or boiling point and is plotted in Fig. 61, which is applicable to petroleum fractions and hydrocarbons containing more than four carbon atoms.

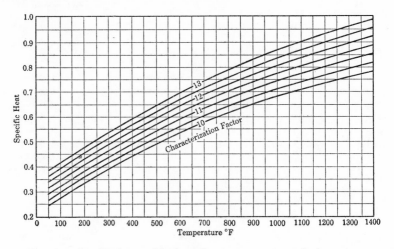

FIG. 61. Specific heats of hydrocarbon gases at atmospheric pressure
Number of carbon atoms per molecule = 4 or more
(Reproduced in *CPP Charts*)

Effect of Pressure on the Heat Capacity of Gases. The effect of pressure on the heat capacity of gases at pressures below one atmosphere and below the critical temperature is negligible. Above atmospheric pressure the heat capacity of a gas at constant pressure increases with increase in pressure, reaching a value of infinity at the critical temperature and pressure. At pressures above the critical pressure the effect of pressure on the heat capacity diminishes with increase in temperature. The general effects of pressure on heat capacities of gases are discussed in Chapter 14.

Special Units for Heat Capacities of Gases. From values of molal heat capacities of gases, values for any other units of mass or volume may be calculated, such as kilocalories per kilogram per Kelvin degree, kilocalories per standard cubic meter per Kelvin degree, Btu per pound per Rankine degree and Btu per standard cubic foot per Rankine degree.

Where heat capacities are based on unit volumes, the basic quantity of gas involved is the mass contained in a unit volume measured at standard conditions of temperature and pressure and not in a unit volume at the existing conditions of temperature and pressure. The heat capacity per standard unit volume refers to the heat capacity of a definite and constant mass of gas, regardless of its temperature and pressure. For example, the heat capacity per standard cubic meter of oxygen at 1000° C signifies the heat capacity at 1000° C of the mass of gas contained in 1 cubic meter at standard conditions, that is, of 1.44 kg of oxygen. It does not signify the heat capacity of the oxygen contained in 1 cubic meter of gas at the given temperature and pressure. This distinction must be made in speaking of the heat capacities of unit volumes of gases at various temperatures.

In the fuel-gas industries a special unit of gas quantity is employed as the standard. This unit is the mass of gas mixture contained in 1000 cu ft, measured at a pressure of 30 in. of mercury, a temperature of 60° F, and saturated with water vapor. This volume of gas corresponds to 2.596 lb-moles of dry gas containing 0.046 lb-mole of water vapor, or 2.642 lb-moles of the mixture (assuming ideal behavior). The heat capacity of the mass of gas equivalent to 1000 cu ft as measured in the gas industry at 30 in. of mercury, saturated, and at 60° F can be obtained by multiplying the mean molal heat capacity of that gas, expressed in Btu per pound-mole per degree Fahrenheit, by the factor 2.642.

Mean Heat Capacities of Gases. The heat-capacity data tabulated in Table 16 represent the values at the stated temperatures. Similarly, the values of heat capacities calculated from the constants a, b, and c of Table 17 give values at temperatures $T°$ K. In heating or cooling a gas from one temperature to another it is convenient to use a mean or average heat capacity over that temperature range, where the mean molal heat capacity at constant pressure is defined by the following equation,

$$c_{pm} = \frac{q_p}{(T_2 - T_1)} = \frac{\int_{T_1}^{T_2} c_p \, dT}{(T_2 - T_1)} \tag{24}$$

where T_1, T_2 are the lower and higher temperatures, respectively.

The total heat required q_p to heat the gas from one temperature to another at constant pressure can be calculated by multiplying the number of moles by the mean molal capacity and by the temperature rise. This method avoids integration of heat-capacity data over the temperature range. Over short temperature ranges the mean heat capacity may be taken as the heat capacity at the average temperature. Even for a gas such as carbon dioxide where the heat capacity increases 5.5% from 0 to 1000° C, the heat capacity at the mean temperature of 500° C is only 0.6% higher than the correct mean heat capacity over that temperature range.

Where an empirical equation for heat capacity is available, the mean molal heat capacity may be calculated by integration, thus

$$c_{pm} = \frac{\int_{T_1}^{T_2} (a + bT + cT^2)dT}{T_2 - T_1} =$$

$$a + \frac{b}{2}(T_2 + T_1) + \frac{c}{3}(T_2{}^2 + T_2T_1 + T_1{}^2) \quad (25)$$

The mean heat capacities of gases over the temperature range from 25 to $t°$ C are tabulated in Table 19 for intervals of 100 C° from 0 to 1200 or 2200° C, calculated from the data of Table 16. The lower limit of 25° C (77° F) has been selected to agree with the reference temperature of thermodynamic data. Mean molal heat capacities are plotted against degrees Fahrenheit in Fig. 62 for values of temperature from 0 to 5000° F.

TABLE 19. MEAN MOLAL HEAT CAPACITIES OF GASES BETWEEN 25 AND
$t°$ C $(p = 0)$

g-cal/(g-mole)(K°)

t	H_2	N_2	CO	Air	O_2	NO	H_2O	CO_2	HCl	Cl_2	CH_4	SO_2	C_2H_4	SO_3	C_2H_6
25	6.894	6.961	6.965	6.972	7.017	7.134	8.024	8.884	6.96	8.12	8.55	9.54	10.45	12.11	12.63
100	6.924	6.972	6.983	6.996	7.083	7.144	8.084	9.251	6.97	8.24	8.98	9.85	11.35	12.84	13.76
200	6.957	6.996	7.017	7.021	7.181	7.224	8.177	9.701	6.98	8.37	9.62	10.25	12.53	13.74	15.27
300	6.970	7.036	7.070	7.073	7.293	7.252	8.215	10.108	7.00	8.48	10.29	10.62	13.65	14.54	16.72
400	6.982	7.089	7.136	7.152	7.406	7.301	8.409	10.462	7.02	8.55	10.97	10.94	14.67	15.22	18.11
500	6.995	7.159	7.210	7.225	7.515	7.389	8.539	10.776	7.06	8.61	11.65	11.22	15.60	15.82	19.39
600	7.011	7.229	7.289	7.299	7.616	7.470	8.678	11.053	7.10	8.66	12.27	11.45	16.45	16.33	20.58
700	7.032	7.298	7.365	7.374	7.706	7.549	8.816	11.303	7.15	8.70	12.90	11.66	17.22	16.77	21.68
800	7.060	7.369	7.443	7.447	7.792	7.630	8.963	11.53	7.21	8.73	13.48	11.84	17.95	17.17	22.72
900	7.076	7.443	7.521	7.520	7.874	7.708	9.109	11.74	7.27	8.77	14.04	12.01	18.63	17.52	23.69
1000	7.128	7.507	7.587	7.593	7.941	7.773	9.246	11.92	7.33	8.80	14.56	12.15	19.23	17.86	24.56
1100	7.169	7.574	7.653	7.660	8.009	7.839	9.389	12.10	7.39	8.82	15.04	12.28	19.81	18.17	25.40
1200	7.209	7.635	7.714	7.719	8.068	7.898	9.524	12.25	7.45	8.94	15.49	12.39	20.33	18.44	26.15
1300	7.252	7.692	7.772	7.778	8.123	7.952	9.66	12.39							
1400	7.288	7.738	7.818	7.824	8.166	7.994	9.77	12.50							
1500	7.326	7.786	7.866	7.873	8.203	8.039	9.89	12.69							
1600	7.386	7.844	7.922	7.929	8.269	8.092	9.95	12.75							
1700	7.421	7.879	7.958	7.965	8.305	8.124	10.13	12.70							
1800	7.467	7.924	8.001	8.010	8.349	8.164	10.24	12.94							
1900	7.505	7.957	8.033	8.043	8.383	8.192	10.34	13.01							
2000	7.548	7.994	8.069	8.081	8.423	8.225	10.43	13.10							
2100	7.588	8.028	8.101	8.115	8.460	8.255	10.52	13.17							
2200	7.624	8.054	8.127	8.144	8.491	8.277	10.61	13.24							

Heat Capacities of Solids

According to the law of Petit and Dulong the atomic heat capacities of the crystalline solid elements are nearly constant and equal to 6.2 cal per g-atom. This rule applies satisfactorily to elements having

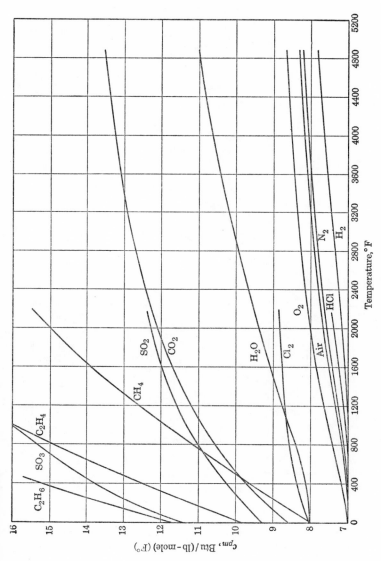

Fig. 62. Mean molal heat capacities of gases at constant pressure
(Mean values from 77° to $t°$ F)

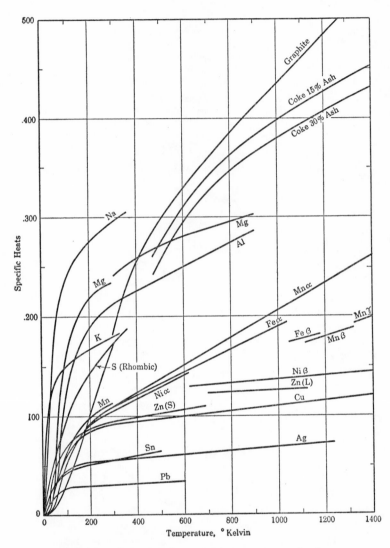

Fig. 63. Specific heats of common elements and cokes

atomic weights above 40 when applied to constant-volume conditions at room temperatures. From kinetic theory, Boltzmann showed that with rise in temperature the atomic heat capacities of the elements at constant volume reach a maximum value of $3R = 5.97$ cal per degree. At room temperature the capacities of the low-atomic-weight elements such as carbon, hydrogen, boron, silicon, oxygen, fluorine, phosphorus, and sulfur are lower than 6.2. At increasing temperatures,

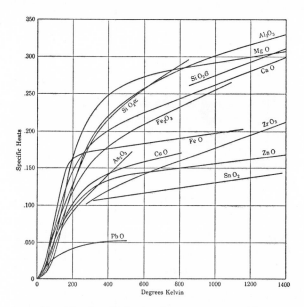

Fig. 64. Specific heats of some common oxides

however, the atomic heat capacities of these elements also approach the value 6.2. The atomic heat capacities of all solid elements decrease greatly with decrease in temperature, approaching a value of zero at absolute zero temperature when in the crystalline state. In general, the heat capacities of compounds are higher in the liquid than in the solid state. At the melting point the two heat capacities are nearly the same.

The heat capacity of a heterogeneous mixture is an additive property, the total heat capacity being equal to the sum of the heat capacities of the component parts. When solutions are formed, this additive property may no longer exist.

The specific heats of various elements and oxides are presented graphically in Figs. 63 and 64. In Fig. 65 are specific heats of a few calcium compounds. The heat capacities of many other common solids are tabulated in Tables 20 and 21. Transition points, *TP*, indicate changes in crystalline structure, and melting points *MP* correspond to abrupt changes in the heat-capacity relationships.

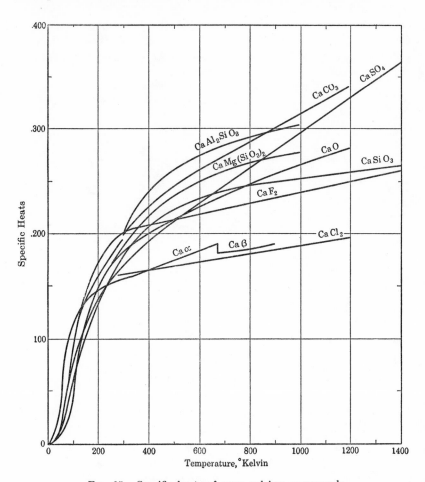

FIG. 65. Specific heats of some calcium compounds

Kopp's Rule. The heat capacity of a solid compound is approximately equal to the sum of the heat capacities of the constituent elements. This generalization was first shown by Kopp to be approximate, provided the following atomic heat capacities are assigned to the elements at 20° C: C 1.8, H 2.3, B 2.7, Si 3.8, O 4.0, F 5.0, P 5.4, all others 6.2. This rule should be used only where experimental values are lacking. Since the heat capacities of solids increase with temper-

TABLE 20.   HEAT CAPACITIES OF SOLID INORGANIC COMPOUNDS
$$C_p = \text{cal}/(\text{gram}) (C°)$$
Values for other elements and compounds given in Figs. 63–65.

Compound	Formula	$t°C$	C_p	Reference
Aluminum sulfate	$Al_2(SO_4)_3$	50	0.184	2
	$Al_2(SO_4)_3 \cdot 18H_2O$	34	0.353	2
Ammonium chloride	NH_4Cl	0	0.357	2
Antimony	Sb	25	0.05	1
Antimony trisulfide	Sb_2S_3	0	0.0830	2
		100	0.0884	2
Arsenic	As	25	0.0796	1
Arsenic oxide	As_2O_3	0	0.117	2
Barium carbonate	$BaCO_3$	100	0.110	2
		400	0.123	2
		800	0.130	2
Barium chloride	$BaCl_2$	0	0.0853	2
		100	0.0945	2
Barium sulfate	$BaSO_4$	0	0.1112	2
		1000	0.1448	2
Bismuth	Bi	25	0.0292	1
Bismuth trioxide	Bi_2O_3	25	0.0584	1
Boron	B	25	0.264	1
Boron oxide	B_2O_3	25	0.2138	1
Cadmium	Cd	25	0.0551	1
Cadmium sulfate	$CdSO_4 \cdot 8H_2O$	0	0.1950	2
Cadmium sulfide	CdS	0	0.0881	2
		100	0.0924	2
Calcium chloride	$CaCl_2$	61	0.164	2
	$CaCl_2 \cdot 6H_2O$	0	0.321	2
Calcium fluoride	CaF_2	0	0.204	2
		40	0.212	2
		80	0.216	2
Calcium sulfate	$CaSO_4 \cdot 2H_2O$	0	0.2650	2
		50	0.198	2
Calcium sulfide	CaS	25	0.157	1
Carbon (diamond)	C	25	0.147	2
Cesium	Cs	25	0.0558	1
Chromium	Cr	25	0.1075	1
Chromium oxide	Cr_2O_3	0	0.168	2
		50	0.188	2
Cobalt	Co	25	0.1045	1
Cupric oxide	CuO	25	0.133	1
Copper sulfate	$CuSO_4$	0	0.148	2
	$CuSO_4 \cdot H_2O$	0	0.1717	2
	$CuSO_4 \cdot 3H_2O$	0	0.2280	2
	$CuSO_4 \cdot 5H_2O$	0	0.2560	2
Ferrous carbonate	$FeCO_3$	54	0.193	2
Ferrous sulfate	$FeSO_4$	45	0.167	2
Gold	Au	25	0.0306	1
Iodine	I	25	0.0518	1
Lead carbonate	$PbCO_3$	32	0.080	2
Lead chloride	$PbCl_2$	0	0.0649	2
		200	0.0704	2
		400	0.0800	2
Lead nitrate	$Pb(NO_3)_2$	45	0.1150	2
Lead sulfate	$PbSO_4$	45	0.0838	2
Lithium	Li	25	0.815	1
Magnesium chloride	$MgCl_2$	48	0.193	2

TABLE 20-*Continued*

Compound	Formula	$t°$ C	C_p	Reference
Magnesium sulfate	$MgSO_4$	61	0.222	2
	$MgSO_4 \cdot H_2O$	9	0.239	2
	$MgSO_4 \cdot 6H_2O$	9	0.349	2
	$MgSO_4 \cdot 7H_2O$	12	0.361	2
Manganese dioxide	MnO_2	0	0.152	2
Manganese oxide	MnO	58	0.158	2
Manganic oxide	Mn_2O_3	58	0.162	2
Mercuric chloride	$HgCl_2$	0	0.0640	2
Mercuric sulfide	HgS	0	0.0506	2
Mercurous chloride	$HgCl$	0	0.0499	2
Molybdenum	Mo	25	0.0585	1
Nickel sulfide	NiS	0	0.116	2
		100	0.128	2
		200	0.138	2
Palladium	Pd	25	0.059	1
Platinum	Pt	25	0.0326	1
Potassium chloride	KCl	0	0.1625	2
		200	0.1725	2
		400	0.1790	2
Potassium chlorate	$KClO_3$	0	0.1910	2
		200	0.2960	2
Potassium chromate	K_2CrO_4	46	0.1864	2
Potassium dichromate	$K_2Cr_2O_7$	0	0.178	2
		400	0.236	2
Potassium nitrate	KNO_3	0	0.214	2
		200	0.267	2
		500	0.292	2
Potassium perchlorate	$KClO_4$	25	0.190	1
Potassium sulfate	K_2SO_4	0	0.1760	2
Selenium	Se	25	0.0755	1
Silver chloride	$AgCl$	0	0.0848	2
		200	0.0974	2
		500	0.101	2
Silver nitrate	$AgNO_3$	50	0.146	2
Sodium borate	$Na_2B_4O_7$	45	0.234	2
(Borax)	$Na_2B_4O_7 \cdot 10H_2O$	35	0.385	2
Sodium carbonate	Na_2CO_3	45	0.256	2
Sodium chloride	$NaCl$	0	0.204	2
		100	0.217	2
		400	0.229	2
		600	0.236	2
Sodium nitrate	$NaNO_3$	0	0.2478	2
		100	0.294	2
		250	0.358	2
Sodium sulfate	Na_2SO_4	0	0.202	2
		100	0.220	2
Sulfur (monoclinic)	S	25	0.1765	1
Titanium	Ti	25	0.1255	1
Titanium dioxide	TiO_2	25	0.165	1
Tungsten	W	25	0.0325	1
Water (ice)	H_2O	-40	0.435	2
		0	0.492	2

Sources: 1. *Selected Values of Chemical Thermodynamic Properties*, as of July 1, 1953, edited by D. D. Wagman, National Bureau of Standards.
2. *International Critical Tables* (1929).

TABLE 21.   HEAT CAPACITIES OF MISCELLANEOUS MATERIALS

$$C_p = \text{cal}/(\text{gram})(\text{C}°)$$

Substance	C_p	Temperature range, ° C
Alundum	0.186	100
Asbestos	0.25	
Asphalt	0.22	
Bakelite	0.3 to 0.4	
Brickwork	0.2 (approx.)	
Carbon (gas retort)	0.204	
Cellulose	0.32	
Cement	0.186	
Charcoal (wood)	0.242	
Chrome brick	0.17	
Clay	0.224	
Coal	0.26 to 0.37	
Coal tar	0.35	40
	0.45	200
Coke	0.265	21–400
	0.359	21–800
	0.403	21–1300
Concrete	0.156	70–312
	0.219	72–1472
Cryolite	0.253	16–55
Fireclay brick	0.198	100
	0.298	1500
Fluorspar	0.21	30
Glass (crown)	0.16 to 0.20	
(flint)	0.117	
(Pyrex)	0.20	
(silicate)	0.188 to 0.204	0–100
	0.24 to 0.26	0–700
(Wool)	0.157	
Granite	0.20	20–100
Magnesite brick	0.222	100
	0.195	1500
Pyrites (copper)	0.131	19–50
(iron)	0.136	15–98
Sand	0.191	
Steel	0.12	

Source: John H. Perry, *Chemical Engineers Handbook*, 3d ed., McGraw-Hill Book Co. (1950), with permission.

ature, it is obvious that the above values do not apply over a wide temperature range.

Heat Capacities of Liquids and Solutions

Few generalizations can be stated regarding the heat capacities of liquids. The heat capacities of most liquids increase with an increase in temperature. The heat capacity of most substances is greater for the liquid state than for either the solid or the gas. Where experi-

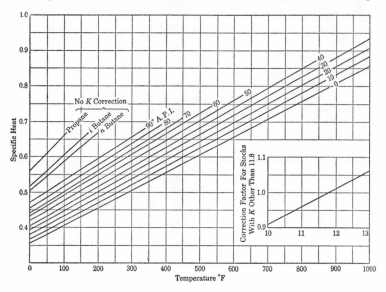

FIG. 66. Specific heat of liquid petroleum oils where $K = 11.8$ (mid-continent stocks). For other stocks multiply by correction factor

(Reproduced in *CPP Charts*)

mental data are lacking Kopp's rule (page 262) may be applied by assigning the following values of atomic heat capacities at room temperature to the atoms of the liquid, according to Wenner[6]: C = 2.8, H = 4.3, B = 4.7, Si = 5.8, O = 6.0, F = 7.0, P = 7.4, S = 7.4, and to most other elements a value of 8.

Specific Heats of the Liquid Hydrocarbons. The following equation was recommended by Fallon and Watson[5] for the specific heats of liquid hydrocarbons and petroleum fractions at temperatures between 0° F and reduced temperatures of 0.85.

$$C_p = [(0.355 + 0.128 \times 10^{-2}\,°\text{API}) + (0.503 + 0.117$$
$$\times 10^{-2}\,°\text{API}) \times 10^{-3}t]\,[0.05K + 0.41] \quad (26)$$

[6] R. R. Wenner, *Thermochemical Calculations*, McGraw-Hill Book Co. (1941), with permission.

where t is in ° F and K is the characterization factor as defined on page 404.

Figure 66 is a plot of this relationship together with curves for the individual light paraffin hydrocarbons as recommended by Holcomb and Brown.[7] The curves on the main plot apply directly to mid-

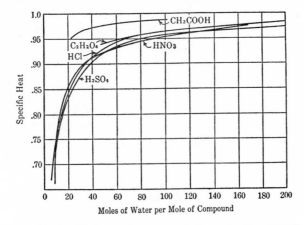

Fɪɢ. 67. Specific heats of aqueous solutions of acids at 20° C

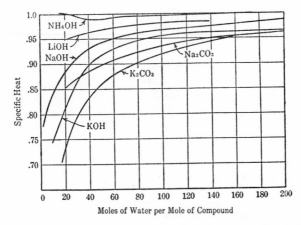

Fɪɢ. 68. Specific heats of aqueous solutions of bases at 20° C

continent stocks whose characterization factors are approximately 11.8. For other stocks the value read from the main plot is multiplied by a correction factor derived as a function of K from the small plot in the lower right-hand corner.

A generalized method of estimating heat capacities of the liquid

[7] D. E. Holcomb and G. G. Brown, *Ind. Eng. Chem*, **34**, 590 (1942), with permission.

state from data on the gaseous state is discussed in Chapter 14. Since heat capacities of gases can be calculated by generalized methods based on spectroscopic data, these methods make it possible to estimate the heat capacity of a fluid at any conditions of the liquid or gaseous state from a minimum of experimental data.

Water has a higher specific heat than any other substance, with the exception of liquid ammonia and a few organic compounds. The heat capacity of water is a minimum at 30° C. The specific heats of aqueous solutions in general decrease with increasing concentration of solute. In dilute solutions the heat capacity of aqueous solutions is nearly equal to the heat capacity of the water present. The heat capacities of some common aqueous solutions of acids, bases, and salts at 20° C are shown graphically in Figs. 67 through 71. In Tables 22 and 23 are values of the specific heats of some common liquids.

TABLE 22. HEAT CAPACITIES OF INORGANIC LIQUIDS

C_p = heat capacity, g-cal/(gram)(C°) at $t°$ C

a = temperature coefficient in the equation $C_p = C_{p0} + at$
over the indicated temperature range.

Liquid	Formula	$t°$ C	C_p	C_{p0}	a	Temperature Range, ° C	Ref.
Ammonia	NH_3	−40	1.051				2
		0	1.098				2
		60	1.215				2
		100	1.479				2
Mercury	Hg	0	0.0335				2
		60	0.0330				2
		100	0.0329				2
		200	0.0329				2
		280	0.0332				2
Nitric acid	HNO_3	25	0.417				1
Silicon tetrachloride	$SiCl_4$	25	0.204				1
Sodium nitrate	$NaNO_3$	350	0.430				2
Sulfuric acid	H_2SO_4			0.339	0.00038	10° to 45° C	2
	H_2SO_4	25	0.369				1
Sulfuryl chloride	SO_2Cl_2	25	0.234				1
Sulfur dioxide	SO_2	−20	0.3130	0.318	0.00028	10° to 140° C	2
Water	H_2O	0	1.008				3
		15	1.000				3
		100	1.006				3
		200	1.061				3
		300	1.155				3

Sources:

1. *Selected Values of Chemical Thermodynamic Properties*, National Bureau of Standards (1953).

2. *International Critical Tables* (1929).

3. *Handbook of Chemistry and Physics*, Chemical Rubber Publishing Co. (1953), with permission.

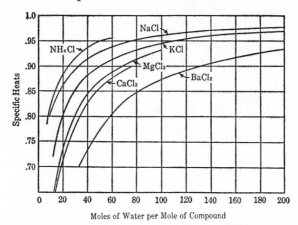

Fig. 69.   Specific heats of aqueous solutions of chlorides at 20° C

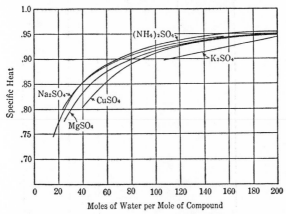

Fig. 70.   Specific heats of aqueous solutions of sulfates at 20° C

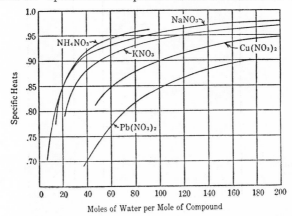

Fig. 71.   Specific heats of aqueous solutions of nitrates at 20° C

TABLE 23. HEAT CAPACITIES OF ORGANIC LIQUIDS

Data from *International Critical Tables* unless otherwise indicated.

C_p = heat capacity, calories per gram per C° at t° C
a = temperature coefficient in equation: $C_p = C_{p0} + at$,
applying over the indicated temperature range

Liquid	Formula	t° C	C_p	C_{p0}	a	Temp. Range
Carbon tetrachloride	CCl_4	20	0.201	0.198	0.000031	0 to 70°
Carbon disulfide	CS_2			0.235	0.000246	−100 to +150°
Chloroform	$CHCl_3$	15	0.226	0.221	0.000330	−30 to +60°
Formic acid	CH_2O_2	0	0.496	0.496	0.000709	40 to 140°
Methyl alcohol	CH_4O	0	0.566			
		40	0.616			
Acetic acid	$C_2H_4O_2$			0.468	0.000929	0 to 80°
Ethyl alcohol	C_2H_6O	−50	0.473			
		0	0.535			
		25	0.580			
		50	0.652			
		100	0.824			
		150	1.053			
Glycol	$C_2H_6O_2$	0	0.544	0.544	0.001194	−20 to +200°
Allyl alcohol	C_3H_6O	0	0.3860			
		21 to 96	0.665			
Acetone	C_3H_6O			0.506	0.000764	−30 to +60°
Propane	C_3H_8	0	0.576	0.576	0.001505	−30 to +20°
Propyl alcohol	C_3H_8O	−50	0.456			
		0	0.525			
		+50	0.654			
Glycerol	$C_3H_8O_3$	−50	0.485			
		0	0.540			
		+50	0.598			
		+100	0.668			
Ethyl acetate	$C_4H_8O_2$	20	0.478			
n-Butane	C_4H_{10}	0	0.550	0.550	0.00191	−15 to +20°
Ether	$C_4H_{10}O$	0	0.529			
		30	0.548			
		120	0.802			
Isopentane	C_5H_{12}	0	0.512			
		8	0.526			
Nitrobenzene	$C_6H_5NO_2$	10	0.358			
		50	0.329			
		120	0.393			
Benzene	C_6H_6	5	0.389			
		20	0.406			
		60	0.444			
		90	0.473			
Aniline	C_6H_7N	0	0.478			
		50	0.521			
		100	0.547			
n-Hexane	C_6H_{14}	20 to 100	0.600			
Toluene	C_7H_8	0	0.386			
		50	0.421			
		100	0.470			
n-Heptane	C_7H_{16}	0 to 50	0.507			
		30	0.518	0.476	0.00142	30 to 80°
Decane (BP 172°)	$C_{10}H_{22}$	0 to 50	0.502			
n-Hexadecane	$C_{16}H_{34}$	0 to 50	0.496			
Stearic acid	$C_{18}H_{36}O_2$	75 to 137	0.550			
Diphenyl*	$C_{12}H_{10}$			0.300	0.00120	80 to 300°

* H. O. Forrest, E. W. Brugmann, and L. W. Cummings, *Ind. Eng. Chem.* **23**, 340 (1931).

Latent Heats

Heat of Fusion. The fusion of a crystalline solid at its melting point to form a liquid at the same temperature is accompanied by an increase in enthalpy corresponding to an absorption of heat. Since the volume changes and hence the external work in fusion are small, this heat of fusion is largely utilized in increasing the internal energy through rearrangement of the atoms. Attempts have been made to establish general relationships between latent heats of fusion and other more easily measured properties. None of these generalizations are accurate. For most elements[6] the ratio of λ_f/T_f varies from 2 to 3, for most inorganic compounds from 5 to 7, and for most organic compounds from 9 to 11, where

λ_f = heat of fusion, calories per gram-formula weight

T_f = melting point, degrees Kelvin

There are a few marked exceptions to these rules.

Values of heats of fusion are given in Table 24.

Heat of Transition. Many crystalline substances exhibit transformation or transition points where changes in crystalline structure take place. The equilibrium temperature of transformation is constant although the actual temperature of transformation is frequently a function of the rate at which the substance is heated or cooled before the transformation. The actual transition usually takes place at a slightly higher temperature when the substance is being heated than when it is being cooled.

Crystalline transformations are accompanied by either an absorption or evolution of heat. The transformation of the phase which is stable at low temperatures into the phase stable at high temperatures requires an absorption of heat.

Data for heats of transition of a few solids are recorded in Table 25.

Heat of Vaporization

The heat required to vaporize a substance consists of the energy absorbed in overcoming the intermolecular forces of attraction in the liquid and the work performed by the vapors in expanding against an external pressure. Molal heats of vaporization are given in Table 26.

As pointed out in Chapter 4, page 78, an exact relationship between heat of vaporization and vapor pressure is expressed by the Clapeyron equation. This equation permits accurate calculations of latent heats of vaporization at any temperature from vapor-pressure data and molal volumes of the liquid and vapor. The necessary data for use of this rigorous equation are available for only relatively few substances.

TABLE 24. HEATS OF FUSION

λ_f = heat of fusion, g-cal per g-atom or g-mole; t_f = melting point, ° C;
T_f = melting point, °K

To convert heats of fusion to Btu per pound-mole, multiply table values by 1.8.

	λ_f	t_f° C	λ_f/T_f	Reference
Elements				
Ag	2,700	961	2.19	1
Al	2,600	660	2.79	1
Cu	3,110	1083	2.29	1
Fe	3,660	1535	2.02	2
Na	629	98	1.70	2
Ni	4,200	1455	2.43	1
Pb	1,220	327	2.03	1
S (rhombic)	300	115	0.77	2
Sn	1,690	232	3.35	1
Zn	1,595	419	2.30	1
Compounds				
H_2O	1,436.3	0.0	5.26	1
Sb_2S_3	11,200	547	13.66	1
CO_2	1,999	−56.2	9.21	1
$CaCl_2$	6,780	782	6.43	1
NaOH	1,700	318	2.88	1
NaCl	6,800	808	6.29	1
Carbon tetrachloride	600	−22.9	2.40	1
Methyl alcohol	757	−98	4.33	1
Acetic acid	2,800	16.6	9.65	1
Ethyl alcohol	1,200	−114.6	7.58	1
Benzene	2,370	5.4	8.51	2
Aniline	1,950	−7.0	7.32	2
Naphthalene	4,550	80	12.88	2
Diphenyl	4,020	71	11.70	2
Stearic acid	13,500	70.5	39.35	2

Sources:
1. *Selected Values of Chemical Thermodynamic Properties*, as of July 1, 1953, edited by D. D. Wagman, National Bureau of Standards.
2. *International Critical Tables* (1929).

Trouton's Rule. According to Trouton's rule, the ratio of the molal heat vaporization λ_b of a substance at its normal boiling point to the absolute temperature T_b is a constant. Thus

$$\frac{\lambda_b}{T_b} = K \qquad (27)$$

where K is termed Trouton's ratio. For many substances this ratio is equal to approximately 21 where the latent heat is expressed in

TABLE 25. HEATS OF TRANSITION

λ_t = heat absorbed in transition, g-cal per g-atom or Chu per lb-atom or mole.
To convert to Btu per lb-mole, multiply by 1.8.
t_t = temperature of transformation, ° C.

Transition	λ_t	t_t, ° C
Sulfur		
Rhombic → monoclinic	7.0	114–151
Iron (electrolytic) (see also Fig. 121, p. 482)		
$\alpha \rightarrow \beta$	363	770
$\beta \rightarrow \gamma$	313	910
$\gamma \rightarrow \delta$	106	1400
Manganese		
$\alpha \rightarrow \beta$	1325	1070–1130
Nickel		
$\alpha \rightarrow \beta$	78	320–330
Tin		
White → gray	530	0

Source: *International Critical Tables* (1929).

calories per gram-mole and the temperature in degrees Kelvin. For nonpolar liquids Trouton's ratio increases slightly as the normal boiling point increases. For polar liquids values of this ratio are much greater than 21.

Kistyakowsky Equation for Nonpolar Liquids. A thermodynamic equation was developed by Kistyakowsky for the calculation of Trouton's ratio at the normal boiling point of nonpolar liquids:

$$\frac{\lambda_b}{T_b} = 8.75 + 4.571 \log_{10} T_b \qquad (28)$$

where λ_b = normal heat of vaporization in calories per gram-mole at T_b° K
T_b = normal boiling point in degrees Kelvin.

This equation is in excellent agreement with experimental results for a wide variety of nonpolar liquids but is inapplicable to polar liquids.

Heats of Vaporization from Vapor-Pressure Data and Compressibility Factors. The molal heat of vaporization of a substance may be calculated from data on its vapor-pressure–temperature relations and specific molal volumes of the liquid and gas by the rigorous equation 3, page 78, which may be written in terms of reduced properties as

$$\lambda = p_c \left(\frac{dp_r}{dT_r}\right) (v_G - v_L) T_r \qquad (29)$$

Any reliable equation for dp_r/dT_r or dp/dT may be used. For example,

TABLE 26. HEATS OF VAPORIZATION

λ = heat of vaporization at $t_b°$ C, g-cal per g-mole

t = temperature, ° C, $\quad t_b$ = normal boiling-point, ° C

To convert heats of vaporization to Btu per lb-mole, multiply table values by 1.8.

Substance	λ	$t_b°$ C	Reference
Ammonia	5,581	−33.4	2
Argon	1,590	−185.8	2
Bromine	7,340	25.0	1
Carbon dioxide	6,030	−78.4	2
Carbon disulfide	6,400	46.25	1
Carbon monoxide	1,444	−191.5	2
Carbon oxysulfide	4,423	−50.2	2
Carbon tetrachloride	7,170	76.7	1
Chlorine	4,878	−34.1	2
Dichlorodifluoromethane	4,850	−30.5	1
Dichloromonofluoromethane (Freon 21)	6,400	8.9	1
Helium	22	−268.9	2
Hydrogen	216	−252.7	2
Hydrogen bromide	4,210	−66.7	2
Hydrogen chloride	3,860	−85.0	2
Hydrogen cyanide	6,027	25.7	2
Hydrogen fluoride	1,800	19.9	1
Hydrogen iodide	4,724	−35.35	1
Hydrogen sulfide	4,463	−60.3	2
Mercury	13,890	356.6	1
Nitric oxide	3,292	−151.7	1
Nitrogen	1,336	−195.8	2
Nitrous oxide	3,950	−88.5	2
Oxygen	1,629	−183.0	2
Silicon tetrafluoride	6,150	−95.5	1
Sulfur	25,000	444.6	1
Sulfur dioxide	5,950	−10.0	1
Sulfur trioxide	9,990	43.3	1
Trichloromonofluoromethane (Freon 11)	5,960	23.6	1
Water	9,717	100.0	1

Sources:

1. *Selected Values of Chemical Thermodynamic Properties*, as of July 1, 1953, edited by D. D. Wagman, National Bureau of Standards.

2. John H. Perry, *Chemical Engineers Handbook*, 3d ed., McGraw-Hill Book Co. (1950).

differentiation of equation 16, page 95, gives

$$\frac{dp_r}{dT_r} = 2.303 p_r \left[\frac{A}{T_r^2} + 40(T_r - b)e^{-20(T_r-b)^2} \right] \tag{30}$$

The compressibility factor z of a substance is defined as $z = pv/RT$. For ideal-gas behavior this factor is unity. Applying this factor to both gas and liquid states and rearranging gives

$$v_G - v_L = (z_G - z_L)\frac{RT}{p} = (z_G - z_L)R\frac{T_rT_c}{p_rp_c} \tag{31}$$

From vapor-pressure data on 95 different compounds Lydersen, Greenkorn, and Hougen[8] found that the term $z_G - z_L$ of a saturated liquid-vapor system is a unique function of reduced pressure nearly independent of the nature of the compound as tabulated in Table 26A.

TABLE 26A.   VALUES OF $z_G - z_L$ AS A FUNCTION OF REDUCED PRESSURE

p_r	$z_G - z_L$	p_r	$z_G - z_L$	p_r	$z_G - z_L$	p_r	$z_G - z_L$
0	1.0	0.15	0.838	0.60	0.542	0.95	0.210
0.01	0.983	0.20	0.802	0.65	0.506	0.96	0.192
0.02	0.968	0.25	0.769	0.70	0.467	0.97	0.170
0.03	0.954	0.30	0.738	0.75	0.426	0.98	0.142
0.04	0.942	0.35	0.708	0.80	0.382	0.99	0.106
0.05	0.930	0.40	0.677	0.85	0.335	1.00	0.000
0.06	0.919	0.45	0.646	0.90	0.280		
0.08	0.899	0.50	0.612	0.92	0.256		
0.10	0.880	0.55	0.578	0.94	0.226		

Combining equations 29, 30, and 31 gives

$$\lambda = 2.303(z_G - z_L)RT_c\left[A + 40T_r{}^2(T_r - b)e^{-20(T_r-b)^2}\right] \tag{32}$$

At high values of T_r and low values of b equation 32 reduces to

$$\lambda = 2.303(z_G - z_L)RT_cA \tag{33}$$

Values of A and b for specific compounds are given in Table 8, page 95, or may be calculated from equation 16, page 95, from accurate measurements of vapor pressure at two temperatures combined with reliable values of p_c and T_c.

Illustration 2.   Estimate the molal heat of vaporization of acetone at pressures of 1.0 and 8.0 atm.

Boiling point at 1 atm   =   56.5° C (329.7° K)

at 8 atm   =   133.1° C (406.3° K)

$T_c = 508.7°$ K; p_c =   46.6 atm

At 1 atm, $T_r = 329.7/508.7 =$ 0.648; $p_r = 1/46.6 = 0.0215$

At 8 atm, $T_r = 406.3/508.7 =$ 0.799; $p_r = 8/46.6 = 0.172$

From Table 26A:

At 1 atm, $z_G - z_L$ =   0.966

At 8 atm, $z_G - z_L$ =   0.822

[8] A. L. Lydersen, R. A. Greenkorn, and O. A. Hougen, Private communication (1954).

From Table 8, $A = 3.0644$; $b = 0.180$

At 1 atm, from equation 33:

$$\lambda = (2.303)(0.966)(1.987)(508.7)(3.0644) = 6890 \text{ cal/g-mole}$$

From equation 32:

$$\lambda = \frac{(6890)}{3.0644} [3.0644 + 40(0.648)^2(0.648 - 0.180)e^{-20(0.467)^2}]$$

$$= (2248)(3.1627) = 7111 \text{ cal/g-mole}$$

Experimental value $\lambda = 7231$ cal/g-mole

At 8 atm, from equation 33, $\lambda = 5683$ cal/g-mole

From equation 32, $\qquad \lambda = 5689$ cal/g-mole

Heats of Vaporization from Reference-Substance Plots. Latent heats of vaporization may be estimated from the slopes of three reference-substance plots for vapor pressures: namely, from plots of equal vapor pressures, plots of equal temperatures, and plots of equal reduced pressures.

From Equal-Vapor-Pressure Reference-Substance Plots. In the Dühring chart, page 81, the temperature of a substance is plotted against the temperature of a reference substance at equal vapor pressures. Nearly straight lines result. The slopes of these lines are related to the ratio of molal latent heats of vaporization of the two substances. Thus, from the Clausius-Clapeyron equation written for the two substances at equal vapor pressures the following equation results:

$$\frac{\lambda}{\lambda'} = \left(\frac{T}{T'}\right)^2 \left(\frac{dT'}{dT}\right) \tag{34}$$

where
$\lambda = $ molal heat of vaporization at temperature T
$\lambda' = $ molal heat of vaporization at temperature T' where both substances are at the same vapor pressure
$dT'/dT = $ slope of Dühring line

From Equal-Temperature Reference-Substance Plots. In the chart of Fig. 15, the vapor pressure of a substance is plotted against the vapor pressure of a reference substance, both at the same temperature. Nearly straight lines result. The slopes of these lines are related to the ratio of the molal heats of vaporization of the two substances. Thus from the Clausius-Clapeyron equation for the two substances at equal temperatures the following equation results:

$$\frac{\lambda}{\lambda'} = \left(\frac{d \ln p}{d \ln p'}\right)_{T=T'} \tag{35}$$

Upon integration,[9]

$$\left(\log p = \frac{\lambda}{\lambda'} \log p' + C\right)_{T=T'} \tag{36}$$

The satisfactory linearity of reference-substance plots up to the critical point appears to result from a fortuitous compensation of errors in the assumptions involved. Even though such a plot may be a straight line, the actual ratio of the latent heats may vary over a wide range and become meaningless as the critical point of either substance is approached. For this reason great care must be exercised in accepting heats of vaporization calculated from these last two plots at elevated pressures. It is recommended that their use be restricted to the low-pressure range and that the methods of the following sections be used for all pressures substantially above atmospheric.

Reduced Reference-Substance Plots. Gordon[10] has pointed out that the accuracy of reference substance plots for the estimation of heats of vaporization may be improved by basing the reference-substance plot on equal reduced conditions. This plot will be accurate at the critical temperature since at this temperature the latent heats of all substances converge to zero. The most convenient form is a logarithmic plot of vapor pressures of the substance in question against that of the reference substance at equal reduced temperatures. It is found that this method of plotting gives nearly straight lines for a wide variety of both polar and nonpolar substances, and over wide ranges of conditions up to the critical.

Since $T = T_r T_c$, the Clausius-Clapeyron equation (page 79) may be written in terms of reduced temperatures:

$$d \ln p = \frac{\lambda dT_r}{R T_r{}^2 T_c} \tag{37}$$

Applying equation 37 to a substance and a reference substance at the same reduced temperature gives

$$\frac{d \ln p}{d \ln p'} = \frac{\lambda}{\lambda'} \frac{T'_c}{T_c} = s_r \tag{38}$$

$$\lambda = s_r \frac{T_c}{T'_c} \lambda' \tag{39}$$

where the primed quantities designate the reference substance. The group of terms $s_r(T_c/T'_c)$ is a constant for any one pair of substances.

[9] D. F. Othmer, *Ind. Eng. Chem.*, **32**, 841–56 (1940).
[10] D. H. Gordon, University of Wisconsin, Ph.D. thesis (1942).

The term s_r is the slope of the line resulting when the logarithm of the vapor pressure of the given substance is plotted against the logarithm of the vapor pressure of a reference substance at the same reduced temperature.

Values of $s_r(T_c/T'_c)$ are given in Table 27 for various refrigerants with water as the reference substance. In Table 28 are given the heats of vaporization of water in Btu per pound-mole at various values of reduced temperature T_r and the corresponding vapor pressures.

TABLE 27.   HEAT OF VAPORIZATION FACTORS AND CRITICAL
TEMPERATURES OF REFRIGERANTS[10]

Reference substance is water.

Refrigerant	Formula	s_r	Critical Temp., ° R	$\dfrac{s_r T_c}{T'_c}$
Ammonia	NH_3	0.933	730.1	0.584
Benzene	C_6H_6	0.923	1012.7	0.803
Butyl alcohol	C_4H_9OH	1.327	1008.3	1.149
Carbon dioxide	CO_2	0.872	547.7	0.410
Carbon disulfide	CS_2	0.839	983.2	0.707
Carbon tetrachloride	CCl_4	0.869	1001.3	0.746
Chlorine	Cl_2	0.776	751.0	0.502
Chlorobenzene	C_6H_5Cl	0.933	1138	0.905
Chloroform	$CHCl_3$	0.916	965	0.749
Ethane	C_2H_6	0.810	549.8	0.382
Ethyl alcohol	C_2H_5OH	1.280	929.3	1.022
Ethyl chloride	C_2H_5Cl	0.885	824	0.625
Ethyl ether	$(C_2H_5)_2O$	0.952	841	0.687
Freon 11	CCl_3F	0.897	848.1	0.654
Freon 12	CCl_2F_2	0.869	692.5	0.516
Freon 21	$CHCl_2F$	0.891	813.0	0.621
Freon 113	$C_2Cl_3F_3$	0.962	877.1	0.724
Isobutane	C_4H_{10}	0.879	734.7	0.551
Methane	CH_4	0.716	343.3	0.211
Methyl alcohol	CH_3OH	1.179	923	0.936
Methyl chloride	CH_3Cl	0.845	749.3	0.543
n-Butane	C_4H_{10}	0.897	765.3	0.591
Nitrogen dioxide	NO_2	0.851	557.4	0.407
n-Propyl alcohol	C_3H_7OH	1.303	766.4	1.083
Propane	C_3H_8	0.851	665.9	0.486
Sulfur dioxide	SO_2	0.949	774.7	0.631
Toluene	C_7H_8	0.956	1069.2	0.877

Illustration 3. Estimate the heat of vaporization of Freon 12 (difluorodichloromethane) at 200° F.

From Table 27 the critical temperature of Freon 12 is 232.5° F, and the value of s_r (T_c/T'_c) is 0.516.   The molecular weight is 121.

$$T_r = \frac{660}{692.5} = 0.951$$

At $T_r = 0.951$ the molal heat of vaporization of water from Table 28 is 7707 Btu per lb-mole.   From equation 39,

$$\lambda = s_r \frac{T_c}{T'_c} \lambda' = (0.516)(7707) = 3977 \text{ Btu per lb-mole}$$

TABLE 28.   Molal Heats of Vaporization and Vapor Pressures of Water

$T_c = 1165.1°$ R, λ = Btu per lb-mole, p = psi abs

T_r	λ	p	T_r	λ	p
0.423	19,370	0.092	0.82	12,970	654.2
0.44	19,170	0.198	0.84	12,400	804.5
0.46	18,940	0.446	0.86	11,770	980.3
0.48	18,700	0.934	0.88	11,070	1184.7
0.50	18,460	1.824	0.90	10,290	1419.3
0.52	18,210	3.365	0.91	9,684	1548.7
0.54	17,960	5.896	0.92	9,401	1688.4
0.56	17,770	9.871	0.93	8,907	1836.0
0.58	17,440	15.871	0.94	8,365	1995.6
0.60	17,170	24.613	0.95	7,774	2164.3
0.62	16,880	36.959	0.96	7,103	2346.1
0.64	16,590	54.006	0.97	6,336	2538.6
0.66	16,280	76.785	0.98	5,393	2746.8
0.68	15,950	106.65	0.99	4,143	2967.5
0.70	15,600	145.08	1.00	0	3206.2
0.72	15,230	193.68			
0.74	14,840	254.1			
0.76	14,420	328.3			
0.78	13,980	418.4			
0.80	13,490	526.1			

Gordon has studied the application of this method to much of the available data on heats of vaporization and found good agreement, with errors generally less than 5%.   This method is particularly convenient to use where a number of values of heats of vaporization are required at different conditions for a single substance.

Another good method for the prediction of heats of vaporization at all conditions has been developed by Meissner.[11]   This method is applicable to all substances and has approximately the same accuracy as the methods herein described.

[11] H. P. Meissner, *Ind. Eng. Chem.*, **33**, 1440 (1941).

Heats of Vaporization of Hydrocarbons. Heats of vaporization of hydrocarbons and petroleum fractions under atmospheric pressure were calculated by Fallon and Watson[5] by differentiating equation 15, page 92, and substituting in the Clapeyron equation 3, page 78. The resultant correlation is plotted in Fig. 72 relating heat of vaporization to boiling point at various parameters of API gravity or molecular

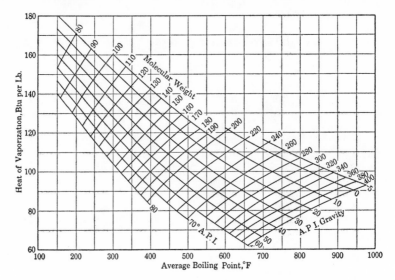

Fig. 72. Heats of vaporization of hydrocarbons and petroleum fractions. Note: Use molecular weights if available rather than API gravities
(Reproduced in *CPP Charts*)

weight. In dealing with pure compounds use of the molecular weight parameter is preferable. Heats of vaporization at other temperatures may be obtained by means of equation 40.

The values of Fig. 72 are in close agreement with the Kistyakowsky equation, page 273, for low-boiling compounds but are considerably higher for high-boiling materials. It is believed that these higher values represent a more reliable extrapolation than that of the Kistyakowsky equation.

Empirical Relationship between Heat of Vaporization and Temperature. The heat of vaporization of a substance diminishes as its temperature and pressure are increased. At the critical temperature, as pointed out in Chapter 4, the kinetic energies of translation of the molecules become sufficiently great to overcome the potential energies of the attractive forces which hold them together, and molecules pass from the liquid to the vapor state without additional energization. At the critical point there is no distinction between the liquid and vapor states, either in enthalpy or in other physical properties, and the heat of vaporization becomes zero.

It was pointed out by Watson[12] that, if values of Trouton's ratio are plotted against reduced temperature, curves of the same shape are obtained for all substances, both polar and nonpolar. These curves may be superimposed by multiplying the ordinates of each by the proper constant factor. It was found that the following empirical equation satisfactorily represents these curves over the entire range of available data:

$$\frac{\lambda_2}{\lambda_1} = \left(\frac{1 - T_{r2}}{1 - T_{r1}}\right)^{0.38} \tag{40}$$

Illustration 4. The latent heat of vaporization of ethyl alcohol is experimentally found to be 204 cal per gram at its normal boiling point of 78° C. Its critical temperature is 243° C. Estimate the heat of vaporization at a temperature of 180° C.

$$T_b/T_c = 351/516 = 0.680 = T_{r1}$$

$$T/T_c = 453/516 = 0.880 = T_{r2}$$

From equation 40,

$$\lambda_2 = 204 \left(\frac{1 - 0.880}{1 - 0.680}\right)^{0.38} = 140 \text{ cal per gram}$$

Evaluation of Enthalpy

As pointed out on page 248, the absolute enthalpy or energy content of matter is unknown. However, the enthalpy of a given substance relative to some reference state can be calculated from its thermophysical properties. This state can be taken arbitrarily as a temperature of 0° C, atmospheric pressure, and the state of aggregation normally existent at this temperature and pressure. The reference state for steam is usually taken as the liquid state under its own vapor pressure of 4.58 mm of mercury at 0° C.

The relative enthalpy of a substance is calculated as the change in enthalpy in passing from the reference state to the existing conditions. As previously pointed out, at constant pressure the increase in enthalpy is equal to the heat absorbed. Ordinarily at moderate pressures the effect of pressure on the enthalpy of liquids and solids may be neglected except when conditions are close to the critical point. This subject is discussed in Chapter 14.

Illustration 5. Calculate the enthalpy of 1 lb of steam at a temperature of 350° F and a pressure of 50 psi, referred to the liquid state at 32° F.

Solution. From the vapor-pressure data of water (Table 5), it is found that

[12] K. M. Watson, *Ind. Eng. Chem.*, **23**, 360 (1931); **35**, 398 (1943).

the saturation temperature under an absolute pressure of 50 psi is 281° F. The steam is therefore superheated 69° F above its saturation temperature. The enthalpy will be the heat absorbed in heating 1 lb of liquid water from 32 to 281° F, vaporizing it to form saturated steam at this temperature, and heating the steam at constant pressure to a temperature of 350° F. The total enthalpy is the sum of the enthalpy of the liquid, the latent heat of vaporization, and the superheat of the vapors. The effect of pressure on the enthalpy of the liquid water is neglected.

The mean specific heat of water between 32 and 281° F is 1.006. The mean heat capacity of water vapor between 281 and 350° F at a pressure of 50 psi is 9.2 Btu per lb-mole per F°. The latent heat of vaporization of water at 281° F is 924.0 Btu per lb.

Enthalpy of liquid water at 281° F $= (281 - 32)\ 1.006 =$ 250.1 Btu per lb

Heat of vaporization at 281° F $=$ 924.0

Superheat of vapor $= (350 - 281)\ \dfrac{9.20}{18}$ $=$ 36.0

         Enthalpy $= \overline{1210.1}$ Btu per lb

Extensive steam tables have been compiled, giving the enthalpies and other properties of steam under widely varying conditions, for both saturated and superheated vapors. In calculating these tables, it is necessary to take into account the variation of the heat capacity with pressure, as discussed in Part II. Tables and charts of enthalpies have been worked out for a number of substances for which frequent thermal calculations are made in engineering practice.

Calculations of enthalpy often include several changes of state. For example, in calculating the enthalpy of zinc vapor at 1000° C and atmospheric pressure, relative to the solid at standard conditions, it is necessary to include the sensible enthalpy of the solid metal at the melting point, the latent heat of fusion, the heat absorbed in heating the liquid from the melting point to the normal boiling point, the latent heat of vaporization, and the heat absorbed in heating the zinc vapor from the boiling point up to 1000° C at constant pressure.

Illustration 6. Calculate the enthalpy of zinc vapor at 1000° C and atmospheric pressure, relative to the solid at 0° C. Zinc melts at 419° C and boils under atmospheric pressure at 907° C.

The mean heat capacities of the solid and liquid may be estimated from Fig. 63, page 260.

         Mean specific heat of solid, 0 to 419° C $= 0.105$

         Mean specific heat of liquid, 419 to 907° C $= 0.12$

From Table 24, page 272, the heat of fusion is 1595 cal per g-atom.

The heat of vaporization at the normal boiling point may be estimated from equation 28:

$$\lambda/1180 = 8.75 + 4.571 \log 1180 = 22.80$$
$$\lambda = 26,900 \text{ cal per g-mole}$$

Since zinc vapor is monatomic, its molal heat capacity at constant pressure is constant and equal to 4.97 cal per g-mole

$$\text{Heat absorbed by solid} = 0.105(419 - 0) \qquad = \quad 44 \text{ cal per gram}$$

$$\text{Heat of fusion} = \frac{1660}{65.4} \qquad\qquad\qquad = \quad 25 \text{ cal per gram}$$

$$\text{Heat absorbed by liquid} = 0.109(907 - 419) \quad = \quad 53 \text{ cal per gram}$$

$$\text{Heat of vaporization} = \frac{26,900}{65.4} \qquad\qquad = 412 \text{ cal per gram}$$

$$\text{Heat absorbed by vapor} = \frac{4.97}{65.4}(1000 - 907) \quad = \quad 7 \text{ cal per gram}$$

$$\text{Total enthalpy} \qquad\qquad\qquad\qquad = \overline{541} \text{ cal per gram}$$

Frequently it is difficult to determine experimentally the individual heats of transition involved in heating a substance. Under such conditions the enthalpy is measured directly and tabulated as such for various temperatures. For example, the enthalpy of steel at various temperatures is determined by cooling in a calorimeter from these initial temperatures. This determination includes all heats of transition undergone in the cooling process.

Enthalpy of Humid Air

The properties of humid air are conveniently expressed on the basis of the weight of humid air that contains either 1 lb or 1 lb-mole of moisture-free air. The enthalpy of the quantity of humid air containing a unit quantity of moisture-free air is the sum of the sensible enthalpy of the dry air and that of the water vapor which is associated with it. The reference states ordinarily chosen are air and liquid water at 0° C. The water vapor in the air may be considered as derived from liquid water at 0° C by the following series of processes:

1. The liquid water is heated to the dew point of the humid air.
2. The water is vaporized at the dew-point temperature to form saturated vapor.
3. The water vapor is superheated to the dry-bulb temperature of the air.

The enthalpy of the water is the sum of the heat absorbed by the liquid, the heat of vaporization at the dew point, and the superheat of the vapor.

Illustration 7. Calculate the enthalpy, per pound of dry air, of air at a pressure of 1 atm and a temperature of 100° F with a percentage humidity of 50.

Solution: From the humidity chart (Fig. 20, page 122) it is seen that air under these conditions contains 0.0345 lb-mole of water per mole of dry air or 0.0345(18)/29 = 0.0215 lb of water per pound of dry air. This corresponds to a dew point of 79° F.

From Fig. 62, the mean molal heat capacity of water vapor between 79 and 100° F is 8.02 and that of air between 32 and 100° F is 6.95.

The heat of vaporization at 79° F may be estimated from Fig. 19 as 18,840 Btu per lb-mole or 1046 Btu per lb.

$$\text{Sensible enthalpy of air} = (100 - 32)\frac{6.95}{29.0} \qquad = 16.3 \text{ Btu}$$

$$\text{Sensible enthalpy of liquid water} = (79 - 32)0.0215 \quad = 1.0$$

$$\text{Latent heat of water} = 1046 \times 0.0215 \qquad\qquad = 22.5$$

$$\text{Superheat of water vapor} = (100 - 79) \times 0.0215 \times \frac{8.02}{18} = 0.2$$

$$\text{Total enthalpy} \qquad\qquad\qquad\qquad = \overline{40.0} \text{ Btu}$$
$$\text{per lb of dry air}$$

Humid Heat Capacity of Air. In dealing with humid air it is convenient to use 1 lb or 1 lb-mole of dry air as the basis of calculations, regardless of the humidity of the air. In problems dealing with the heating or cooling of air where no change in moisture content takes place, the total change in enthalpy is equal to the sum of the change in the sensible enthalpy of the dry air and the change in sensible enthalpy of the water vapor. For example, in heating 1 lb of dry air associated with H lb of water vapor from t_1 to t_2° F, the total heat q required is given by the equation

$$q = C_{pa}(t_2 - t_1) + H(C_{pw})(t_2 - t_1) \tag{41}$$

where

C_{pa} = the mean heat capacity of air at constant pressure

C_{pw} = the mean heat capacity of water vapor at constant pressure

Instead of considering the air and water vapor separately, it is convenient to employ a heat-capacity term which combines the two. Thus,

$$q = S(t_2 - t_1) \tag{42}$$

where S = heat capacity of 1 lb of dry air and of the water associated with it, expressed in Btu per pound of dry air per degree Fahrenheit. Combining equations 41 and 42 gives

$$S = C_{pa} + HC_{pw} \tag{43}$$

The combined heat capacity S is termed the *humid heat capacity* of the air. Over the low-temperature range from 30 to 180° F, the mean heat capacity of dry air is 0.240 Btu per lb and that of water vapor is 0.446 Btu per lb, from Fig. 62. Accordingly the humid heat capacity of air when expressed in Btu per pound of air per degree Fahrenheit is given by the equation

$$S = 0.240 + 0.446H \tag{43a}$$

Adiabatic Humidification. In the discussion of the humidity chart (Chapter 5, page 126) it was pointed out that for the water vapor–air system the wet-bulb temperature lines and lines of adiabatic cooling are nearly identical. This identity is fortuitous for the water vapor–air system only but not valid for other known systems. The identity of the two lines even for the air–water vapor system begins to break down at temperatures above 150° F. The derivation of the adiabatic cooling lines for the air–water vapor system follows:

When air is cooled by the adiabatic vaporization of water into it, sensible heat is derived from the humid air to supply the heat necessary in vaporizing the water at the wet-bulb temperature and in heating the evolved vapor to the existing dry-bulb temperature. Since the total enthalpy of the system remains constant, the heat lost by the humid air must equal that gained by the water in vaporization and superheating. This equality may be expressed mathematically for the evaporation of dH moles of water into humid air containing 1 mole of dry air. Thus,

$$dH[\lambda + c_{pw}(t - t_w)] = -S\,dt \tag{44}$$

where H = molal humidity
$\quad \lambda$ = molal heat of vaporization at temperature t_w
$\quad c_{pw}$ = mean molal heat capacity of water vapor
$\quad t$ = dry-bulb temperature
$\quad t_w$ = temperature of adiabatic evaporation
$\quad S$ = mean molal humid heat capacity of air

Assuming that the wet-bulb temperature remains constant, as humidification proceeds the final dry-bulb temperature reached by the entire weight of air will be the wet-bulb temperature t_w, corresponding to saturation and a humidity H_w. In the temperature range from 32 to 200° F the molal heat capacities of air and water vapor may be taken from Fig. 60 as constant at 6.95 and 8.04, respectively. Then, from equation 43, $S = 6.95 + 8.04H$. Substituting these values in equation 44 and rearranging gives

$$\frac{dH}{6.95 + 8.04H} = -\frac{dt}{\lambda + 8.04(t - t_w)} \tag{45}$$

Integrating between the limits H, t and H_w, t_w,

$$\frac{1}{8.04}\ln\frac{6.95 + 8.04H}{6.95 + 8.04H_w} = \frac{1}{8.04}\ln\frac{\lambda}{\lambda + 8.04(t - t_w)}$$

or $\quad 6.95\lambda + 8.04\lambda H + 8.04(t - t_w)(6.95 + 8.04H) = 6.95\lambda + 8.04\lambda H_w$

or

$$t = \frac{(H_w - H)\lambda}{6.95 + 8.04H} + t_w \tag{46}$$

The temperature t_w of adiabatic evaporation corresponds to the experimental value of wet-bulb temperature, provided evaporation from the wet-bulb thermometer proceeds adiabatically, that is, with no gain or loss of heat by radiation, and also provided the actual vapor-pressure equilibrium is established at the liquid-air interface. The first condition is realized where the air is passed rapidly over the wet-bulb thermometer so that radiation errors become negligible; the second condition is true where the rate of evaporation keeps pace with the rate of heat transfer by convection. This latter condition is realized without appreciable error at temperatures below 150° F. *The lines of adiabatic evaporation are therefore commonly referred to as wet-bulb temperature lines.*

The adiabatic-cooling lines of Fig. 19 were constructed from equation 46. Corresponding to a selected value of t_w, values of dry-bulb temperature were calculated to correspond to various humidities, thus establishing a complete curve. The adiabatic cooling temperature lines of Fig. 19 which apply to gases of appreciable carbon dioxide content were constructed from a similar equation in which the effect of the carbon dioxide on the humid heat capacity of the gas was considered. The molal heat capacity of carbon dioxide may be assumed to be 9.3 (Fig. 62). Then,

$$S = 6.95(1 - x) + 9.3x + 8.04H \tag{47}$$

where x = mole fraction of CO_2 in the dry gas. With this modification equation 46 becomes

$$t = \frac{(H_w - H)\lambda}{6.95(1 - x) + 9.3x + 8.04H} + t_w \tag{48}$$

This equation permits calculations of adiabatic-cooling lines to apply to combustion gases or other mixtures containing appreciable amounts of carbon dioxide.

In using Figs. 19 and 20 it is assumed that the wet-bulb temperature lines coincide with the adiabatic-cooling lines even in the presence of carbon dioxide.

Enthalpy of Humid Air. The ordinary psychrometric chart is limited to direct use at an atmospheric pressure of 29.92 in. of mercury. For other pressures different sets of wet-bulb temperature and percentage (or relative) humidity lines are required. Goodman[13] has designed a psychrometric chart which covers a range of atmospheric pressures from 22 to 32 in. of mercury as shown in Fig. 73. In this chart lines of constant enthalpy have been constructed instead of the usual wet-bulb temperature lines, where enthalpy of the humid air is expressed on the basis of one pound of dry air. The horizontal lines in this chart represent absolute humidities, the diagonal lines constant enthalpies, the nearly vertical lines dry-bulb temperatures, and the curved lines humidities at saturation for various constant atmospheric pressures. The temperature lines are given a slight slope to allow

[13] W. Goodman, *Air Conditioning Analysis*, Macmillan Co. (1943), with permission.

for the increase in heat capacity of air and water vapor with temperature and thus avoid curvature in the constant enthalpy lines.

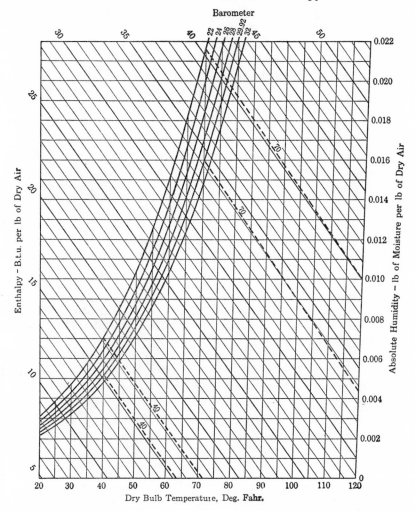

Fig. 73.   Enthalpy chart for water vapor–air system (From W. Goodman, *Air Conditioning Analysis*, Macmillan, 1943, with permission)

Reference states: Dry air at 0° F, liquid water at 32° F

(Reproduced in *CPP Charts*)

In adiabatic evaporation the wet-bulb temperature is constant, and the system including the humid air and the water to be evaporated is hence at constant enthalpy. At low temperatures the enthalpy of the liquid water to be evaporated is negligible, so that the enthalpy of the humid air is nearly equal to that of the system, and hence the slope of the constant-enthalpy lines is nearly equal to that of the constant wet-bulb lines. The constant-enthalpy lines instead of wet-bulb termperature lines have the advantage that the humidity lines are nearly independent of atmospheric pressure and may be used directly in establishing heat requirements in air-conditioning problems. In Fig. 73 wet-bulb temperature lines have been constructed for 70 and for 40° F. It will be observed that the wet-bulb temperature lines have nearly the same slope as the enthalpy lines, and become more nearly the same as the temperature is lowered. It will also be observed that the location of a given wet-bulb temperature line depends on the atmospheric pressure. The slopes of these particular lines were obtained from equation 46; other lines can be drawn in similarly, or the slopes may be estimated from the existing 40 and 70° F wet-bulb temperature lines.

Problems

1. (a) From the data of Table 17, page 255, calculate the mean heat capacity of one of the following gases: oxygen, hydrogen, water vapor, sulfur dioxide, ammonia:

    (1) In kilocalories per kilogram per degree centigrade from 0 to $t°$ C.
    (2) In Chu per pound per degree centigrade from 0 to $t°$ C.
    (3) In kcal per standard cubic meter per degree centigrade from 0 to $t°$ C.
    (4) In kcal per kilogram per degree centigrade from 1000 to 2000° C.
    (5) In Btu per pound-mole per degree Fahrenheit from 32 to $t°$ F.
    (6) In Btu per standard cubic foot per degree Fahrenheit from 32 to $t°$ F.
    (7) In Btu per pound per degree Fahrenheit from 1000 to 2000° F.

    (b) Calculate the heat capacity of the assigned gas in kcal per kilogram per degree centigrade at 1500° C.

2. From the experimental data for the molal heat capacities of oxygen at various temperatures, derive the constants in the following types of empirical equations over the temperature range from 0 to 1500° C:

$$c_p = a + bT + cT^2$$

$$c_p = a + \frac{b}{T} + \frac{c}{T^2}$$

$$c_p = a + \frac{b}{\sqrt{T}} + \frac{c}{T}$$

3. Calculate the amount of heat given off when 1 cubic meter of air (standard conditions) cools from 500 to −100° C at a constant pressure of 1 atm, assuming the heat-capacity formula of Table 17, page 255, to be valid over this temperature range. **Ans.** 195.2 kcal.

4. Calculate the number of kilocalories required to heat, from 500 to 1500° C, 1 cubic meter (standard conditions) of a gas having the following composition by volume:

CO_2	70%
N_2	27
O_2	2
H_2	1
	100%

5. Calculate the number of Btu required to heat 1 lb each of the following liquids from a temperature of 32 to 100° F:

(a) Acetone. **Ans.** 35.4 Btu.
(b) Carbon tetrachloride. **Ans.** 13.5 Btu.
(c) Ether. **Ans.** 37.3 Btu.
(d) Propyl alcohol. **Ans.** 38.7 Btu.

6. Calculate the number of calories required to heat 1000 grams of each of the following aqueous solutions from 0 to 100° C.

(a) 5% NaCl by weight.
(b) 20% NaCl by weight.
(c) 20% H_2SO_4 by weight.
(d) 20% KOH by weight.
(e) 20% NH_4OH by weight.
(f) 20% Pb $(NO_3)_2$ by weight.

7. From Fig. 63 determine the heat required to raise 1 lb of graphite from 32 to 1450° F. Show how the graph is utilized to determine the quantity of heat required. **Ans.** 482 Btu.

8. Calculate the specific heat at 20° C of ZnO, Al_2O_3, SnO_2, FeO, As_2O_3 from Kopp's rule, and compare with the experimental values.

9. Calculate the heat equivalent in Btu of the external work of vaporization of 1 lb of water at a temperature of 80° F, assuming that water vapor follows the ideal-gas law. **Ans.** 59.1 Btu.

10. Calculate the heat of vaporization in Btu per pound of carbon tetrachloride at its normal boiling temperature by the following methods:

(a) From the equation of Kistyakowsky 28.
(b) From equation 32.
(c) From equation 39.

11. Obtaining the necessary boiling-point and critical data from Fig. 15, estimate the heat of vaporization, in Btu per pound of n-butane (C_4H_{10}) at a pressure of 200 psi.

(a) By equations 28 and 40. **Ans.** 118.7 Btu per lb.
(b) By equation 39 and Tables 27 and 28. **Ans.** 120.9 Btu per lb.

12. Using Fig. 15 for the necessary boiling-point and critical data, estimate the heat of vaporization, in Btu per pound of carbon disulfide (CS_2) when under a pressure of 77.5 psi, using the two methods of problem 11.

13. Cyclohexane (C_6H_{12}) has a normal boiling point of 80.8° C. Estimate:

(a) The critical temperature. **Ans.** 281° C.

(b) The heat of vaporization at 10 atm pressure, expressed as Btu per pound. **Ans.** 117.2 Btu per lb.

14. Utilizing the thermal data for diphenyl, ($C_6H_5 \cdot C_6H_5$) tabulated below and vapor pressure data from a handbook, estimate the following:

(a) Critical temperature.

(b) Boiling point at 25 psi.

(c) Heat of vaporization at 25 psi, as Btu per pound.

(d) Enthalpy of 1 lb of saturated diphenyl vapor at 25 psi relative to solid diphenyl at 32° F.

Data for diphenyl

Normal boiling point	255° C
Melting point	71° C
Specific heat of solid diphenyl	0.385 Btu per lb per °F
Heat of fusion	46.9 Btu per lb
Specific heat of liquid diphenyl	$C_p = 0.300 + 0.00120t°$ C

15. Calculate the heat of vaporization, in calories per gram, of water at a temperature of 100° C by means of the Clapeyron equation. At this temperature $dp/dt = 27.17$, where p is the vapor pressure in millimeters of mercury and t is the temperature in degrees centigrade. **Ans.** 548 cal per gram, using equation 4, page 79; 538 cal per gram, using equation 3, page 78, and specific volume data from steam tables.

16. The vapor pressure of zinc in the range from 600 to 985° C is given by the equation

$$\log p = -\frac{6160}{T} + 8.10$$

where p = vapor pressure, millimeters of mercury
T = temperature, degrees Kelvin

Estimate the heat of vaporization of zinc at 907° C, the normal boiling point. Compare this result with that calculated from equation 28.

17. In a cracking operation a petroleum refinery produces 500,000 cu ft per hr of volatile gases analyzing by volume

H_2	10.0%
CH_4	40.0
C_2H_6	28.0
C_3H_8	20.0
C_3H_6	2.0
	100 %

measured at 80 psi absolute pressure and 400° F. In order to cool this stream to 90° F, before introduction into an absorption tower, the gases are passed countercurrently through a heat exchanger in which crude petroleum feed possessing a specific heat of 0.80 Btu/(lb)(° F), a specific gravity of 0.85, and an initial temperature of 70° F is used as the cooling medium. The crude petroleum leaves the heat exchanger at 300° F. Calculate:

(a) Average molecular weight of the gas mixture.

(b) Specific volume of the gas mixture at 80 psi absolute pressure and 400° F.

(c) Weight of gas circulated through the heat exchanger in pounds per hour.

(d) Mean molal heat capacity of the gas mixture at constant pressure.

(e) Weight of crude petroleum circulated through the heat exchanger in pounds per hour.

(f) Volume of crude petroleum circulated through the heat exchanger in cubic feet per hour.

(g) Number of pounds per hour of C_3H_8 that can be removed in the absorption tower if 85% of the total C_3H_8 is removed.

(h) Volume analysis of the gases leaving the absorption tower on the assumption that the only gas removed is C_3H_8 and this only to the extent of 85% of the total amount present in the gas stream.

18. Calculate the enthalpy in kilocalories per kilogram, referred to the solid at 0° C, of molten zinc at a temperature of 750° C.

19. Obtaining the latent-heat data from the steam tables, calculate the enthalpy in Btu per pound, relative to the liquid at 32° F, of steam at a temperature of 500° F superheated 200° F above its saturation point. **Ans.** 1272.3 Btu per lb.

20. Calculate the enthalpy in Btu per pound, relative to 32° F, of pure molten iron at a temperature of 3000° F. When iron is heated from 32° F to its melting point, it undergoes three transformations, from α to β, from β to γ, and from γ to δ forms.

21. Using the latent-heat data calculated in problem 14, calculate the enthalpy in Btu per pound relative to the solid at 32° F, of saturated diphenyl vapors under a pressure of 40 psi. **Ans.** 439 Btu per lb.

22. Calculate the enthalpy in Btu per pound of dry air, relative to air and liquid water at 32° F, of humid air at a temperature of 120° F, a pressure of 1 atm, and a percentage humidity of 60.

23. Humid air at a pressure of 1 atm has a dry-bulb temperature of 180° F and a wet-bulb temperature of 120° F. This air is cooled to a dry-bulb temperature of 115° F. Calculate the heat evolved, in Btu per pound of dry air. **Ans.** 17.4 Btu per lb.

24. Hot gases are passing through a chimney at a rate of 10,000 cu ft per min, measured at the existing conditions of 800° C and a pressure of 740 mm Hg. The gases have the following composition by volume on the dry basis:

CO_2	14%
N_2	80
O_2	6
	100%

The dew point of the gases is 50° C, and they contain 20 grams of carbon soot per cubic meter measured at the chimney conditions. Calculate the enthalpy of the material passing through the chimney per minute in Btu relative to gases, solid carbon, and liquid water at 25° C.

25. The gas feeder main to a small petroleum refinery delivers 200,000 cu ft per hr of gases at 400° F and 18 psi absolute pressure to a gas-treatment plant. The gas analyzes as follows by volume:

H_2	5.0%
CH_4	50.0
C_2H_6	20.0
C_3H_8	15.0
C_4H_{10}	10.0
	100.0%

Before introduction of these gases into the debutanizing unit, they are compressed to 150 psi absolute pressure and are then cooled at constant pressure from 380° F (temperature after compression in water-cooled pumps) to 200° F in a counter-current-flow heat exchanger cooled by cold crude oil entering at 80 and leaving at 280° F; they are cooled still further to 100° F in a countercurrent-flow heat exchanger cooled by water entering at 60 and leaving at 160° F. The gases then pass through the plate-absorption unit (or debutanizing tower) where 90% of the C_4H_{10} and 80% of the C_3H_8 are dissolved in the absorption oil, and the residual gases, now at 100 psi absolute pressure and 130° F, are piped to the power plant of the refinery. Calculate:

(a) Number of pounds per hour of crude oil circulated through the first heat exchanger. Assume a specific heat of 0.85 Btu per lb per F° for crude oil.

(b) Number of cubic feet per hour of cooling water circulated through the second heat exchanger.

(c) Number of cubic feet per hour of debutanized or residual gas leaving debutanizing tower.

(d) Number of pounds per hour of C_3H_8 and C_4H_{10} absorbed in the debutanizing tower.

(e) Volume analysis of the gases piped to the power plant.

26. A thermal unit in a petroleum refinery is fed with a gas stream containing steam, propane and butane. The hourly feed rate is 50 lb steam, 800 lb propane and 1200 lb butane. Water and a liquid mixture of propane and butane in the foregoing proportions are passed into a vaporizer at 120° F and 300 psia. The liquid materials are vaporized and leave the vaporizer at 350° F and 300 psia. The heating medium in the vaporizer is a liquid asphalt stock having a specific heat of 0.60 Btu/(lb)(F°). The asphalt undergoes a temperature drop of 200 F° in passing through the vaporizer. The quantity of heat lost from the vaporizer to its surroundings amounts to 40% of the heat transmitted to the fluid being vaporized and heated. Compute the following:

(a) Volumetric analysis of the gas.

(b) Molecular weight of the gas.

(c) Volume in cubic feet of the gas leaving the vaporizer.

(d) Number of pounds per hour of asphalt stock circulated through the vaporizer.

9

Thermochemistry

All chemical reactions are accompanied by either an absorption or an evolution of energy, which usually manifests itself as heat. The science of thermochemistry deals with the changes of energy in chemical reactions.

As discussed in Chapter 8, the internal energy of a given substance is dependent on its temperature, pressure, and state of aggregation and is independent of the means by which this state was brought about. Likewise the change in internal energy ΔU of a system that results from any physical change or chemical reaction depends only on the initial and final state of the system and is independent of the path taken during the reaction. The total change in internal energy will be the same, whether or not energy is absorbed or evolved in the form of heat, radiant energy, electric energy, mechanical work, or other forms.

For a flow reaction proceeding with negligible changes in kinetic energy and potential energy and with no electrical work and no mechanical work beyond that required for flow, the heat added is equal to the increase in enthalpy of the system,

$$q = \Delta H \tag{1}$$

For a nonflow reaction proceeding at constant pressure the heat added is also equal to the gain in enthalpy,

$$q = \Delta H \tag{2}$$

For a nonflow reaction proceeding at constant volume, the heat added is equal to the gain in internal energy of the system,

$$q = \Delta U \tag{3}$$

Standard Heat of Reaction. The heat of a chemical reaction is the heat absorbed in the course of the reaction, or, in a more general sense it is equal to the change in enthalpy of the system for the reaction proceeding at constant pressure. This heat of reaction is dependent not only on the chemical nature of each reacting material and product but also on their physical states. For purposes of organizing thermochemical data it is necessary to define a standard heat of reaction which may be recorded as a characteristic property of the reaction and from which heats of reaction under other conditions may be calculated. The *standard heat of reaction* is defined as the *change in enthalpy resulting from the procedure of the reaction under a pressure of 1 atm, starting and ending with all materials at a constant temperature of 25° C.*

For example, 1 g-atom (65.38 grams) of zinc may be allowed to react with 2073 grams of 1.0 molal aqueous hydrochloric acid containing 2.0 g-moles of HCl. The reaction may be carried out in a calorimeter under atmospheric pressure with all reactants at an initial temperature of 25° C. During the course of the reaction the system will become heated, hydrogen gas will be evolved, and a 0.5 molal solution of zinc chloride will be formed. When the reaction is completed, the resultant solution and the hydrogen gas may be cooled to 25° C. If no evaporation of water takes place, it will be found that 34,900 cal will be evolved by the system. The net result of the reaction is the conversion of 2 moles of hydrochloric acid in aqueous solution into 1 mole of zinc chloride in aqueous solution and 1 mole of hydrogen gas at atmospheric pressure, all at a temperature of 25° C. The measured amount of heat absorbed represents the standard heat of reaction for this particular reaction, proceeding under atmospheric pressure in an aqueous solution of the specified concentration.

Exactly the same net result is obtained by allowing the above reaction to proceed in an electrolytic cell in which one electrode is zinc and the other platinum. An electric motor might be connected to the cell and be permitted to do work as the reaction proceeds. In this case the amount of heat evolved will be less than 34,900 cal by the heat equivalent of the electric energy produced by the cell. However, the heat of reaction is the same and is equal to the algebraic sum of the amounts of energy absorbed as heat and as electric energy.

Conventions and Symbols. As pointed out in the preceding section, the heat of reaction accompanying a chemical change is dependent on the physical state of each reactant and product, as well as on its chemical nature. For this reason, in order to define a heat of reaction it is necessary to specify completely the nature and state of each material involved. The following system of conventions and symbols, to be used in conjunction with the conventional chemical equation, is adopted for this purpose.

The formula of a substance appearing in an equation designates not only the nature of the substance but also the quantity that is involved in the reaction. Thus, H_2SO_4 indicates 1 mole of sulfuric acid, and $1\frac{1}{2}N_2$ indicates $1\frac{1}{2}$ moles of molecular nitrogen. All equations are written with the reactants on the left and the products on the right side.

The value of the heat of reaction accompanying an equation is the heat of reaction resulting from the procedure of the reaction from the left to the right of the equation as written. If the reaction proceeds in the reverse direction the heat of reaction is of opposite sign.

Unless otherwise specified it is assumed that each reactant or product is in its normal state of aggregation at a temperature of 25° C and a pressure of 1 atm.

The state of aggregation of a substance is indicated by a letter in parentheses following its chemical formula. Thus (g) indicates the gaseous state, (l) the liquid, and (s) the solid.

Additional information may accompany these letters in parentheses. Thus, S (rhombic) and C (diamond) indicate sulfur in the rhombic state and carbon as diamond, respectively, while S (monoclinic) and C (graphite) indicate monoclinic sulfur and solid graphitic carbon. For a gas the pressure may be specified. Thus, CO_2 $(g, 2 \text{ atm})$ indicates gaseous carbon dioxide under a pressure of 2 atm.

The concentration of a substance in aqueous solution is indicated by its molality (m), by the number of moles of solvent (n_1) per mole of solute, or by the mole fraction of the solute (N_2). Thus, $(m = 0.1)$ following a chemical formula indicates that the substance is in aqueous solution with a molality of 0.1. The symbol $(n_1 = 200)$ indicates an aqueous solution with 200 moles of water per mole of solute. The symbol $(N_2 = 0.55)$ indicates an aqueous solution in which the mole fraction of the solute is 0.55. If the aqueous solution is highly dilute, such that additional dilution produces no thermal effect, the symbol (aq) follows the formula of the solute.

The concentration of a substance in nonaqueous solution is indicated by the above symbols, accompanied in parentheses by the formula of the solvent. Thus, $(C_2H_6O, N_2 = 0.55)$ indicates that a substance is in alcoholic solution with a mole fraction of 0.55.

Ionic reactions are indicated in the usual manner, for example, H^+ and Ca^{++} for the positive hydrogen and calcium ions and Cl^- and SO_4^- for the negative chloride and sulfate ions, respectively.

As previously pointed out, positive values of q represent an absorption of heat by the system under consideration, that is, an increase in enthalpy of the system, $q = \Delta H$. Where the initial and final temperatures of the system are the same, a subscript may be used to designate this temperature; thus ΔH_{25} is the heat of reaction, or change in enthalpy, at 25° C and 1 atm pressure.

When heat is evolved in a reaction, corresponding to a decrease in enthalpy, the reaction is termed *exothermic;* when heat is absorbed the reaction is said to be *endothermic.*

With the aid of the above symbols, the states of components participating in a chemical reaction are indicated by the following:

$$Zn(s) + 2HCl(m = 1.0) = ZnCl_2(m = 0.5) + H_2(g, 1.0 \text{ atm})$$
$$\Delta H_{25} = -34,900 \text{ cal per g-mole}$$

This equation designates the changes occurring in the reaction described in the preceding section.

Heat of Formation.   The *heat of formation* of a chemical compound is a special case of the standard heat of a chemical reaction wherein the reactants are the necessary elements and the compound in question is the only product formed.   Heats of formation are always expressed with reference to a standard state.   The molal heat of formation of a compound represents, unless otherwise stated, the heat of reaction, ΔH_f, when 1 mole of the compound is formed from the elements in a reaction beginning and ending at 25° C and at a pressure of 1 atm, with the reacting elements originally in the states of aggregation which are stable at these conditions of temperature and pressure.   The heat of formation of a compound is positive when its formation from the elements is accompanied by an increase in enthalpy.   A compound whose heat of formation is negative is termed an *exothermic compound*. If the heat of formation is positive it is called an *endothermic compound*.

For example, the molal heat of formation of liquid water is $-68,317.4$ cal per g-mole.   This means that, when 2.016 grams of hydrogen gas combine with 16 grams of molecular oxygen at a temperature of 25° C and a pressure of 1 atm to form 18.016 grams of liquid water at the same temperature, the heat given off to the surroundings is 68,317.4 cal, and the enthalpy of the system is decreased by 68,317.4 cal.   It is obvious that this reaction will not proceed at a constant temperature but during its progress will be at a very high temperature, and the product formed will be temporarily in the vapor state.   However, upon cooling to 25° C this sensible and latent heat appearing temporarily in the system itself is evolved and included in the heat of formation.   If water vapor were the final product at 25° C, the heat of formation would be numerically less by an amount equal to the heat of vaporization of water at 25° C.   The heat of vaporization of water at 25° C is 10,519.5 cal per g-mole.   Therefore, the heat of formation of water vapor at 25° C is $-68,317.4 + 10,519.5 = -57,797.9$ cal per g-mole.

The basic thermochemical data of inorganic compounds are presented in terms of standard heats of formation.   In *Selected Values of Chemical Thermodynamic Properties*, National Bureau of Standards, are extensive tables giving the heats of formation of a great variety of inorganic compounds, both in pure states and in solutions of various concentrations.

TABLE 29. HEATS OF FORMATION AND SOLUTION

Reference Conditions: 25° C (298.16° K), 1 atm pressure, gaseous substances in ideal state.

$\Delta H°_f$ = standard heat of formation, kcal per g-mole
$\Delta H°_s$ = standard integral heat of solution, kcal per g-mole

Multiply values by 1000 to obtain g-cal per g-mole, or kcal per kg-mole.
Multiply values by 1800 to obtain Btu per lb-mole.

Source: *Selected Values of Chemical Thermodynamic Properties*, as of July 1, 1953, edited by D. D. Wagman, National Bureau of Standards.

Abbreviations

c = crystalline state
l = liquid state
g = gaseous state
dil = in dilute aqueous solution

∞ = infinite dilution
ppt = precipitated solid
$amorph$ = amorphous state

Compound	Formula	State	$\Delta H°_f$, Heat of Formation	Moles of Water	$\Delta H°_s$, Heat of Solution
Acetic acid	CH_3COOH	l	−116.4	∞	−0.343
Aluminum chloride	$AlCl_3$	c	−166.2	600	−79.3
Aluminum hydroxide	$Al(OH)_3$	$amorph$	−304.2		
Aluminum oxide	Al_2O_3	Corundum, c	−399.09		
Aluminum silicate	Al_2SiO_5	Sillimanite, c	−648.9		
Aluminum sulfate	$Al_2(SO_4)_3$	c	−820.98	∞	−76.12
Ammonia	NH_3	g	−11.04	∞	−8.28
Ammonia	NH_3	l	−16.06	∞	−3.26
Ammonium carbonate	$(NH_4)_2CO_3$	dil	−225.11		
Ammonium bicarbonate	$(NH_4)HCO_3$	c	−203.7	∞	+6.78
Ammonium chloride	NH_4Cl	c	−75.38	∞	+3.62
Ammonium hydroxide	NH_4OH	$in\ 1\ H_2O$	−87.64		
Ammonium nitrate	NH_4NO_3	c	−87.27	∞	+6.16
Ammonium oxalate	$(NH_4)_2C_2O_4$	c	−268.72	2100	+8.12
Ammonium sulfate	$(NH_4)_2SO_4$	c	−281.86	∞	+1.48
Ammonium acid sulfate	$(NH_4)HSO_4$	c	−244.83	800	−0.76
Antimony trioxide	Sb_2O_3	c	−168.4	∞	+1.9
Antimony pentoxide	Sb_2O_5	c	−234.4	∞	+8.0
Antimony sulfide	Sb_2S_3	c	−43.5		
Arsenic acid	H_3AsO_4	c	−215.2	∞	+0.4
Arsenic trioxide	As_2O_3	$monoclinic, c$	−156.4	∞	+6.7
Arsenic pentoxide	As_2O_5	c	−218.6	∞	−6.0
Arsine	AsH_3	g	41.0		
Barium acetate	$Ba(C_2H_3O_2)_2$	c	−355.1	400	−6.4
Barium carbonate	$BaCO_3$	c	−291.3		
Barium chlorate	$Ba(ClO_3)_2$	c	−181.7	∞	+6.1
Barium chloride	$BaCl_2$	c	−205.56	∞	−3.16
Barium chloride	$BaCl_2 \cdot 2H_2O$	c	−349.35	∞	+4.00
Barium hydroxide	$Ba(OH)_2$	c	−226.2	∞	−12.38
Barium oxide	BaO	c	−133.4	∞	−36.9
Barium peroxide	BaO_2	c	−150.5		
Barium silicate	$BaSiO_3$	c	−359.5		
Barium sulfate	$BaSO_4$	c	−350.2	∞	+4.63
Barium sulfide	BaS	c	−106.0	∞	−12.4
Bismuth oxide	Bi_2O_3	c	−137.9		
Boric acid	H_3BO_3	c	−260.2	∞	+5.0
Boron oxide	B_2O_3	c	−302.0	∞	−3.45
Bromine chloride	$BrCl$	g	+3.51		
Cadmium chloride	$CdCl_2$	c	−93.0	∞	−4.39
Cadmium oxide	CdO	c	−60.86		

TABLE 29—*Continued*

Compound	Formula	State	ΔH°_f, Heat of Formation	Moles of Water	ΔH°_s, Heat of Solution
Cadmium sulfate	$CdSO_4$	c	-221.36	∞	-12.84
Cadmium sulfide	CdS	c	-34.5		
Calcium acetate	$Ca(C_2H_3O_2)_2$	c	-355.0	∞	-7.5
Calcium aluminate	$CaO \cdot Al_2O_3$	$glass$	-545		
Calcium aluminate	$2CaO \cdot Al_2O_3$	$glass$	-695		
Calcium aluminate	$3CaO \cdot Al_2O_3$	$glass$	-848		
Calcium aluminum silicate	$3CaO \cdot Al_2O_3 \cdot 2SiO_2$	c	-1303		-7.5
Calcium aluminum silicate	$CaO \cdot Al_2O_3 \cdot 6SiO_2$	c	-1828		
Calcium carbide	CaC_2	c	-15.0		
Calcium carbonate	$CaCO_3$	Calcite, c	-288.45		
Calcium chloride	$CaCl_2$	c	-190.0	∞	-19.82
Calcium chloride	$CaCl_2 \cdot 6H_2O$	c	-623.15	∞	$+3.43$
Calcium fluoride	CaF_2	c	-290.3		
Calcium hydroxide	$Ca(OH)_2$	c	-235.80	∞	-3.88
Calcium nitrate	$Ca(NO_3)_2$	c	-224.0	∞	-4.51
Calcium oxalate	$CaC_2O_4 \cdot H_2O$	c	-399.1		
Calcium oxide	CaO	c	-151.9	∞	-19.46
Calcium phosphate	$Ca_3(PO_4)_2$	c, α	-986.2		
Calcium silicate	$CaSiO_3$	(Wallastonite, c)	-378.6		
Calcium silicate	Ca_2SiO_4	c, β	-538.0		
Calcium sulfate	$CaSO_4$	Anhydrite, c	-342.42	∞	-4.25
Calcium sulfide	CaS	c	-115.3	∞	-4.5
Carbon graphite	C	c	0		
Diamond	C	c	$+0.4532$		
Amorphous (in coke)	C	$amorph$	$+2.6$		
Carbon monoxide	CO	g	-26.4157		
Carbon dioxide	CO_2	g	-94.0518	∞	-4.64
Carbon disulfide	CS_2	g	$+27.55$		
Carbon disulfide	CS_2	l	$+21.0$		
Carbon tetrachloride	CCl_4	g	-25.50		
Carbon tetrachloride	CCl_4	l	-33.34		
Chloric acid	$HClO_3$	dil	-23.50		
Chromium chloride	$CrCl_3$	c	-134.6		
Chromium chloride	$CrCl_2$	c	-94.56	∞	-18.64
Chromium oxide	Cr_2O_3	c	-269.7		
Chromium trioxide	CrO_3	c	-138.4	80	-2.5
Cobalt oxide	CoO	c	-57.2		
Cobalt oxide	Co_3O_4	c	-210		
Cobalt chloride	$CoCl_2$	c	-77.8		
Cobalt sulfide	CoS	ppt	-21.4		
Copper acetate	$Cu(C_2H_3O_2)_2$	c	-213.2	∞	-2.7
Copper carbonate	$CuCO_3$	c	-142.2		
Copper chloride	$CuCl_2$	c	-52.3	in aq. HCl	-6.3
Copper chloride	$CuCl$	c	-32.5		
Copper nitrate	$Cu(NO_3)_2$	c	-73.4	800	-10.4
Copper oxide	CuO	c	-37.1		
Copper oxide	Cu_2O	c	-39.84		
Copper sulfate	$CuSO_4$	c	-184.00	∞	-17.51
Copper sulfide	CuS	c	-11.6		
Copper sulfide	Cu_2S	c	-19.0		
Cyanogen	C_2N_2	g	$+73.60$		
Hydrobromic acid	HBr	g	-8.66	∞	-20.24
Hydrochloric acid	HCl	g	-22.063	∞	-17.960
Hydrocyanic acid	HCN	g	$+31.2$	∞	-6.0
Hydrofluoric acid	HF	g	-64.2	∞	-14.46

TABLE 29–*Continued*

Compound	Formula	State	ΔH°_f, Heat of Formation	Moles of Water	ΔH°_s, Heat of Solution
Hydriodic acid	HI	*g*	+6.20	∞	**−19.57**
Hydrogen oxide	H_2O	*g*	−57.7979		
Hydrogen oxide	H_2O	*l*	−68.3174		
Hydrogen oxide (heavy water)	D_2O	*g*	−59.5628		
Hydrogen oxide (heavy water)	D_2O	*l*	−70.4133		
Hydrogen peroxide	H_2O_2	*l*	−44.84	∞	−0.84
Hydrogen sulfide	H_2S	*g*	−4.815	∞	−4.58
Iron acetate	$Fe(C_2H_3O_2)_3$	*in 1800* H_2O	−353.8		
Iron carbide	Fe_3C	*c*	+5.0		
Iron carbonate	$FeCO_3$	*c*	−178.70		
Iron chloride	$FeCl_2$	*c*	−81.5	∞	−19.5
Iron chloride	$FeCl_3$	*c*	−96.8	∞	−31.1
Iron hydroxide	$Fe(OH)_2$	*c*	−135.8		
Iron hydroxide	$Fe(OH)_3$	*c*	−197.0		
Iron nitride	Fe_4N	*c*	−2.55		
Iron oxide	FeO	*c*	−64.3		
Iron oxide	$Fe_{0.95}O$	Wustite, *c*	−63.7		
Iron oxide	Fe_3O_4	*c*	−267.0		
Iron oxide	Fe_2O_3	*c*	−196.5		
Iron silicate	$FeO \cdot SiO_2$	*c*	−276		
Iron silicate	$2FeO \cdot SiO_2$	*c*	−343.7		
Iron sulfate	$Fe_2(SO_4)_3$	*in 3000* H_2O	−653.62		
Iron sulfate	$FeSO_4$	*c*	−220.5	200	−15.5
Iron sulfide	FeS	*c*	−22.72		
Iron sulfide	FeS_2	Pyrites, *c*	−42.52		
Lead acetate	$Pb(C_2H_3O_2)_2$	*c*	−230.5	∞	−2.1
Lead carbonate	$PbCO_3$	*c*	−167.3		
Lead chloride	$PbCl_2$	*c*	−85.85	∞	+6.20
Lead nitrate	$PbNO_3$	*c*	−107.35	∞	+9.00
Lead oxide (yellow)	PbO	*c*	−52.07		
Lead peroxide	PbO_2	*c*	−66.12		
Lead suboxide	Pb_2O	*c*	−51.2		
Lead sesquioxide	Pb_3O_4	*c*	−175.6		
Lead sulfate	$PbSO_4$	*c*	−219.50		
Lead sulfide	PbS	*c*	−22.54		
Lithium chloride	LiCl	*c*	−97.70	∞	−8.877
Lithium hydroxide	LiOH	*c*	−116.45	∞	−5.061
Magnesium carbonate	$MgCO_3$	*c*	−266		
Magnesium chloride	$MgCl_2$	*c*	−153.40	∞	−37.06
Magnesium hydroxide	$Mg(OH)_2$	*c*	−221.00		
Magnesium oxide	MgO	*c*	−143.84		
Magnesium silicate	$MgSiO_3$	*c*	−357.9		
Magnesium sulfate	$MgSO_4$	*c*	−305.5	∞	−21.81
Manganese carbonate	$MnCO_3$	*c*	−213.9		
Manganese carbide	Mn_3C	*c*	−1		
Manganese chloride	$MnCl_2$	*c*	−115.3	400	−16.7
Manganese oxide	MnO	*c*	−92.0		
Manganese oxide	Mn_3O_4	*c*	−331.4		
Manganese oxide	Mn_2O_3	*c*	−232.1		
Manganese dioxide	MnO_2	*c*	−124.5		
Manganese dioxide	MnO_2	*amorph*	−117.0		
Manganese silicate	$MnO \cdot SiO_2$	*c*	−302.5		
Manganese silicate	$MnO \cdot SiO_2$	*glass*	−294.0		
Manganese sulfate	$MnSO_4$	*c*	−254.24	∞	−14.96
Manganese sulfide	MnS	*c*	−48.8		

TABLE 29–*Continued*

Compound	Formula	State	$\Delta H°_f$, Heat of Formation	Moles of Water	$\Delta H°_s$, Heat of Solution
Mercury acetate	Hg(C$_2$H$_3$O$_2$)$_2$	c	-199.4	∞	$+4.2$
Mercury bromide	HgBr$_2$	c	-40.5	∞	$+3.4$
Mercury chloride	HgCl$_2$	c	-55.0	∞	$+3.2$
Mercury chloride	Hg$_2$Cl$_2$	c	-63.32		
Mercury nitrate	Hg(NO$_3$)$_2$	dil	-58.0		
Mercury nitrate	Hg$_2$(NO$_3$)$_2$·2H$_2$O	c	-206.9		
Mercury oxide	HgO	red, c	-21.68		
Mercury oxide	HgO	yellow, c	-21.56		
Mercury oxide	Hg$_2$O	c	-21.8		
Mercury sulfate	HgSO$_4$	c	-168.3		
Mercury sulfate	Hg$_2$SO$_4$	c	-177.34		
Mercury sulfide	HgS	Cinnabar, c	-13.90		
Mercury thiocyanate	Hg(CNS)$_2$	c	$+48.0$		
Molybdenum oxide	MoO$_2$	c	-130		
Molybdenum oxide	MoO$_3$	c	-180.33	∞	-7.77
Molybdenum sulfide	MoS$_2$	c	-55.5		
Nickel chloride	NiCl$_2$	c	-75.5	10,000	-19.63
Nickel cyanide	Ni(CN)$_2$	c	$+27.1$		
Nickel hydroxide	Ni(OH)$_3$	c	-162.1		
Nickel hydroxide	Ni(OH)$_2$	c	-128.6		
Nickel oxide	NiO	c	-58.4		
Nickel sulfide	NiS	c	-17.5		
Nickel sulfate	NiSO$_4$	c	-213.0	∞	-19.2
Nitrogen oxide	NO	g	$+21.600$		
Nitrogen oxide	N$_2$O	g	$+19.49$		
Nitrogen oxide	NO$_2$	g	$+8.091$		
Nitrogen pentoxide	N$_2$O$_5$	g	$+3.6$	∞	-34.03
Nitrogen pentoxide	N$_2$O$_5$	c	-10.0	∞	-20.43
Nitrogen tetroxide	N$_2$O$_4$	g	$+2.309$		
Nitrogen trioxide	N$_2$O$_3$	g	$+20.0$		
Nitric acid	HNO$_3$	l	-41.404	∞	-7.968
Oxalic acid	H$_2$C$_2$O$_4$·2H$_2$O	c	-340.9	2100	$+8.70$
Oxalic acid	H$_2$C$_2$O$_4$	c	-197.6	2100	$+2.03$
Perchloric acid	HClO$_4$	l	-11.1	∞	-20.31
Phosphoric acid (meta)	HPO$_3$	c	-228.2	∞	-6.7
Phosphoric acid (ortho)	H$_3$PO$_4$	c	-306.2	3000	-3.2
Phosphoric acid (pyro)	H$_4$P$_2$O$_7$	c	-538.0	∞	-7.9
Phosphorous acid (hypo)	H$_3$PO$_2$	l	-143.2	∞	-2.4
Phosphorous acid (ortho)	H$_3$PO$_3$	l	-229.1	∞	-3.1
Phosphorus trichloride	PCl$_3$	g	-73.22		
Phosphorus pentoxide	P$_2$O$_5$	c	-360.0		
Platinum chloride	PtCl$_4$	c	-62.9	∞	-19.5
Platinum chloride	PtCl	c	-17.7		
Potassium acetate	KC$_2$H$_3$O$_2$	c	-173.2	∞	-3.68
Potassium carbonate	K$_2$CO$_3$	c	-273.93	1000	-7.63
Potassium chlorate	KClO$_3$	c	-93.50	∞	$+9.96$
Potassium chloride	KCl	c	-104.175	∞	$+4.115$
Potassium chromate	K$_2$CrO$_4$	c	-330.49	∞	$+4.49$
Potassium cyanide	KCN	c	-26.90	∞	$+2.80$
Potassium dichromate	K$_2$Cr$_2$O$_7$	c	-485.90	2000	$+17.20$
Potassium fluoride	KF	c	-134.46	∞	-4.24
Potassium nitrate	KNO$_3$	c	-117.76	∞	$+8.35$
Potassium oxide	K$_2$O	c	-86.4	∞	-75.28
Potassium sulfate	K$_2$SO$_4$	c	-342.66	∞	$+5.68$
Potassium sulfide	K$_2$S	c	-100	400	-9.9
Potassium sulfite	K$_2$SO$_3$	c	-266.9	∞	-2.2
Potassium thiosulfate	K$_2$S$_2$O$_3$	aq	-274		

TABLE 29–*Continued*

Compound	Formula	State	$\Delta H°_f$, Heat of Formation	Moles of Water	$\Delta H°_s$, Heat of Solution
Potassium hydroxide	KOH	c	−101.7S	∞	−13.22
Potassium nitrate	KNO₃	c	−117.76	∞	+8.348
Potassium permanganate	KMnO₄	c	−194.4	4000	+10.5
Selenium oxide	SeO₂	c	−55.0	∞	+0.93
Silicon carbide	SiC	c	−26.7		
Silicon tetrachloride	SiCl₄	l	−153.0		
Silicon tetrachloride	SiCl₄	g	−145.7		
Silicon dioxide	SiO₂	Quartz, c	−205.4		
Silver bromide	AgBr	c	−23.78		
Silver chloride	AgCl	c	−30.362		
Silver nitrate	AgNO₃	c	−29.43	∞	+5.37
Silver sulfate	Ag₂SO₄	c	−170.50	∞	+4.50
Silver sulfide	Ag₂S	α, c	−7.60		
Sodium acetate	NaC₂H₃O₂	c	−169.8	∞	−4.322
Sodium arsenate	Na₃AsO₄	c	−365	500	−16.5
Sodium tetraborate	Na₂B₄O₇	c	−777.7	900	−10.2
Sodium borate	Na₂B₄O₇·10H₂O	c	−1497.2	900	+26.1
Sodium bromide	NaBr	c	−86.030	∞	−0.15
Sodium carbonate	Na₂CO₃	c	−270.3	400	−5.6
Sodium carbonate	Na₂CO₃·10H₂O	c	−975.6	400	+16.5
Sodium bicarbonate	NaHCO₃	c	−226.5	∞	+4.0
Sodium chlorate	NaClO₃	c	−85.73	∞	+4.95
Sodium chloride	NaCl	c	−98.232	∞	+0.930
Sodium cyanide	NaCN	c	−21.46	200	+0.26
Sodium fluoride	NaF	c	−136.0	∞	+0.06
Sodium hydroxide	NaOH	c	−101.99	∞	−10.246
Sodium iodide	NaI	c	−68.84	∞	−1.81
Sodium nitrate	NaNO₃	c	−101.54	∞	−5.111
Sodium oxalate	NaC₂O₄	c	−314.3	600	+3.8
Sodium oxide	Na₂O	c	−99.4	∞	−56.8
Sodium triphosphate	Na₃PO₄	c	−460	1000	−13.9
Sodium diphosphate	Na₂HPO₄	c	−417.4	1000	−6.04
Sodium monophosphate	NaH₂PO₄	in 300 H₂O	−367.7		
Sodium phosphite	Na₂HPO₃	c	−338	800	−9.5
Sodium selenate	Na₂SeO₄	c	−258	∞	−1.7
Sodium selenide	Na₂Se	c	−63.0	∞	−19.9
Sodium sulfate	Na₂SO₄	c	−330.90	∞	−0.56
Sodium sulfate	Na₂SO₄·10H₂O	c	−1033.48	∞	+18.85
Sodium bisulfate	NaHSO₄	c	−269.2	200	−1.4
Sodium sulfide	Na₂S	c	−89.2	800	−15.16
Sodium sulfide	Na₂S·4½H₂O	c	−416.9	800	+5.11
Sodium sulfite	Na₂SO₃	c	−260.6	∞	−3.2
Sodium bisulfite	NaHSO₃	dil	−206.6		
Sodium silicate	Na₂SiO₃	glass	−360		
Sodium silicofluoride	Na₂SiF₆	c	−677	600	+5.8
Sulfur dioxide	SO₂	g	−70.96	10,000	9.90
Sulfur trioxide	SO₃	g	−94.45	∞	−54.13
Sulfuric acid	H₂SO₄	l	−193.91	∞	−22.99
Tellurium oxide	TeO₂	c	−77.69	∞	+1.31
Tin chloride	SnCl₄	l	−130.3	aq HCl	−29.9
Tin chloride	SnCl₂	c	−83.6	aq HCl	−0.4
Tin oxide	SnO₂	c	−138.8		
Tin oxide	SnO	c	−68.4		
Titanium oxide	TiO₂	amorph	−207		
Titanium oxide	TiO₂	Rutile, c	−218.0		
Tungsten oxide	WO₂	c	−136.3		
Vanadium oxide	V₂O₅	c	−373		

TABLE 29–*Concluded*

Compound	Formula	State	$\Delta H^\circ{}_f$, Heat of Formation	Moles of Water	$\Delta H^\circ{}_s$, Heat of Solution
Zinc acetate	$Zn(C_2H_3O_2)_2$	c	−258.1	800	−9.8
Zinc bromide	$ZnBr_2$	c	−78.17	∞	−16.06
Zinc carbonate	$ZnCO_3$	c	−194.2		
Zinc chloride	$ZnCl_2$	c	−99.40	∞	−17.08
Zinc hydroxide	$Zn(OH)_2$	c	−153.5		
Zinc iodide	ZnI_2	c	−49.98	∞	−13.19
Zinc oxide	ZnO	c	−83.17		
Zinc sulfate	$ZnSO_4$	c	−233.88	∞	−19.45
Zinc sulfide	ZnS	c	−48.5		
Zirconium oxide	ZrO_2	c	−258.2		

In Table 29 are listed selected values of heats of formation. It will be observed in these tables that the heat of formation is made synonymous with increase in enthalpy. This results in a negative sign for all exothermic compounds. This practice is in agreement with the nomenclature of the American Standards Association which is here adopted. The opposite sign for heats of formation are to be found in some older references.

When a compound is hydrated, its heat of formation in Table 29 includes the heat of formation of the water forming the hydrate. For example, the heat of formation of solid $CaCl_2 \cdot 6H_2O$ is given as −623,150. This represents the heat of reaction accompanying the formation at 25° C of 1 mole of $CaCl_2 \cdot 6H_2O$ from solid calcium and gaseous chlorine, hydrogen, and oxygen.

In the sixth column of Table 29 are values of the heats of solution of a few compounds. The heat of solution represents the change in enthalpy resulting from the formation of a solution of the specified concentration from 1 g-mole of the compound and the number of gram-moles of liquid water indicated in the fifth column. By means of these values of the total heat of formation of a dissolved material may be calculated. For example, the heat of solution of 1 g-mole of $CaCl_2 \cdot 6H_2O$ in an infinite amount of water is +3430 cal. The heat of formation of the undissolved $CaCl_2 \cdot 6H_2O$ is −623,150 cal. The combined heat of formation and solution of 1 g-mole of $CaCl_2 \cdot 6H_2O$ in an infinite amount of water is the algebraic sum of these two values, −623,150 + 3430 = −619,720 cal.

If an element normally exists in more than one allotropic form at 25° C and atmospheric pressure, one of these forms is selected to serve as the basis of heats of formation throughout the table. This is equivalent to assigning a heat of formation of zero to the element in the particular form selected. For example, carbon may exist as graphite, diamond, or amorphous carbon. The β-graphite form has been selected as the basis of the heats of formation of all carbon compounds. This is indicated by assigning a value of zero to the heat of formation of β-graphite. On this basis all other forms of elementary carbon have positive heats of formation. The heat of formation, for example, of barium carbonate is the heat of reaction accompanying the formation of the compound from graphite and the other necessary elements. When more than one allotropic form of an element exists, the particular form on which the tables are based can be identified as the form for which the heat of formation is given as zero.

Laws of Thermochemistry. At a given temperature and pressure the quantity of energy required to decompose a chemical compound into its elements is precisely equal to that evolved in the formation of that compound from its elements. This principle was first formulated by Lavoisier and Laplace in 1780. For example, the heat of formation of sodium chloride is $-98,232$ cal. The same amount of energy is required to decompose sodium chloride into sodium and chlorine.

A corollary of this first principle of thermochemistry is known as the law of constant-heat summation, which states that the net heat evolved or absorbed in a chemical process is the same whether the reaction takes place in one or in several steps. The total change in enthalpy of a system is dependent on the temperature, pressure, state of aggregation, and state of combination at the beginning and at the end of the reaction and is independent of the number of intermediate chemical reactions involved. This principle is known as the law of Hess, formulated in 1840.

By means of this principle it is possible to calculate the heat of formation of a compound from a series of reactions not involving the direct formation of the compound from the elements. The majority of chemical compounds cannot be prepared in the pure state directly from the elements. For example, the heat of formation of carbon monoxide cannot be measured directly because it cannot be prepared in a pure state from the elements without the concomitant formation of carbon dioxide. However, pure carbon dioxide may be formed from its elements and the heat of reaction measured. Also, pure carbon monoxide may be oxidized to form carbon dioxide and the heat of this reaction measured.

Thus, at 25° C,

$$C(\beta) + O_2(g) = CO_2(g), \qquad \Delta H_f = -94,051.8 \text{ cal} \qquad (4)$$

$$CO(g) + \tfrac{1}{2}O_2(g) = CO_2(g), \qquad \Delta H_f = -67,636.1 \text{ cal} \qquad (5)$$

From the law of Hess, the heat of formation of carbon monoxide is the same as the net heat of reaction accompanying (a) the formation of carbon dioxide from the elements and (b) the decomposition of this carbon dioxide into carbon monoxide and oxygen. The first step is represented by equation 4 with a heat of reaction of $-94{,}051.8$. The second step is the reverse of the reaction of equation 5 with a heat of reaction of $+67{,}636.1$. The net heat of reaction of the two processes is $-94{,}051.8 + 67{,}636.1 = -26{,}415.7$ cal, the heat of formation of carbon monoxide.

The result of this application of the law of Hess is exactly the same as that obtained by treating equations 4 and 5 as algebraic equalities and combining them as such. If equation 5 is subtracted from equation 4,

$$\text{C (graphite)} + \tfrac{1}{2}O_2 = CO, \qquad \Delta H_f = -26{,}415.7 \text{ cal} \qquad (6)$$

All thermochemical equations may be treated in this manner and combined with each other according to the rules of algebra. In this way it is possible to use the principle of Hess effectively in calculating the heat of a reaction from a series of intermediate reactions. The heat of formation of any compound may be calculated if the heat of any one reaction into which it enters is known together with the heat of formation of each of the other compounds present in the reaction.

For example, it is practically impossible to measure directly the heats of formation of the hydrocarbons. However, a hydrocarbon may be oxidized completely to carbon dioxide and water and this heat of reaction measured. The heat of formation of the hydrocarbon may then be calculated from the heats of formation of the other compounds present in the reaction, namely, carbon dioxide and water. Thus,

$$CH_4(g) + 2O_2(g) = CO_2(g) + 2H_2O\,(l), \quad \Delta H_1 = -212{,}798 \text{ cal} \qquad (7)$$

$$C(\beta) + O_2(g) = CO_2(g), \qquad\qquad\qquad \Delta H_2 = -94{,}051.8 \text{ cal} \qquad (8)$$

$$2H_2(g) + O_2(g) = 2H_2O\,(l), \qquad\qquad \Delta H_3 = -136{,}634.8 \text{ cal} \qquad (9)$$

Equation 8 + equation 9 − equation 7 gives

$$C(\beta) + 2H_2(g) = CH_4(g) \qquad\qquad\qquad\qquad\qquad (10)$$

$$\Delta H_f = \Delta H_2 + \Delta H_3 - \Delta H_1 = -94{,}051.8 - 136{,}634.8 - (-212{,}798)$$
$$= -17{,}888.6 \text{ cal}$$

By the principle of constant-heat summation it is thus possible to calculate heats of reaction which cannot be determined by direct measurement. For example, the oxidation of linseed oil and the souring of milk are reactions that proceed very slowly. By measuring the heats of combustion of the initial reactants and final products, it is possible to calculate the desired heat of reaction. Similarly, the heat of transition of a compound or element from one allotropic form to another such as graphite to diamond may be impossible to measure directly. However, the difference between the heats of combustion of these two forms of carbon will give the desired heat of transition. Since this combustion method involves taking the difference between two large numbers, small errors in either number may result in a large error in the difference.

Standard Heat of Combustion. The heat of combustion of a substance is the heat of reaction resulting from the oxidation of the substance with molecular oxygen. Thermochemical data on organic compounds are ordinarily expressed in terms of the heats of combustion. These data are not necessarily the results of direct-combustion experiments but may be indirectly obtained from measurements of other heats of reaction which lead to greater accuracy. Calculated heats of combustion data are available for many substances which are noncombustible under ordinary circumstances, as, for example, carbon tetrachloride.

The assignment of negative values to heats of combustion is consistent with the use of changes of enthalpy as synonymous with heats of formation, heats of vaporization, and so forth. Since combustion proceeds with a reduction in the enthalpy of the system, the value of ΔH must be negative, and hence the heat of combustion is also negative.

The usually accepted *standard heat of combustion* is that resulting from the combustion of a substance, in the state that is normal at 25° C and atmospheric pressure, with the combustion beginning and ending at a temperature of 25° C. The data ordinarily presented for standard heats of combustion correspond to final products of combustion that are in their normal states at a temperature of 25° C and at a pressure of 1 atm. The major final products are generally gaseous carbon dioxide and liquid water.

The standard heat of combustion of a substance is dependent on the extent to which oxidation is carried. Unless otherwise specified, a value of standard heat of combustion corresponds to complete oxidation of all carbon to carbon dioxide and of all hydrogen to liquid water. Where other oxidizable elements are present, it is necessary to specify the extent to which the oxidation of each is carried in designating a heat of combustion. For example, in oxidizing an organic chlorine compound, either gaseous chlorine or hydrochloric acid may be formed, depending on the conditions of combustion. If sulfur is present, its final form may be SO_2, SO_3, or the corresponding acids.

The situation is further complicated by the fact that such products may form aqueous solutions with the water. Standard heats of combustion of compounds that contain elements such as S, Cl, I, Br, N, and F must always be accompanied by complete specification of the final state of each product.

In Table 30 are a few selected values of standard heats of combus-

TABLE 30. STANDARD HEATS OF COMBUSTION

Reference conditions: 25° C (298.16° K), 1 atm pressure, gaseous
substances in ideal state

$\Delta H^°_c$ = standard heat of combustion, kcal per g-mole

Multiply values by 1000 to obtain g-cal per g-mole, or kcal per kg-mole.
Multiply values by 1800 to obtain Btu per lb-mole.

Abbreviations

s = solid l = liquid g = gaseous

Hydrocarbons

Final Products: $CO_2(g)$, $H_2O(l)$

Compound	Formula	State	$-\Delta H^°_c$
Carbon (graphite)	C	s	94.0518
Carbon monoxide	CO	g	67.6361
Hydrogen	H₂	g	68.3174
Methane	CH₄	g	212.798
Ethyne (acetylene)	C₂H₂	g	310.615
Ethene (ethylene)	C₂H₄	g	337.234
Ethane	C₂H₆	g	372.820
Propyne (allylene, methylacetylene)	C₃H₄	g	463.109
Propene (propylene)	C₃H₆	g	491.987
Propane	C₃H₈	g	530.605
1,2-Butadiene	C₄H₆	g	620.71
2-Methylpropene (isobutylene, isobutene)	C₄H₈	g	646.134
2-Methylpropane (isobutane)	C₄H₁₀	g	686.342
n-Butane	C₄H₁₀	g	687.982
1-Pentene (amylene)	C₅H₁₀	g	806.85
Cyclopentane	C₅H₁₀	l	786.54
2,2-Dimethylpropane (neopentane)	C₅H₁₂	g	840.49
2-Methylbutane (isopentane)	C₅H₁₂	g	843.24
n-Pentane	C₅H₁₂	g	845.16
Benzene	C₆H₆	g	789.08
Benzene	C₆H₆	l	780.98
1-Hexene (hexylene)	C₆H₁₂	g	964.26
Cyclohexane	C₆H₁₂	l	936.88
n-Hexane	C₆H₁₄	l	995.01
Methylbenzene (toluene)	C₇H₈	g	943.58
Methylbenzene (toluene)	C₇H₈	l	934.50
Cycloheptane	C₇H₁₄	l	1086.9
n-Heptane	C₇H₁₆	l	1151.27
1,2-Dimethylbenzene (o-xylene)	C₈H₁₀	g	1098.54
1,2-Dimethylbenzene (o-xylene)	C₈H₁₀	l	1088.16
1,3-Dimethylbenzene (m-xylene)	C₈H₁₀	g	1098.12
1,3-Dimethylbenzene (m-xylene)	C₈H₁₀	l	1087.92
1,4-Dimethylbenzene (p-xylene)	C₈H₁₀	g	1098.29
1,4-Dimethylbenzene (p-xylene)	C₈H₁₀	l	1088.16
n-Octane	C₈H₁₈	l	1307.53
1,3,5-Trimethylbenzene (mesitylene)	C₉H₁₂	l	1241.19
Naphthalene	C₁₀H₈	s	1231.6
n-Decane	C₁₀H₂₂	l	1620.06
Diphenyl	C₁₂H₁₀	s	1493.5
Anthracene	C₁₄H₁₀	s	1695
Phenanthrene	C₁₄H₁₀	s	1693
n-Hexadecane	C₁₆H₃₄	l	2557.64

Table 30–*Continued*

Alcohols

Final Products: $CO_2(g)$, $H_2O(l)$

Compound	Formula	State	$-\Delta H^\circ_c$
Methyl alcohol	CH_4O	g	182.59
Methyl alcohol	CH_4O	l	173.65
Ethyl alcohol	C_2H_6O	g	336.82
Ethyl alcohol	C_2H_6O	l	326.70
Ethylene glycol	$C_2H_6O_2$	l	284.48
Allyl alcohol	C_3H_6O	l	442.3
n-Propyl alcohol	C_3H_8O	g	494.26
n-Propyl alcohol	C_3H_8O	l	483.56
Isopropyl alcohol	C_3H_8O	g	493.02
Isopropyl alcohol	C_3H_8O	l	481.11
Glycerol	$C_3H_8O_3$	l	396.27
n-Butyl alcohol	$C_4H_{10}O$	g	649.98
n-Butyl alcohol	$C_4H_{10}O$	l	638.18
Amyl alcohol	$C_5H_{12}O$	l	786.7
Methyl-diethyl carbinol	$C_6H_{14}O$	l	926.9

Acids

Final Products: $CO_2(g)$, $H_2O(l)$

Compound	Formula	State	$-\Delta H^\circ_c$
Formic (monomolecular)	CH_2O_2	g	75.70
Formic	CH_2O_2	l	64.57
Oxalic	$C_2H_2O_4$	s	58.82
Acetic	$C_2H_4O_2$	g	219.82
Acetic	$C_2H_4O_2$	l	208.34
Acetic anhydride	$C_4H_6O_3$	g	432.34
Acetic anhydride	$C_4H_6O_3$	l	426.00
Glycolic	$C_2H_4O_3$	s	166.54
Propionic	$C_3H_6O_2$	g	378.36
Propionic	$C_3H_6O_2$	l	365.41
Lactic	$C_3H_6O_3$	s	325.8
d-Tartaric	$C_4H_6O_6$	s	274.9
n-Butyric	$C_4H_8O_2$	l	520
Citric (anhydrous)	$C_6H_8O_7$	s	474.3
Benzoic	$C_7H_6O_2$	s	771.5
o-Phthalic	$C_8H_6O_4$	s	770.8
Phthalic anhydride	$C_8H_4O_3$	s	781.4
o-Toluic	$C_8H_8O_2$	s	928.6
Palmitic	$C_{16}H_{32}O_2$	s	2379
Stearolic	$C_{18}H_{32}O_2$	s	2628
Elaidic	$C_{18}H_{34}O_2$	s	2663
Oleic	$C_{18}H_{34}O_2$	l	2668
Stearic	$C_{18}H_{36}O_2$	s	2697

Carbohydrates, Cellulose, Starch, etc.

Final Products: $CO_2(g)$, $H_2O(l)$

Compound	Formula	State	$-\Delta H^\circ_c$
d-Glucose (dextrose)	$C_6H_{12}O_6$	s	673
l-Fructose	$C_6H_{12}O_6$	s	675
Lactose (anhydrous)	$C_{12}H_{22}O_{11}$	s	1350.1
Sucrose	$C_{12}H_{22}O_{11}$	s	1348.9

	g-cal per gram
Starch	4177
Dextrin	4108
Cellulose	4179
Cellulose acetate	4495

TABLE 30–*Concluded*

Other CHO Compounds

Final Products: $CO_2(g)$, $H_2O(l)$

Compound	Formula	State	$-\Delta H^\circ_c$
Formaldehyde	CH_2O	g	134.67
Acetaldehyde	C_2H_4O	g	284.98
Acetone	C_3H_6O	g	435.32
Acetone	C_3H_6O	l	427.79
Methyl acetate	$C_3H_6O_2$	g	397.5
Ethyl acetate	$C_4H_8O_2$	g	547.46
Ethyl acetate	$C_4H_8O_2$	l	538.76
Diethyl ether	$C_4H_{10}O$	l	652.59
Diethyl ketone	$C_5H_{10}O$	l	738.05
Phenol	C_6H_6O	g	747.55
Phenol	C_6H_6O	l	731.46
Pyrogallol	$C_6H_6O_3$	s	639
Amyl acetate	$C_7H_{14}O_2$	l	1040
Camphor	$C_{10}H_{16}O$	s	1411

Nitrogen Compounds

Final Products: $CO_2(g)$, $N_2(g)$, $H_2O(l)$

Urea	CH_4N_2O	s	151.05
Cyanogen	C_2N_2	g	261.70
Trimethylamine	C_3H_9N	l	578.4
Pyridine	C_5H_5N	l	660
Trinitrobenzene (1,3,5)	$C_6H_3N_3O_6$	s	664.0
Trinitrophenol (2,4,6)	$C_6H_3N_3O_7$	s	620.0
o-Dinitrobenzene	$C_6H_4N_2O_4$	s	703.2
Nitrobenzene	$C_6H_5NO_2$	l	739
o-Nitrophenol	$C_6H_5NO_3$	s	689
o-Nitroaniline	$C_6H_6N_2O_2$	s	766
Aniline	C_6H_7N	l	812
Trinitrotoluene (2,4,6)	$C_7H_5N_3O_6$	s	821
Nicotine	$C_{10}H_{14}N_2$	l	1428

Halogen Compounds

Final Products: $CO_2(g)$, $H_2O(l)$, dil.sol. of HCl

Carbon tetrachloride	CCl_4	g	92.01
Carbon tetrachloride	CCl_4	l	84.17
Chloroform	$CHCl_3$	g	121.8
Chloroform	$CHCl_3$	l	114.3
Methyl chloride	CH_3Cl	g	182.81
Chloracetic acid	$C_2H_3ClO_2$	s	172.24
Ethylene dichloride	$C_2H_4Cl_2$	l	296.77
Ethyl chloride	C_2H_5Cl	g	339.66

Sulfur Compounds

Final Products: $CO_2(g)$, $SO_2(g)$, $H_2O(l)$

Carbonyl sulfide	COS	g	132.21
Carbon disulfide	CS_2	g	263.52
Carbon disulfide	CS_2	l	256.97
Methyl mercaptan	CH_4S	g	298.68
Dimethyl sulfide	C_2H_6S	g	457.12
Dimethyl sulfide	C_2H_6S	l	450.42
Ethyl mercaptan	C_2H_6S	l	448.0

References:

1. Selected Values of Physical and Thermodynamic Properties of Hydrocarbons and Related Compounds, *Am. Petroleum Inst. Research Proj.* **44**, edited by F. D. Rossini, Carnegie Institute of Technology (1952).

2. *International Critical Tables*, vol. V (1929). The values taken from this source were converted to a reference temperature of 25° C.

3. John H. Perry, *Chemical Engineers Handbook*, 3d ed., McGraw-Hill Book Co. (1950).

tion of organic compounds. These values, in all cases, correspond to the formation of gaseous carbon dioxide from the carbon present in the compound. The hydrogen in the original compound forms liquid water or may be utilized, in part, to form mineral acids when such elements as Cl, S, or N are present. The final products of combustion which are formed from other elements are specifically designated. For example, the heat of combustion of gaseous chloroform is given as $-121,800$ cal per g-mole. In the heading of the section of the table dealing with halogen derivatives the final state of chlorine is specified as HCl in dilute aqueous solution. Therefore, the heat of combustion of chloroform corresponds to the heat of the following reaction:

$$CHCl_3(g) + \tfrac{1}{2}O_2 + H_2O(aq) = CO_2 + 3HCl(aq)$$

$$\Delta H_c = -121,800 \text{ cal} \tag{11}$$

Heats of Combustion of Hydrocarbons. Average values of heats of combustion of petroleum fractions and hydrocarbons are plotted in Fig. 74 as a function of API gravity and characterization factor as defined on page 404. These are total heating values, corresponding to the formation of liquid water at 60° F.

Heats of Formation Calculated from Heats of Combustion. From the heat of combustion of a substance its heat of formation may be calculated if the heat of formation of each of the other products entering into the combustion reaction is known. Thus, in order to calculate the heat of formation of chloroform from equation 11 it is necessary to know the heats of formation of CO_2, $H_2O(l)$, and $HCl(aq)$. These values may be obtained from Table 29.

$$H_2(g) + \tfrac{1}{2}O_2(g) = H_2O(l), \qquad \Delta H_1 = -68,317.4 \text{ cal} \tag{12}$$

$$C(\beta) + O_2(g) = CO_2(g), \qquad \Delta H_2 = -94,051.8 \text{ cal} \tag{13}$$

$$\tfrac{1}{2}H_2(g) + \tfrac{1}{2}Cl_2(g) = HCl(aq), \qquad \Delta H_3 = -40,023 \text{ cal} \tag{14}$$

Equation 13 + 3(equation 14) − equation 11 − equation 12 gives

$$C(\beta) + \tfrac{1}{2}H_2(g) + 1\tfrac{1}{2}Cl_2(g) = CHCl_3(g) \tag{15}$$

or

$$\Delta H_f = \Delta H_2 + 3\Delta H_3 - \Delta H_c - \Delta H_1$$

$$\Delta H_f = (-94,051.8) + 3(-40,023) - (-121,800) - (-68,317.4)$$
$$= -24,003 \text{ cal}$$

Thus the heat of formation of $CHCl_3(g)$ is $-24,003$ cal per g-mole.

In this manner a general equation may be derived for use in calculating the heat of formation of a compound $C_aH_bBr_cCl_dF_eI_fN_gO_hS_i$ from its heat of combustion. If ΔH_c is the heat of combustion of this compound corresponding to the final products, $CO_2(g)$, $H_2O(l)$, $Br(l)$, $Cl_2(g)$, $HF(aq)$, $I(s)$, $N_2(g)$, $SO_2(g)$, and ΔH_f its heat of formation, then,

$$\Delta H_f = -\Delta H_c - 94{,}051.8a - 34{,}158.7b - 44{,}501e - 70{,}960i \quad (16)$$

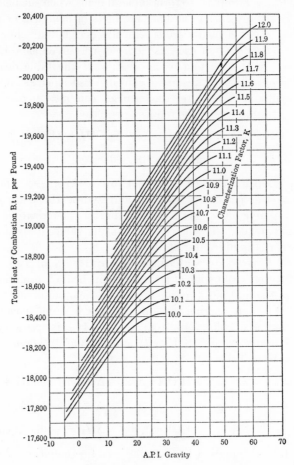

Fɪɢ. 74.   Total heats of combustion of liquid petroleum hydrocarbons
(Reproduced in *CPP Charts*)

If $\Delta H'_c$ is the heat of combustion of this compound corresponding to the final products $CO_2(g)$, $H_2O(l)$, $Br_2(g)$, $HCl(aq)$, $I(s)$, $HNO_3(aq)$, $H_2SO_4(aq)$, then,

$$\Delta H_f = -\Delta H'_c - 94{,}051.8a - 34{,}158.7b + 3670c - 5864.3d - 44{,}501e \\ -15{,}213.3g - 148{,}582.6i \quad (17)$$

The coefficient of c in equation 17 is the heat of vaporization per g-atom of liquid bromine.

If neither equation 16 nor 17 is applicable, the heat of formation of any compound may be calculated from its heat of combustion by the method demonstrated above in the algebraic combination of equations 11, 12, 13, and 14.

Calculation of the Standard Heat of Reaction from Heats of Formation. The standard heat of reaction accompanying any chemical change may be calculated if the heats of formation of all compounds involved in the reaction are known. If the reference state of enthalpy for a compound at 25° C and 1 atm pressure is taken as its separate component elements at 25° C and 1 atm pressure and in their normal states of aggregation, then the relative enthalpy of the compound is equal to its heat of formation. Thus, the standard heat of reaction, or enthalpy change, is equal to the algebraic sum of the standard heats of formation of the products less the algebraic sum of the standard heats of formation of the reactants. Thus,

$$[\Delta H_{\text{reaction}} = \sum \Delta \text{H}_{f(\text{products})} - \sum \Delta \text{H}_{f(\text{reactants})}]_{25° C} \qquad (18)$$

When an element enters into a reaction its heat of formation is zero if its state of aggregation is that selected as the basis for the heats of formation of its compounds.

Illustration 1. Calculate the standard heat of reaction of the following:

$$\text{HCl}(g) + \text{NH}_3(g) = \text{NH}_4\text{Cl}(s)$$

From Table 29 the standard heats of formation are for

$\text{HCl}(g)$,	$\Delta \text{H}_f = -22{,}063$ cal
$\text{NH}_3(g)$,	$\Delta \text{H}_f = -11{,}040$ cal
$\text{NH}_4\text{Cl}(s)$,	$\Delta \text{H}_f = -75{,}380$ cal

Substituting these values in equation 18 gives

$$\Delta H_{25} = + (-75{,}380) - (-22{,}063) - (-11{,}040)$$
$$\text{or } \Delta H_{25} = -42{,}277 \text{ cal}$$

Illustration 2. Calculate the standard heat of reaction, ΔH_{25} of the following:

$$\text{CaC}_2(s) + 2\text{H}_2\text{O}(l) = \text{Ca(OH)}_2(s) + \text{C}_2\text{H}_2(g)$$

From Table 29,

$\text{CaC}_2(s)$,	$\Delta \text{H}_f = -15{,}000$ cal
$\text{H}_2\text{O}(l)$,	$\Delta \text{H}_f = -68{,}317.4$ cal
$\text{Ca(OH)}_2(s)$,	$\Delta \text{H}_f = -235{,}800$ cal

The heat of formation of acetylene, $\text{C}_2\text{H}_2(g)$, is calculated from the heat of combustion data of Table 30 by means of equation 17:

$$\text{C}_2\text{H}_2(g), \qquad \Delta \text{H}_f = +54{,}194 \text{ cal}$$

$$\Delta H_{25} = (-235{,}800) + (54{,}194) - (-15{,}000) - 2(-68{,}317.4)$$
$$\text{or } \Delta H_{25} = -29{,}971 \text{ cal}$$

Illustration 3. Calculate the standard heat of reaction ΔH_{25} of the following:

$$2FeS_2(s) + 5\tfrac{1}{2}O_2(g) = Fe_2O_3(s) + 4SO_2(g)$$

The standard heats of formation of the compounds are obtained from Table 29 for

$$\begin{aligned}
FeS_2(s), &\qquad \Delta_{H_f} = -42{,}520 \text{ cal} \\
Fe_2O_3(s), &\qquad \Delta_{H_f} = -196{,}500 \text{ cal} \\
SO_2(g), &\qquad \Delta_{H_f} = -70{,}960 \text{ cal}
\end{aligned}$$

$$\Delta H_{25} = (-196{,}500) + 4(-70{,}960) - 2(-42{,}520) = -395{,}300 \text{ cal}$$

Calculation of the Standard Heat of Reaction from Heats of Combustion. For a reaction between organic compounds the basic thermochemical data are generally available in the form of standard heats of combustion. The standard heat of reaction where organic compounds are involved can be conveniently calculated by using directly the standard heats of combustion instead of standard heats of formation. An energy balance is again employed, but in this case the standard reference state is not the elements but the products of combustion at 25° C and 1 atm pressure and in the state of aggregation specified by the heat of combustion data. For example, the enthalpy of methane relative to its products of combustion, gaseous CO_2 and liquid H_2O, is equal to the negative value of its standard heat of combustion, or $+212{,}798$ cal per g-mole. Therefore, in any equation involving combustible materials the formula of a compound may be replaced by its enthalpy relative to its products of combustion. The enthalpy of the products minus the enthalpy of the reactants is then equal to the standard heat of reaction, or the standard heat of combustion of the reactants minus the standard heat of combustion of the products is equal to the standard heat of reaction. Thus,

$$[\Delta H_{\text{reaction}} = \sum \Delta_{H_{c(\text{reactants})}} - \sum \Delta_{H_{c(\text{products})}}]_{25° C} \qquad (19)$$

Illustration 4. Calculate the standard heat of reaction ΔH_r of the following:

$$\underset{\text{(ethyl alcohol)}}{C_2H_5OH(l)} + \underset{\text{(acetic acid)}}{CH_3COOH(l)} = \underset{\text{(ethyl acetate)}}{C_2H_5OOCCH_3(l)} + H_2O(l)$$

From Table 30, heats of combustion are as follows:

$$\begin{aligned}
C_2H_5OH, &\qquad \Delta_{H_c} = -326{,}700 \text{ cal} \\
CH_3COOH, &\qquad \Delta_{H_c} = -208{,}340 \text{ cal} \\
C_2H_5OOCCH_3, &\qquad \Delta_{H_c} = -538{,}760 \text{ cal}
\end{aligned}$$

Since the heat of reaction is the difference between the heats of combustion of the reactants and the products,

$$\Delta H_{25} = (-326{,}700) + (-208{,}340) - (-538{,}760) = 3720 \text{ cal}$$

In general, the heats of formation of organic compounds are small in comparison to the heats of combustion. Similarly, the heats of reaction are small in systems involving only combinations of organic compounds. Since these relatively small quantities can be determined only by the differences between the large heats of combustion, it follows that they are rarely known with a high degree of accuracy. For example, the small heat of reaction which is the final result of illustration 4 is uncertain. An error of only 0.2% in determining the heat of combustion of ethyl acetate results in an error of 40% in the value of this heat of reaction.

When both organic and inorganic compounds appear in a reaction, it is best to obtain the heat of reaction by means of heat of formation data. The heats of formation of the organic compounds may be calculated from their heats of combustion by means of equation 16 or 17. The procedure is then the same as that demonstrated in illustration 2.

Illustration 5. Calculate the standard heat of reaction ΔH_{25} of the following:

$$CH_3Cl(g) + KOH(s) = CH_3OH(l) + KCl(s)$$

The heats of formation of the inorganic compounds are obtained from Table 29.

$$KOH(s), \qquad \Delta_H{}_f = -101,780 \text{ cal}$$
$$KCl(s), \qquad \Delta_H{}_f = -104,175 \text{ cal}$$

The heats of combustion of the organic compounds are obtained from Table 30 and their heats of formation calculated by means of equation 17; thus for $CH_3Cl(g)$

$$\Delta_{H_c} = -182,810 \text{ cal}$$

$$\Delta_H{}_f = 182,810 - 94,051.8 - 3(34,158.7) - 5864.3 = -19,582 \text{ cal}$$

For $CH_3OH(l)$, $\qquad \Delta_{H_c} = -173,650 \text{ cal}$

$$\Delta_H{}_f = 173,650 - 94,051.8 - 4(34,158.7) = -57,037 \text{ cal}$$

$$\Delta H_{25} = (-57,037) + (-104,175) - (-19,582) - (-101,780)$$

$$\Delta H_{25} = -39,850 \text{ cal}$$

Heats of Neutralization of Acids and Bases. The neutralization of a dilute aqueous solution of NaOH with a dilute solution of HCl may be represented by the following thermochemical equation:

$$NaOH(aq) + HCl(aq) = NaCl(aq) + H_2O(l) \qquad (20)$$

The heat of reaction ΔH_{25} may be calculated from the respective heats of formation in Table 29. Thus,

$$NaOH(aq), \qquad \Delta_H{}_f = -112,236 \text{ cal}$$
$$HCl(aq), \qquad \Delta_H{}_f = -\ 40,023 \text{ cal}$$
$$NaCl(aq), \qquad \Delta_H{}_f = -\ 97,302 \text{ cal}$$
$$H_2O(l), \qquad \Delta_H{}_f = -\ 68,317 \text{ cal}$$

From an energy balance

$$\Delta H_{25} = (-97{,}302) + (-68{,}317) - (-112{,}236) - (-40{,}023)$$
$$\Delta H_{25} = -13{,}360 \text{ cal}$$

Heats of neutralization may be determined by direct calorimetric measurements in a series of solutions of finite concentrations of progressively increasing dilution extrapolated to infinite dilution. Following are heats of neutralization based on such measurements.

$$\text{HCl}(aq) + \text{LiOH}(aq) = \text{LiCl}(aq) + \text{H}_2\text{O}, \qquad \Delta H = -13{,}360$$
$$\text{HNO}_3(aq) + \text{KOH}(aq) = \text{KNO}_3(aq) + \text{H}_2\text{O}, \qquad \Delta H = -13{,}355$$
$$\tfrac{1}{2}\text{H}_2\text{SO}_4(aq) + \text{KOH}(aq) = \tfrac{1}{2}\text{K}_2\text{SO}_4(aq) + \text{H}_2\text{O}, \qquad \Delta H = -13{,}357$$

It will be noted that the heat of neutralization of strong acids with strong bases in dilute solution is practically constant when 1 mole of water is formed in the reaction. The explanation of this fact is that strong acids and bases and their salts are completely dissociated into their respective ions when in dilute aqueous solution. From this viewpoint, a dilute solution of hydrochloric acid consists of only hydrogen and chlorine ions in aqueous solution, and similarly a dilute solution of sodium hydroxide consists of sodium and hydroxyl ions in aqueous solution. Upon neutralization, the resulting solution contains only sodium and chlorine ions. The reaction may be looked upon as ionic, and equation 20 may be rewritten in ionic form:

$$\text{Na}^+(aq) + \text{OH}^-(aq) + \text{H}^+(aq) + \text{Cl}^-(aq) =$$
$$\text{Na}^+(aq) + \text{Cl}^-(aq) + \text{H}_2\text{O}(l) \quad (21)$$

Canceling the similar terms from equation 21 gives

$$\text{OH}^-(aq) + \text{H}^+(aq) = \text{H}_2\text{O}, \qquad \Delta H_{25} = -13{,}360 \text{ cal} \quad (22)$$

Thus, the actual net result of the neutralization of dilute solutions of strong acids and bases is the production of water from hydrogen and hydroxyl ions. The accepted average value of the heat of neutralization is $-13{,}360$ cal per g-mole of water formed. This corresponds to the heat of reaction of equation 22, representing the formation of water from its ions.

In the neutralization of dilute solutions of weak acids and weak bases, the heat given off is less than 13,360 cal. For example, in the neutralization of hydrocyanic acid with sodium hydroxide, the heat evolved is only 2481 cal per g-mole of water formed. The unevolved heat may be considered as the heat required to complete the dissociation of hydrogen cyanide into hydrogen and cyanide ions as neutralization proceeds. As a hydrogen ion from hydrogen cyanide is neutralized by a hydroxyl ion, more hydrogen cyanide ionizes until neutralization is complete. This ionization of hydrogen cyanide requires the absorption of heat at the expense of the heat evolved in the union of hydrogen and hydroxyl ions.

Thermoneutrality of Salt Solutions. When dilute aqueous solutions of two neutral salts are mixed, there is no thermal effect provided there is no precipitation, or evolution of gas. However, upon evaporation of a mixture of such solutions, four crystalline salts will be found, indicating that double decomposition or metathesis has taken place. For example,

$$NaCl(aq) + KNO_3(aq) = NaNO_3(aq) + KCl(aq), \qquad \Delta H = 0 \quad (23)$$

In dilute aqueous solutions it may be considered that each of these four salts is completely ionized and equation 23 written in ionic form,

$$Na^+(aq) + Cl^-(aq) + K^+(aq) + NO_3^-(aq) = Na^+(aq) +$$
$$NO_3^-(aq) + K^+(aq) + Cl^-(aq), \qquad \Delta H = 0 \quad (24)$$

From this viewpoint it is evident that mixing such systems actually produces no change, the initial and final solutions consisting of the same four ions. It is only upon concentration of the solution and reassociation of the ions that the metathesis leads to a definite change in the nature of the system.

The experimentally observed fact that dilute solutions of neutral salts of strong acids and bases may be mixed without thermal effect is termed the *thermoneutrality of salt solutions*.

Heats of Formation of Ions. Equation 22 represents the formation of one mole of water from the combination of hydrogen and hydroxyl ions. The average value of this heat of reaction has been determined as $-13,360$ cal per g-mole. The heat of formation of water from gaseous oxygen and hydrogen is given in Table 29 as $-68,317.4$ cal per g-mole.

$$H_2(g) + \tfrac{1}{2}O_2(g) = H_2O(l), \qquad \Delta H_a = -68,317.4 \text{ cal} \quad (a)$$
$$OH^-(aq) + H^+(aq) = H_2O(l), \qquad \Delta H_b = -13,360 \text{ cal} \quad (b)$$

Equation a − equation b gives

$$H_2(g) + \tfrac{1}{2}O_2(g) = H^+(aq) + OH^-(aq)$$
$$\Delta H_f = -68,317.4 - (-13,360) = -54,957 \text{ cal} \quad (25)$$

Thus, the combined heats of formation of the hydrogen and hydroxyl ions from elementary hydrogen and oxygen are $-54,957$ cal. By arbitrarily assigning a zero value to the heat of formation of the hydrogen ion the heat of formation of the hydroxyl ion is obtained. Thus, by definition,

$$\tfrac{1}{2}H_2(g) = H^+, \qquad\qquad \Delta H_f = 0 \qquad\qquad (26)$$

$$\tfrac{1}{2}H_2(g) + \tfrac{1}{2}O_2(g) = OH^-(aq), \qquad \Delta H_f = -54{,}957 \text{ cal} \qquad (27)$$

On this basis the relative heats of formation of the other ions of highly dissociated acids and bases may be calculated. For example, from Table 29, the heat of formation of $NaOH(aq)$ is $-112{,}236$ cal per g-mole. Since sodium hydroxide is completely dissociated into sodium and hydroxyl ions when in dilute solution, the formation of 1 mole of $NaOH(aq)$ from the elements is, in actual effect, the formation of 1 mole of sodium and 1 mole of hydroxyl ions. Thus,

$$Na(s) + \tfrac{1}{2}O_2(g) + \tfrac{1}{2}H_2(g) = Na^+(aq) + OH^-(aq)$$
$$\Delta H_f = -112{,}236 \text{ cal} \qquad\qquad (28)$$

Combining equations 28 and 27 gives

$$Na(s) = Na^+, \qquad \Delta H_f = -57{,}279 \text{ cal} \qquad (29)$$

Therefore, the heat of formation of the sodium ion is $-57{,}279$ cal per g-atom. In a similar manner the heats of formation of other ions may be calculated, based on the assignment of a value of zero to hydrogen. Heats of formation of a few common ions, taken from *Selected Values of Chemical Thermodynamic Properties*, National Bureau of Standards, are given in Table 31.

The heat of formation in dilute aqueous solution of any compound that is completely dissociated under these conditions is equal to the sum of the heats of formation of its ions. From the data of Table 31 the heats of formation of such compounds may be predicted.

Illustration 6. Calculate, from the data of Table 31, the heat of formation of barium chloride in dilute solution.

Since $BaCl_2$ may be assumed to be completely dissociated in dilute solution, its heat of formation in dilute solution is equal to the sum of the heats of formation of one barium ion and two chlorine ions. From Table 31,

$$Ba^{++}; \qquad \Delta H_f = -128{,}670 \text{ cal}$$
$$Cl^-; \qquad \Delta H_f = -40{,}023 \text{ cal}$$
$$BaCl_2(aq); \qquad \Delta H_f = (-128{,}670) + 2(-40{,}023)$$

or
$$\Delta H_f = -208{,}716 \text{ cal}$$

The heat of formation of a substance that is soluble and highly ionized in water may be calculated from its heat of formation in dilute solution from the heat of formation of its ions, provided its standard heat of solution is known. Since experimental determinations of the heats of solution are available, a method is provided for estimating

Table 31. Heats of Formation of Ions

$\Delta H^\circ{}_f$ = standard heat of formation at 25° C, kcal per g-mole

Cations			Anions		
Ion	Formula	$\Delta H^\circ{}_f$	Ion	Formula	$\Delta H^\circ{}_f$
Aluminum	Al^{+++}	-125.4	Acetate	CH_3COO^-	-116.843
Ammonium	$NH_4{}^+$	-31.74	Bicarbonate	$HCO_3{}^-$	-165.18
Barium	Ba^{++}	-128.67	Bisulfate	$HSO_4{}^-$	-211.70
Calcium	Ca^{++}	-129.77	Bisulfite	$HSO_3{}^-$	-150.09
Hydrogen	H^+	Zero by	Bromide	Br^-	-28.90
		definition	Carbonate	$CO_3{}^{--}$	-161.63
Iron	Fe^{++}	-21.0	Chloride	Cl^-	-40.023
Iron	Fe^{+++}	-11.4	Fluoride	F^-	-78.66
Lithium	Li^+	-66.554	Hydroxide	OH^-	-54.957
Magnesium	Mg^{++}	-110.41	Iodide	I^-	-13.37
Manganese	Mn^{++}	-52.3	Nitrite	$NO_2{}^-$	-25.4
Manganese	Mn^{+++}	-24	Nitrate	$NO_3{}^-$	-49.372
Potassium	K^+	-60.04	Phosphate	$PO_4{}^{---}$	-306.9
Sodium	Na^+	-57.279	Sulfate	$SO_4{}^{--}$	-216.90
Zinc	Zn^{++}	-36.43	Sulfite	$SO_3{}^{--}$	-149.2

Source: *Selected Values of Chemical Thermodynamic Properties,* as of July 1, 1953, edited by D. D. Wagman, National Bureau of Standards.

the heat of formation of many inorganic compounds. The heat of formation in infinitely dilute solution is calculated from the ionic heats of formation, and the heat of solution is subtracted from this value. For example, the standard heat of solution of $BaCl_2(s)$ in an infinite amount of water is -3160 cal per g-mole. From the result of illustration 6, the heat of formation of $BaCl_2(s)$ is $-208,716 + 3160$ or $-205,556$ cal per g-mole.

Chemical reactions which take place between strong acids and bases and their salts in dilute aqueous solution may be treated as ionic, and heats of reaction may be calculated directly from ionic heats of formation. This method is particularly desirable in dealing with analytical data for complex solutions. By treating the reactions as ionic it is unnecessary to formulate hypothetical combinations of the analytically determined ionic quantities.

Heats of Formation of Atoms. At high temperatures the elementary gases are known to dissociate into their atomic states with absorption of great amounts of energy. Upon cooling, these monatomic gases rapidly recombine to form the original molecular gas. An interesting phenomenon of this type is the association of monatomic hydrogen, $H(g)$, to form the molecular gas, $H_2(g)$ in the hydrogen welding process.

Molecular dissociations in the gaseous state are accompanied by large absorptions of energy. Conversely, the reassociation is accompanied by a great evolution of energy. In Table 32 are data for the heats of dissociation of several gases into their atomic forms.

TABLE 32. HEATS OF FORMATION OF ATOMS

$$\frac{1}{2}O_2(g) = O(g), \qquad \Delta H°_{25} = +59,159 \text{ g-cal}$$
$$\frac{1}{2}H_2(g) = H(g), \qquad \Delta H°_{25} = +52,089 \text{ g-cal}$$
$$\frac{1}{2}N_2(g) = N(g), \qquad \Delta H°_{25} = +85,566 \text{ g-cal}$$
$$\frac{1}{2}F_2(g) = F(g), \qquad \Delta H°_{25} = +32,250 \text{ g-cal}$$
$$\frac{1}{2}Cl_2(g) = Cl(g), \qquad \Delta H°_{25} = +29,012 \text{ g-cal}$$
$$\frac{1}{2}Br_2(g) = Br(g), \qquad \Delta H°_{25} = +23,040 \text{ g-cal}$$
$$\frac{1}{2}I_2(g) = I(g), \qquad \Delta H°_{25} = +18,044 \text{ g-cal}$$

Source: *Selected Values of Chemical Thermodynamic Properties*, as of July 1, 1953, edited by D. D. Wagman, National Bureau of Standards.

Thermochemistry of Solutions

The enthalpy change accompanying the dissolution of a substance is termed its *heat of solution* or, better, its heat of dissolution. If chemical combination takes place between the solvent and the substance being dissolved, the heat of dissolution will include the heat of solvation or the heat of hydration accompanying this combination. If ionization takes place, the heat of solution will also include the energy of ionization. The heat of solution of a neutral, nondissociating salt is generally positive; that is, heat is absorbed from the surroundings in the isothermal formation of the solution, or the solution cools if dissolution proceeds adiabatically. The dissolution of such a material is analogous to the evaporation of a liquid in that the result is the breaking down of a condensed structure into a state of great dispersion. Thus, energy is absorbed in overcoming the attractive forces between the particles of the condensed state.

Heats of solvation, especially in aqueous systems, are generally negative and relatively large. For this reason the heat of solution of a substance that forms a solvate or hydrate has generally a large negative value, indicating the evolution of heat when the unhydrated substance is dissolved.

The enthalpy change when a substance is dissolved depends on the amounts and natures of the solute and solvent, on the temperature, and on the initial and final concentrations of the solution. The numerical value of the heat effect, therefore, requires an exact and complete statement of all reference conditions.

Standard Integral Heats of Solution. Arbitrarily the *standard integral heat of solution* is defined as the change in enthalpy of the system when one mole of solute is dissolved in n_1 moles of solvent with the temperature maintained at 25° C and the pressure at one atmosphere.

The numerical value of the integral heat of solution depends on the value of n_1. As successive equal increments of solvent are added to a given mass of solute, the heat evolved with each addition progressively diminishes until a high dilution is attained, usually about 100 or 200 moles of solvent per mole of solute, following which no further heat effect is perceptible. The integral heat of solution approaches a maximum numerical value at infinite dilution. This limiting value is termed the integral heat of solution at infinite dilution.

In Figs. 75 through 79 are presented the standard integral heats of solution of common acids, bases, and salts in water. Integral heats of solution are determined calorimetrically by first measuring the heat of solution of a solute in enough solvent to form a relatively concentrated solution. The heat of dilution accompanying the addition of solvent to this concentrated solution is then measured. The integral heat of solution at any desired concentration is obtained by adding, algebraically, the observed heats of initial solution and of subsequent dilutions.

It is evident that the integral heat of solution as defined above is the enthalpy of a solution containing 1.0 mole of solute, relative to the pure solute and solvent at the same temperature and pressure. Thus, the enthalpy of a solution at temperature T relative to the pure solute and solvent at temperature T_0 is expressed thus,

$$H_s = n_1 \text{H}_1 + n_2 \text{H}_2 + n_2 \Delta \text{H}_{s2} \qquad (30)$$

where H_s = enthalpy of $n_1 + n_2$ moles of solution of components 1 and 2 at temperature T relative to temperature T_0

H_1, H_2 = molal enthalpies of pure components 1 and 2 at temperature T relative to temperature T_0

ΔH_{s2} = integral heat of solution of component 2 at temperature T

Heat of Formation of a Compound in Solution. By combination of the data of Table 29 with those of Figs. 75 to 79 it is possible to calculate the heat of formation of a compound in an aqueous solution of specified concentration. This total standard heat of formation is the sum of the standard heat of formation of the solute ΔH_f and its standard integral heat of solution ΔH_{s2}, at the specified concentration.

Illustration 7. (a) Calculate the heat of formation of H_2SO_4 to form an aqueous solution containing 5 moles of water per mole of H_2SO_4.

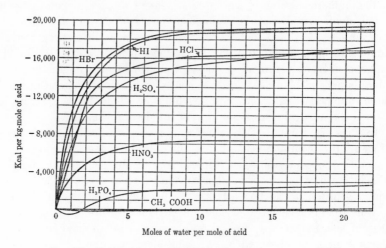

Fig. 75.   Integral heats of solution of acids in water at 25° C

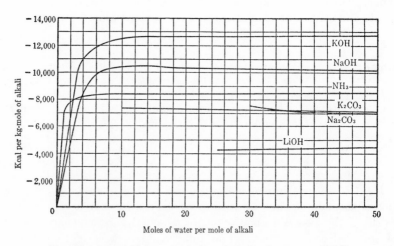

Fig. 76.   Integral heats of solution of alkalies in water at 25° C

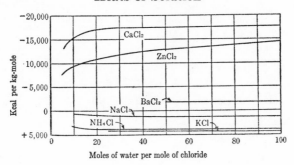

Fig. 77.   Integral heats of solution of chlorides in water at 25° C

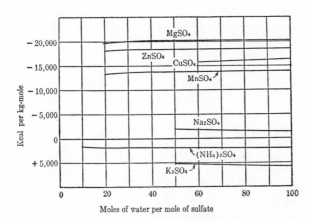

Fig. 78.   Integral heats of solution of sulfates in water at 25° C

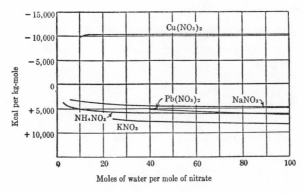

Fig. 79.   Integral heats of solution of nitrates in water at 25° C

(b) Calculate the heat of dilution of the solution of part a to a concentration of 1 mole of H_2SO_4 in 20 moles of water.

(a) From Table 29,

$$H_2SO_4(l): \qquad \Delta H_f = -193,910 \text{ cal}$$

From Fig. 75,

$$H_2SO_4(n_1 = 5.0), \qquad \Delta H_s = -13,600 \text{ cal}$$
$$H_2(g) + S(s) + 2O_2(g) + 5H_2O = H_2SO_4(n_1 = 5)$$
$$\Delta H_{25} = -193,910 - 13,600 = -207,510 \text{ cal}$$

Since the same amount of water appears on both sides of the equation, its heat of formation does not appear.

(b) From Fig. 75,

$$H_2SO_4(n_1 = 20), \qquad \Delta H_s = -17,200 \text{ cal}$$
$$H_2SO_4(n_1 = 5), \qquad \Delta H_s = -13,600 \text{ cal}$$
$$H_2SO_4(n_1 = 5) + 15H_2O = H_2SO_4(n_1 = 20)$$
$$\Delta H_{25} = -17,200 - (-13,600)$$
$$\Delta H_{25} = -3600 \text{ cal}$$

Heat of Solution of Hydrates. If a solute forms a hydrate, the standard heat of solution of the hydrate is the difference between the heat of solution of the anhydrous substance and its heat of hydration. The heat of hydration is calculated from the data of Table 29 as the difference between the heat of formation of the hydrated compound and the sum of the heats of formation of the anhydrous substance and of the water of hydration.

Illustration 8. Calculate the standard heat of solution of $CaCl_2 \cdot 6H_2O$ to form a solution containing 10 moles of water per mole of $CaCl_2$.

From Table 29,

$$CaCl_2: \qquad \Delta H_f = -190,000 \text{ cal}$$
$$H_2O: \qquad \Delta H_f = -68,317.4 \text{ cal}$$
$$CaCl_2 \cdot 6H_2O: \quad \Delta H_f = -623,150 \text{ cal}$$

The enthalpy change accompanying hydration is then

$$\Delta H_{25} = (-623,150) - (-190,000) - 6(-68,317.4)$$

or

$$\Delta H_{25} = -23,246 \text{ cal per g-mole}$$

This heat of reaction represents the molal enthalpy of $CaCl_2 \cdot 6H_2O$ relative to $CaCl_2$ and H_2O at 25 C. The enthalpy of $CaCl_2$ ($n_1 = 10$) relative to the same reference substances and state is obtained from Fig. 77:

$$CaCl_2(n_1 = 10), \qquad \Delta H_s = -15,500 \text{ cal per g-mole}$$

For the reaction

$$CaCl_2 \cdot 6H_2O + 4H_2O \rightarrow CaCl_2(n_1 = 10)$$
$$\Delta H_{25} = -15,500 - (-23,246)$$
$$\Delta H_{25} = 7746 \text{ cal per g-mole}$$

Heats of Mixing. Heats of solution in a system in which both solute and solvent are liquids are termed heats of mixing. Heats of mixing are frequently expressed on a unit weight rather than a molal basis. If 10 grams of glycerin are mixed with 90 grams of water, the heat evolved is 191 cal. Therefore, the heat of mixing water and glycerin to form a solution containing 10% glycerin is -1.91 cal per gram of solution formed. The integral heat of solution of either component may be calculated from heat of mixing data.

Illustration 9. The heat of mixing of water and glycerin to form a solution containing 40% glycerin is -4.50 cal per gram of solution. Calculate the integral heats of solution of glycerin and of water at this concentration.

Basis: 100 grams of solution.

Heat of mixing $= -450$ cal per 100 grams of solution

$$\text{Heat of solution of glycerin in water} = \frac{-450}{40} = -11.25 \text{ cal per gram}$$
$$\text{of glycerin}$$

$$\text{Heat of solution of water in glycerin} = \frac{-450}{60} = -7.5 \text{ cal per gram}$$
$$\text{of water}$$

Enthalpy-Concentration Charts. McCabe[1] has called attention to the usefulness of the enthalpy-concentration diagram for binary solutions. In such charts the enthalpy per unit weight of solution is plotted against concentration for a series of constant-temperature and constant-pressure lines. Once such a diagram for a given binary solution has been constructed, calculations of the heat effects involved in changing the concentrations and temperatures of the solution become simple and rapid. The expenditure of the time required in constructing such a diagram is well justified in case of specialization on a given system.

In constructing an enthalpy-concentration chart it is convenient to choose as reference states the pure components each at a specified temperature, pressure, and physical state. With heats of solution measured at 25° C it is convenient to establish the first enthalpy curve at this temperature and subsequently from this base line derive all other constant-temperature lines. This is done directly from equation 30, employing standard integral heats of solution together with heat-capacity data on both the pure components and the solutions.

In Figs. 80, 81, 82, and 83 are enthalpy-concentration charts for aqueous systems of hydrochloric acid, sulfuric acid, calcium chloride and ethanol.

---

[1] W. L. McCabe, *Trans. Am. Inst. Chem. Engrs.*, **31**, 129–62 (1935).

Illustration 10. Calculate the final temperature when 5.0 lb of 10% HCl at 60° F are mixed with 8 lb of 30% HCl at 100° F.   From Fig. 80,

$$\begin{aligned}
\text{Enthalpy of 10\% HCl at 60° F} &= (5)(-60) &&= -\ 300\\
\text{Enthalpy of 30\% HCl at 100° F} &= (8)(-175) &&= -1400\\
\text{Total enthalpy of mixture} & &&= -1700\\
\text{Enthalpy per lb of final mixture} &= \frac{-1700}{13} &&= -131\\
\text{Final concentration} &= (0.5 + 2.4)/13 &&= 22.3\% \text{ HCl}
\end{aligned}$$

From Fig. 80 this enthalpy corresponds to a temperature of 90° F.

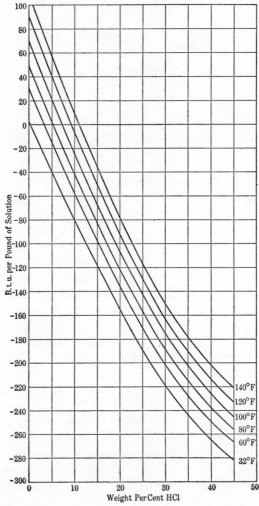

Fɪɢ. 80. Enthalpy-concentration chart of hydrochloric acid solutions relative to pure HCl(g) and pure H$_2$O(l) at 32° F and 1 atm

In the enthalpy-concentration diagrams for the systems sulfuric acid–water, and calcium chloride–water (Figs. 81 and 82), steep sloping lines are shown in the two-phase region of vapor and solution. These lines represent the enthalpies of the entire system in Btu per pound,

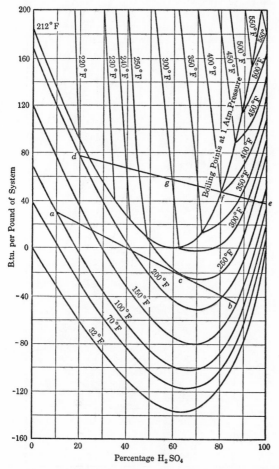

Fig 81. Enthalpy-concentration of sulfuric acid–water system relative to pure components (water and H_2SO_4 at 32° F and own vapor pressures)

(Reproduced in *CPP Charts*)

including the liquid and vapor phases for isothermal conditions at one atmosphere pressure and meet the saturation curves at the normal boiling points of the solutions. In the absence of other gases no vapor phase exists when the total vapor pressure of the system is less than the pressure imposed.

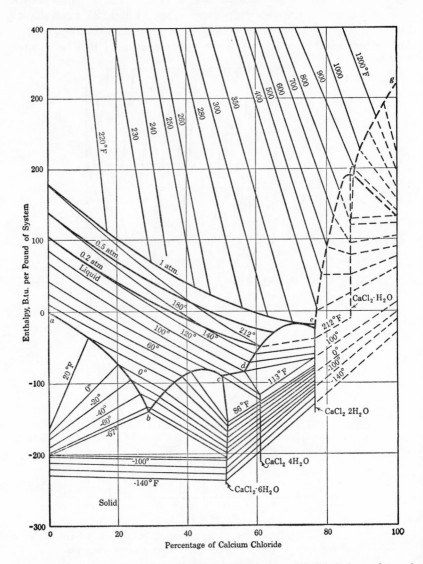

FIG. 82. Enthalpy of calcium chloride–water system (Modified from data of Bošnjakovic, *Technische Thermodynamik*, Theodor Steinkopff, Dresden and Leipzig, 1937)

For the illustrated temperature-concentration range of the sulfuric acid system the vapors are essentially pure water. Thus, a point on the chart in the two-phase region indicates a mixture of water vapor

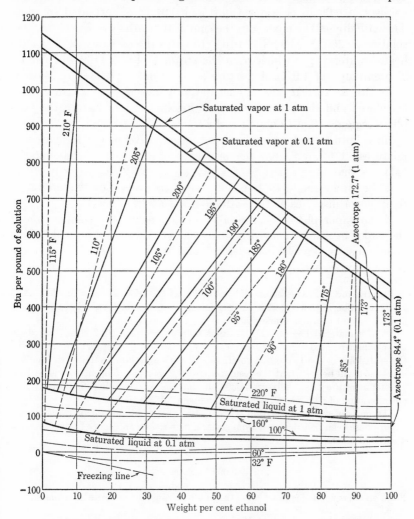

Fig. 83. Enthalpy-composition diagram of ethanol-water system. Reference states, pure liquid alcohol and pure liquid water, each at 32° F saturated

and solution saturated at the indicated temperature and atmospheric pressure. The enthalpy of the mixture is the sum of the enthalpies of the saturated solution and the water vapor.

For example, a mixture at 250° F with an over-all composition of 45% H_2SO_4 consists of two phases, water vapor and a saturated solution containing 48% H_2SO_4. From a material balance 1 lb of the mixture consists of 0.06 lb of water vapor and 0.94 lb of saturated solution. The enthalpy of the mixture is the sum of the enthalpy of the saturated solution at 250° F which is 8 Btu per lb plus the enthalpy of the super-heated water vapor which from the steam tables is 1169 Btu per lb. The enthalpy of 1.0 lb of mixture is then $(0.06)(1169) + (0.94)(8.0)$ = 78 Btu per lb. In this manner the enthalpies of other mixtures were calculated and the constant temperature lines established in the two-phase regions of both Figs. 81 and 82.

Since both enthalpies and masses are additive in the formation of mixtures, it follows from the principles of the energy and material balances that the properties of a mixture of two solutions or mixtures must lie on a straight tie line connecting the properties of the original solutions or mixtures on the enthalpy-concentration chart.

If the vapor phase consists of two components, the composition of the vapor in equilibrium with the liquid solution at its normal boiling point must be known in order to establish the vapor lines and the composition of both phases.

Illustration 11. One pound of pure H_2SO_4 at 150° F is mixed with 1 lb of 20% H_2SO_4 solution initially at 200° F. Calculate the temperature of adiabatic mixing and the weight of water evaporated.

From a material balance the resultant mixture contains 60% H_2SO_4 based on the combined liquid and vapor phases. By constructing a tie line de connecting the enthalpies of the two initial solutions and noting its intersection with the 60% abscissa it will be seen that the resultant enthalpy per pound of mixture is 58 Btu. The temperature of the mixture is 300° F, and the corresponding liquid phase has a composition of 63% H_2SO_4 with a boiling point of 300° F.

The mass of water evaporated y (neglecting the small amount of H_2SO_4 in the vapor phase) can be calculated from either a material or an energy balance. From a material balance,

$$0.80 = y + 0.37(2 - y)$$

$$y = 0.095 \text{ lb water vapor}$$

From an energy balance in which enthalpies at a constant pressure are equated,

Enthalpy of H_2SO_4 at 150° F	=	40 Btu per lb
Enthalpy of 20% H_2SO_4 at 200° F	=	76 Btu per lb
Enthalpy of water vapor at 300° F and 1 atm	=	1194 Btu per b
Enthalpy of 63% solution at 300° F	=	1 Btu per lb

$$116 = y(1194) + (2 - y)$$

$$y = 0.095 \text{ lb water vapor}$$

Maximum Temperature in Mixing Solutions. The maximum temperature attainable in mixing two solutions of the same components but of different temperatures and concentrations may be readily obtained from an enthalpy-concentration diagram by constructing a tie line connecting the enthalpies corresponding to the two initial solutions. The point of meeting or tangency of this line with the highest isotherm on the diagram is the desired maximum temperature for adiabatic mixing. For example, in Fig. 81, a tie line ab is constructed joining the enthalpy a of a 10% H_2SO_4 solution at 100° F and the enthalpy b of an 87% solution at 150° F. The maximum temperature attainable occurs at point c, corresponding to the 250° F isotherm, and at a composition of 63%. The relative weights of the two solutions to give this concentration are obtained from a material balance. Thus, for 1 lb of 10% H_2SO_4 solution 2.21 lb of an 87% solution are required to produce 3.21 lb of a 63% solution.

If the tie line crosses the region of partial vaporization the maximum temperature corresponds to the higher temperature of intersection with the saturation line. Thus in Fig. 81 the tie line de connects enthalpy d of a 20% solution at 200° F with the enthalpy e of pure H_2SO_4 at 150° F. The maximum temperature occurs at point f corresponding to 400° F and a solution of 80% H_2SO_4. The relative amount of the two solutions to give this concentration is obtained from a material balance. Thus, for 1 lb of 20% solution, 3.0 lb of H_2SO_4 are required to give 4.0 lb of solution having a concentration of 80%.

Solid-Liquid Systems. The enthalpy composition chart of a complex system is shown in Fig. 82 for the calcium chloride–water system. In this diagram line $abcde$ represents the solubility of the various hydrates of calcium chloride at various temperatures with corresponding enthalpies. Line efg represents the melting points and corresponding enthalpies of the various solid hydrates.

a = freezing point of water, 32° F
b = eutectic of ice and $CaCl_2 \cdot 6H_2O$, −67° F
c = transition of $CaCl_2 \cdot 6H_2O$ to $CaCl_2 \cdot 4H_2O$ at 86° F
d = transition of $CaCl_2 \cdot 4H_2O$ to $CaCl_2 \cdot 2H_2O$ at 113° F
e = transition of $CaCl_2 \cdot 2H_2O$ to $CaCl_2 \cdot H_2O$
f = transition of $CaCl_2 \cdot H_2O$ to $CaCl_2$
g = melting point of $CaCl_2$

In the two-phase region containing vapor, the sloping lines starting at the line marked 1 atm represent the enthalpies of the liquid-vapor system at the stated temperatures and compositions of combined phases. For example, 1 lb of a vapor-liquid system containing 40% $CaCl_2$ at

250° F has an enthalpy of 95 Btu and consists of (0.40/0.42) lb of a 42% solution, boiling at 250° F at 1 atm pressure and [1 − (0.40/0.42)] lb of water vapor at 250° F in equilibrium with the boiling solution. The value of 42% is obtained as the intersection of the 250° F line with the 1 atm line.

The sloping lines running upward from the dotted melting-point line *efg* represent the enthalpies of the liquid-vapor system at the stated temperature and composition of the combined phases. For example, 1 lb of liquid-vapor system containing 80% $CaCl_2$ at 800° F has an enthalpy of 220 Btu and consists of (0.8/0.84) lb of liquid containing 84% $CaCl_2$ and [1 − (0.8/0.84)] lb of water vapor at 800° F in equilibrium with the solid. The vapor pressures corresponding to the latter sloping lines in the two-phase region are not given.

Illustration 12. To 200 lb of anhydrous $CaCl_2$ at 100° F are added 500 lb of a solution containing 20% $CaCl_2$ at 80° F.

(a) What is the temperature of the final mixture?

(b) How much heat must be removed to start crystallization?

From Fig. 82,

Enthalpy of $CaCl_2$ at 100° F = (200)(12) = 2400 Btu

Enthalpy of $CaCl_2$ solution at 80° F = (500)(−15) = −7500 Btu

Total enthalpy = −5100 Btu

$$\text{Final composition} = \frac{(200 + 100)(100)}{700} = 42.8\% \ CaCl_2$$

$$\text{Final enthalpy} = \frac{-5100}{700} = -7.3 \ \text{Btu/lb}$$

(a) Final temperature is hence 190° F.

(b) This solution must be cooled to 70° F before crystallization will begin.

$$\Delta H = 700[-83 - (-7.3)] = -53,000 \ \text{Btu}$$

This problem can also be solved by the construction of a tie line as demonstrated in illustration 11.

Partial Properties. In dealing with the extensive properties of solutions, it is convenient to consider separately the contribution attributable to each component present. For example, the total volume of a solution represents the sum of the contributions of all components present. In ideal systems, such as a gas at low pressure, these contributions are the same as the properties of the pure components existing separately at the temperature and pressure of the solution. However, in many systems this additivity of properties does not exist, and the volume of a solution is quite different from the sum of the pure-component volumes. That portion of the total volume of a solution which is attributable to the presence of a particular component is termed the *partial volume* of that component. Other partial extensive properties may be similarly defined such as *partial enthalpy* and *partial heat capacity*.

The partial volume of a component in a solution of given composition may be determined experimentally by adding a unit mass of the pure component to a quantity of solution so large that only a negligible change in composition results. The increase in total volume under these conditions is termed the *partial specific volume* of the component added. Similarly, the *partial molal volume* of the component may be determined.

As the composition of a solution approaches 100% of the component under consideration, the partial molal volume approaches the molal volume of the pure component. In nonideal solutions the partial molal volume of each component is a function of composition.

Mathematically, the partial molal volume of a component of a solution may be defined[2] as

$$\bar{v}_1 = \left(\frac{\partial V}{\partial n_1}\right)_{pTn_2n_3\cdots} \tag{31}$$

where　　　　$\bar{v}_1$ = partial molal volume of component 1
　　　　　　V = total volume of solution
　　　　　　n_1 = moles of component 1

The partial derivative implies constancy of all intensive properties. Equation 31 is the formal expression for the experimental measurement, described above, of the change in volume per mole of component added at conditions of constant composition, temperature, and pressure.

Where z is a single-valued, continuous function of several independent variables x, y, u, the total differential of z can be expressed in terms of its partial derivative with respect to its independent variables, thus:

$$dz = \frac{\partial z}{\partial x}\,dx + \frac{\partial z}{\partial y}\,dy + \frac{\partial z}{\partial u}\,du \tag{32}$$

This principle is illustrated in Fig. 84 for two independent variables x and y. Applying this mathematical principle to the volume of a solution in terms of the independent variables of composition gives

$$dV = \frac{\partial V}{\partial n_1}\,dn_1 + \frac{\partial V}{\partial n_2}\,dn_2 + \frac{\partial V}{\partial n_3}\,dn_3 + \cdots \tag{33}$$

[2] G. N. Lewis and M. Randall, *Thermodynamics*, McGraw-Hill Book Co. (1923).

or, combining equations 31 and 32,

$$dV = \bar{v}_1 dn_1 + \bar{v}_2 dn_2 + \bar{v}_3 dn_3 + \cdots \tag{34}$$

A general integration of equation 34 requires a knowledge of partial molal volumes as a function of concentration. This difficulty can be avoided by integration at constant composition since then the partial molal properties are constants, thus:

$$V = \bar{v}_1 n_1 + \bar{v}_2 n_2 + \bar{v}_3 n_3 + \cdots \tag{35}$$

This procedure is justified by the following argument. The total volume in mixing n_1, n_2, etc. moles of components 1, 2, etc. at constant

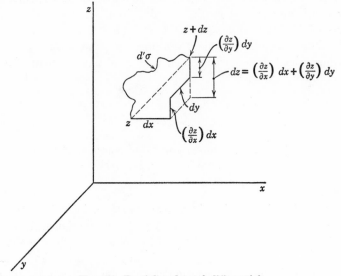

Fig. 84. Partial and total differentials

temperature and pressure is dependent only on the initial and final composition and is independent of order of mixing. Therefore, the volume obtained by mixing at constant composition according to equation 35 must also give the correct value.

Equation 35 may be differentiated to obtain

$$dV = \bar{v}_1 dn_1 + n_1 d\bar{v}_1 + \bar{v}_2 dn_2 + n_2 d\bar{v}_2 + \bar{v}_3 dn_3 + n_3 d\bar{v}_3 + \cdots \tag{36}$$

Combining equations 34 and 36 gives

$$n_1 d\bar{v}_1 + n_2 d\bar{v}_2 + n_3 d\bar{v}_3 + \cdots = 0 \tag{37}$$

Equation 37 relates the changes in the partial molal volume of one component to changes in the partial molal volumes of the others.

In Fig. 85 are shown the specific volumes of alcohol-water solutions and the partial specific volumes of the separate components. The departure of the specific volume of a solution from a straight-line relationship indicates that the volume of the solution is not an additive property but that shrinkage in volume occurs when the two components are mixed. At alcohol concentrations above 28% by weight, each component contributes less than the pure specific volume; below 28%, water contributes more than its own pure specific volume and alcohol less.

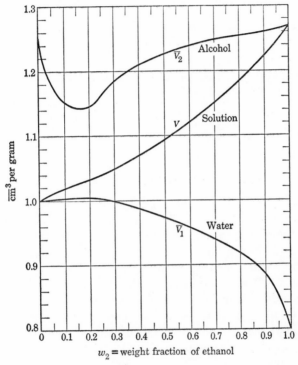

Fig. 85. Partial and total specific volumes of the ethanol-water system at 20° C

Equations 33 to 37 were developed for volumes because of the ease with which this property may be visualized. Similar equations may be developed relating any extensive property to the contributions of the separate components. Thus, the total molal enthalpy of a solution is equal to the sum of the products of the partial enthalpies of the components times their mole fractions, or the total enthalpy per gram is equal to the sum of the products of the partial enthalpies per gram times the weight fractions.

It is evident that partial extensive properties per mole or unit weight of component are themselves intensive properties, independent of the mass of solution under consideration but functions of composition as well as of temperature and pressure.

Partial Enthalpies. The enthalpy of a solution may be calculated either from integral heat of solution data combined with the enthalpies of its pure components or directly as the sum of the partial enthalpies of the components. Thus,

$$H_s = n_1 H_1 + n_2 H_2 + n_2 \Delta H_{s2} = n_1 \bar{H}_1 + n_2 \bar{H}_2 \qquad (38)$$

The use of partial molal properties is frequently the more convenient and more nearly accurate, particularly where small changes in composition occur.

Integral heats of solution may also be expressed in terms of differential heats of solution, thus:

$$(n_2 \Delta H_{s2} = n_1 \Delta \bar{H}_{s1} + n_2 \Delta \bar{H}_{s2})_T \qquad (39)$$

where $\Delta \bar{H}_{s1}$, $\Delta \bar{H}_{s2}$ = differential heats of solution of components 1 and 2.

Where arbitrary values of zero are assigned to

H₁ and H₂ at temperature T, $\Delta \bar{H}_{s1} = \bar{H}_1$ and $\Delta \bar{H}_{s2} = \bar{H}_2$.

Lewis and Randall[2] present several methods of calculating partial properties; two of these are illustrated by determining in one case partial enthalpies per pound and in the other partial molal enthalpies of components in solutions. The same procedures are used in calculating other partial extensive properties.

Method of Tangent Slope. If the total enthalpy of a solution is plotted as ordinates against the moles of solute, component 2, per fixed quantity of all other components, it follows from an equation similar to equation 31 that the partial molal enthalpy of component 2 is represented by the slope of this curve. If the solution is binary, the partial molal enthalpy of the solvent component then may be calculated from an equation similar to 38.

For most accurate results covering the entire range of composition of a binary solution two such plots should be constructed, one covering the concentration range of solute from 0 to 50% on the basis of one mole or unit weight of solute and the other covering the concentration range of solvent from 0 to 50% constructed on the basis of one mole or unit weight of solvent.

Illustration 13. From the following data for the heats evolved when water and glycerin are mixed to form 1 gram of solution, calculate the partial enthalpies at 25° C of water and glycerin, per gram of each component, at each of the designated concentrations. Plot curves relating the partial enthalpies of glycerin and of water to percentage of glycerin by weight. As the reference state of zero enthalpy use the pure components at 25° C, the temperature of the solutions.

Solution: Method of Tangent Slope

% Glycerin by Weight	$\Delta H_m = H$	w_2	H_1	w_1	H_2	$\bar{H}_2$	$\bar{H}_1$
10	−1.91	111	−2120	....	......	−13.5	−0.6
20	−3.21	250	−4020	....	......	−10.5	−1.4
30	−3.90	429	−5560	....	......	−8.1	−2.1
40	−4.50	668	−7500	....	......	−6.2	−3.4
50	−4.50	1000	−9000	1000	−9000	−4.2	−4.8
60	−4.21	....	......	668	−7000	−2.65	−6.5
70	−3.70	....	......	429	−5290	−1.7	−8.3
80	−2.61	....	......	250	−3660	−0.8	−11.4
90	−1.79	....	......	111	−1990	−0.4	−14.3

ΔH_m = heat of mixing, $= H$, total enthalpy, calories per gram of solution

w_2 = grams of glycerin per 1000 grams of water

w_1 = grams of water per 1000 grams of glycerin

H_1 = total enthalpy of solution per 1000 grams of water

H_2 = total enthalpy of solution per 1000 grams of glycerin

$\bar{H}_2$ = partial enthalpy of glycerin, calories per gram

$\bar{H}_1$ = partial enthalpy of water, calories per gram

The values of H_1, the enthalpy of solution per 1000 grams of water, are obtained by multiplying the heat of solution, per gram of solution, by the number of grams of the solution ($w_2 + 1000$). For a 40% solution of glycerin,

$$H_1 = (668 + 1000)(-4.50) = -7500 \text{ cal}$$

Values of w_2 are plotted as abscissas and values of H_1 as ordinates in Fig. 86. The slope of a tangent to this curve is the partial enthalpy per gram of glycerin $\bar{H}_2$ in a solution of concentration corresponding to the abscissa of the point of tangency. For a 40% solution,

$$\text{Slope of tangent} = \bar{H}_2 = \frac{-9600 + 3400}{1000} = -6.2$$

The corresponding values of $\bar{H}_1$, the partial enthalpy of water, are calculated from the following equation, which is similar to equation 38:

$$(w_1 + w_2)H = w_1\bar{H}_1 + w_2\bar{H}_2 \qquad (a)$$

For a 40% solution,

Basis: 100 grams of solution.

$$\bar{H}_1 = \frac{100H}{w_1} - \frac{w_2\bar{H}_2}{w_1} = \frac{-450 + (40)(6.2)}{60} = -3.4$$

The concept of partial enthalpies may be clearly visualized by inspection of Fig. 86. Point *a* represents the total enthalpy of a solution containing 668 grams

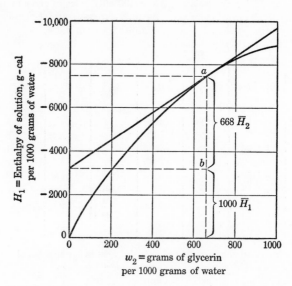

Fig. 86. Calculation of partial enthalpies by the tangent slope method

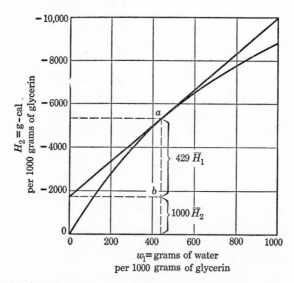

Fig. 87. Calculation of partial enthalpies by the tangent slope method

of glycerin and 1000 grams of water. This total enthalpy is the sum of the partial enthalpy of the 668 grams of glycerin, equal to $a - b$ on the diagram, plus the partial enthalpy of the 1000 grams of water, equal to b.

It will be noted that at the higher concentrations of glycerin the slope of the curve of Fig. 86 becomes small, and it is difficult to determine a value of $\bar{H}_2$ from it with sufficient accuracy to permit a reliable calculation of $\bar{H}_1$. For this reason it is inadvisable to use this curve in the range of concentrations above 50% glycerin. In this range more accurate graphical results may be obtained by plotting total enthalpies for solutions containing 1000 grams of glycerin against the weight of water dissolved. From this curve values of $\bar{H}_1$ are determined directly. The corresponding values of $\bar{H}_2$ are calculated from equation a. In Fig. 87 values of w_1, grams of water per 1000 grams of glycerin, are plotted as abscissas against H_2, total enthalpy per 1000 grams of glycerin. The slope of a tangent to this curve is the partial enthalpy of water per gram in a solution of concentration corresponding to the abscissa at the point of tangency. For a solution containing 70% glycerin,

$$\text{Slope of tangent} = \bar{H}_1 = \frac{-10,000 + 1700}{1000} = -8.3 \text{ cal per gram of water}$$

From equation a, on the basis of 100 grams of solution,

$$\bar{H}_2 = \frac{100H}{w_2} - \frac{w_1\bar{H}_1}{w_2} = \frac{-370 - (-8.3 \times 30)}{70} = -1.7 \text{ cal per gram of glycerin}$$

In Fig. 87 point a represents the total enthalpy of a solution containing 429 grams of water and 1000 grams of glycerin. This total is the sum of the partial enthalpy of the water, equal to $a - b$ plus the partial enthalpy of the glycerin, equal to b.

Method of Tangent Intercepts. In the method of tangent intercepts the total enthalpy of the solution based on unit quantity of solution instead of unit quantity of either component is plotted against the fraction of either component covering the range of concentrations from zero to unity. If a tangent is drawn to this curve at a concentration corresponding to a desired fraction of solute, then the intercepts of this tangent line with the ordinates corresponding to zero and unit fractions of solute give the respective partial enthalpies of the solvent and solute.

This method may be used for obtaining partial enthalpies per unit weight of either component if compositions are expressed in weight fractions and if the total enthalpies are expressed on the basis of a unit weight of solution. Partial molal enthalpies are obtained by plotting total enthalpy per mole of solution against mole fraction.

A proof of this relationship based on molal units follows:

Let $H =$ enthalpy of a solution containing n_1 moles of solvent and n_2 moles of solute relative to the pure components (the enthalpies of the pure solvent and pure solute are each arbitrarily taken as zero at the temperature of the solution).

$\text{H} =$ enthalpy per mole of solution

N_1 = mole fraction of solvent
N_2 = mole fraction of solute

$$\text{H} = \frac{H}{n_1 + n_2} \tag{40}$$

$$N_2 = \frac{n_2}{n_1 + n_2} \tag{41}$$

On curve III of Fig. 88 values of enthalpy ʜ in Btu per pound-mole of solution are plotted against mole fraction N_2 of the solute. A tangent is drawn to curve III at point E corresponding to a mole fraction of N_2 equal to C. This tangent intercepts the ordinate $N_2 = 0$ at point A and the ordinate $N_2 = 1$ at point F. Ordinate OB represents the

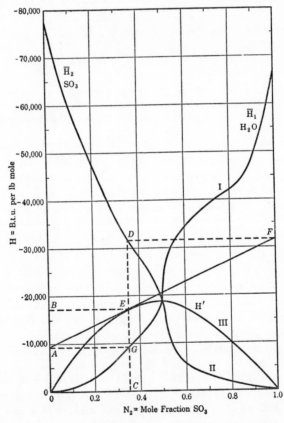

Fɪɢ. 88. Calculation of differential heats of solution by the method of tangent intercepts, SO_3—H_2O system

value of H at a mole fraction $N_2 = C$. The slope of the tangent at E is equal to $\dfrac{d\text{H}}{dN_2}$; also

$$OA = OB - AB = \text{H} - N_2 \frac{d\text{H}}{dN_2} \tag{42}$$

Variation in the mole fraction of solute N_2 may be produced by addition or removal of solvent, n_2 being kept constant. Differentiating equation 40 with respect to n_1, keeping n_2 constant, gives

$$d\text{H} = \frac{dH}{n_1 + n_2} - H\frac{dn_1}{(n_1 + n_2)^2} \tag{43}$$

Differentiating equation 41 with respect to n_1, keeping n_2 constant, yields

$$dN_2 = -\frac{n_2 dn_1}{(n_1 + n_2)^2} \tag{44}$$

Combining equations 43 and 44,

$$\frac{d\text{H}}{dN_2} = \frac{-dH(n_1 + n_2)}{n_2 dn_1} + \frac{H}{n_2} \tag{45}$$

Substituting equation 45 in equation 42,

$$OA = \text{H} + \frac{N_2 dH(n_1 + n_2)}{n_2 dn_1} - \frac{N_2 H}{n_2} \tag{46}$$

Substituting equations 40 and 41 in equation 46,

$$OA = \text{H} + \frac{dH}{dn_1} - \text{H} = \frac{dH}{dn_1} \tag{47}$$

Since n_2 was taken as constant in all differentiations,

$$\frac{dH}{dn_1} = \left(\frac{\partial H}{\partial n_1}\right)_{n_2} = OA = \bar{\text{H}}_1 \tag{48}$$

which by definition is the partial molal enthalpy of component 1. Similarly, it may be shown that

$$OF = \bar{\text{H}}_2 \tag{49}$$

which is the partial molal enthalpy of the solute.

In Fig. 88 the total and partial enthalpies at 25° C, referred to the pure components at 25° C, of liquid SO_3 dissolved in liquid H_2O are plotted. The partial enthalpies for this system were calculated by

Morgen[3] from the data of Bichowsky and Rossini[4] for values of SO_3 up to $N_2 = 0.5$ and from Herrmann[5] for values of SO_3 from $N_2 = 0.5$ to 1.0.

The use of integral heat of solution data or an enthalpy concentration chart is sound, but more nearly accurate results are obtained from partial enthalpy data, particularly where slight changes in concentration are involved. The heat evolved in any process involving changes in concentration can be obtained by the usual energy-balance method where the enthalpy of the initial system plus the heat absorbed is equal to the enthalpy of the final system. The following illustration is taken from Morgen.[3]

Illustration 14. It is desired to increase the strength of a 23.2% H_2SO_4 solution (19% SO_3) to 80.6% H_2SO_4 (65.6% SO_3) with oleum containing 41.2% free SO_3 (89.2% SO_3).

Calculate the heat evolved at 25° C on the basis of 100 lb of weak acid.

From a material balance of SO_3,

$$\text{Weak acid} + \text{oleum} = \text{strong acid}$$
$$(0.19)(100) + x(0.892) = (100 + x)(0.656)$$
$$x = 197.5 \text{ lb oleum}$$
$$100 + x = 297.5 \text{ lb strong acid}$$

	Weak Acid	Oleum	Strong Acid
Lb SO_3	19	176	195
Lb H_2O	81	21	102
Lb total	100	197.5	297.5
Lb-moles SO_3	0.238	2.202	2.44
Partial molal enthalpy, $SO_3 \times 10^{-3}$	−67.1	−4.68	−35.82
Lb-moles H_2O	4.50	1.185	5.686
Partial molal enthalpy of $H_2O \times 10^{-3}$	−0.14	−35.10	−7.16

From an energy balance of the process using partial enthalpies the following results are obtained.

$$\frac{\Delta H}{1000} + \overbrace{0.238(-67.1)}^{SO_3} + \overbrace{4.50(-0.14)}^{H_2O} + \overbrace{2.202(-4.68)}^{SO_3} + \overbrace{1.185(-35.10)}^{H_2O}$$

(Weak acid: SO_3, H_2O; Oleum: SO_3, H_2O)

$$= \overbrace{2.440(-35.82)}^{SO_3} + \overbrace{5.686(-7.16)}^{H_2O}$$

(Strong acid: SO_3, H_2O)

$$\Delta H = -59,620 \text{ Btu per 100 lb weak acid}$$

or   Heat evolved = 59,620 Btu

[3] R. A. Morgen, *Ind. Eng. Chem.*, **34**, 571 (1942), with permission.

[4] F. R. Bichowsky and F. D. Rossini, *The Thermochemistry of the Chemical Substances*, Reinhold Publishing Corp. (1936), with permission.

[5] C. V. Herrmann, *Ind. Eng. Chem.*, **33**, 898 (1941).

Incomplete and Successive Reactions

In the preceding sections consideration was given to the thermal effects accompanying chemical reactions which went to completion and in which the reactants were present in stoichiometric proportions. In industrial processes, excess reactants are nearly always present, and the reactions seldom go to completion.

Frequently the chemical transformations of the same reactant proceed in successive steps or in divergent stages, and the quantities of the different products that are formed bear no stoichiometric relationship to one another but are dependent on the relative rates of the different reactions involved.   For example, in the combustion of carbon, both carbon monoxide and carbon dioxide are formed in variable proportions.   In such reactions the standard heat of reaction is calculated by subtracting the heats of formation of the reactants which actually react from the heats of formation of the products actually formed. Or the heats of combustion of the products actually formed are subtracted from the heats of combustion of the reactants transformed to give the standard heat of reaction.

Illustration 15.   In the production of metallic manganese, 10 kg of manganese oxide, Mn_3O_4, are heated in an electric furnace with 3.0 kg of amorphous carbon (coke).   The resulting products are found to contain 4.8 kg of manganese metal and 2.6 kg of manganous oxide, MnO, as slag.   The remainder of the products consists of unconverted charge and carbon monoxide gas.   Calculate the standard heat of reaction of this process for the entire furnace charge.

Solution

$$\text{Initial } Mn_3O_4 = 10.0 \text{ kg or } 10/229 \qquad = 0.0437 \text{ kg-mole}$$
$$\text{Initial } C = 3.0 \text{ kg or } 3.0/12 \qquad = 0.250 \text{ kg-atom}$$
$$\text{Mn formed} = 4.8 \text{ kg or } 4.8/55 \qquad = 0.0874 \text{ kg-atom}$$
$$\text{MnO formed} = 2.6 \text{ kg or } 2.6/71 \qquad = 0.0366 \text{ kg-mole}$$

$$\text{Unconverted } Mn_3O_4 = 0.0437 - \frac{0.0874 + 0.0366}{3} = 0.0024 \text{ kg-mole}$$

$$\text{O in final CO} = 4(0.0437 - 0.0024) - 0.0366 \qquad = 0.1286 \text{ kg-atom}$$
$$\text{CO formed} \qquad = 0.1286 \text{ kg-mole}$$
$$\text{Unconverted } C = 0.250 - 0.1286 \qquad = 0.1214 \text{ kg-mole}$$

MATERIAL BALANCE

Materials Entering		Materials Leaving		
Mn_3O_4	= 10.0 kg	Mn		= 4.8 kg
C	= 3.0	MnO		= 2.6
	Total = 13.0 kg	Mn_3O_4	= 0.0024 × 229	0.55
		CO	= 0.1286 × 28	3.6
		C	= 0.1214 × 12	1.45
				13.00 kg

Heats of formation of active reactants (data from Table 29):

$$
\begin{aligned}
Mn_3O_4 &= (-331,400)(0.0437 - 0.0024) &&= -13,700 \text{ kcal}\\
C(coke) &= (2600)(0.250 - 0.1214) &&= 335\\
&\quad\text{Total} &&= -14,035 \text{ kcal}
\end{aligned}
$$

Heats of formation of products actually formed:

$$
\begin{aligned}
MnO &= -92,000 \times 0.0366 &&= -3365 \text{ kcal}\\
CO &= -26,416 \times 0.1286 &&= -3400\\
Mn &&&= 0\\
&\quad\text{Total} &&= -6765 \text{ kcal}
\end{aligned}
$$

Standard heat of reaction $= -6765 - (-14,035) = +7270$ kcal

Effect of Pressure on Heat of Reaction

The definition of standard heat of reaction includes the specification that the reaction be carried out at a constant pressure of 1 atm and a temperature of $298.16°$ K. The relation between the heats of re-

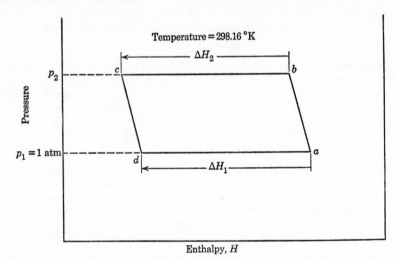

Fig. 89.   Effect of pressure on heat of reaction

action carried out at two different constant pressures may be developed with the aid of Fig. 89. Point a indicates the enthalpy of the reactants at $298.16°$ K and 1 atm, whereas point b indicates the enthalpy of the reactants at the same temperature but at the higher pressure p_2. Likewise, points d and c represent the respective enthalpies of the products at pressures p_1 and p_2.

Since enthalpy is a point function, the total change for a process is governed solely by the terminal conditions, and not by the path followed.

Accordingly, the total enthalpy change in going from a to c along path abc is exactly equal to the total enthalpy change in going from a to c along the path adc. This principle results in the equation

$$H_{R2} - H_{R1} + \Delta H_2 = \Delta H_1 + H_{P2} - H_{P1} \tag{50}$$

where subscripts R and P refer to reactants and products, respectively. Rearranging gives

$$\Delta H_2 = \Delta H_1 + (H_{P2} - H_{P1}) - (H_{R2} - H_{R1}) \tag{51}$$

This equation is subject to considerable simplification if the pressures are such that ideal behavior for gaseous reactants and products may be assumed. For an ideal gas, enthalpy is independent of pressure; therefore, for the gaseous reactants, $H_{R2} = H_{R1}$, and, for the gaseous products, $H_{P2} = H_{P1}$. As far as any solid or liquid reactants and products are concerned, pressure has but little effect on enthalpy. For example, in the reaction $C(s) + O_2(g) = CO_2(g)$, the enthalpy change in compressing 1 g-atom of graphite from 1 to 5 atm is 0.52 g-cal. In most thermochemical calculations an item of such magnitude is negligible. Therefore, it is usually permissible to assume that for solid or liquid reactants $H_{R2} = H_{R1}$ and for solid or liquid products $H_{P2} = H_{P1}$. It may, therefore, be concluded that for moderate pressure differences,

$$\Delta H_2 = \Delta H_1 \tag{52}$$

If the pressure differences are great, the assumption of ideal behavior is no longer justified, and all terms of equation 51 must be evaluated. For gaseous reactants and products, the enthalpy change in shifting from the lower to the higher pressure is evaluated as explained in Chapter 14. For liquids and solids, it is usually satisfactory to assume constancy of internal energy and volume with change of pressure and calculate the enthalpy change as equal to $V\Delta p$.

Heats of Reaction at Constant Pressure and at Constant Volume

The definition of standard heat of reaction includes the requirement that the reaction be carried out at a constant pressure of 1 atm. Actually, in much calorimetric work, particularly in the determination of heats of combustion of solid or liquid substances, the experiments are conducted under constant-volume conditions rather than under conditions of constant pressure. It therefore is necessary to convert q_v, the experimentally determined absorption of heat under constant-volume conditions, into q_p, the absorption of heat under constant-pressure conditions.

By applying the first law to constant-volume conditions and recognizing that, at constant volume, work of expansion is zero, the following equation is obtained:

$$q_V = \Delta U_V \tag{53}$$

The same reaction may be carried out under constant pressure conditions, with the pressure equal to the initial pressure for the constant-volume case, and with the same terminal temperatures as for the constant-volume conditions. The first law applied to this case gives

$$q_p = \Delta U_p + w_{ep} = \Delta U_p + p\Delta V \tag{54}$$

Subtracting equation 53 from equation 54 yields

$$q_p - q_V = \Delta U_p - \Delta U_V + p\Delta V \tag{55}$$

If ideal behavior of gaseous reactants and products is assumed, the internal energy of such components will be unaffected by pressure. Furthermore, for liquids and solids, it is known that the internal energy is principally dependent on temperature and is but slightly affected by pressure. Hence, under these circumstances, $\Delta U_p = \Delta U_V$.

In evaluating $p\Delta V$ in equation 55, the assumption is made that the volume changes due to the consumption or formation of solids or liquids by the reaction are very small and may be ignored. If gases are involved in the reaction, it is assumed that the ideal-gas law is obeyed, and, accordingly,

$$p\Delta V = p\Delta V_g = \Delta n_g RT \tag{56}$$

where ΔV_g is the volume change of the gaseous materials participating in the process and Δn_g represents the change in the number of moles of gaseous materials. For example, in the reaction

$$C(s) + \tfrac{1}{2}O_2(g) = CO(g), \qquad \Delta n_g = 1 - \tfrac{1}{2} = \tfrac{1}{2}$$

The final working equation thus becomes

$$q_p = q_V + \Delta n_g RT \tag{57}$$

In terms of the heat evolutions Q_p and Q_V, the equation becomes

$$Q_p = Q_V - \Delta n_g RT \tag{58}$$

Illustration 16. A fuel oil analyzed 87% carbon and 13% hydrogen. Its heating value was determined in an oxygen-bomb calorimeter, with all water formed from the combustion of the hydrogen going to the liquid state. The calorimeter temperature was approximately 25° C. Duplicate determinations gave values of 19,450 and 19,410 Btu per lb. Calculate the difference, as Btu per pound, between Q_v, the experimentally determined heating value under constant-volume conditions, and Q_p, the heating value under constant-pressure conditions.

Reactions

C (in fuel oil) + $O_2(g)$ = $CO_2(g)$, $\Delta n_g = 1 - 1 = 0$

H_2 (in fuel oil) + $\frac{1}{2}O_2(g)$ = $H_2O(l)$, $\Delta n_g = -\frac{1}{2}$ mole per mole hydrogen

Basis: 1 lb fuel oil

$$\text{Lb-moles hydrogen} = \frac{1 \times 0.13}{2.016} = 0.06448$$

Substituting in equation 6 gives

$$(Q_p - Q_v) = -(-\tfrac{1}{2})(0.06448)(1.987)(1.8)(298.16) = 34.4 \text{ Btu per lb}$$

In technical calorimetry, using standard ASTM methods, duplicate determinations are to be within 0.3% of the mean value. With this particular fuel oil, the permissible spread between the duplicate determinations is 117 Btu. Hence, in this instance, the correction to constant-pressure conditions is of no significance. However, in precision calorimetry with pure compounds, the correction to constant-pressure conditions usually is of sufficient magnitude in relation to the accuracy of the experimental method to warrant its application.

Effect of Temperature on Heat of Reaction

Standard heats of reaction represent the enthalpy changes during a reaction at constant pressure in which all reactants are initially at a selected standard temperature and all products are finally existent at that same temperature. Such conditions are rarely encountered in industrial reactions. Various reactants may enter at different temperatures, and the various products may each leave at still different temperatures. The heat effects of such reactions may be calculated from data on standard heats of reaction and thermophysical properties.

Kirchhoff's Equation. An analytical relation for the effect of temperature on the heat of reaction may be derived for the special case of reactions that begin and end with all materials at the same temperature. Let

$$\Delta H_T = \text{heat of reaction at temperature } T$$
$$\Delta H_T + d\Delta H = \text{heat of reaction at temperature } T + dT$$

Since the enthalpy change in going from reactants at T to products at $T + dT$ is independent of path,

$$dH_R + \Delta H_T + d\Delta H = \Delta H_T + dH_P \tag{59}$$

where dH_R and dH_P are the changes in enthalpies of the reactants and products, respectively, corresponding to temperature change dT.

This can be shown diagrammatically in Fig. 90.

From Fig. 90 at temperature T the enthalpy of reactants is H_R and of products is H_P. At temperature $T + dT$ the enthalpy of reactants is H_R' and of products is H_P'.

$$H_R' - H_R = dH_R = C_p\, dT \tag{60}$$

where C_p = total heat capacity of reactants. Similarly,

$$H_P' - H_P = dH_P = C_p'\, dT \tag{61}$$

where C_p' = total heat capacity of products.

$$\Delta H_T = H_P - H_R \tag{62}$$

$$\Delta H_{T+dT} = H_P' - H_R' = \Delta H_T + d\Delta H \tag{63}$$

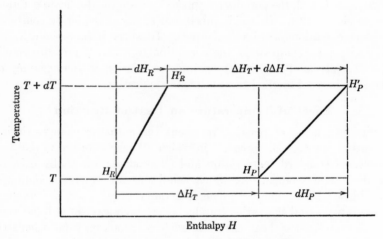

Fig. 90.   Effect of temperature on heat of reaction

ΔH_T is the same whether the reaction goes directly from H_R to H_P or through the path H_R, H_R', H_P' to H_P.

Equation 59 becomes

$$d\Delta H = (C_p' - C_p)\, dT = \Delta C_p\, dT \tag{64}$$

where ΔC_p = change in heat capacity of entire system at constant pressure.

Equation 64, known as Kirchhoff's equation, may be applied to the following general reaction at constant pressure and temperature:

$$n_b B + n_c C + \cdots = n_r R + n_s S + \cdots$$

where n_b, n_c, n_r, n_s = number of moles of components $B, C, R, S \cdots$.

For this reaction,

$$\Delta C_p = n_r c_{pr} + n_s c_{ps} + \cdots - n_b c_{pb} - n_c c_{pc} \cdots \tag{65}$$

Where the molal heat capacity of a substance may be represented by an empirical equation of the form

$$c_p = a + bT + cT^2 \tag{66}$$

each heat capacity term in equation 65 may be replaced by an empirical equation of type 66. Then,

$$\Delta C_p = \Delta a + \Delta bT + \Delta cT^2 \tag{67}$$

where

$$\Delta a = n_r a_r + n_s a_s \cdots - n_b a_b - n_c a_c \cdots \tag{68}$$

$$\Delta b = n_r b_r + n_s b_s \cdots - n_b b_b - n_c b_c \cdots \tag{69}$$

$$\Delta c = n_r c_r + n_s c_s \cdots - n_b c_b - n_c c_c \cdots \tag{70}$$

Substituting equation 67 in equation 64 and integrating yields

$$\Delta H_T = I_H + \Delta aT + \tfrac{1}{2}\Delta bT^2 + \tfrac{1}{3}\Delta cT^3 \tag{71}$$

The constant of integration I_H of equation 71 may be evaluated from a single value of ΔH_T. This is usually done at the standard conditions of 25° C. The other constants of equation 71 are obtained directly from the empirical heat-capacity equations for the reactants and products.

Equation 71 is of thermodynamic importance in determining the effect of temperature on chemical equilibria. The utility of this equation is limited to reactions that begin and end at the same temperature where no changes in phase are involved in going from the given reference temperature to temperature T and where heat capacities are expressed by the given empirical equations.

Enthalpy Changes in Reactions with Different Temperatures. The enthalpy change accompanying any reaction may be expressed in terms of an over-all energy balance. Thus, in constant pressure or flow processes where changes in kinetic, potential, and surface energies are negligible and no work is performed,

$$q = \Delta H = \sum H_{P'} - \sum H_R \tag{72}$$

where

$\sum H_R$ = sum of enthalpies of all materials entering the reaction relative to the reference state for standard heats of reaction, conveniently 25° C and 1.0 atm

$\sum H_{P'}$ = sum of enthalpies of all materials leaving the reaction, referred to the form of chemical combination in which they entered the reaction at the standard reference conditions

A substance that is produced by chemical reaction has an enthalpy at 25° C referred to the reactants at 25° C which is by definition equal to the standard heat of reaction. Thus,

$$H_{P'} = H_P + \Delta H_{25} \qquad (73)$$

where

H_P = enthalpy of the product, referred to its standard state at 25° C
ΔH_{25} = standard heat of reaction

Combining equations 72 and 73 gives

$$q = \Delta H = \sum H_P + \sum \Delta H_{25} - \sum H_R \qquad (74)$$

In using equation 74, standard heats of reaction are included in the summations of the equation only for the products actually resulting from reactions taking place in the process and to the extent that they are formed in the process. This term becomes zero for all materials passing through the process without chemical change.

Illustration 17. Carbon monoxide at 200° C is burned under atmospheric pressure with dry air at 500° C in 90% excess of that theoretically required. The products of combustion leave the reaction chamber at 1000° C. Calculate the heat evolved in the reaction chamber in kilo calories per kilogram-mole of CO burned, assuming complete combustion.

Basis: 1.0 kg-mole of CO.

$$CO(g) + \tfrac{1}{2} O_2(g) = CO_2(g)$$

From the data of Table 30, page 306,

$$\begin{aligned}
\Delta H_{25} &= -67,636 \text{ kcal} \\
O_2 \text{ required} &= 0.5 \text{ kg-mole} \\
O_2 \text{ supplied} = 0.5 \times 1.90 &= 0.95 \text{ kg-mole} \\
\text{Air supplied} = 0.95/0.21 &= 4.52 \text{ kg-moles} \\
N_2 \text{ present} = 4.52 - 0.95 &= 3.57 \text{ kg-moles} \\
\text{Unused } O_2 = 0.95 - 0.50 &= 0.45 \text{ kg-mole}
\end{aligned}$$

Enthalpy of reactants (H_R) relative to standard state at 25° C where c_{pm} = mean molal heat capacity between 25 and $t°$ C:

CO:

$$c_{pm} = 7.017$$
$$\text{Enthalpy} = (7.017)(1.0)(200 - 25) = 1227 \text{ kcal}$$

Air:

$$c_{pm} = 7.225$$
$$\text{Enthalpy} = (7.225)(4.52)(500 - 25) = 15,520 \text{ kcal}$$
$$\sum H_R = 15,520 + 1,227 = 16,747 \text{ kcal}$$

Enthalpy of products H_p relative to standard state at 25° C:

CO_2:

$$c_{pm} = 11.92$$
$$\text{Enthalpy} = (11.92)(1.0)(1000 - 25) = 11,630 \text{ kcal}$$

O_2:

$$c_{pm} = 7.941$$
$$\text{Enthalpy} = (7.941)(0.45)(1000 - 25) = 3482 \text{ kcal}$$

N_2:

$$c_{pm} = 7.507$$
$$\text{Enthalpy} = (3.57)(7.507)(1000 - 25) = 26,150 \text{ kcal}$$
$$\sum H_P = 11,630 + 3,482 + 26,150 = 41,262 \text{ kcal}$$

From equation 74,

$$\Delta H = 41,262 - 67,636 - 16,747 = -43,121 \text{ kcal}$$

Since the reaction proceeds at constant pressure, the enthalpy change is equal to the heat absorbed, or there is an evolution of heat of 43,121 kcal.

Enthalpy Terms in Energy Balances. As has been pointed out on page 247, enthalpy is a convenient property combining internal energy and the product pV. Although enthalpy has energy units, it cannot in the general case be considered as solely energy. For example, the enthalpy of an incompressible liquid may be increased by merely increasing the pressure without doing any work or changing the energy of the system. In such a case enthalpy is not a measure of energy, and there can be no generalization of "conservation of enthalpy" parallel to the principle of conservation of energy. In the general case the enthalpy input and output items of a process do not necessarily balance, and true energy balances should include only energy items.

It is proper to include enthalpy terms in energy balances under two conditions:

1. In a reversible flow process, the enthalpy is the sum of its internal energy and the flow work, and as such is properly included in the general energy balance represented by equation 1, page 246.

2. In a non-flow process at constant pressure where work is performed only by expansion, enthalpy is properly included as a term in the general energy balance of the form of equation 12, page 247.

Fortunately the great majority of processes of industrial importance fall in one or the other of these classifications.

In setting up an energy balance, it is common practice to tabulate all energy input items in one column and all output items in another. For convenience in adding, all entries in each column are arranged to be of positive sign. Thus, in such a tabulation any work done on the system will appear as an input item while work done by the system will appear as an output item. Similarly, heat absorbed is an input item, and heat evolved an output item.

For all flow processes and for nonflow processes at constant pressure, the enthalpies of all entering materials are entered as input items, and those of the product materials, referred to their own standard states, as output items. In such processes enthalpy changes equal heat absorptions or evolutions and may be entered in the energy balance as such. Standard heats of reaction are input items if they are negative, representing a contribution of heat to the process and output items if positive. The over-all enthalpy changes of the process are entered as output heat items if negative and as input items if positive.

Since energy balances of this type are entirely valid for the specified types of processes, all forms of energy may be properly included along with the work, heat, and enthalpy items. Where potential, kinetic, and surface energies are not negligible, these input and output items are entered as part of the general balance.

Illustration 18. From the results of illustration 17 prepare an energy balance for the combustion process under consideration.

Since this is a flow process, enthalpies may be included in the energy balance, and enthalpy changes are equal to heat absorptions and liberations.

Input		Output	
Enthalpy of CO	1,227 kcal	Enthalpy of CO_2	11,630 kcal
Enthalpy of air	15,520	Enthalpy of O_2	3,482
Standard reaction,		Enthalpy of N_2	26,150
heat added	67,636	Heat evolved	43,121
	84,383 kcal		84,383 kcal

Temperature of Reaction

Adiabatic Reactions. If a reaction proceeds without loss or gain of heat and if all the products of the reaction remain together in a single mass or stream of materials, these products will assume a definite temperature known as the adiabatic-reaction temperature. In this particular case, for a flow process or a nonflow process at constant pressure where the only energy terms involved are internal energy and flow work, the enthalpy change is zero, and it follows from equation 74 that the sum of the enthalpies of the reactants must equal the sum of the standard heat of reaction and the enthalpies of all the products. The temperature of the products which corresponds to this total enthalpy may be calculated by expressing the enthalpy of the products as a function of temperature and solving the resultant equation for temperature T. This requires data on the heat capacities and latent heats of all products.

The products considered in calculating an adiabatic-reaction temperature must include all materials actually present in the final system; inerts and excess reactants as well as the new compounds formed must be included. If the reaction is incomplete, only the standard heat of reaction resulting from the degree of completion actually obtained is considered, and the products will include some of each of the original reactants.

The enthalpy $\sum H_P$ of n_i moles of any product material at a temperature $T°$ K, referred to a temperature of 298° K is expressed by

$$\sum H_P = \sum n_i \int_{298}^{T} c_{p_i}\, dT + \sum n_i \lambda_i \tag{75}$$

The integration term refers to the change in enthalpy of component i for the temperature change of each phase, and $\sum n_i \lambda_i$ refers to all latent heats of all phase changes of component i in being heated from its reference state to the adiabatic-reaction temperature at constant pressure.

Where the molal heat capacity c_{p_i} is expressed as a quadratic function of temperature,

$$c_{p_i} = a_i + b_i T + c_i T^2 \tag{76}$$

upon integration,

$$\sum H_P = \sum n_i \left[a_i(T - 298) + \frac{b_i}{2}(T^2 - 298^2) + \frac{c_i}{3}(T^3 - 298^3) \right] + \sum n_i \lambda_i \tag{77}$$

The adiabatic-reaction temperature is then obtained by solving equation 77 for T. The equation is most easily solved by assuming values of T until the equality is satisfied.

The above calculation procedure presupposes a knowledge of the exact composition of the products and an assumption that the normal values of heat capacities are valid. For highly exothermic reactions extremely high temperatures may be obtained which result in the formation of free radicals and free atoms. The heat of formation of these species must be included, and the composition of such products can be calculated only for equilibrium conditions.

The above calculation procedures also fall short where normal heat-capacity equations are invalid. For example in extremely rapid reactions as in the explosion of gases the early temperatures will be higher than calculated from normal heat-capacity values. A small fraction of a second is required to establish an equilibrium distribution of energy in the gaseous products. Where the energy is initially absorbed only as kinetic energy of translation, extremely high temperatures will result. An equilibrium temperature will be attained only after the various degrees of molecular motion of rotation and vibration have attained equilibrium values.

Illustration 19. For the production of sulfuric acid by the contact process, iron pyrites, FeS_2, is burned with air in 100% excess of that required to oxidize all iron to Fe_2O_3 and all sulfur to SO_2. It may be assumed that the combustion of the pyrites is complete to form these products and that no SO_3 is formed in the burner. The gases from the burner are cleaned and passed into a catalytic converter in which 80% of the SO_2 is oxidized to SO_3 by combination with the oxygen present in the gases. The gases enter the converter at a temperature of 400° C.

Assuming that the converter is thermally insulated so that heat loss is negligible, calculate the temperature of the gases leaving the converter.

$$4FeS_2 + 11O_2 = 2Fe_2O_3 + 8SO_2$$
$$SO_2 + \tfrac{1}{2}O_2 = SO_3$$

Basis: 4.0 g-moles of FeS_2

Oxygen supplied for 100% excess = 11.0×2.0 =	22 g-moles
Air introduced = 22/0.21	= 104.8 g-moles
N_2 introduced = 104.8 − 22	= 82.8 g-moles
Excess O_2 in burner gases = 22 − 11	= 11.0 g-moles
SO_2 in burner gases	= 8.0 g-moles

Gases entering converter:

$$
\begin{aligned}
SO_2 &= 8.0 \text{ g-moles} \\
O_2 &= 11.0 \\
N_2 &= 82.8 \\
\hline
\text{Total} &= 101.8 \text{ g-moles}
\end{aligned}
$$

SO_3 formed in converter = 8.0(0.8) = 6.4 g-moles
O_2 consumed in converter = 6.4/2 = 3.2 g-moles

Gases leaving converter:

$$
\begin{aligned}
SO_3 &= 6.4 \text{ g-moles} \\
SO_2 = 8.0 - 6.4 &= 1.6 \\
O_2 = 11.0 - 3.2 &= 7.8 \\
N_2 &= 82.8 \\
\hline
\text{Total} &= 98.6 \text{ g-moles}
\end{aligned}
$$

Mean molal heat capacities of gases entering converter (mean values between 25 and 400° C are taken from Table 19, page 258):

$$
\begin{aligned}
SO_2 &= (8.0)(10.94) &= 87 \\
O_2 &= (11.0)(7.406) &= 81 \\
N_2 &= (82.8)(7.089) &= 587 \\
\hline
\text{Total} & &= 755
\end{aligned}
$$

Enthalpy $\sum H_R$ of gases entering converter = $(755)(400 - 25)$ = 283,000 cal
Standard heat of reaction ΔH_{298} (heats of formation from Table 29, page 301),

$$
\begin{aligned}
SO_2: \quad \Delta_{Hf} &= -70{,}960 \text{ cal} \\
SO_3: \quad \Delta_{Hf} &= -94{,}450 \text{ cal}
\end{aligned}
$$

$$\Delta H_{298} = (6.4)[(-94{,}450) - (-70{,}960)] = -150{,}400 \text{ cal}$$

From equation 74, since $\Delta H = 0$,

$$\sum H_p = \sum H_R - \sum \Delta H_{298} = (283{,}000) - (-150{,}400) = 433{,}400$$

In order to solve for the temperature of the products, $\sum H_P$ is expressed as a function of temperature by the constants of Table 17, page 255. Integrating these equations between the limits of 298 and $T°$ K gives expressions for the enthalpies of the individual components.

O_2:

$$H = 47.7(T - 298) + 0.01236(T^2 - \overline{298}^2) - 2.612(10^{-6})(T^3 - \overline{298}^3)$$

N_2:

$$H = 535(T - 298) + 0.0575(T^2 - \overline{298}^2) - 1.904(10^{-6})(T^3 - \overline{298}^3)$$

SO_2:

$$H = 11.10(T - 298) + 0.00801(T^2 - \overline{298}^2) - 2.022(10^{-6})(T^3 - \overline{298}^3)$$

SO_3:

$$H = 47.6(T - 298) + 0.0612(T^2 - \overline{298}^2) - 14.14(10^{-6})(T^3 - \overline{298}^3)$$

Adding these equations gives

$$\sum H_P = 641.4(T - 298) + 0.1391(T^2 - \overline{298}^2) - 20.68(10^{-6})(T^3 - \overline{298}^3)$$
$$= 641.4T + 0.1391T^2 - 20.68(10^{-6})T^3 - 203,100$$

Equating the heat input to the enthalpy of the products yields

$$641.4T + 0.1391T^2 - 20.68(10^{-6})T^3 - 203,100 = 433,400$$

This equation is solved graphically or by substituting values of T until the equation is satisfied. Solving,

$$T = 855° \text{ K or } 582° \text{ C}$$

A shorter, more direct procedure involves assuming a final temperature, and then using corresponding values of mean heat capacity from Table 19 to compute the final temperature. For example, assuming a final temperature of 590° C as a first approximation, the mean molal heat capacities over the temperature range from 25 to 590° C are obtained from Table 19, page 258. Then

$$O_2 = 7.606$$
$$N_2 = 7.222$$
$$SO_2 = 11.43$$
$$SO_3 = 16.28$$

$$\sum H_P = [(7.8)(7.606) + (82.8)(7.222) + (1.6)(11.43) + (6.4)(16.28)](T - 298)$$
$$= (779.4)(T - 298) = 433,400$$
$$T = 854° \text{ K}(581° \text{ C})$$

If the calculated final temperature is considerably different from that assumed as a first approximation for obtaining mean heat capacities, the resultant error may be reduced by repeating the calculation with heat-capacity data based on the temperature calculated as a first approximation. Since mean heat capacities do not vary greatly with temperature, successive approximations of this type rapidly approach the correct solution.

Nonadiabatic Reactions. If a reaction does not proceed adiabatically, the amount of heat gained or lost from the system during the reaction must be known in order to calculate the reaction temperature.

This loss or gain of heat may be obtained experimentally by calorimetric measurements or estimated from the laws of heat transfer by conduction, convection, and radiation.

If a flow process or a nonflow process at constant pressure is under consideration the heat absorbed by the system is equal to ΔH of equation 74. Using this value of ΔH, the enthalpy and temperature of the products are calculated as in illustration 19.

Theoretical Flame Temperatures

The temperature attained when a fuel is burned in air or oxygen without gain or loss of heat is termed the *theoretical flame temperature.* The methods developed in the preceding paragraphs may be used to calculate the theoretical flame temperature of a gaseous, atomized liquid, or powdered solid fuel when burned with air or oxygen in any desired proportions.

The assumption is made that no mechanical work is involved and that the only energy terms present are internal energy and flow work. The same limitations are involved in calculating flame temperature or other reactions by these methods. The actual composition of the products must be known, including the presence of unreacted reactants, free radicals, and free atoms, and the method is not applicable to the first fraction of a second required to attain equilibrium values of heat capacities.

The maximum adiabatic flame temperature is attained when the fuel is burned with the theoretically required amount of pure oxygen. The maximum adiabatic flame temperature *in air* corresponds to combustion with the theoretically required amount of air and is obviously much lower than the maximum flame temperature in pure oxygen. Because of the necessity of using excess air in order to insure complete combustion, the adiabatic flame temperatures of actual combustions are always less than the maximum values.

Illustration 20. Calculate the theoretical flame temperature of a gas containing 20% CO and 80% N_2 when burned with 100% excess air, both air and gas initially being at 25° C.

Basis: 1.0 g-mole of CO.

$$N_2 \text{ in original gas} = \frac{1}{0.20} \times 0.80 \qquad\qquad = 4.0 \text{ g-moles}$$

$$O_2 \text{ supplied} = 0.5 \times 2 \qquad\qquad\qquad = 1.0 \quad \text{g-mole}$$

$$N_2 \text{ from air} = \frac{1.0}{0.21} \times 0.79 \qquad\qquad = 3.76 \text{ g-moles}$$

Total $N_2 = 3.76 + 4.0$ $= 7.76$ g-moles

Moles of original N_2, O_2, $CO = 7.76 + 1.0 + 1.0 = 9.76$ g-moles

Combustion products:

CO$_2$ formed $= 1.0$ g-mole

O_2 remaining $= 1.0 - 0.5$ $= 0.5$ g-mole

N_2 $= 7.76$ g-moles

Enthalpy of products, $\sum H_P$ (referred to 25° C)

CO_2:

$$H = 1.0 \left[6.339(T - 298) + \frac{0.01014}{2}(T^2 - \overline{298}^2) - \frac{3.415}{3}(10^{-6})(T^3 - \overline{298}^3) \right]$$

O_2:

$$H = 0.5 \left[6.117(T - 298) + \frac{0.003167}{2}(T^2 - \overline{298}^2) - \frac{1.005}{3}(10^{-6})(T^3 - \overline{298}^3) \right]$$

N_2:

$$H = 7.76 \left[6.457(T - 298) + \frac{0.001389}{2}(T^2 - \overline{298}^2) - \frac{0.069}{3}(10^{-6})(T^3 - \overline{298}^3) \right]$$

Adding:

$$\sum H_P = 59.504(T - 298) + 0.01125(T^2 - \overline{298}^2) - 1.484(10^{-6})(T^3 - \overline{298}^3)$$
$$\sum H_P = 59.504T + 0.01125T^2 - 1.484 \times 10^{-6}T^3 - 18{,}703$$

Since $\sum H_R = 0$ and $\Delta H = 0$, from equation 74, $\sum H_P = -\Delta H_{298}$ or

$$59.504 + 0.01125T^2 - 1.484 \times 10^{-6}T^3 - 18{,}703 = 67{,}636$$

Solving this equation graphically, $T = 1216°$ K or 943° C.

A more direct solution may be obtained by assuming the final temperature and then using the corresponding values of mean heat capacities from Table 19, page 258. Thus the mean heat capacities between 25 and 943° C are

$$CO_2 = 11.82, \qquad O_2 = 7.903, \qquad N_2 = 7.471$$

Then,

$$\sum H_P = [(1)(11.82) + (0.5)(7.903) + (7.76)(7.471)](T - 298) = 67{,}636$$

or $T = 1215°$ K (942° C)

The adiabatic flame temperature of a fuel is dependent on the initial temperature of both the fuel and the air with which it is burned. By preheating either the fuel or the air or both the total heat input is increased and the flame temperatures are correspondingly increased.

Illustration 21. Calculate the effect on the theoretical flame temperature of illustration 20 of preheating both the gas and air to 1000° C before combustion.

Basis: Same as illustration 20.

Enthalpy of reactants at 1000° C relative to 25° C:

$$CO = (1.0)(7.587)(1000 - 25) = 7,397$$
$$O_2 = (1.0)(7.941)(1000 - 25) = 7,742$$
$$N_2 = (7.76)(7.507)(1000 - 25) = \underline{56,798}$$
$$H_R = \overline{71,937}$$

Using the values of ΔH_{298} and the equation for $\sum H_P$ from illustration 20,

$$67,636 + 71,937 = 59.504T + 0.01125T^2 - 1.484(10^{-6})T^3 - 18,703$$
$$59.904T + 0.01125T^2 - 1.484(10^{-6})T^3 = 158,276$$

Solving this equation gives

$$T = 2070° \text{ K} \ (1797° \text{ C})$$

Solving by the use of mean heat-capacity data for the products of combustion yields

$$T = 2076° \text{ K} \ (1803° \text{ C})$$

In Table 37, page 409, are values of maximum adiabatic flame temperatures of various hydrocarbon gases when burned with air at 25° C, assuming complete combustion to CO_2 and H_2O. Equilibrium degrees of conversion including the presence of free radicals and free atoms have been neglected, and normal values of heat capacities are assumed.

Actual Flame Temperatures

The adiabatic flame temperature, assuming complete combustion, is always higher than can be obtained by actual combustion under the same specified initial conditions. There is always loss of heat from the flame, and it is impossible to obtain complete combustion at high temperatures. The partial completion of these reactions results from the establishment of definite equilibrium conditions between the products and reactants. For example, at high temperatures an equilibrium is established among carbon monoxide, carbon dioxide, and oxygen, corresponding to definite proportions of these three gases. Combustion of carbon monoxide will proceed only to the degree of completion that will give a mixture of gases in proportions corresponding to these equilibrium conditions. Furthermore, the presence of free radicals and elements must be included in calculating both heats of reaction and energy contents. Any energy expended in performing mechanical work, increasing the external kinetic energy and elevation of the gas, will reduce the temperature correspondingly.

In Table 37[6] are experimentally observed values of the maximum

[6] G. W. Jones, B. Lewis, J. B. Friauf, G. St. J. Perrott, *J. Am. Chem. Soc.*, **53**, 869 (1931).

flame temperatures of various hydrocarbon gases when burned with air at 25° C. Values of such calculated adiabatic flame temperatures, assuming complete combustion to CO_2 and H_2O, are included in Table 37.

It will be noted from Table 37 that the adiabatic flame temperatures of the various gases vary but little. For example, although pentane has 12 times the heating value of hydrogen, its adiabatic flame temperature is lower by only 110° C. This results from the fact that, in the combustion of the gases of high heating values, correspondingly large quantities of combustion products including nitrogen are present with high total heat capacities.

Nuclear Reactions.[7] Nuclear reactions are emerging rapidly as sources of energy for power purposes. A single illustration to indicate the magnitude of this energy source compared with ordinary chemical reactions and a simple procedure of calculations will be presented.

An atom is composed of a *nucleus* surrounded by negatively charged *electrons* of relatively small mass. The atomic nucleus is built of primary particles called *protons* and *neutrons*. The term *nucleon* includes both protons and neutrons. Each proton carries a single positive charge and is thus identical with the hydrogen ion or its equivalent, the nucleus of the *hydrogen atom*. The neutron carries no electric charge. Relative to the value of 16.00000 for the atomic mass of the main isotope of oxygen, the masses of these building units of the atom are as follows:

Hydrogen atom	1.00813 amu	(no charge)
Proton	1.00758 amu	(positive charge)
Electron	0.00055 amu	(negative charge)
Neutron	1.00898 amu	(no charge)

where amu = atomic mass unit. It will be observed that the mass of the neutron is greater than that of either the hydrogen atom or the proton.

The number of protons present in an atomic nucleus is designated as its *atomic number Z*. The *mass number A* of an element is equal to the sum of the protons and neutrons in the nucleus.

Z = atomic number = number of protons in the nucleus

A = mass number = number of nucleons

$A - Z$ = number of neutrons.

The mass number A of an element corresponds to the integer nearest the atomic weight of that element. The chemical nature of an element depends on the atomic number Z, that is, on the number of protons or electrons present, and is independent of the number of neutrons

[7] R. G. Taecker, private communication (1954).

present.　Atoms having the same number of protons Z but different number of neutrons $A - Z$ are designated as *isotopes*.

Some naturally and artificially produced isotopes are *radioactive*, that is, they undergo spontaneous change accompanied by the emission of electrically charged particles called *alpha* and *beta* particles.　The alpha particle corresponds to the *helium nucleus* and the beta particle to the electron.　Beta particles may carry either a positive or negative charge.　Some product nuclei resulting from nuclear fission emit *gamma* rays before reaching a stable state.　In unstable nuclei a neutron is also subject to spontaneous conversion into a proton and a negative beta particle, thus:

	Neutron	→	proton	+	negative beta particle	
Charge	0	→	+1		−1	$\Delta Z = +1$
Mass	1	→	1		0	$\Delta A = 0$

In this nuclear reaction the atomic number of the product nucleus increases by one (one proton is formed) and the mass number remains unchanged.

The decomposition of a proton may also lead to stability by formation of a neutron and a positive beta particle, thus:

	Proton	→	neutron	+	positive beta particle	
Charge	+1	→	0		+1	$\Delta Z = -1$
Mass	1	→	1		0	$\Delta A = 0$

In this nuclear change the product nucleus decreases in *atomic number* by one (one proton disappears) and the mass number remains the same.

If no nuclear energy changes were involved, the mass of the nucleus would be equal to the sum of the masses of the protons and neutrons, and the total mass of an atom, to the sum of the masses of its protons, neutrons, and electrons.　In the assembly of an atom, however, the mass of the resulting atom is less than the mass of the constituent protons, neutrons, and electrons by an amount designated as the *mass defect*.　This *mass defect*, designated also as *binding energy*, represents the mass converted into energy by this assembly.　In terms of energy units this binding energy is represented by the Einstein equation

$$E = \Delta m c^2 \tag{78}$$

where　Δm = mass defect, grams
　　　　c = velocity of light, $2.998(10^{10})$ cm per sec
　　　　$E = \Delta m\ 8.988(10^{20})$ ergs $= \Delta m\ 2.147(10^{13})$ g-cal

Uranium 235 is capable of undergoing fission by neutron absorption with the formation of two lighter nuclei or fission fragments from each uranium nucleus. This particular fission is accompanied by an instantaneous release of energy, gamma radiation, and two or three neutrons. The resultant fission fragments are unstable and *decay* to form stable products by further emission of beta particles, gamma rays, and energy.

Neutrons of any energy level can cause U235 to fission, but the fast neutrons must be slowed down or *moderated* if a self-sustaining chain reaction is to be attained within a reasonable mass of the element. In order for a chain reaction to be self-sustaining, a so-called *critical mass* of the element must be present. For power purposes uranium 235 fuel can be in any chemical form or phase. This fuel requires reprocessing by periodic removal of fission products which retard the fissioning reaction.

Uranium 235 occurs in nature associated with its isotope U238 in the ratio of 1 to 140.

Illustration 22. Calculate the amount of uranium 235 required per day for operating a commercial power unit generating 100 Mw of heat, and compare this with the amount of carbon required to release the same energy by ordinary combustion. It will be assumed that the nuclear reaction proceeds by the nuclear fission of uranium 235 by neutron bombardment to form the isotopes molybdenum 42 and xenon 54. This chain reaction results in the emission of two neutrons and four beta particles for each nucleus of uranium, thus:

$$_{92}U^{235} + {}_0n^1 \rightarrow {}_{42}Mo^{100} + {}_{54}Xe^{134} + 2({}_0n^1) + 4({}_{-1}\beta^0) + E$$

where

$_0n^1$ = neutron (zero atomic number, mass number 1) = 1.00898 amu

$_{-1}\beta^0$ = beta particle (negative charge, mass number 0) = 0.00055 amu

Mass of Reactants		Mass of Products	
$_{92}U^{253}$	235.11240 grams	$_{42}Mo^{100}$	99.94686 grams
$_0n^1$	1.00898	$_{54}Xe^{131}$	133.96517
		$2({}_0n^1)$	2.01796
		$4({}_{-1}\beta^0)$	0.00220
	236.12138 grams		235.93219 grams

Mass defect = 236.12138 − 235.93219 = 0.18919 gram

The calculation of atomic weights to this high degree of precision is given by the methods of Fermi.[8]

The energy equivalent to this mass defect is given by equation 78; thus

$$E = \Delta mc^2 = (0.18919)(2.998 \times 10^{10})^2 = 1.700(10^{20}) \text{ ergs}$$
$$= 4.063(10^{12}) \text{ cal}$$

One hundred megawatts are equivalent to $20.65(10^{11})$ cal per day. The mass of U235 required to give this energy is hence

$$\frac{20.65(10^{11})}{4.063(10^{12})} \times 235.11240 = 119.5 \text{ grams per day}$$

[8] E. Fermi, *Nuclear Physics*, University of Chicago Press (1950).

The heat of combustion of carbon at 25° C = −94.0518 kcal per g-mole. The mass of carbon burned to produce 100 Mw or 20.65 (10^{11}) cal per day is hence

$$\frac{20.65(10^{11})}{94.0518} \times 12.010 = 26.37(10^7) \text{ grams per day}$$

On the basis of this nuclear reaction 1 lb of uranium is equivalent in energy output to 2,207,000 lb of carbon. On the basis of mass defect, 1 lb-mass defect is equivalent to

$$2{,}207{,}000 \times \frac{235.1124}{0.18919} = 2.744 \times 10^9 \text{ lb of carbon burned} = 3.866 \times 10^{13} \text{ Btu}$$

Problems

1. Calculate the heat of formation, in calories per gram-mole, of $SO_3(g)$ from the following experimental data on standard heats of reaction:

$$PbO(s) + S(s) + \tfrac{3}{2}O_2(g) = PbSO_4(s), \qquad \Delta H = -167{,}430 \text{ cal}$$
$$PbO(s) + H_2SO_4 \cdot 5H_2O(l) = PbSO_4(s) + 6H_2O(l), \qquad \Delta H = -27{,}967 \text{ cal}$$
$$SO_3(g) + 6H_2O(l) = H_2SO_4 \cdot 5H_2O(l), \qquad \Delta H = -45{,}013 \text{ cal}$$

2. From heat of formation data, calculate the standard heats of reaction of the following, in kilocalories per kilogram-mole:

    (a) $SO_2(g) + \tfrac{1}{2}O_2(g) + H_2O(l) = H_2SO_4(l)$.
    (b) $CaCO_3(s) = CaO(s) + CO_2(g)$.
    (c) $CaO(s) + 3C(\text{graphite}) = CaC_2(s) + CO(g)$.
    (d) $2AgCl(s) + Zn(s) + aq = 2Ag(s) + ZnCl_2(aq)$.
    (e) $CuSO_4(aq) + Zn(s) = ZnSO_4(aq) + Cu(s)$.
    (f) $N_2(g) + 3H_2(g) = 2NH_3(g)$.
    (g) $N_2(g) + O_2(g) = 2NO(g)$.
    (h) $2NaClO_3(s) \rightarrow 2NaCl(s) + 3O_2(g)$.
    (i) $CuO(s) + H_2(g) \rightarrow H_2O(l) + Cu(s)$.
    (j) $Ca_3(PO_4)_2(s) + 3SiO_2(s) + 5C(s) \rightarrow 3[CaO \cdot SiO_2](s) + 5CO(g) + 2P(s)$.
    (k) $4NH_3(g) + 5O_2(g) \rightarrow 4NO(g) + 6H_2O(g)$.
    (l) $3NO_2(g) + H_2O(l) \rightarrow 2HNO_3(aq) + NO(g)$.
    (m) $CaF_2(s) + H_2SO_4(l) \rightarrow CaSO_4(s) + 2HF(l)$.
    (n) $P_2O_5(s) + 3H_2O(l) \rightarrow 2H_3PO_4(aq)$.
    (o) $CaC_2(s) + 5H_2O(g) \rightarrow CaO(s) + 2CO_2(g) + 5H_2(g)$.
    (p) $2AsH_3(g) + 3O_2(g) \rightarrow As_2O_3(s) + 3H_2O(l)$.

3. Calculate the heats of formation of the following compounds from the standard heat of combustion data:

    (a) Benzene (C_6H_6) (l).
    (b) Ethylene glycol ($C_2H_6O_2$) (l).
    (c) Oxalic acid $(COOH)_2(s)$.
    (d) Aniline ($C_6H_5NH_2$) (l).
    (e) Carbon tetrachloride (CCl_4) (l).
    (f) Propane (g).
    (g) Phenol (s).
    (h) Carbon disulfide (l).

(i) Urea (s).
(j) Chloroform (l).

Note: The heat of combustion data given in the table for halogen compounds are based on having all chlorine converted into $HCl(aq)$, by hydrolysis. Accordingly, the actual reaction to be considered is one of combined oxidation and hydrolysis. For CCl_4 it is all hydrolysis and no oxidation.

4. Calculate the standard heats of reaction of the following reactions, expressed in calories per gram-mole:

(a) $(COOH)_2(s) = HCOOH(l) + CO_2(g)$.
(oxalic acid)   (formic acid)

(b) $C_2H_5OH(l) + O_2(g) = CH_3COOH(l) + H_2O(l)$.
(ethyl alcohol)   (acetic acid)

(c) $2CH_3Cl(g) + Zn(s) = C_2H_6(g) + ZnCl_2(s)$.
(methyl chloride)   (ethane)

(d) $3C_2H_2(g) = C_6H_6(l)$.
(acetylene)   (benzene)

(e) $(CH_3COO)_2Ca(s) = CH_3COCH_3(l) + CaCO_3(s)$.
(calcium acetate)   (acetone)

(f) $CH_3OH(l) + \frac{1}{2}O_2(g) = HCHO(g) + H_2O(l)$.
(methyl alcohol)   (formaldehyde)

(g) $2C_2H_5Cl(l) + 2Na(s) = C_4H_{10}(g) + 2NaCl(s)$.
(ethyl chloride)   (n-butane)

(h) $C_2H_2(g) + H_2O(l) = CH_3CHO(g)$.
(acetylene)   (acetaldehyde)

(i) $C_6H_5NO_2(l) + 3Fe(s) + 6HCl(aq) = C_6H_5NH_2(l) + 3FeCl_2(aq) + 2H_2O(l)$.
(nitrobenzene)   (aniline)

5. The integral heat of solution of LiCl in water to form a solution of infinite dilution is -8877 cal per g-mole. Calculate the heat of formation of $LiCl(s)$ from the data of Table 29, **page 299**, and Table 31, page 317.

6. (a) Calculate the number of Btu evolved at 25° C when 80 lb of $ZnCl_2$ are added to 200 lb of water.

(b) Calculate the number of Btu evolved when 40 lb of $CaCl_2$ are added to 200 lb of an aqueous solution containing 10% $CaCl_2$ by weight at 25° C.

7. Calculate the heat evolved, expressed as Btu, when the following materials are mixed at 25° C:

(a) 50 lb H_2SO_4 and 50 lb H_2O. **Ans.** 12,770 Btu.

(b) 50 lb H_2SO_4 and 200 lb of a solution of sulfuric acid and water, which contains 50% by weight of H_2SO_4. **Ans.** 8380 Btu.

(c) 50 lb H_2O and 200 lb of a solution of sulfuric acid and water, which contains 50% by weight of H_2SO_4. **Ans.** 2020 Btu.

(d) 60 lb of $Na_2SO_4 \cdot 10H_2O$ and 100 lb of water. **Ans.** 4350 Btu.

8. An aqueous sulfuric acid solution contains 60% H_2SO_4 by weight. To 500 grams of this solution are added 700 grams of a solution containing 95% H_2SO_4 by weight. Calculate the quantity of heat evolved.

9. One hundred pounds of an oleum solution containing 15.4% free SO_3 are to be diluted with water to make a 30.8% solution of H_2SO_4. Calculate the heat evolved. **Ans.** 39,500 Btu.

10. Five hundred grams of oleum containing 30.0% free SO_3 are diluted with a large volume of water to infinite dilution. Calculate the heat evolved.

11. One ton of 19.1% H_2SO_4 is to be concentrated to 91.6% H_2SO_4 at 70° F. Steam is available to heat the acid to 302° F, and a vacuum can be maintained equivalent to the vapor pressure of 91.6% H_2SO_4 at 150° C, i.e., 14 mm. Calculate the heat to be supplied. **Ans.** 1,971,000 Btu.

12. A batch of dilute sulfuric acid at 80° F, which weighs 500 lb and contains 10% H_2SO_4, is to be brought to 50% strength by the addition of 98% acid, which is at 70° F. How much heat is abstracted by the cooling system if the temperature of the final acid is 100° F?

By means of a sketch, show how Fig. 81 was used in the solution of this problem.

13. In a continuous concentrating system, dilute acid (60% H_2SO_4) is concentrated to 95% strength. The dilute acid enters the system at 70° F, while the water vapor and the concentrated acid leave the system at the boiling temperature of the latter. How many Btu are needed to concentrate 1000 lb of dilute acid? **Ans.** 704,000 Btu.

By means of a sketch, show how Fig. 81 was used in the solution of this problem.

14. Hydrochloric acid, $G(60°/60° F) = 1.20$, is prepared by absorbing HCl gas at 80° F in water that is admitted to the absorption system at 50° F. If the final acid leaves the system at 80° F, how much heat is withdrawn from the equipment per 1500 lb of acid produced?

By means of a sketch, show how Fig. 80 was used in the solution of this problem.

15. Assume that pure sulfuric acid and water, both at 70° F, are mixed under adiabatic conditions. If the acid is gradually added to the water, what is the maximum temperature that can be attained? **Ans.** 350° F.

By means of a sketch, show how Fig. 81 was used in the solution of this problem.

16. Two sulfuric acid solutions at 25° C of 5% and 80% concentrations are to be mixed under adiabatic conditions. If the stronger solution is gradually added to the weaker, what is the maximum temperature that can be attained?

By means of a sketch, show how Fig. 81 was used in the solution of this problem.

17. Calculate the heat evolved when 5 lb HCl gas at 80° F are dissolved in 20 lb of 10% HCl at 60° F, to form a solution at 60° F (Fig. 80). **Ans.** 3655 Btu.

18. Calculate the resultant temperature when 20 lb of water at 100° F are added to 10 lb of 40% HCl at 60° F (Fig. 80).

19. Calculate the heat required to concentrate 40 lb of 5% HCl at 120° F to 8 lb of 20% HCl at 120° F with the vapors leaving at 120° F (Fig. 80). **Ans.** 32,760 Btu.

20. Calculate the final temperature when 100% H_2SO_4 at 60° F is diluted with a solution containing 20% H_2SO_4 at 100° F to form a solution containing 50% H_2SO_4 (Fig. 81), assuming adiabatic mixing.

21. In the following table are values of the standard integral heats of solution ΔH_s in calories per gram-mole, of liquid acetic acid in water to form solutions containing m moles of water per mole of acid (from *International Critical Tables*, Vol. V, page 159).

m	ΔH_s	m	ΔH_s
0.25	+ 70	5.00	+ 24
0.58	+126	6.19	− 13
1.11	+149	30.00	− 92
1.42	+149	63.3	−107
1.95	+130		

(a) Using the method of tangent slopes, calculate the differential molal heat of solution of acetic acid in a solution containing 15% acetic acid by weight. **Ans.** −34.3 cal per g-mole acetic acid.

(b) Using the method of tangent intercepts, calculate the differential molal heats of solution of acetic acid and of water in a solution containing 50% acetic acid by weight. **Ans.** 198.8 cal per g-mole acetic acid, −34.8 cal per g-mole water.

22. From the data of problem 21 calculate the heat, in Btu, evolved when 50 lb of acetic acid are added to 1000 lb of a solution containing 50% acetic acid by weight. Perform the calculation both from integral heat of solution data and from differential heat of solution data derived in problem 21. Assume constant temperature at 25° C.

23. From the integral heat of solution data of problem 21 calculate the heat, in calories, that is evolved at 25° C when 1 liter of aqueous acetic acid containing 75% acid by weight is diluted to 2 liters by the addition of water. **Ans.** 1844 cal.

24. The heats of mixing, in calories per gram-mole of solution, of carbon tetrachloride (CCl_4) and aniline ($C_6H_5NH_2$) at 25° C are given in the following table (from the *International Critical Tables*, Vol. V, page 155).

Mole Fraction CCl_4	ΔH	Mole Fraction CCl_4	ΔH
0.0942	98	0.6215	288
0.1848	169	0.7175	270
0.3005	237	0.7888	246
0.4152	282	0.8627	188
0.4827	291	0.9092	149
0.5504	298		

(a) Calculate the integral and differential molal heats of solution of each component in a solution containing 40% CCl_4 by weight.

(b) Calculate the heat evolved, in Btu, when 1 lb of aniline is added to a large quantity of solution containing 50% CCl_4 by weight.

25. (a) For one of the binary systems from the following list, determine the differential heat of solution of each component at each of the concentrations listed (from *International Critical Tables*, Vol. V).

Carbon Tetrachloride Benzene		Carbon Disulfide Acetone		Chloroform Acetone	
Weight %		Weight %		Weight %	
CCl_4	ΔH	CS_2	ΔH	$CHCl_3$	ΔH
10	0.301	10	5.78	10	− 4.77
20	0.598	20	11.80	20	− 9.83
30	0.816	30	16.45	30	−14.31
40	0.963	40	19.92	40	−19.38
50	1.030	50	20.93	50	−23.27
60	1.030	60	20.80	60	−25.53
70	0.912	70	17.62	70	−25.07
80	0.699	80	16.15	80	−21.55
90	0.452	90	10.80	90	−13.56
$t = 18°$ C		$t = 16°$ C		$t = 14°$ C	

ΔH = heat of mixing, as joules per gram of solution (1 joule = 0.2389 cal)

(b) Prepare a plot showing the differential heats of solution (calories per gram of component) and the heats of mixing (calories per gram of solution) as a function of the composition, expressed as weight per cent.

(c) Calculate the integral molal heat of solution of each component for a solution containing 50 weight per cent of each component.

(d) Calculate the Btu evolved when 1 lb of the first-named component is dissolved in a large volume of solution containing 60% by weight of the first-named component.

26. Two hundred grams of 20% $CaCl_2$ at 40° C and 300 grams of $CaCl_2 \cdot 6H_2O$ at 20° C are mixed.

(a) Estimate the temperature of the final mixture.

(b) How much heat must be removed to solidify the mixture completely?

27. Calculate the standard heat of reaction, in kilocalories, accompanying the reduction of 20 kg of Fe_2O_3 by carbon (coke) to form 12 kg of $Fe(s)$. The only other products leaving the process are $FeO(s)$ and $CO(g)$.

28. The heat requirements are to be estimated for a low-temperature reduction process applied to a magnetite ore. The ore averages 90% Fe_3O_4; the remainder is inert gangue, largely SiO_2. The process is to be conducted in an externally heated retort which is closed except for the two openings that are to serve as entry for the charge and as exit for the product. The opening for admitting the charge also serves as a vent for the escape of the gas formed by the reaction between the magnetite and the reducing agent. On the basis of laboratory tests, it is anticipated that 95% of the iron will be reduced to the metallic state, the remainder being reduced to FeO. The reducing agent to be used is a metallurgical coke containing 85% carbon and 15% ash, the latter being largely SiO_2. The coke is to be charged 300% in excess of the theoretical demand for complete reduction. The solid discharged is thus a mixture of Fe, FeO, unused coke, and SiO_2 from the ore and from the coke used up in the reduction process. The gas escaping from the retort will be practically pure CO, formed by the reaction between the magnetite and the coke.

Analyses

| Coke: Carbon | 85% | | Ore: | Fe_3O_4 | 90% |
| Ash | 15% | | | SiO_2 | 10% |

Temperatures			Specific Heats		
Entering ore	200° C		Fe_3O_4:	0° C	0.151
Entering coke	200° C			100° C	0.179
Leaving solids	950° C			200° C	0.203
Leaving gas	950° C			300° C	0.222

Calculate the material balance for the process, on the basis of 2000 lb of metallic iron produced.

Estimate the heat requirements for the process, as Btu per 2000 lb of metallic iron produced.

29. By the combustion at constant volume of 2.0 grams of $H_2(g)$ to form liquid water at 17° C, 67.45 kcal are evolved. Calculate the quantity of heat that would be evolved were the reaction conducted under a constant pressure at 17° C.

30. When 1.0 gram of naphthalene ($C_{10}H_8$) is burned in a bomb calorimeter, the water formed being condensed, 9621 cal are evolved at 25° C. Calculate the heat of combustion at constant pressure and 25° C, the water vapor remaining uncondensed. **Ans.** 9302 cal per gram.

31. A fuel oil analyzes 88% C and 12% H by weight. The standard heat of combustion of this oil is determined in an oxygen-bomb calorimeter. Calculate the correction that must be applied to get the heat of reaction at constant pressure. Which is the greater, the heat of reaction at constant pressure or the heat of reaction at constant volume?

32. Calculate the actual heat of reaction in gram-calories per gram-mole for each of the following reactions. Reactants and products are at a constant pressure of 1 atm. The temperature of each reactant and product is as indicated. Assume the mean specific heat of $NaHSO_4$ (0 to 250° C) to be 0.23 and Na_2SO_4 (0 to 600° C) to be 0.26.

$$SO_2(g) + \tfrac{1}{2}O_2(g) \rightarrow SO_3(g)$$
$$\text{450° C} \quad\quad \text{450° C} \quad\quad \text{450° C}$$

$$NaHSO_4(s) + NaCl(s) \rightarrow Na_2SO_4(s) + HCl(g)$$
$$\text{250° C} \quad\quad \text{30° C} \quad\quad \text{600° C} \quad\quad \text{300° C}$$

$$SiO_2(s) + 3C\ (\text{coke}) \rightarrow SiC(s) + 2CO(g)$$
$$\text{25° C} \quad\quad \text{25° C} \quad\quad \text{1800° C} \quad\quad \text{1200° C}$$

$$2NaCl(s) + SO_2(g) + \tfrac{1}{2}O_2(g) + H_2O(g) \rightarrow Na_2SO_4(s) + 2HCl(g)$$
$$\text{400° C} \quad\quad \text{400° C} \quad\quad \text{400° C} \quad\quad \text{400° C} \quad\quad \text{400° C} \quad\quad \text{400° C}$$

33. Derive a general equation for each of the following reactions which will express the heat of reaction as a function of temperature, with the reactants and products at the same temperature. Base all equations on the assumption that H_2O is in the vapor state.

$$CO(g)\ \ + \tfrac{1}{2}O_2(g) \rightarrow CO_2(g)$$
$$H_2O(g) + CO(g) \rightarrow H_2(g) + CO_2(g)$$
$$N_2(g)\ \ + 3H_2(g) \rightarrow 2NH_3(g)$$
$$SO_2(g) + \tfrac{1}{2}O_2(g) \rightarrow SO_3(g)$$
$$H_2(g)\ \ + Cl_2(g) \rightarrow 2HCl(g)$$

34. Sulfur dioxide gas is oxidized in 100% excess air with 80% conversion to SO_3. The gases enter the converter at 400° C and leave at 450° C. How many kilocalories are absorbed in the heat interchanger of the converter per kilogram-mole of SO_2 introduced? **Ans.** 21,412 kcal.

35. Calculate the heat of neutralization in calories per gram-mole of NaOH $(n_1 = 5)$ with $HCl(n_1 = 7)$ at 25° C where $n_1 =$ moles H_2O per mole of solute.

36. A bed of petroleum coke (pure carbon) weighing 3000 kg, at an initial temperature of 1300° C, has saturated steam at 100° C blown through it until the temperature of the bed of coke has fallen to 1000° C. The average temperature of the gases leaving the generator is 1000° C. The analysis of the gas produced is CO_2 3.10%, CO 45.35%, H_2 51.55% by volume, dry basis. How many kilograms of steam are blown through the bed of coke to reduce the temperature to 1000° C? Neglect loss of heat by radiation, and assume that no steam passes through the process undecomposed. **Ans.** 241 kg.

37. If, in problem 36, 20% of the steam passes through the coke undecomposed, how much steam is blown through the bed of coke to reduce its temperature to 1000° C? Neglect loss of heat by radiation.

38. Steam at 200° C, 50° superheat, is blown through a bed of coke initially at 1200° C. The gases leave at an average temperature of 800° C with the following composition by volume on the dry basis:

H_2	53.5%
CO	39.7
CO_2	6.8
	100.0%

Of the steam introduced 30% passes through undecomposed. Calculate the heat of reaction in kilocalories per kilogram-mole of steam introduced. Mean specific heat of coke (0 to 1200° C) = 0.35. **Ans.** 22,147 kcal.

39. Calculate the number of Btu required to calcine completely 100 lb of limestone containing 80% $CaCO_3$, 11% $MgCO_3$, and 9% H_2O. The lime is withdrawn at 1650° F, and the gases leave at 400° F. The limestone is charged at 70° F.

40. Limestone, pure $CaCO_3$, is calcined in a continuous vertical kiln by the combustion of producer gas in direct contact with the charge. The gaseous products of combustion and calcination rise vertically through the descending charge. The limestone is charged at 25° C, and the calcined lime is withdrawn at 900° C. The producer gas enters at 600° C and is burned with the theoretically required amount of air at 25° C. The gaseous products leave at 200° C. The analysis of the producer gas by volume is as follows:

CO_2	9.21%
O_2	1.62
CO	13.60
N_2	75.57
	100.00%

Calculate the number of cubic meters (0° C, 760 mm Hg) of producer gas, required to burn 100 kg of limestone, neglecting heat losses and the moisture contents of the air and producer gas. **Ans.** 103.9 cubic meters.

41. A fuel gas of the following composition at 1600° F is burned in a copper melting furnace with 25% excess air at 65° F:

CH_4	40%
H_2	40
CO	4
CO_2	3
N_2	11
O_2	2
	100%

The copper is charged at 65° F and poured at 2000° F. The gaseous products leave the furnace at an average temperature of 1000° F. How much copper is melted by burning 4000 cu ft (32° F; 29.92 in. Hg; dry) of the above gas, assuming that the heat lost by radiation is 50,000 Btu and neglecting the moisture contents of the fuel gas and air?

42. One thousand cubic meters of gas, measured at standard conditions (0° C, 760 mm and dry), containing 20 grams of ammonia per cubic meter (as measured above) are passed into an ammonia absorption tower at a temperature of 40° C, saturated with water vapor, and at a total pressure of 740 mm. The gas is passed upward countercurrent to a descending stream of water which absorbs 95% of the incoming ammonia. The gas leaves the tower at 38° C, saturated with water vapor. Six hundred kilograms of water enter the top of the tower at 20° C. What is the temperature of the solution leaving the tower, neglecting heat losses? Assume the mean molal heat capacity of moisture-free, ammonia-free gas to be 7.2. **Ans.** 44.8° C.

43. Pure HCl gas comes from a Mannheim furnace at 300° C. This gas is cooled to 60° C in a silica coil and is then completely absorbed by being passed counter-current to a stream of aqueous hydrochloric acid in a series of Cellarius vessels and absorption towers. The unabsorbed gas from the last Cellarius vessel enters the first absorption tower at 40° C. Fresh water is introduced in the last absorption tower at 15° C and leaves the first absorption tower at 30° C containing 31.45% HCl (20° Bé acid). This acid is introduced into the last Cellarius vessel and leaves the first vessel at 30° C containing 35.21% HCl (22° Bé acid). There are produced in this system 9000 lb of 22° Bé acid in 10 hr. Calculate separately the heat removed in the cooling coil, Cellarius vessels, and absorption towers, neglecting the presence of water vapor in the gas stream and assuming complete absorption of HCl.

44. The integral heats of solution of HF, HCl, and NH_3 in n moles of water may be represented in kilocalories per gram-mole of solute by the empirical equation

$$\Delta H = \frac{an}{n + b}$$

where,   for HF, $a = -11.71$ kcal per g-mole, $b = 0.15$
            for HCl, $a = -17.53$ kcal per g-mole, $b = 0.90$
            for NH_3, $a = -8.49$ kcal per g-mole, $b = 0.164$

For a 10% solution by weight calculate:

(a) Integral heat of solution per kilogram of solute, per kilogram of solvent, per kilogram of solution.

(b) Differential heat of solution per kilogram of solute, per kilogram of solvent.

45. Calculate the energy released in calories per gram of uranium in the following nuclear reactions resulting from neutron bombardment of U235 to form different fission products:

(a) $_{92}U^{235} + {_0}n^1 \rightarrow {_{62}}Sm^{152} + {_{36}}Kr^{82} \quad + 6(_{-1}\beta^0) \quad + 2(_0n^1) + E$

(b) $_{92}U^{235} + {_0}n^1 \rightarrow {_{62}}Sm^{152} + {_{34}}Sc^{82} \quad + 4(_{-1}\beta^0) \quad + 2(_0n^1) + E$

(c) $_{92}U^{235} + {_0}n^1 \rightarrow {_{40}}Zr^{90} \quad + {_{60}}Nd^{144} + 8(_{-1}\beta^0) \quad + 2(_0n^1) + E$

Atomic Weights

$$_{92}U^{235} \;\;= 235.11240 \qquad\qquad _{34}Sc^{82} \;\;= \;\;81.95079$$
$$_{62}Sm^{152} = 151.97761 \qquad\qquad _{40}Zr^{90} \;\;= \;\;89.94454$$
$$_{26}Kr^{82} \;\;= \;\;81.94508 \qquad\qquad _{60}Nd^{144} = 143.96895$$

The adsorption of gases by solids is of industrial importance in the purification and drying of gases, in the recovery of solvent vapors and casing-head gasoline, in the fractional separation of gases by hypersorption, and in gaseous reactions catalyzed by solid surfaces. In one method of air conditioning, dehumidification is accomplished by intermittent adsorption and desorption of water vapor by a solid desiccant.

Many forces of attraction exist between a gas and a solid, depending on the complex physical structure of the solid as well as on the chemical nature of both gas and solid. Inasmuch as the adsorption of a gas takes place below its saturation pressure, it is evident that the physical force of attraction of a gas with a solid adsorbent is greater than the force of attraction bringing about normal condensation of the gas. Where a single molecular layer of adsorbed gas is formed, the force of attraction between the gas and the solid surface progressively diminishes with the extent of surface covered due to interaction between adsorbed molecules and the nonuniformity of the surface. In the adsorption of a gas the freedom of molecular translation is reduced from three dimensions in space to two dimensions on a surface. Where multiple molecular layers of gas are adsorbed, the forces of attraction fall off with successive layers. In a porous solid of capillary structure the number of molecular layers that build upon the walls is limited by the diameter of the capillary pore or the distance between capillary walls. In submicroscopic capillary pores, the vapor pressure of the condensed phase progressively diminishes with decreased diameter of capillary, the force of attraction following the complex capillary structure of the solid. Physical adsorption may also result in an orientation of the adsorbed molecule. Polarization is induced in the adsorbed molecule by the polar nature of the solid resulting from the unequal distribution of electric charges where covalent atomic bonds are not equally shared. By this induction, orientation is extended to nonpolar as well as to polar gases.

Physical and Chemical Adsorption. The foregoing forces of attraction are characteristic of *physical* or *van der Waals adsorption*.

Besides these physical forces, stronger forces of attraction between a gas and a solid result in the formation of surface compounds in variable stoichiometric ratios. This latter type of adsorption is designated as chemisorption. Chemisorption is highly specific in nature, may be restricted to definite molecular sites on the surface, and proceeds slowly at low temperatures. Chemisorption should in turn be distinguished from the strong stoichiometric chemical reactions between a gas and a solid that result in the disappearance of the original solid. A comparison between physical adsorption and chemisorption is given in Table 33. For example, the heats of condensation for the three types of adsorption of nitrogen gas on an iron catalyst are as follows:

Normal liquefaction $\Delta_H = -1360$ g-cal per g-mole
Physical adsorption $\Delta_H = -2000$ to -3000 g-cal per g-mole
Chemisorption $\Delta_H = -35,000$ g-cal per g-mole

TABLE 33.   COMPARISON OF PHYSICAL ADSORPTION AND CHEMISORPTION

	Physical Adsorption	Chemisorption
Heat of adsorption	Small, same order as normal liquefaction	Large, many times greater than the heat of normal liquefaction
Rate of adsorption	Controlled by resistance to mass transfer	Controlled by resistance to surface reaction
	Rapid rate at low temperatures	Negligible rate at low temperatures
Specificity	Low. Entire surface available for physical adsorption	High. Chemisorption limited to active sites on the surface
Surface coverage	Complete and extendable to multilayers	Incomplete and limited to a unimolecular layer
Adsorption above critical temperature	None	No restriction
Adsorption at low partial pressures	Small	Large
Adsorption at high partial pressures	Large, may extend to infinite amounts	Slight increase
Activation energy	Low, nearly negligible	High, corresponding to a chemical reaction
Quantities adsorbed per unit mass	High	Low

Equilibrium Adsorption. The amount of gas adsorbed by a solid under equilibrium conditions may be expressed either as a percentage by weight or as the mass of gas or volume of gas (reduced to standard conditions, 0° C, 760 mm Hg pressure), both on the basis of unit mass of gas-free adsorbent. Three methods of graphical representation of equilibrium adsorption data are in common use as illustrated in Figs. 91, 92, and 93 for the adsorption of ammonia gas on charcoal taken from the data of Titoff.[1] In *adsorption isotherms* (Fig. 91) the volume of gas adsorbed is plotted against partial pressure at various parameters of *constant temperature*. In *adsorption isobars* (Fig. 92), the volume

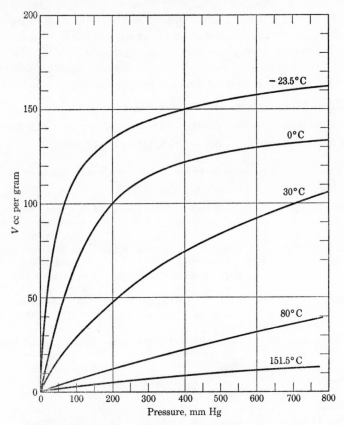

Fɪɢ. 91. Adsorption isotherms for ammonia gas on charcoal[1]

of gas adsorbed is plotted against temperature at various parameters of *constant partial pressure*. The isobaric plot is a cross plot of the isothermal curves. In *adsorption isosteres* (Fig. 93), the partial pressure of the adsorbed gas is plotted against temperature at various parameters of *constant volume*. The isostere diagram is a cross plot of the isobaric curves. The adsorption isotherms represent the most generally useful plot in adsorption studies. The adsorption isostere is of particular

[1] A. Titoff, *Z. Phys. Chem.*, **74**, 641 (1910).

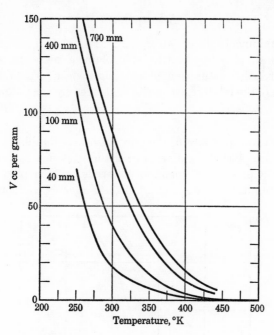

FIG. 92. Adsorption isobars for ammonia gas on charcoal[1]

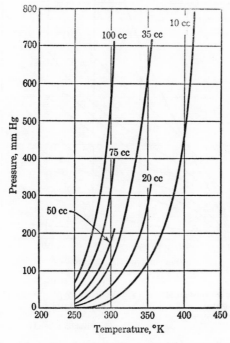

FIG. 93. Adsorption isosteres for ammonia gas on charcoal[1]

371

value for obtaining heats of adsorption. The plot of adsorption isotherms can be simplified by plotting the volume or mass of gas adsorbed against the relative saturation of the gas instead of against partial pressure, where the relative saturation is defined by the following ratio:

$$x = p/p_s \tag{1}$$

and p_s is the vapor pressure of the pure adsorbate liquid.

In Fig. 94 are plotted equilibrium moisture contents against relative humidity for a number of common materials. When plotted in this manner, the resultant curves are but slightly affected by temperature.

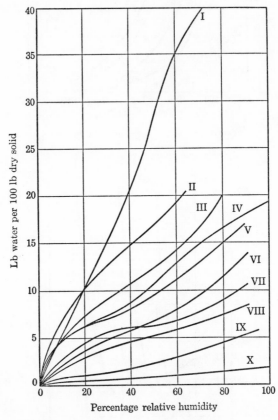

I.	Silica gel	VI.	Cotton cloth
II.	Leather, chrome-tanned	VII.	Sulfite pulp, fresh, unbleached
III.	Wool, worsted	VIII.	Bond paper
IV.	Activated alumina	IX.	Cellulose acetate silk, fibrous
V.	Viscose	X.	Kaolin (Florida)

Fɪɢ. 94. Equilibrium moisture content of various substances at 77° F

Equilibrium moisture content is of importance in drying operations. Water will not evaporate from an adsorbent solid into a gas whose relative humidity is higher than that corresponding to the equilibrium moisture content of the solid. The equilibrium moisture content represents the minimum moisture content to which a material can be dried by a gas of a given relative humidity. Continued passage of gas will not result in further drying, even though the gas is far from a state of normal saturation.

For silica gel the commercially dry basis is used in expressing moisture content. The commercially dry gel contains 5% water in chemical combination which is not removed in desorption or regeneration operations and hence is not included in reporting moisture content. If this chemically adsorbed water were removed, the adsorptive capacity of the gel would disappear. For a particular silica gel the useful range of moisture content for temperatures up to 100° F may be represented approximately by the simple equation:

$$w_e = 0.55 \, x_r = 55 \, (p/p_s) \tag{2}$$

where

x_r = relative humidity, per cent

w_e = equilibrium moisture content, pounds per 100 lb of commercially dry gel

p = partial pressure of water vapor in air

p_s = vapor pressure of pure water at temperature of air

The advantage of a plot of equilibrium adsorption content against relative saturation instead of as isotherms is apparent. It should be noted that relative saturation should be employed and not percentage saturation. Relative saturation is independent of total pressure, as shown on page 372.

Adsorption Isotherms. Brunauer[2] classifies adsorption isotherms into five general types as represented by the diagrams of Fig. 95 where at constant temperature the volume of gas adsorbed (at 0° C, 760 mm Hg. pressure) per unit mass of adsorbate-free solid is plotted against the relative saturation of the adsorbate gas.

In type I the adsorption isotherm is hyperbolic, approaching a constant asymptotic value at a relative saturation of unity. The explanation of this perfomance is that adsorption is restricted to a surface layer one molecule thick. Type I is represented by the adsorption of ethyl chloride on carbon black at 0° C, of oxygen on carbon black at −183° C, and of ammonia on charcoal (Figs. 91, 92, 93).

[2] S. Brunauer, *The Adsorption of Gases and Vapors*, Princeton University Press (1945), with permission.

In type II the adsorption curve is S-shaped and increases to infinity as the relative saturation approaches unity. The explanation of this behavior is the formation of a multimolecular layer of indefinite thickness. Type II is represented by the adsorption of water vapor on carbon black at 29° C and of nitrogen on an iron catalyst at −195° C.

In type III the curvature of the adsorption isotherm is convex toward the side of abscissas. The second derivative of the curve is thus always positive. The amount of gas adsorbed increases without limit as its

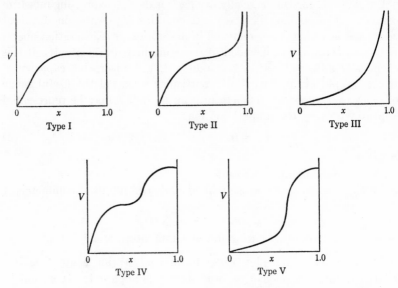

FIG. 95. Types of adsorption isotherms

relative saturation approaches unity. In this case as in type II an infinite molecular layer is possible. The convex curvature can be accounted for by heats of adsorption of the first layer becoming less than the heat of normal condensation, owing to interaction taking place in the first layer. The adsorption of ethyl chloride on charcoal at −78° C, and of bromine and iodine vapors on silica gel at 19° C, is representative of this type.

Type IV is similar to type II in the low and intermediate ranges of relative saturation, but the values of V approach a maximum finite value at a relative saturation of unity. The high values of V are accounted for by capillary condensation, the maximum value corresponding to complete filling of the capillaries. Type IV is represented by the adsorption of water vapor on charcoal at 29° C and of benzene vapors on ferric oxide gel at 50° C.

Type V is similar to type III at low and intermediate ranges of relative saturation but approaches a maximum finite value for V at a relative saturation of unity.   This behavior is accounted for by capillary condensation and the building up of a layer of finite thickness at saturation. The initial convex curvature can be accounted for by the heat of adsorption of the first layer becoming less than the heat of normal condensation due to interaction.   Type V is represented by the adsorption of water vapor on activated carbon at 100° C and of nitrogen on silica-alumina gel.

Equations for Adsorption Isotherms.   No general equations suitable for all types of adsorption isotherms have been developed from theoretical considerations.   Numerous empirical equations for specific types of adsorption have been formulated.

In many cases the adsorption isotherm is satisfactorily represented by the empirical equation proposed by Freundlich

$$w = kp^n \tag{3}$$

where k and n are empirical constants.

Brunauer, Emmett and Teller[3-7] have started with simple equations suitable for type I and progressively built up additional equations for the succeeding four types of adsorption.

The simplest type of adsorption isotherm, type I, occurs where adsorption is restricted to a single molecular layer.   On a theoretical basis, Langmuir developed an equation for this adsorption isotherm, assuming that at any pressure less than saturation the amount of gas adsorbed is proportional to the partial pressure of the gas and to the fraction of the surface left uncovered.   He further assumed in developing equation 4 that the adsorbed gas did not dissociate nor interact. On this basis,

$$V = \frac{V_m b p}{1 + bp} \tag{4}$$

where V = volume of gas (0°, 760 mm Hg) adsorbed per unit mass of adsorbent

V_m = volume of gas (0°, 760 mm Hg) adsorbed per unit mass of adsorbent to cover surface with a layer one molecule thick

b = empirical constant, in reciprocal pressure units

[3] S. Brunauer, P. H. Emmett, and E. Teller, *J. Am. Chem. Soc.*, **60**, 309 (1938).

[4] S. Brunauer, L. S. Deming, and E. Teller, *J. Am. Chem. Soc.*, **62**, 1723 (1940).

[5] P. H. Emmett, *Advances in Catalysis*, Academic Press (1948).

[6] P. H. Emmett, *Ind. Eng. Chem.*, **31**, 639 (1945).

[7] P. H. Emmett and S. Brunauer, *J. Am. Chem. Soc.*, **59**, 1553 (1937).

Equation 4 is applicable to adsorption isotherms of type I over the entire range of partial pressures. As p approaches infinity, equation 4 becomes $V = V_m$ corresponding to complete surface coverage. As p approaches zero, equation 4 becomes $V = V_m b p$ which holds fairly well for all types of adsorption as the partial pressure approaches zero. On the basis of its theoretical development, V_m should be nearly independent of pressure. The constant b was assumed to be independent of pressure but highly dependent on temperature by the relation $b = e^{-A/RT}$ where A is equal to the heat of adsorption. Actually V_m as calculated by equation 4 is not a reliable measure of surface coverage, and usually b is a function of pressure as well as temperature, indicating that the heat of adsorption is not constant but diminishes as the extent of surface covered diminishes. For molecular dissociation of a diatomic gas into two atoms upon adsorption, the Langmuir equation is reduced to the form

$$V = \frac{V_m \sqrt{bp}}{1 + \sqrt{bp}} \tag{5}$$

Otherwise, the same limitations and restrictions apply to both equations 4 and 5.

For adsorption isotherms of types I, II, and III Brunauer, Emmett, and Teller proposed the following equation:

$$V = \frac{V_m C x}{(1 - x)[1 + (C - 1)x]} \tag{6}$$

where V_m and C are empirical constants and $x = p/p_s$. The constant C is related to heats of adsorption by the following:

$$C = e^{(E_1 - E_L)/RT} \tag{7}$$

where E_1 = heat of adsorption of first layer
E_L = heat of normal condensation

Equation 6 is based on adsorption on a free surface without capillary condensation but applicable to adsorption in multimolecular layers of indefinite thickness. The constant V_m represents the volume of gas required to cover completely the surface with a unimolecular layer. Equation 6 is more reliable than equation 4 in the calculation of V_m. Equation 6 is based on the formation of an infinite number of layers for types II and III at $x = 1.0$. This equation was developed, assuming that the heat of adsorption E_1 was constant over the entire surface coverage of the first layer and that the heat of condensation for all subsequent layers is equal to E_L, the normal heat of condensation. Equation 6 does not hold for adsorbents where capillary condensation occurs. It fits types I and II for values of C greater than 1 and for values of E_1 greater than E_L and holds for type III where C is equal or less than 1 $(C \gtreqless 1)$ and where E_1 is equal to or less than $E_L (E_1 \gtreqless E_L)$.

Where C is great compared to 1, equation 6 reduces to the Langmuir equation 4 where $C = bp_s$. Equation 6 is suitable for types II and III for the range of values of $x = 0.05$ to $x = 0.5$.

Where the number of layers formed is finite but without capillary condensation, Brunauer, Emmett, and Teller have expanded equation 6 to include the number n of layers formed to give

$$V = \frac{V_m C x}{1 - x} \left[\frac{1 - (n + 1)x^n + nx^{n+1}}{1 + (C - 1)x - Cx^{n+1}} \right] \tag{8}$$

where n = number of layers formed. Equation 8 holds for types II and III for values of x over the entire range. Equation 8 reduces to equation 6 in the limiting case where $n = \infty$ and reduces to the equation 4 where $n = 1$. Equation 8 holds for types I, II, and III under the following conditions:

For Type I, $C > 1$ and $n = 1$
For Type II, $C > 1$ and $n > 1$
For Type III, $C \gtreqless 1$ and $n > 1$

Equation 8 is suitable for all five types of adsorption for values of x from 0.05 to 0.7.

Adsorption types IV and V are characterized by capillary condensation to form finite molecular layers. In capillary condensation the vapor pressure is lowered by the curvature of the capillaries, and the heats of adsorption are greater than the normal heat of condensation.

For complete filling of the space between capillary walls the final layer adsorbed has a special heat of adsorption of Q. The resultant equation including this additional term gives

$$V = \frac{V_m x C}{1 - x} \left[\frac{1 + (0.5ng - 0.5n)x^{n-1} - (ng + 1)x^n + (0.5ng + 0.5)x^{n+1}}{1 + (C - 1)x + (0.5Cg - 0.5C)x^n - (0.5Cg + 0.5C)x^{n+1}} \right] \tag{9}$$

where $g = e^{Q/RT}$

Equation 9 describes types IV and V over the entire range of x from 0.05 to 1.0. It gives a finite value of V at $x = 1.0$.

For type IV, $C > 1$
For type V, $C \gtreqless 1$

Equation 9 reduces to equation 8 when $Q = 0$ and $g = 1$, corresponding to weak capillary forces, reduces to equation 6 where n is ∞, and reduces to equation 4 where $n = 1$.

Illustration 1. For the absorption of argon on an alumina-iron catalyst at $-183°$ C Emmett and Brunauer[7] obtained the following experimental data:

$p,$ mm Hg	V (0° C, 760 mm Hg) cc per gram
21	70
45	93
90	120
175	135
245	155
330	175
405	200
485	220
540	245

From these data calculate the constants V_m, C, and n. At $-183°$ C, $p_s = 1026$ mm Hg.

The values of V_m and C are most easily obtained from equation 6 by using the data at low values of x.

By rearrangement of equation 6,

$$\frac{x}{V(1-x)} = \frac{1}{V_m C} + \frac{C-1}{V_m C} x$$

At $-183°$ C, $x = p/1026$.

p	x	$1-x$	V	$\dfrac{x}{V(1-x)}$
21	0.0185	0.9815	70	0.00030
45	0.044	0.956	93	0.00049
90	0.088	0.912	120	0.00081
175	0.170	0.830	135	0.00152
245	0.239	0.761	155	0.00202
330	0.322	0.678	175	0.00271
405	0.394	0.616	200	0.00320
485	0.472	0.529	220	0.00407
540	0.525	0.475	245	0.00450

From a plot of $x/V(1-x)$ against x in Fig. 96 a straight line results (except for high values of x), with an intercept

$$\frac{1}{V_m C} = 0.00015$$

and slope

$$\frac{C-1}{V_m C} = 0.0079$$

Solving simultaneously gives

$$C = 54, \qquad V_m = 124 \ \overline{\text{cm}}^3 \text{ per gram}$$

The number of molecular layers n can be then obtained from the data at high pressure, using the three-term equation 8. By substituting values of $C = 54$ and $V_m = 124$ into equation 8 it will be found by trial and error that $n = 6$.

Capillary Condensation. Adsorbents of industrial importance are generally substances of highly porous structure, which expose enormous

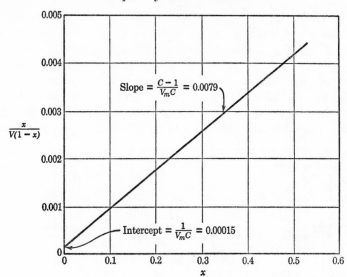

Fɪɢ. 96. Evaluation of V_m and C in the adsorption of argon on an
alumina-iron catalyst

interior surfaces. Activated alumina, charcoal, and silica gel are
familiar examples. In silica gel, for instance, the capillary pores are
about 4×10^{-7} cm in diameter, only 10 times the diameter of simple
molecules, and comprise about 50% of the total volume. The total
interior area is about one acre per cubic inch. In such materials capillary condensation is of great importance.

Capillary condensation can occur only when the solid is wetted by
the condensate, resulting in concave surfaces of the condensed liquid.
The equilibrium vapor pressure of a liquid having a concave surface is
less than the normal value by an amount depending on the radius of
curvature. In the submicroscopic capillary pores of a wetted solid the
condensation of liquid will produce concave liquid surfaces of extremely
small radii of curvature and correspondingly low vapor pressures. For
this reason vapors that are at partial pressures much less than the
normal saturation value are condensed, augmenting the adsorption
normally taking place on flat surfaces. The adsorbing capacity of a
material possessing submicroscopic capillarity is considerably greater
than for one having the same surface area but having no capillary
structure.

As the partial pressure of a gas is increased, adsorption progressively
increases. At low partial pressures capillary condensation does not
take place. When a pressure sufficient to produce condensation in the

smallest capillaries is reached, capillary condensation begins, and the capillaries fill to levels of greater diameter at higher vapor pressures. The relationship between pressure and amount adsorbed is, therefore, dependent on the size distribution of the capillary pores, as well as on the area of exposed surface and the nature of both adsorbent and gas.

Adsorption Calculations. In Fig. 97 the isotherms of benzene vapor adsorbed on activated charcoal from the data of Coolidge[8] are shown. The adsorbed quantity x in milliliters of vapor, measured at

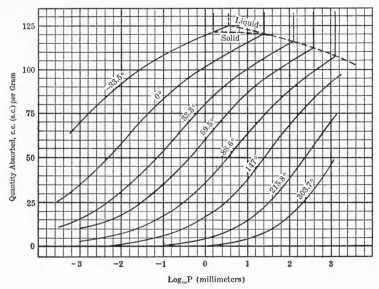

FIG. 97. Adsorption of benzene on activated coconut charcoal
(outgassed at 550° C)

standard conditions, adsorbed per gram of gas-free or "outgassed" charcoal is plotted as ordinate. Logarithms of the pressure of benzene vapor in millimeters of mercury are plotted as abscissa. The logarithmic scale is desirable because of the wide range of pressures required. The characteristic shape of these curves when plotted in linear coordinates is indicated by type II, Fig. 95. The charcoal used in the experiments on which Fig. 97 is based had been outgassed by heating to 550° C at reduced pressure. The data of Fig. 97 are rigorously applicable only to the particular charcoal for which they were determined. The quantity adsorbed is dependent on the source of charcoal, the method of its preparation, and its subsequent treatment. The same limitations apply to adsorption data for other materials.

[8] A. S. Coolidge, *J. Am. Chem. Soc.*, **46**, 596 (1924).

Illustration 2. (a) Estimate the number of pounds of benzene that may be absorbed by 1 lb of the activated charcoal of Fig. 97 from a gas mixture at 20° C in which the partial pressure of benzene is 30 mm Hg.

(b) Calculate the percentage of the adsorbed vapor of part a that would be recovered by passing superheated steam at a pressure of 5 psi and a temperature of 200° C through the adsorbent until the partial pressure of benzene in the steam leaving is reduced to 10.0 mm Hg.

(c) Calculate the residual partial pressure of benzene in a gas mixture treated with the freshly stripped charcoal of part b at a temperature of 20° C.

Solution: From Fig. 97,

(a) C_6H_6 adsorbed at 20° C, 30 mm Hg = 110 cc per gram

$$\text{or} \frac{110}{22,400} \times 78 = 0.382 \text{ gram per gram or lb per lb}$$

(b) C_6H_6 adsorbed at 200° C, 10.0 mm Hg = 22 cc per gram
Benzene recovered = 110 − 20   = 90 cc per gram

$$\% \text{ recovery} = \frac{90}{110} \qquad\qquad = 82\%$$

(c) Pressure of benzene in equilibrium at 20° C with charcoal containing 20 cc of benzene per gram = antilog of $\overline{4}.9 = 8 \times 10^{-4}$ mm Hg, the residual partial pressure of benzene.

Reversibility of Adsorption: Stripping. Van der Waals adsorption is a reversible process, and such an adsorbed gas is vaporized if its partial pressure in the gas phase is reduced below its vapor pressure in the adsorbed phase. The recovery or stripping of adsorbed gases may be accomplished in the following ways:

(a) The temperature of the solid may be raised until the vapor pressure of the adsorbed gas exceeds atmospheric pressure. The adsorbate vapor will then be evolved and may be collected at atmospheric pressure.

(b) The adsorbed gas may be withdrawn by applying a vacuum lowering the total pressure below the vapor pressure of the adsorbate. Enough heat should be supplied to prevent a drop in temperature as a result of evaporation. The undiluted adsorbate vapor may then be collected at this low pressure.

(c) A stream of an inert, condensible gas may be blown through the adsorbent, keeping the partial pressure of the adsorbate gas in the gas stream below the equilibrium pressure of the adsorbate in the solid. The adsorbate vapor will be evolved in admixture with the inert gas. By using an easily condensible vapor for stripping, such as superheated steam, the adsorbed material may be easily recovered by condensing the stripping vapor only or by condensing the entire mixture and separating by decantation, provided the two condensed vapors are immiscible.

(d) The adsorbed vapors may be displaced by treatment with some other vapor which is preferentially adsorbed.

The various desiccants used for drying gases are usually regenerated by blowing hot air through the spent adsorbent. For example, silica gel is regenerated by hot gases at 250 to 350° F and activated alumina by hot gases from 350 to 600° F.

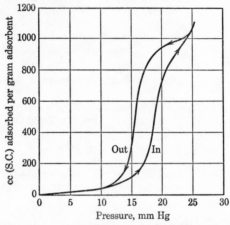

FIG. 98.   Adsorption hysteresis of water vapor on pine charcoal at 25° C

[A. J. Allmand, P. G. T. Hand, J. E. Manning, and D. O. Shiels, *J. Phys. Chem.*, **33**, 1682 (1929)]

Adsorption Hysteresis. The adsorption isotherm of Fig. 98 indicates a behavior that is typical of certain adsorbents which possess a high degree of capillarity. As indicated, the quantity adsorbed in equilibrium with a selected partial pressure is dependent on the direction from which equilibrium is approached. If equilibrium is reached by the evolution of adsorbate, the upper or "out" curve is applicable. If the equilibrium conditions are reached by adsorption, with a continually increasing concentration, the lower or "in" curve applies. This behavior is known as *adsorption hysteresis* and is exhibited by many capillary adsorbents. In calculations dealing with such systems data for both "in" and "out" curves are required.

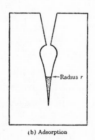

FIG. 99.   Adsorption hysteresis

This phenomenon can be explained by assuming a particular geometric shape of pore spaces within the solid. In Fig. 99 is shown a pore space inside a solid undergoing desorption in one case and adsorption in the other. If desorption is started with a full pore space and unsaturated gas is passed over the surface with a partial pressure of adsorbate gas of p_a, evaporation continues until the curvature attained in the upper capillary corresponds to the equilibrium vapor pressure at the level of liquid indicated. In adsorption, starting with an adsorbate-free solid, adsorption stops when the liquid level has reached the point indicated in b, corresponding to the same curvature as in a and exerting a vapor pressure equal to p_a. In each case equilibrium is attained, in desorption with a high liquid content and in adsorption with a low liquid content as indicated by shaded portions in the two figures. The situation is exaggerated in the illustration. With interconnecting pore spaces open to the atmosphere the forces produced by small capillaries will cause liquid to flow from large openings into the finer capillaries, thus emptying the large pore spaces. A discussion of these forces is given by Ceaglske and Hougen.[9] Other types of adsorption hysteresis are explained by Emmett and Dewitt.[10]

Preferential Adsorption. Preferential adsorption has already been referred to as one of the methods of removing an adsorbate gas. In illustration 2 it was assumed that the presence of the gases from which the benzene was adsorbed had no effect on the equilibrium between the benzene vapor and the charcoal. This assumption is not necessarily true because charcoal adsorbs considerable quantities of all the ordinary gases as well as benzene vapors. It would be expected that, when charcoal is exposed to a mixture of gases, a complicated equilibrium would be reached between each of the gases and its adsorbed quantity. Few quantitative data are available on the adsorption of mixtures of gases and vapors, but it is apparent that, when several gases are adsorbed, the presence of each must affect the equilibrium concentration of others.

It is an experimentally observed fact that, in general for van der Waals adsorption, a gas of high molecular weight, high critical temperature, and low volatility is adsorbed in preference to a gas of low molecular weight, low critical temperature, and high volatility. Such a preferentially adsorbed gas or vapor will displace other gases which have already been adsorbed. The chemical nature of the gas also plays an important part, but ordinarily it may be assumed that a heavy vapor of low volatility will displace a light gas of high volatility and similar chemical type. In the experiments of Coolidge it was found that exposure of the outgassed charcoal to air before treatment with benzene vapors had no apparent effect on the final equilibrium.

[9] N. H. Ceaglske and O. A. Hougen, *Trans. Am. Inst. Chem. Engrs.*, **33**, 283 (1937).
[10] P. H. Emmett and T. W. Dewitt, *J. Am. Chem. Soc.*, **65**, 1253 (1943).

In the absence of definite data it may ordinarily be assumed that, when an adsorbent is treated with a vapor of low volatility such as water or benzene in admixture with a very volatile gas such as air, the adsorption of the gas will exert only a negligible influence on the normal

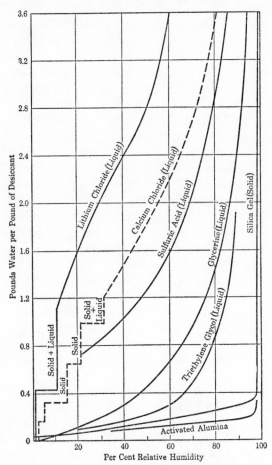

FIG. 100. Equilibrium moisture content of various desiccants

equilibrium between the vapor and the adsorbent. When a gaseous mixture having several components of similar volatility is treated with an adsorbent, equilibrium quantities of these components are adsorbed, and general predictions of the equilibrium conditions cannot be made without specific data.

The preferential adsorptive properties that many substances exhibit are of great industrial importance in the selective separation of components of gaseous mixtures.   An important application of adsorbents is in the drying of gases.   In Fig. 100 are the equilibrium moisture contents of silica gel and activated alumina compared with various solid and liquid chemical desiccants.   It will be observed that the solid desiccants compare unfavorably with the liquid desiccants in adsorption capacity.   However, both silica gel and activated alumina are capable of drying air to much lower dew points than homogeneous solutions of lithium chloride or calcium chloride.

Othmer Chart for Adsorbents.   A useful way of presenting the vapor-pressure relations of adsorbed water is by means of the Othmer chart wherein, for the same temperature, the vapor pressure of adsorbed water is plotted against the vapor pressure of pure water for lines of constant moisture content.   Straight lines generally result on a logarithmic plot, as shown in Fig. 101 for silica gel.   The relative humidity at any

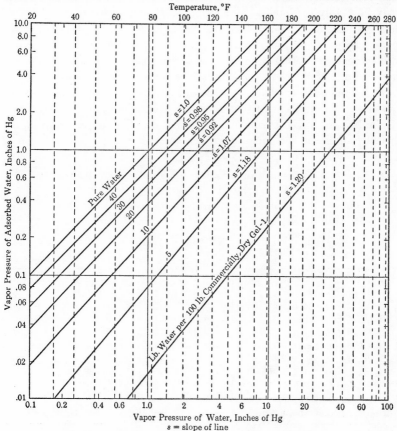

FIG. 101.   Vapor pressure of water adsorbed on silica gel

temperature and composition is obtained directly from the ratio p/p_s, where p_s is the vapor pressure of pure water. This chart is also useful in estimating heats of adsorption.

It should be noted that all the discussion in this chapter is restricted to equilibrium conditions. Calculation of rates of sorption requires consideration of particle size, diffusion, rates of gas flow, and other factors.

Heat of Wetting

When a solid surface is brought into contact with a liquid in which it is insoluble, the liquid will spread in a thin film over the surface of the solid, provided the solid is wettable by the liquid. This implies that the surface tension of the liquid relative to air is less than the adhesion tension between the liquid and the solid. The liquid film may be highly compressed as a result of attractive forces, or chemical binding may occur. The formation of such films of liquids is accompanied by an evolution of heat.

The heat of wetting per unit area of interfacial surface is small, and hence the heat of wetting per unit mass of solid is negligible unless a large wettable interfacial area is present as in a fine powder or a porous material. In addition to the extent of interfacial area the magnitude of the heat of wetting depends on the nature of both solid and liquid. Because of the uncertainty of surface areas of porous materials, heats

TABLE 34.   HEATS OF COMPLETE WETTING OF POWDERS DRIED AT 100° C

Experiments at 12 to 13° C

Unit: cal per gram of the dry material.

Liquid	ΔH, Clay	ΔH, Amorphous Silica	ΔH, Starch	ΔH, Sugar Charcoal
H_2O (water)	-12.6	-15.3	-20.4	-3.9
CH_3OH (methyl alcohol)	-11.0	-15.3	-5.6	-11.5
C_2H_5OH (ethyl alcohol)	-10.8	-14.7	-4.9	-6.8
$C_5H_{11}OH$ (amyl alcohol)	-10.0	-13.5	-7.0	-5.6
CH_3COOH (acetic acid)	-9.3	-13.5	-3.0–4.0	-6.0
CH_3COCH_3 (acetone)	-8.0	-13.5	-2.0	-3.6
$C_2H_5OC_2H_5$ (ether)	-5.8	-8.4	-2.2	-1.2
C_6H_6 (benzene)	-5.8	-8.1	-1.2	-4.2
CCl_4 (carbon tetrachloride)	-1.8	-8.1	-1.7	-1.5
CS_2 (carbon disulfide)	-1.7	-3.6	-0.5	-4.0
C_5H_{12}, C_6H_{14} (pentane, hexane)	-1.2	-3.1	-0.3	-0.4

Source: *International Critical Tables*, vol. V, p. 142 (1929).

of wetting are expressed on the basis of unit mass of solid rather than
on the basis of unit interfacial area. The *heat of complete wetting* is
the change in enthalpy when unit mass of solid is wetted with sufficient
liquid so that further addition of liquid produces no additional thermal
effect. The magnitude of the heat of wetting is proportional to the
surface area exposed by a given mass of solid and is dependent on the
quantity of liquid initially present in the solid substance.

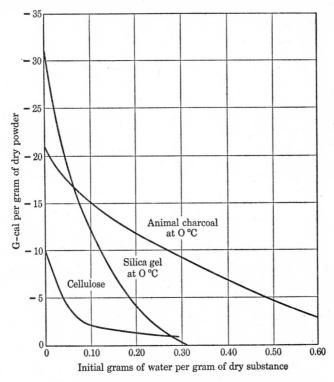

Fig. 102. Heats of complete wetting with water

In Table 34 are experimental values of heats of wetting of dry clay,
silica gel, starch, and charcoal when wetted by various liquids. The
chemical and physical structures of these substances are not specified.
It must be understood that these data may be used only for prediction
of the order of magnitude of the effect to be expected in a specific case.

As previously mentioned, the heat given off in the complete wetting
of a material is diminished if the solid contains initially some of the
same liquid. This effect is shown graphically in Fig. 102 where heats
of complete wetting are plotted, per gram of dry material, against the

grams of liquid originally present per gram of dry solid. The points at which the extrapolated curves reach the axis correspond to zero heat of wetting and indicate the minimum quantities of liquid required to produce complete wetting. The heat of wetting of an already partially wetted solid is the difference between the heats of complete wetting at the final and at the initial concentrations.

A more convenient method of presenting heat of wetting data is as *integral* and *differential heats of wetting*. These terms are analogous to integral and differential heats of solution and are similarly defined.

The enthalpy relationships involved in wetting are as follows:

$$\Delta H = (w_2 + 1)H - w_2 H_2 - H_1 = w_2 \bar{H}_2 + \bar{H}_1 - w_2 H_2 - H_1 \quad (10)$$

where

H = total enthalpy of wetted solid, Btu per pound of wet solid
ΔH = integral heat of wetting, Btu per pound of dry solid
$\bar{H}_1$ = partial enthalpy of gel, Btu per pound of dry solid
$\bar{H}_2$ = partial enthalpy of liquid, Btu per pound of liquid
w_2 = pounds of liquid per pound of dry solid
H_2, H_1 = enthalpy of pure liquid and dry solid, respectively, at the temperature of the system, in Btu per pound (Both values are given as zero in Fig. 103.)

When all enthalpies and heats of wetting are taken at the same temperature H_1 and H_2 become 0 and equation 10 becomes

$$\Delta H_w = (w_2 + 1)H = \Delta \bar{H}_1 + w_2 \Delta \bar{H}_2 = \bar{H}_1 + w_2 \bar{H}_2 \quad (11)$$

where the integral heat of wetting becomes identical with the enthalpy per unit mass of wetted solid and the differential heat of wetting for each component $\Delta \bar{H}_i$ becomes identical with its partial enthalpy $\bar{H}_i$.

Values of integral and differential heats of wetting of silica gel with water are shown in Fig. 103 from the data of Ewing and Bauer.[11] The values are reported on the basis of one pound of commercially dry silica gel.

It will be observed that the value of $\bar{H}_2$ is -400 Btu per lb of water when adsorbed on dry gel but rapidly falls off when the water content rises, reaching zero at 40% water (commercially dry basis). The value of $\bar{H}_1$ changes in a reverse manner, being zero for dry gel and becoming equal to the integral heat of wetting at a water content of 40% (commercially dry basis). In commercially dry silica gel approximately 5% of water of constitution is present. If chemically bound water were

[11] D. T. Ewing and G. T. Bauer, *J. Am. Chem. Soc.*, **59**, 1548 (1937).

removed, the gel would lose its capacity for adsorption. The removal of combined water would result in evolution of heat rather than in absorption of heat, as indicated by that portion of the ΔH curve of Fig. 103 below zero.

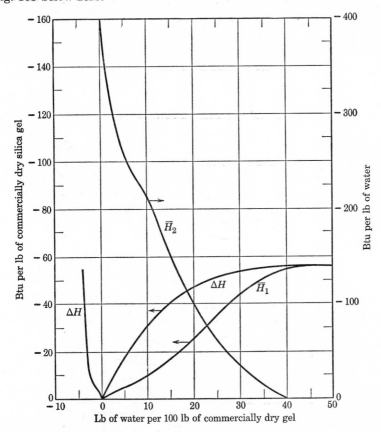

FIG. 103. Integral and differential heats of wetting of silica gel at 25° C[11]

Illustration 3. One hundred pounds of commercially dry silica gel are wetted with water from a content of 5 to 20 lb. Estimate the heat evolved from (a) integral and (b) differential heat of wetting data.

(a) From integral heat of wetting data (Fig. 103) the enthalpy change is equal to

$$100[-46 - (-18)] = -2800 \text{ Btu}$$

(b) From the partial enthalpy data (Fig. 103) the enthalpy change is equal to

$$100[-26 + 0.20(-100)] - 100[-5 + 0.05(-260)] = -2800 \text{ Btu}$$

Therefore, the heat evolved is 2800 Btu.

Enthalpies of Wetted Systems. In Fig. 104 the enthalpy of the silica gel–water system is shown at various temperatures and compositions relative to dry silica gel at 32° F and liquid water at 32° F. For convenience, the abscissas represent pounds of water per 100 lb of

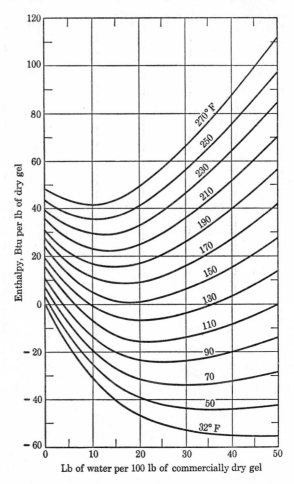

Fɪɢ. 104. Enthalpy of silica gel–water system per pound of commercially dry gel
Reference state: Commercially dry gel and liquid water at 32° F

commercially dry gel. In constructing this chart, the 70° F isotherm was derived first since heats of wetting are known at this temperature. Thus,

$$(w_2 + 1)H = H_1 + H_2w_2 + \Delta H_w \tag{12}$$

where

w_2 = pounds of water per pound of dry gel
H_1 = enthalpy of dry gel at 70° F, Btu per pound
H_2 = enthalpy of liquid water at 70° F, Btu per pound
ΔH_w = heat of wetting when w_2 lb of water are added
　　　　to one lb of dry gel at 70° F
H = enthalpy of wet gel at 70° F, Btu per pound

The enthalpy at any other temperature is then obtained from

$$H = H_{70} + \int_{70}^{t} C_p dt \tag{13}$$

where C_p is the specific heat of the system.

From a combination of the enthalpy chart for air and that of silica gel, the thermal effects in· drying air by silica gel may be readily calculated.

Illustration 4. Calculate the heat to be removed when 100 lb of air (dry basis) at 70° F and 0.010 humidity are dried to a final humidity of 0.001 at 70° F and 1 atm pressure starting with 10 lb of dry gel at 70° F. The average temperature of the gel is initially at 70° F and finally at 80° F.

Average composition of gel after adsorption $= \dfrac{100(0.010 - 0.001)}{10} = 0.09$

From Fig. 73,　　　Heat removed from the air $= 100(27.7 - 18) = $ 970 Btu
From Fig. 104,　　Heat removed from the gel $= 10[6 - (-14)] = $ 200
　　　　　　　　　Total heat removed　　　　　　　　$= $ 1170 Btu

Heat of Adsorption

The heat of adsorption of a gas caused by van der Waals forces of attraction and capillarity is the sum of the heat of normal condensation plus the heat of wetting. In activated adsorption the gas is adsorbed by formation of a surface compound at temperatures even above the critical temperature of the gas. The corresponding heat of adsorption greatly exceeds that of normal condensation. The difference between heat of adsorption and heat of normal condensation represents the heat of wetting in van der Waals adsorption and the heat of formation of a surface compound in activated adsorption.

Data on heats of adsorption are presented either as integral or as differential values. The integral heat of adsorption is the change in enthalpy per unit weight of adsorbed gas when adsorbed on gas-free or "outgassed" adsorbent to form a definite concentration of adsorbate.

The integral heat of adsorption varies with the concentration of the adsorbate, diminishing with an increase in concentration.

The differential heat of adsorption of a gas is the change in enthalpy when a unit quantity of the gas is adsorbed by a relatively large quantity of adsorbent on which a definite concentration of the adsorbed gas already exists. The differential heat of adsorption is also a function of concentration, diminishing with an increase in concentration. As complete saturation of an adsorbent is approached, the differential heat of adsorption approaches that of normal condensation. In activated adsorption the differential heat of adsorption decreases as the more active centers of the surface become covered. In capillary adsorption the heat of adsorption decreases as the capillaries become progressively filled. When the gas begins to condense at its normal saturation pressure, the heat of adsorption assumes the value of its normal heat of condensation.

Integral and differential heats of adsorption bear the same relationship to each other as do integral and differential heats of solution and may be calculated in a similar manner from total and partial enthalpy relationships. Heat effects in adsorption are calculated more accurately from differential heat of adsorption data because concentration changes are usually small and average values of the differential heat of adsorption may be used.

Experimental measurements have been made of the heats of adsorption of many of the more common gases on the important adsorbents such as charcoal, silica gel, and various solid catalysts. In Fig. 105 are values of differential heats of adsorption in calories per gram-mole adsorbed at 0° C, plotted against concentrations in cubic centimeters of adsorbed gas, measured at standard conditions of 0° C and 760 mm of mercury, per gram of adsorbent.

Illustration 5. An adsorber for the removal of water vapor from air contains 250 lb of silica gel on which is initially adsorbed 28.0 lb of water. Calculate the heat evolved per pound of water adsorbed at this concentration, assuming that the characteristics of the gel are similar to that on which the data of Fig. 105 are based.

Solution: Concentration of H_2O in gel $= \dfrac{28}{250} = 0.112$ gram per gram of gel

or $\qquad\qquad \dfrac{0.112}{18} \times 22{,}400 = 139$ cc per gram of gel

From Fig. 105,

$$\Delta_H = -12{,}500 \text{ cal per g-mole} \quad \text{or} \quad -22{,}500 \text{ Btu per lb-mole}$$

Heat evolved per lb of water adsorbed $= \dfrac{22{,}500}{18} = 1250$ Btu

The effect of temperature on heat of adsorption is given by the Clapeyron equation (page 78) and may be evaluated by an equal-temperature reference-substance method of plotting equilibrium adsorption pressure similar in principle to that used for normal vapor pressures

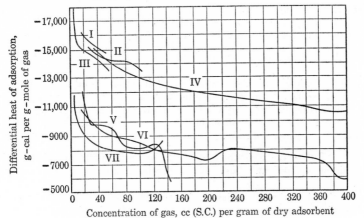

I. C_6H_6, 0° C, on inactive coconut carbon, outgassed at 350° C
II. C_6H_6, 0° C, on active coconut carbon, outgassed at 350° C
III. C_2H_5OH, 0° C, on active coconut carbon, outgassed at 350° C
IV. H_2O, 0° C, on silica gel dried at 300° C for 2 hr and outgassed at 250° C, containing 3.5 to 5.5% H_2O
V. SO_2, 0° C, on same silica gel
VI. SO_2, −10° C, on blood carbon (puriss. Merck), outgassed at 450° C ($d = 1.63$)
VII. NH_3, 0° C, on coconut carbon, heated to 550° C, outgassed at 400° C ($d = 1.86$)

FIG. 105. Differential heats of adsorption

and heats of vaporization. If the heat of adsorption of a vapor is referred to the normal heat of condensation of the same vapor at the same temperature, equation 35, page 276, becomes

$$\lambda_1 = \lambda \frac{d \ln p_1}{d \ln p} \tag{14}$$

where at the same temperatures for both systems

λ = normal heat of condensation
λ_1 = heat of adsorption
p = normal vapor pressure of the condensed vapor
p_1 = actual equilibrium pressure of the adsorbed vapor

Problems

1. Activated charcoal similar to that of Fig. 97 is to be used for the removal of benzene vapors from a mixture of gases at 20° C and a pressure of 1 atm. The relative saturation of the gases with benzene is 83%.

(a) Calculate the maximum weight of benzene that may be adsorbed per pound of charcoal.

(b) The adsorbed benzene is to be removed by stripping with superheated steam at a temperature of 180° C. Calculate the final partial pressure to which the benzene in the steam leaving the stripper must be reduced in order to remove 90% of the adsorbed benzene.

(c) If the adsorbent is so used that the benzene-bearing gases always come into equilibrium with freshly stripped charcoal of part b before leaving the process, calculate the loss of benzene in these treated gases, expressed as percentage of the total benzene entering the process.

2. Fifty pounds of unsized cotton cloth containing 20% total moisture are hung in a room of 4000 cu ft capacity. The initial air is at a temperature of 100° F, at a relative humidity of 20%, and a barometric pressure of 29.92 in. Hg. The air is kept at 100° F with no fresh air admitted and no air vented. Neglect the space occupied by the contents of room.

(a) Calculate the moisture content of the cloth and the relative humidity of the air at equilibrium. **Ans.** 72.5% relative humidity, 0.103 lb water per pound.

(b) Calculate the equilibrium moisture content of the cloth and the corresponding relative humidity of the air if 100 lb of wet cloth instead of 50 lb are hung in the same room. **Ans.** 94.0% relative humidity, 0.145 lb water per pound.

(c) What is the final pressure in the room under part a? **Ans.** 30.9 in. Hg.

3. Air at atmospheric pressure is to be dried at 80° F from 70% to 10% relative humidity by mixing with silica gel, initially dry, and at 80° F. What is the final equilibrium moisture content of the gel, if a constant temperature of 80° F is assumed?

4. If air were to be dried to a dew point of 0° F, what would be the equilibrium water contents of the following desiccants, assuming exit temperature of air and desiccant to be 70° F?

Activated alumina	**Ans.**	0.03
Silica gel	**Ans.**	0.03
Glycerin	**Ans.**	0.01
Sulfuric acid	**Ans.**	0.56
Calcium chloride solution	**Ans.**	0.34
Lithium chloride solution	**Ans.**	0.43
Triethylene glycol	**Ans.**	0.05

5. How would it be possible to dry air to a dew point of 0° F with a homogeneous LiCl solution?

6. Activated charcoal is used for the recovery of benzene (C_6H_6) vapors from a mixture of inert gases. Calculate the heat evolved in Btu per pound of benzene adsorbed on a large quantity of charcoal at 0° C, when the charcoal contains 0.25 lb of benzene per pound of charcoal.

7. Estimate the heat evolved in Btu in completely wetting 10 lb of dried clay with water. **Ans.** 227 Btu.

8. Silica gel contains 12% H_2O by weight. Calculate the heat, in calories, evolved when 2.0 kg of this material at 0° C are completely wetted with water. **Ans.** 12,850 cal.

9. Calculate the heat evolved, expressed in Btu, when 100 lb of silica gel (commercially dry basis) adsorbs 25 lb of water at 70° F.

10. Using the data in Table 34 and Fig. 102, calculate the heat of complete wetting, expressed in Btu, for the following solids when they are completely wetted with the liquid specified.

    (a) 100 lb of dry starch, wetted with water. **Ans.** 3672 Btu.

    (b) 100 lb of dry clay, wetted with ethyl alcohol. **Ans.** 1944 Btu.

    (c) 100 lb of animal charcoal containing 8% water when wetted completely with water. **Ans.** 2550 Btu.

11. From *International Critical Tables* obtain the following data:

    (a) Total heat evolved in the adsorption of 65 cc of CO_2 on 1 gram of outgassed coconut charcoal at 0° C, in calories.

    (b) Differential heat of adsorption of CH_3OH vapors on activated charcoal containing 100 cc of vapor per gram of charcoal at 0° C, in calories per gram-mole.

    (c) Heat of wetting of dried bone charcoal with gasoline, in calories per gram of charcoal.

12. Estimate the final temperature when 100 lb of air (dry basis) initially at 70° F and a humidity of 0.010 lb per lb of dry air are mixed with 10 lb of silica gel, initially commercially dry and at 70° F. The final humidity of the air = 0.001. Assume no heat losses from the system.

13. Calculate the heat removed when 100 lb of air (dry basis) initially at 70° F and a humidity of 0.010 lb per lb of dry air are mixed with 10 lb of silica gel, initially commercially dry and at 70° F. The final humidity of the air is 0.001. The gel and air are kept at 70° F. **Ans.** 1223 Btu.

14. Calculate the heat removed when 100 lb of air (dry basis) initially at 70° F and 70% relative humidity are mixed with 10 lb of silica gel, initially commercially dry at 70° F. The final relative humidity of the air is 10%, and the final temperature of the air and gel is 80° F.

15. Solve illustration 4, assuming that the average temperature of air is 85° F and the average temperature of gel is 95° F. (*Note:* Actually beds of silica gel are operated under intermittent-flow adiabatic conditions. The treatment of the adiabatic case presents formidable difficulties because of the fluctuating temperature waves in the bed.) **Ans.** 780 Btu.

16. Krieger[12] obtained the following data for the adsorption of nitrogen on activated alumina powder at 77.3° K. ($p_s = 759$ mm Hg)

p, mm Hg	V', g-moles N_2 per gram Alumina
31.7	0.000831
40.1	0.000853
56.6	0.000890
64.5	0.000903
82.7	0.000953
96.7	0.000985
112.4	0.001015
128.8	0.001045
148.6	0.001081
169.3	0.001118

[12] K. A. Krieger, *J. Am. Chem. Soc.* **63**, 2712 (1941).

From these data and equation 6 calculate the constants C and V_m. From the value of V_m calculate the surface area of the activated alumina in square meters per gram.

For nitrogen gas the area A_m covered per molecule = 16.2 (10^{-20}) square meter per molecule.

The surface area per gram $A_S = \dfrac{V_m N A_m}{22,414}$

$$N = 6.023 \ (10^{23}) \text{ molecules per g-mole}$$

11

Fuels and Combustion

Because of the universal use of the combustion of fuels for the generation of heat and power, the development of special methods and techniques has been justified for establishing the material and energy balances of such processes. Combustion calculations should be pursued as rigorously as the available data permit and should be based on stoichiometric principles rather than on empirical equations.

Heating Values of Fuels. The heating value of a fuel is numerically equal to its standard heat of combustion but of opposite sign. This property is usually determined by direct experimental measurements; methods are also given for its estimation from composition and rank.

Two methods of expressing heating values are in common use, differing in the state selected for the water present in the system after combustion. The *total heating value* of a fuel is the heat *evolved* in its complete combustion under constant pressure at a temperature of 25° C when all the water initially present as liquid in the fuel and that present in the combustion products are condensed to the liquid state. The *net heating value* is similarly defined, except that the final state of the water in the system after combustion is taken as vapor at 25° C. The total heating value is also termed the "higher" or "gross" heating value; the net is also termed the "lower" heating value. The net heating value is obtained from total heating value by subtracting the latent heat of vaporization at 25° C of the water *formed and vaporized* in the combustion.

Coke and Carbon. The combustible constituents of cokes and charcoals are chiefly carbon. The heating value of such fuels may be predicted with accuracy sufficient for most purposes by simply multiplying its carbon content by the heating value per unit weight of carbon.

In Table 29, page 298, it will be noted that heats of formation of carbon compounds are based on a value of zero assigned to the heat of formation of graphite. On this basis various other forms of carbon have positive heats of formation. The heat of formation of carbon in the form of the diamond is accurately known, and is equal to 0.4532 kcal per g-atom.

The published values for the heats of formation of amorphous carbon from different sources show considerable variation, with most of the values ranging from approximately 1.25 to 3.0 kcal per g-atom. These differences arise in part from differences in allotropic forms and in the surface energy of carbon resulting from different states of subdivision and porosity, and in part from the presence of hydrocarbon compounds of high molecular weights and low hydrogen contents. An average value of 2.6 kcal per g-atom will be used in this text as the heat of formation of amorphous carbon.

The heats of combustion of the various forms of amorphous carbon differ by the same amounts as do their heats of formation. For combustion calculations the value of the heat of combustion of carbon is taken as $-96,650$ cal per g-atom or $-14,490$ Btu per lb. This value is the difference between the heat of formation of carbon dioxide and that of carbon in coke, or $-94,050 - 2600 = -96,650$ cal.

Coal Analyses. Coal consists chiefly of organic matter of vegetable origin which has been altered by decomposition resulting from compression and heating during long ages of submergence in the earth's crust. In addition to organic matter, it contains mineral constituents of the plants from which it was formed and inclusions of other inorganic materials entrained during its geological formation.

Two types of analysis are in common use for expressing the composition of coal. In an *ultimate analysis*, determination is made of each of the major chemical elements. In a *proximate analysis* four arbitrarily defined groups of constituents are determined, namely, *moisture, volatile matter, fixed carbon*, and *ash*. The sum of the fixed carbon and volatile matter of a coal is termed the *combustible*. Following are the ultimate and proximate analyses of a typical Illinois coal:

Ultimate		Proximate	
Moisture	9.61%	Moisture	9.61%
Ash (corrected)	9.19	Ash	9.37
Carbon	66.60	Volatile matter	30.68
Net hydrogen	3.25	Fixed carbon	50.34
Sulfur	0.49		100.00%
Nitrogen	1.42		
Combined H_2O	9.44		
	100.00%		

The proximate analysis of coal should be carried out according to arbitrarily standardized procedures which have been recommended by the United States Bureau of Mines. The details of these methods are described in most books on methods of technical analysis. These determinations may be rapidly and easily carried out. Contracts and specifications for the purchase of coal are frequently based on this analysis. The tedious methods of ultimate analysis are carried out when necessary to serve as a basis for energy- and material-balance calculations. The sulfur content is of particular interest, and a separate determination of this element frequently is desirable.

In both schemes of analysis "moisture" represents the loss in weight on heating a sample of the finely divided coal at 105° C for one hour. The material termed "ash" in the proximate analysis is the residue formed in the ordinary combustion of coal. However, the ash determined in this manner does not accurately represent the mineral content of the original coal because of the changes that take place during combustion. A mineral component of many coals is in pyrites, FeS_2.

In combustion the FeS_2 is oxidized to Fe_2O_3 and SO_2:

$$2FeS_2 + 5\tfrac{1}{2}O_2 \rightarrow Fe_2O_3 + 4SO_2$$

The Fe_2O_3 is weighed in the residual ash. In the oxidation of pyrites 4 g-atoms (128 grams) of sulfur are replaced by 3 g-atoms (48 grams) of oxygen, a loss in weight equal to $\frac{5}{8}$ times the weight of pyritic sulfur present.

Thus,

Mineral content $\qquad\qquad$ = ash as weighed $+ \frac{5}{8}$ pyritic sulfur

Mineral content excluding S = ash as weighed $- \frac{3}{8}$ pyritic sulfur

To determine the actual mineral content, not including the pyritic sulfur, a correction equal to $\frac{3}{8}$ of the pyritic sulfur must be subtracted from the ash as weighed. Other less important corrections may also be applied to the ash. Unless otherwise designated, "ash" refers to ash as weighed.

To obtain the ultimate analysis, direct determinations are made of carbon, sulfur, nitrogen, and hydrogen by the usual analytical methods. The moisture and ash are determined by the standardized procedures of the proximate analysis. The percentage oxygen content is then taken as the difference between 100 and the sum of the percentages of carbon, hydrogen, sulfur, nitrogen, and corrected sulfur-free ash. It is recommended that, for this calculation, the corrected ash be estimated by assuming that all sulfur in the coal is present in the pyritic form. On this basis,

$$\% \text{ corrected ash} = \% \text{ ash as weighed} - \tfrac{3}{8}(\% \text{ S})$$
$$= \% \text{ mineral content} - \% \text{ S}$$

where $\% \text{ S} = $ percentage sulfur content of coal This correction represents only an approxmation, since not all sulfur is pyritic and other changes in the mineral constituents may take place in combustion. More refined methods for estimating oxygen content are not ordinarily justified.

In reporting the ultimate analysis it is convenient to consider that all oxygen is in combination with hydrogen to form moisture and "combined water." The surplus hydrogen above that required to combine with the oxygen is termed "net" or "available" hydrogen. This represents the hydrogen present in the form of hydrocarbons and available for further oxidation.

The Bureau of Mines has published extensive tables[1] of the ultimate analyses of coals representing hundreds of coal deposits throughout the United States. If the source of a coal is known, the ultimate analysis of its combustible matter can be obtained with fair reliability from these tables since the composition of combustible material in any one coal bed is nearly constant. In every coal sample it is necessary, however, to make separate determinations of ash and moisture contents.

Rank of Coal. The *fuel ratio* of a coal is defined as the ratio of its percentage of fixed carbon to that of volatile matter. The *rank* of the coal, whether bituminous or anthracite, may be estimated from its fuel ratio. The generally accepted classification of coals and the corresponding ranges of fuel ratios are given in Table 35. Fuels of lower rank than bituminous, namely, subbituminous and lignite, may have fuel ratios within the bituminous range but are characterized by higher water or oxygen contents.

TABLE 35. RANK OF COALS

Rank	Fuel Ratio
Anthracite	between 10 and 60
Semianthracite	between 6 and 10
Semibituminous	between 3 and 7
Bituminous	between $\frac{1}{2}$ and 3

The classification of coals on the basis of fuel ratio is not entirely satisfactory for many purposes. Other methods[2] have been developed which give better differentiation.

Heating Value of Coal. The total heating value of a coal may be determined by direct calorimetric measurement and is usually expressed in Btu per pound. The net heating value is obtained by subtracting from the total heating value the heat of vaporization at 25° C of the water present in the coal and that formed by the oxidation of the

[1] *U. S. Bur. Mines Bull.* **123.**

[2] R. T. Haslam and R. P. Russell, *Fuels and Their Combustion*, McGraw-Hill Book Co. (1926).

available hydrogen. Thus,

$$\text{Net H.V.} = \text{total H.V.} - 8.94 \times \text{H} \times 1050 \qquad (1)$$

where H.V. = heating value, Btu per pound
　　H = weight fraction of total hydrogen, including available hydrogen, hydrogen in moisture, and hydrogen in combined water

When the heating value of a coal is determined by its combustion in a calorimeter, the sulfur is oxidized to form sulfuric acid. Normally the sulfur in coal burns to form sulfur dioxide only, so that a subtraction to the calorimetric value should be made for the heat evolved in forming sulfuric acid from sulfur dioxide and water.

Many attempts have been made to develop a method of calculating the heating value of coal from its proximate analysis. None of these methods are sufficiently reliable to justify their use except as approximations.

A fair approximation to the heating value of a coal may be obtained by considering that each of the combustible constituents, carbon, available hydrogen, and sulfur, is present in its elementary state. On the basis of this assumption the heating value is the sum of the quantities of heat evolved in the combustion of each of these elements, using for carbon the heating value of amorphous carbon and for sulfur the heating value of FeS_2. The respective heats of combustion, in Btu per pound, may be calculated from the data of Table 29, page 297. It is assumed that the sulfur and the iron in FeS_2 are burned to SO_2 and Fe_2O_3.

Element	Heating Value, Btu per lb
Carbon	14,490
Hydrogen (total)	61,000
Hydrogen (net)	51,610
Sulfur (as FeS_2)	5,550

Then,

$$\text{Total H.V.} = 14{,}490\text{C} + 61{,}000\text{H}_a + 5550\text{S} \qquad (2)$$

where H.V. = heating value, Btu per pound
　　C, H_a, S = weight fractions of carbon, available hydrogen, and sulfur, respectively

Equation 2 is known as Dulong's formula. It is not theoretically sound because it neglects the heats of formation of the compounds of carbon, sulfur, and hydrogen that exist in the coal. However, as previously pointed out, the heats of formation of hydrocarbon compounds are small in comparison to their heats of combustion, and the results of the above equation are rarely in error by more than 3%. The experimentally observed total heating value of the coal whose analysis is given above, page 398, was 11,725 Btu per lb. Applying equation 2, the heating value would be predicted as $(14,490 \times 0.666) + (61,000 \times 0.0325) + (5550 \times 0.0049) = 11,660$ Btu per lb, an error of -0.6%.

Because the heating value of coal is more easily determined than its ultimate analysis, the use of equation 2 for calculation of heating value is rarely advantageous. It is more useful as a means of predicting the available hydrogen content from experimentally determined values of carbon and sulfur contents and heating value.

A useful relationship has been pointed out by Uehling[3] between the heating value of coal *per pound of total carbon* and its rank. It was found that, for each rank of coal, the heating value per pound of total carbon is nearly constant, rarely varying by more than 2%. It was also found that the weight of available hydrogen, per pound of total carbon varies but little among coals of the same rank. On this basis, standard average heating values and available hydrogen contents, per pound of total carbon, were established for the different ranks of fuel. These values, contained in Table 36, were based on the published results of a large number of analyses carried out by the United States Bureau of Mines.

TABLE 36. STANDARD HEATING VALUES AND
NET HYDROGEN CONTENTS OF COAL

H.V.$'$ = heating value, Btu per lb of total carbon
H$'_a$ = available hydrogen content, lb per lb of total carbon

Rank	H.V.$'$	H$'_a$
Coke	14,490	0.0
Anthracite	16,100	0.029
Semibituminous	17,400	0.049
Bituminous	17,900	0.054
Sub-bituminous	17,600	0.045
Lignite	17,100	0.037

From the data of Table 36, the heating value of the bituminous coal whose analysis is given on page 398 would be predicted to be $17,900 \times 0.666 = 11,920$ Btu per lb, in error by only 1.7% as compared to the experimentally determined value of 11,725. The available hydrogen content would be predicted as $0.666 \times 0.054 \times 100 = 3.6\%$ as compared to the experimentally observed value of 3.25%.

[3] E. A. Uehling, *Heat Loss Analysis*, McGraw-Hill Book Co. (1929), with permission.

Because of the relative difficulty of the total carbon determination as compared to the calorimetric determination of heating value, the relationships pointed out by Uehling have their greatest value in predicting the ultimate analysis from the experimentally determined heating value. From only the heating value and the data of Table 36, good approximations to the total carbon and available hydrogen content of a coal may be predicted. In view of the fact that the composition of a coal sample will frequently vary by as much as 5% from the true average composition of the coal from which it was taken, the accuracy of these predictions is often as great as is justifiable for calculations of energy and material balances.

Petroleum

Petroleums are complex mixtures of hydrocarbons including four significant series of compounds: paraffins, naphthenes, olefins, and aromatics. These compounds differ in hydrogen content in this order, paraffins having the highest hydrogen content and aromatics the lowest. In naturally occurring petroleums the first two series predominate; in *cracked* products formed by decomposition of natural oils, large quantities of olefins and aromatics may also be present. In addition to hydrocarbons, varying quantities of sulfur, oxygen, and nitrogen compounds are generally present.

Because of the complexity of petroleum fractions determination of the actual compounds present is generally impracticable. Elementary analyses may be made, determining carbon, hydrogen, sulfur, and nitrogen as for coal. Data of this type are available in the publications of the United States Bureau of Mines for many naturally occurring petroleums. However, such analyses give little indication of the actual character of an oil and its thermal properties. Approximate methods have been developed whereby much of this information may be estimated from easily determined physical properties such as distillation or boiling range, specific gravity, and viscosity.

Characterization of Petroleum. For the general correlation of the average physical properties of petroleum stocks of different types, it is necessary to develop a means of quantitatively expressing the general character of the oil. Paraffin hydrocarbons of maximum hydrogen content may be considered as one extreme and aromatic materials of minimum hydrogen content as the other.

To serve as a quantitative index to this property, which may be termed paraffinicity, the U. O. P. (Universal Oil Products Company) characterization factor[4] has been developed and empirically related to

[4] K. M. Watson and E. F. Nelson, *Ind. Eng. Chem.*, **25**, 880 (1933), with permission.

six commonly available laboratory inspections. This factor is not a true measure of chemical type and does not show constancy in a homologous series. These disadvantages are offset by simplicity and convenience.

The definition of the U. O. P. characterization factor arose from the observation that, when a crude oil of supposedly uniform character is fractionated into narrow cuts, the specific gravity of each cut is approximately proportional to the cube roots of its absolute boiling points at 1 atm pressure. The proportionality factor may then be taken as an index of the paraffinicity of the stock. Thus

$$K = \frac{\sqrt[3]{T_B}}{G} \tag{3}$$

where K = U. O. P. characterization factor
T_B = average boiling point, degrees Rankine at 1 atm pressure
G = specific gravity at 60° F

In dealing with mixtures of wide boiling range, a special method of obtaining the average boiling point as described by Watson and Nelson[4] must be used. For narrower cuts the 50% point of the Engler distillation may be taken as the average boiling point.

The characterization factor shows fair constancy throughout the boiling range of a number of crude oils and for others may either increase or decrease in the higher boiling range. In the paraffin series fair constancy for the average of the reported isomers exists up to a boiling temperature of 700° F. Values of the characterization factor range as follows:

Pennsylvania stocks	12.2–12.5
Midcontinent stocks	11.8–12.0
Gulf Coast stocks	11.0–11.5
Cracked gasolines	11.5–11.8
Cracking-plant combined feeds	10.5–11.5
Recycle stocks	10.0–11.0
Cracked residuums	9.8–11.0

The characterization factor is readily calculated from values of specific gravity and average boiling point according to equation 3 or it may be read directly from API gravity and average boiling point by interpolation between the curves of Fig. 106. In this figure, API gravities are plotted as ordinates and average boiling points as abscissas with lines of constant K from equation 3. The relationship between specific gravity and degrees API is shown by Fig. B in the appendix.

It has also been found that a fair empirical correlation exists between the characterization factor and the viscosity-gravity relationship at a given temperature. Paraffinic stocks have high viscosities compared with aromatic materials of the same gravities.

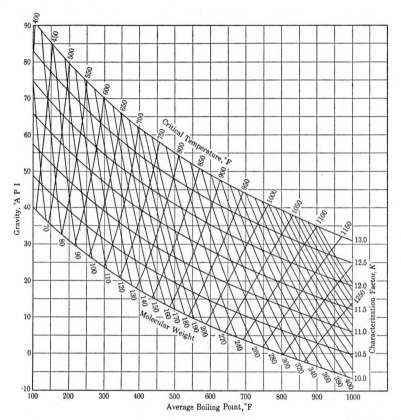

FIG. 106. Molecular weights, critical temperatures, and characterization factors of petroleum fractions
(Reproduced in *CPP Charts*)

Because of uncertainties of molecular aggregation at low temperatures, the viscosity measurements used for physical correlations should be made at as high a temperature as possible. In Fig. 107 viscosity in centistokes at 122° F is plotted against API gravity for stocks of constant characterization factors. By use of the centistoke scale of viscosity the entire range of fractions from light gasolines to heavy residues is covered in a single relationship. A chart for conversion of common viscometer readings into centistokes is included in Fig. C, appendix.

Lines of constant boiling point are plotted on Fig. 107 resulting from combination of the relationships between characterization factor from boiling-point and viscosity data. These lines permit an approximation to the boiling point from only viscosity and gravity data. This rela-

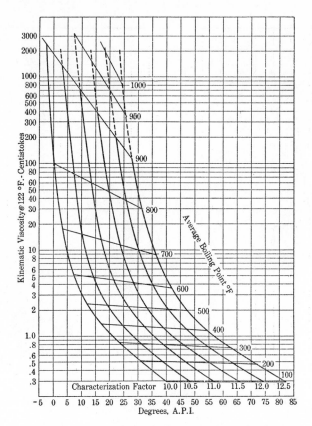

Fig. 107.   Characterization factor from viscosity at 122° F

(Reproduced in *CPP Charts*)

tionship is particularly useful for heavy stocks on which boiling-point data can be obtained only under high vacuum. However, because of the large change in viscosity with a slight change in the gravity of heavy stocks, boiling points estimated in this way may be considerably in error, sometimes as much as 50° F for the heavier residues. Similar charts were developed on the basis of viscosities at other temperatures.[5]

[5] K. M. Watson, E. F. Nelson, and G. B. Murphy, *Ind. Eng. Chem.*, **27**, 1460 (1935).

Molecular Weights of Petroleum Fractions.   The average molecular weights of petroleum fractions may be satisfactorily estimated from average boiling point and gravity.   Aromatic stocks of low characterization factors have lower molecular weights than paraffinic materials of the same average boiling points.

The relationships among molecular weight, characterization factor, boiling point, and API gravity are included in the curves of Fig. 106. By interpolation between these curves, molecular weights may be estimated, with errors rarely exceeding 5%.   If boiling-point data are not available, the boiling point may be estimated from viscosity, using Fig. 107.

Critical Properties.   The critical temperature and molecular weight curves of Fig. 106 are in satisfactory agreement with the existing data on petroleum.

Critical temperatures estimated from Fig. 106 are applicable with little error to pure hydrocarbons, narrow petroleum cuts, or wide-boiling mixtures if a proper method of obtaining average boiling point is used. Correct methods of averaging have been developed by Smith and Watson.[6]   Critical pressures may be estimated by the methods described in Chapter 4.

Hydrogen Content.   The curves of Fig. 108 represent a relationship between hydrogen content and characterization factor for materials of constant boiling points.

Figure 108 combined with the preceding charts permits estimation of hydrogen content from a knowledge of only the specific gravity and one other property.   Ordinarily the error will be less than 0.5%, based on the total weight of the oil, except for highly aromatic, low-boiling materials.

Petroleum oils ordinarily contain little ash and in the absence of specific data may be assumed to be 97% carbon and hydrogen with the remainder oxygen, nitrogen, sulfur, and ash.   This assumption is unsatisfactory for oils of high sulfur content, such as certain California or Mexican stocks, or where salts are present with water in partial solution and suspension.   Specific data should be obtained on such stocks.

Illustration 1.   A fuel oil has an API gravity of 14.1 and a viscosity of 150 Saybolt Furol seconds at 122° F.   Estimate the characterization factor, average boiling point, hydrogen content, specific heat at 200° F, heating value, and average molecular weight of this oil.

From the conversion chart, Fig. C in the appendix, it is found that 150 Saybolt Furol seconds is equivalent to 320 centistokes.

[6] R. L. Smith and K. M. Watson, *Ind. Eng. Chem.*, **29**, 1408 (1937).

(a) Characterization factor (Fig. 107)   = 11.35
(b) Average boiling point (Fig. 106 or 107)   = 880° F
(c) Hydrogen content (Fig. 108)   = 11.5
(d) Specific heat 200° (Fig. 66) = 0.485 × 0.975   = 0.473
(e) Average molecular weight (Fig. 106)   = 410
(f) Heating value (Fig. 74)   = 18,825 Btu per lb

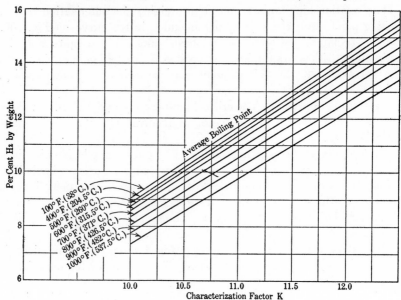

FIG. 108.   Characterization factor *vs.* weight % H_2
(Reproduced in *CPP Charts*)

Fuel Gas

The standard basis that has been adopted for the expression of the total heating value of a fuel gas is the number of Btu evolved when one cubic foot of the gas, at a temperature of 60° F, a pressure of 30 in. of mercury, and saturated with water vapor, is burned with air at the same temperature, and the products are cooled to 60° F, the water formed in the combustion being condensed to the liquid state. This unit is widely used as a basis for specifications and legal standards.

Since the vapor pressure of water at a temperature of 60° F is 0.52 in. of mercury, the heating value per standard cubic foot represents the heating value of 1 cu ft of moisture-free gas under a pressure of 30 − 0.52 or 29.48 in. of mercury and a temperature of 60° F. The number of moles of moisture-free gas in the standard cubic foot is equal to $1.0 \times \dfrac{491.7}{519.7} \times \dfrac{29.48}{29.92} \times \dfrac{1}{359.05} = 0.0025963$ lb-mole. Conversely,

$\dfrac{1}{0.0025963}$ or 385.2 standard cu ft of fuel gas contain 1 lb-mole of moisture-free gas if the gas behaves ideally.

The heating value of a fuel gas of known composition may be calculated as the sum of the heats of combustion of its components. The necessary data may be obtained from Table 29, page 297. The total heating values of the common combustible gases are also contained in Table 37, expressed in Btu per standard cubic foot.

TABLE 37. HEATING VALUES AND FLAME TEMPERATURES OF GASES

H.V. = total heating value, Btu per standard cu ft, measured at 60° F, 30 in. Hg and saturated with water vapor (assuming ideal-gas behavior)

Gas	Formula	H.V.	Maximum Flame Temperatures with Dry Air at 25° C		
			Theoretical (assuming complete combustion), ° C	Calculated* (allowing for equilibrium conditions), ° C	Experimental,* ° C
Carbon monoxide	CO	316	2342		
Hydrogen	H_2	319	2217		
Paraffins:					
Methane	CH_4	994	2012	1918	1885
Ethane	C_2H_6	1742	2065	1949	1900
Propane	C_3H_8	2479	2356	1967	1930
n-Butane	C_4H_{10}	3215	2084	1973	1905
n-Pentane	C_5H_{12}	3949	2087		
Olefins:					
Ethylene	C_2H_4	1576	2250	2072	1980
Propylene	C_3H_6	2299	2180	2050	1940
Butylene	C_4H_8	3036	2158	2033	1935
Amylene	C_5H_{10}	3770	2204		
Acetylene	C_2H_2	1451	2586		
Aromatics:					
Benzene	C_6H_6	3687	2211		
Toluene	$C_6H_5CH_3$	4409	2187		

* G. W. Jones, B. Lewis, J. B. Friauf, and G. St. J. Perrott, *J. Am. Chem. Soc.*, **53**, 869 (1931).

Fuel gases generally contain complex mixtures of both saturated and unsaturated hydrocarbons. The individual analytical determination of each component of these mixtures is not feasible for ordinary industrial purposes. However, Watson and Ceaglske[7] have described a simple scheme of industrial gas analysis which yields data suitable for ordinary

[7] K. M. Watson and N. H. Ceaglske, *Ind. Eng. Chem., Anal. Ed.* (Jan., 1932).

combustion calculations. In this scheme carbon monoxide and hydrogen are separately determined and reported as such. The saturated paraffin hydrocarbon gases are reported in terms of a hypothetical compound C_nH_{2n+2}, representing the average composition of the mixture of paraffins in the gas. Similarly, the unsaturated hydrocarbons or illuminants are reported in terms of a hypothetical compound of average composition, C_aH_b. For example, the analysis of a gas might be: CO 40%, H_2 42%, $C_{2.5}H_{4.2}$ (illuminants) 7%, $C_{1.2}H_{4.4}$ (paraffins) 11%.

An analysis of this type may be used as effectively for stoichiometric calculations as though all components were individually determined. The heating value of the gas may also be calculated by means of the following approximate formulas for the total heating values of mixtures of paraffins and of unsaturated hydrocarbons.

Paraffin hydrocarbons, C_nH_{2n+2}:

$$\text{Cal per g-mole} = 158,100n + 54,700$$

$$\text{Btu per cu ft at } 60° \text{ F, 30 in., sat.} = 745n + 258 \qquad (4)$$

Unsaturated hydrocarbons, C_aH_b:

$$\text{Cal per g-mole} = 98,200a + 28,200b + 28,800$$

$$\text{Btu per cu ft at } 60° \text{ F, 30 in., sat.} = 459a + 132b + 135 \qquad (5)$$

If the analysis of a gas is carried out carefully, its heating value may ordinarily be predicted by means of these equations with an error of less than 2%. Larger errors arise if large quantities of acetylene are present in the gas.

Illustration 2. A city gas has the following composition by volume:

CO_2	2.6%
$C_{2.73}H_{4.72}$ (unsaturateds)	8.4
O_2	0.7
H_2	39.9
CO	32.9
$C_{1.14}H_{4.38}$ (paraffins)	10.1
N_2	5.4
	100.0%

(a) Calculate the theoretical number of moles of oxygen that must be supplied for the combustion of 1 mole of the gas.

(b) Calculate the heating value of the gas in calories per gram-mole and Btu per standard cubic foot.

Solution

(a) *Basis:* 100 g-moles of gas

Oxygen required for:

Unsaturateds = 8.4(2.73 + 4.72/4)	= 32.8
Hydrogen = 39.9/2	= 19.95
Carbon monoxide = 32.9/2	= 16.45
Paraffins = 10.1 (1.14 + 4.28/4)	= 22.3
Total	= 91.5

Oxygen to be supplied per mole of gas = 0.915 − 0.007 = 0.908 mole.

(b)   *Basis:* 1.0 g-mole of gas.

Heating value of:

Hydrogen = 0.399 × 68,317	= 27,259 cal
Carbon monoxide = 0.329 × 67,636	= 22,252
Unsaturateds	
= 0.084 [(2.73 × 98,200) + (4.72 × 28,200) + 28,800] =	36,120
Paraffins = 0.101 [(1.14 × 158,100) + 54,700]	= 23,730
Total	= 109,361 cal

Btu per standard cu ft $= \dfrac{109,361 \times 1.8}{385.2} = 511$

Incomplete Combustion of Fuels.   The standard heating values of fuels correspond to conditions of complete combustion of all carbon to carbon dioxide gas, hydrogen to liquid water, and sulfur to sulfur dioxide gas.   If a fuel is burned in such a manner that complete combustion does not result, the standard heat of reaction may be calculated by subtracting from its standard heat of combustion the standard heats of combustion of the combustible products formed.

Illustration 3.   A coal having a heating value of 12,180 Btu per lb and containing 68.1% total carbon is burned to produce gases having the following composition by volume on the moisture-free basis.

CO_2	12.4%
CO	1.2
O_2	5.4
N_2	81.0
	100.0%

Calculate the standard heat of reaction in Btu per pound of coal burned.

Basis: 1.0 lb-mole of flue gas.

C in CO_2 = 0.124 lb-atom or	1.49 lb
C in CO = 0.012 lb-atom or	0.14 lb
Total carbon = 1.49 + 0.14	= 1.63 lb
Coal burned = 1.63 ÷ 0.681	= 2.39 lb
Heating value of coal = 2.39 × 12,180	= 29,100 Btu
Heat of combustion of CO	
= 0.012(−67,636 × 1.8)	= −1460 Btu
Standard heat of reaction	
= −29,100 − (−1460) = −27,640 Btu or	
−27,640 ÷ 2.39	= −11,560 Btu per lb of coal

Material and Energy Balances

In establishing an energy balance all sources of thermal energy are entered on the input side of the balance and all items of heat utilization and dissipation on the output side. It is ordinarily desirable to base all thermal quantities on a reference temperature of 25° C, thus permitting direct use of standard thermochemical data. Other reference temperatures may be used if desired, but in any event it is necessary that each complete balance be based on a single constant-reference temperature.

Where a fuel is used in an industrial reaction two different points of view are emphasized in expressing an energy balance, depending on whether or not the fuel is intended primarily as a source of heat or principally as a reducing agent. In the first instance the heating value of the fuel is listed on the input side of the balance and the heating value of the products resulting from the partial combustion of the fuel and its reaction with the charge on the output side. In this instance the utilization of the heating value of the fuel is of principal interest for heating purposes or in producing a fuel gas that is subsequently to be used for heating. In the second instance, where fuel is used primarily as a reducing agent, as in the reduction of ores, the principal interest is in the products of reduction and not in the heating value of the fuel or of the products of reaction. In this latter instance it is customary to include on the input side of the energy balance the heat evolved in the partial combustion of the fuel, which represents the difference between the heating value of the fuel and the heating value of the combustible products resulting from the incomplete combustion of that fuel.

With these two points of view in mind, the input and output items of an energy balance of a chemical process, based on a reference temperature of 25° C, are distributed in the following classification:

CASE I. PROCESSES WHERE THE FUEL IS USED PRIMARILY FOR ITS HEATING VALUE

Input Items

Group 1. Enthalpy of each material entering the process.
Group 2. Total heating value of fuel.
Group 3. Heat evolved by the separate exothermic reactions other than the combustion of fuel.
Group 4. Energy supplied to the process from external sources such as by the input of heat, electric energy, radiant energy, and mechanical work.

Output Items

Group 1. Enthalpy of each material leaving the process.

Group 2.   Total heating value of combustible products resulting from the incomplete combustion of the fuel and from the reaction of the fuel with the charge.

Group 3.   Heat absorbed by the separate endothermic reactions involved.

Group 4.   All heat transferred from the processes for useful purposes such as for the generation of steam in a boiler furnace.

Group 5.   All energy lost from the process as heat, electric energy, radiant energy, and mechanical work.

CASE II.   PROCESSES WHERE THE FUEL IS USED PRIMARILY AS A CHEMICAL REAGENT

Input Items

All entries are the same as for Case I except group 2.

Group 2.   Total heating value of fuel minus the total heating value of combustible products resulting from the incomplete combustion of the fuel and from the reaction of the fuel with the charge.

Output Items

All entries are the same as for Case I except there are no entries under group 2.

An energy balance is useful in showing how much energy is consumed by necessary endothermic reactions, how much is transferred to a heat interchanger or stored in a fluid used for supplying useful heat or power, and how much heat is wasted owing to incomplete combustion of fuels, to overheating of products, and to inadequate thermal insulation.

For example, in a coal-fired boiler furnace an energy balance indicates the distribution of the chemical energy of the coal into the enthalpy of steam, the heat lost in the gaseous products due to the presence of combustible gases and to sensible heat, the heat loss due to incomplete combustion of coal as represented by the unburned coke and coal in the refuse, and the loss of heat by radiation and conduction through the boiler setting. The justification of further insulation, of increasing the size of the combustion space, of increasing the draft, and of using automatic stoking can be answered, at least in part, from the study of such an energy balance.

Enthalpy of Water Vapor. The enthalpy of superheated water vapor referred to the liquid at 25° C is the sum of three separate items:

1. The sensible enthalpy of the water in the liquid state at the saturation temperature of the vapor. This item may be either positive or negative, depending on whether the saturation temperature is above or below 25° C.

2. The heat of vaporization of the water at the saturation temperature.

3. The sensible enthalpy of the water vapor referred to the saturation temperature.

Where water vapor is highly superheated, as in flue gases, it is general practice to simplify this calculation by assuming that the *enthalpy of water vapor is equal to the sum of the heat of vaporization at 25° C plus the sensible enthalpy of the vapor referred to 25° C.* In effect, this is assuming a saturation temperature of 25° C for the water vapor. This assumption introduces negligible errors where the partial pressure of the water vapor is small, of the order of one atmosphere or less.

Thermal Efficiency

The thermal efficiency of a process may be defined as the percentage of the heat input that is effectively utilized in a desired manner. As thus defined, the thermal efficiency is arbitrary, depending on both the designation of heat input and the heat that is effectively utilized. A numerical value of thermal efficiency is indefinite unless both these terms are specified.

The heat-input basis may be taken as the sum of the input items of the energy balance. This is a logical basis, and, unless otherwise specified, this basis will be used. Other bases are in common use to fit the needs of particular processes. For example, the thermal efficiency of the combustion of a fuel may be based on its *total heating value* or on its *net heating value.* The net heating value gives a higher value of efficiency and is preferred by some for this reason. This basis is undesirable where combustion apparatus is capable of recovering some of the latent heat of the water vapor from the gaseous products, since such apparatus might give an efficiency value above 100%.

Percentage efficiency is also dependent on the quantity of heat that is designated as effectively utilized. For example, a furnace and steam-boiler unit used in domestic heating might be considered as effectively utilizing only the heat represented by the energy added to the water in forming steam. On the other hand, it might also be logical to include as effective heat the radiation from the furnace itself where this radiant heat is used in heating the surrounding room.

A gas producer or water-gas generator produces a combustible gas at a relatively high temperature. If this gas can be utilized while hot, its sensible enthalpy as well as its heating value should be included in the effectively utilized heat of the producer. The efficiency of the producer on this basis is termed the *hot thermal efficiency.* If the gas is cooled before use, its sensible heat is not available, and only its heating value can be classed as heat effectively utilized in the producer. The efficiency on this basis is termed the *cold thermal efficiency.*

An itemized energy balance gives a complete picture of energy distribution and utilization. From this balance any desired intrepretation can be made relative to thermal efficiency.

Combustion of Fuels

In calculating the material and energy balances of processes involving the partial or complete combustion or decomposition of fuels, the same principles are employed whether such fuels are gases, liquids, or solids. The material balance of a simple combustion process includes the weights of fuel and air supplied and the weights of refuse and gases produced. This material balance can be calculated from a knowledge of the chemical composition of the four items mentioned without any direct measurements of the weights except the weight of fuel consumed. The weights (or volumes) of air and gaseous products are usually not measured because of the great difficulties involved and because these can generally be calculated indirectly with greater accuracy than by direct measurement.

In the burning of coal on a grate as in a boiler furnace the weight and composition of fuel used and composition of gaseous products are measured directly. The chemical analysis of the fuel should include the percentages of carbon, hydrogen, oxygen, moisture, nitrogen, and ash. It may not be necessary to have a complete ultimate analysis, but in any event the carbon, moisture, and ash content should be known. A complete analysis of the dry gaseous products is always necessary. The moisture content in the gaseous products can be calculated, provided the hydrogen content of the fuel is known, or can be determined by measuring the dew point of the gas.

The refuse from the furnace may be considered as consisting of ash, coked carbon, and unchanged combustible matter from the coal. The composition in terms of these constituents may be estimated from a determination of ash, fixed carbon, and volatile matter in the refuse, using the standard scheme of proximate analysis. The weight of refuse actually formed per unit weight of coal should be calculated on the basis of the *ash* contents of refuse and coal, as reported in the proximate analysis, and not on the basis of the *corrected ash* reported in the ultimate analysis.

The air entering the furnace may be assumed to be of average atmospheric composition and its humidity determined by a psychrometric method. The analysis of the flue gases is ordinarily determined by the Orsat type of apparatus, yielding the percentages of carbon dioxide, carbon monoxide, oxygen, and nitrogen in the *moisture-free* gases. For more nearly accurate work, determinations of methane and hydrogen should also be made.

Material and energy balances of combustion processes are based either on a unit weight of fuel or on the weight of fuel used in a given cycle or unit time of operation. For example, in boiler furnaces operating continuously the analysis may be based on a period of 24 hr or reduced to a basis of 1 lb of coal consumed. In operating a ceramic kiln of the batch type the analysis should be conducted over a complete cycle of operation including time of preheating and firing and the final results based on the entire cycle of operation.

When sufficient experimental data are collected, the same scheme of calculations may be employed for all problems in combustion. However, complete information is seldom available, and it becomes necessary to devise methods of circumventing these limitations. Extensive information can often be built up from but few data, and complete material and energy balances established from a few temperature measurements, the proximate analysis of coal, and an Orsat analysis of gas.

The various calculations that follow illustrate the modifications in procedure necessary to make the best use of data available and also to deal with the special variations in combustion processes represented in four special cases:

Case 1. Combustion of coal in a boiler furnace where

   (a) Complete ultimate analysis of fuel is known.
   (b) No uncoked coal appears in refuse.
   (c) Tar and soot are negligible.
   (d) Sulfur is negligible.

Case 2. Combustion of coal where

   (a) Hydrogen and nitrogen contents are unknown.
   (b) Uncoked coal drops into refuse.

Case 3. Combustion of coal where sulfur is not negligible.

Case 4. Partial combustion of fuel, as in a gas producer, where

   (a) Steam is admitted.
   (b) Tar and soot are not negligible.

Case 1. Combustion of Coal in Boiler Furnace. The simplest problem in coal combustion calculations exists where complete information is available or can be directly estimated on the ultimate analysis of coal, where tar and soot in the gases are negligible, where the sulfur content of the fuel is negligible, and where no uncoked coal drops into the refuse. The methods employed in this illustration are general and may be similarly applied to all problems in the combustion or partial combustion of a carbonaceous fuel whether solid, liquid, or gaseous.

The material balance of a furnace is represented by the following items:

Input

1. Weight of fuel charged.
2. Weight of dry air supplied.
3. Weight of moisture in air supplied.

Output

1. Weight of dry gaseous products.
2. Weight of water vapor in gaseous products.
3. Weight of refuse.

The method of calculating each of these items will be discussed in detail. General methods for such calculations have already been discussed in Chapters 2 and 7.

Illustration 4. From a 12-hr test conducted on a coal-fired steam generating plant the following data were obtained:

Data on Coal Fired

Ultimate analysis:

Carbon	65.93%
Available hydrogen	3.50
Nitrogen	1.30
Combined water	6.31
Free moisture	4.38
Ash	18.58
Total	100.00%

Total heating value	= 11,670 Btu per lb
Total weight of coal fired	= 119,000 lb
Average temperature of coal	= 65° F

Data on Refuse Drawn from Ash Pit

Ash content	= 87.4%
Carbon content	= 12.6%
Average temperature	= 255° F
Mean specific heat from 73 to 255° F	= 0.23

(Estimated from Fig. 64, page 261)

Data on Flue Gas

Orsat analysis:

Carbon dioxide	11.66%
Oxygen	6.52
Carbon monoxide	0.04
Nitrogen	81.78
Total	100.00%
Average temperature	= 488° F

Data on Air

Average dry-bulb temperature	= 73.0° F
Average wet-bulb temperature	= 59.4° F
Average barometric pressure	= 29.08 in. Hg

Data on Steam Generated

Average feed water temperature	= 193° F
Weight of water evaporated	= 1,038,400 lb
Average steam pressure	= 137.4 psi gage
Quality of steam	= 98.3%

Calculate the material and energy balances for the entire plant.

Material Balance

All calculations are based upon 100 lb of coal as fired.

1. Weight of Refuse Formed. Where the refuse is not weighed directly, its weight can be readily calculated from its ash content and that of the coal.

Ash content of coal	= 18.58 lb
Ash content per lb of refuse	= 0.8740 lb

$$\text{Weight of refuse formed} = \frac{18.58}{0.8740} = 21.2 \text{ lb}$$

2. Weight of Dry Gaseous Products

A direct measurement of the weight of gaseous products from a combustion process is seldom made because of the many difficulties involved. Pitot tubes measure inaccurately low velocities such as are encountered in chimneys and flues. Orifice and Venturi meters are similarly unreliable because of low-pressure drops encountered and because of soot accumulation. Electric flow meters read inaccurately if the composition of the gas varies with respect to carbon dioxide or water vapor. In any case the direct measurement of gas streams is extremely difficult because of variation in temperature and velocity across each section of the stream. Any accurate measurement must give a correct integrated value of velocity and temperature over the entire cross section. Because of these uncertainties and troubles it becomes easier and more nearly accurate to calculate the weight of gaseous products from the stoichiometric relationships of combustion.

The complete analysis of the gaseous products includes the percentages of carbon dioxide, carbon monoxide, oxygen, methane, ethane, hydrogen, and nitrogen present. Moisture content is not revealed in the usual gas analysis because the entire analysis is conducted with the gas sample saturated with water vapor at a constant temperature and pressure.

The general rule is recommended that the *weight of dry gaseous products should be calculated from a carbon balance.* A carbon balance is selected as the basis of this calculation for two reasons. In the first place, carbon is determined with a higher degree of precision in both fuel and gaseous products than any other element present. Second, carbon is the chief constituent in both fuel and gaseous products so that a slight error in its determination will not be magnified in subsequent calculations. To calculate the weight of gaseous products from a material balance of any other element would invite many additional sources of error. For example, the hydrogen balance would be out of the question because of the many sources of hydrogen, its relatively low percentage content, its several outlets, and its various methods of combination.

Carbon Balance

Carbon gasified. *Basis:* 100 lb coal fired.

Carbon in coal $= 100 \times 0.6593$ $\qquad = 65.93$ lb or 5.49 lb-atoms
Carbon in refuse $= 21.2 \times 0.1260$ $\qquad = \underline{\ 2.67}$ lb or 0.22 lb-atoms
Carbon entering stack gases $\qquad\qquad\qquad = 63.26$ lb or 5.27 lb-atoms

Carbon in stack gases. *Basis:* 1.0 lb-mole of gas.

Carbon in CO_2 $\qquad\qquad\qquad\qquad = 0.1166$ lb-atom
Carbon in CO $\qquad\qquad\qquad\qquad\ = \underline{0.0004}$
$\qquad$ Total carbon $\qquad\qquad\qquad = 0.1170$ lb-atom

Moles of dry stack gas per 100 lb coal fired $= 5.27/0.1170 = 45.1$ lb-moles

Total dry gaseous products. *Basis:* 100 lb coal fired.

$CO_2 = 45.1 \times 0.1166 = 5.26$ lb-moles or $\times 44$ $\quad = 231.5$ lb
$CO = 45.1 \times 0.0004 = 0.018$ lb-moles or $\times 28$ $\quad = 0.560$
$O_2 = 45.1 \times 0.0652 = 2.94$ lb-moles or $\times 32$ $\quad = 94.2$
$N_2 = 45.1 \times 0.8178 = \underline{36.90}$ lb-moles or $\times 28.2$ $\ = \underline{1041}$
$\qquad$ Total $\qquad\qquad = 45.1$ lb-moles $\qquad\qquad = 1367$ lb

Average molecular weight $= 1367/45.1$ $\qquad\qquad\quad = 30.3$

3. Weight of Dry Air Supplied

Direct measurement of the weight or volume of air used in combustion is accompanied by the same difficulties as direct measurement of gaseous products. Furthermore, air is usually drawn through the grate by chimney draft so that there is no need for confining the supply of air in ducts and there is no opportunity for direct measurement of its flow. The dry air used in combustion consists of oxygen and inert gases, chiefly nitrogen. These inert gases also include argon and traces of rare gases, but because of the small amounts present it is customary to include all the inert gases as nitrogen and assign a *molecular weight of 28.2 to atmospheric "nitrogen."* This nitrogen passes through the furnace unchanged and appears entirely in the gaseous products. Any nitrogen present in the fuel burned will also appear in the flue gases. The nitrogen in ordinary solid and liquid fuels burned will usually be negligible or very small. However, in the combustion of gases a considerable portion of the nitrogen appearing in the flue gases may come from the gaseous fuel.

The composition of dry air may ordinarily be taken as constant, containing 21.0% oxygen and 79.0% nitrogen by volume, the nitrogen content including the argon present. The carbon dioxide content of air is 0.03%, and shows no variations which are detectable with the analytical equipment ordinarily used. In combustion calculations, the carbon dioxide of the air may be neglected. The moisture content of air is subject to extreme variations depending on weather conditions, so that its separate determination is invariably necessary.

Because of the constancy of composition of dry air it is possible to calculate readily the weight of air used in a combustion process from a knowledge of the nitrogen contents of the gaseous products and of the fuel used. Accordingly, the general rule is expressed that the *weight of dry air actually used in a combustion process is calculated from a nitrogen balance*. The chief objection to the use of the nitrogen-balance basis is that, in gas analysis, errors resulting accumulate in the nitrogen determination since this is always found by difference.

Nitrogen Balance. Basis: 100 lb coal fired.

Nitrogen in gaseous products	= 36.90	lb-moles
Nitrogen from coal = 1.30/28.0	= 0.0464	
Nitrogen from air	= 36.85	lb-moles
Dry air supplied = 36.85/0.79	= 46.6	lb-moles
or 46.6 × 29	= 1354 lb	

It will be noted that the nitrogen content of the coal might be neglected without introducing a serious error.

4. Weight of Moisture in Air

The weight of moisture per mole of dry air depends on the temperature, pressure, and relative humidity of the air. From the dew point the partial pressure of the water vapor is determined, and the moisture content of the air may be calculated by the methods explained in Chapter 5.

Dry-bulb temperature	= 73.0° F
Wet-bulb temperature	= 59.4° F

From Fig. 20,

Molal humidity of air	= 0.012
Water supplied with air = 46.6 × 0.012	= 0.559 lb-mole

5. Total Volume of Wet Air Introduced

Basis: 100 lb coal fired.

Total moles of moist air $= 46.6 + 0.559$ $\qquad = 47.2$ lb-moles

Volume at 73° F, 29.08 in. Hg $= 47.2 \times 359 \times \dfrac{29.92}{29.08} \times \dfrac{533}{492} = 18{,}870$ cu ft

6. Weight of Moisture in Gaseous Products

To complete the material balance it is necessary to know the weight of moisture in the gaseous products since this is not obtained by the ordinary gas analysis. Direct measurement of the moisture content is difficult. It can also be calculated if the composition of the dry flue gases and the moisture content of the air used and the hydrogen and moisture in the fuel burned are known. As a general rule, it may be stated that the *moisture content of the gaseous products is calculated from a hydrogen balance.*

Hydrogen Balance.   Basis: 100 lb coal fired.

Input

From moisture introduced with dry air	$= 0.559$ lb-mole
From combined water in coal $= 6.31/18$	$= 0.351$
From free moisture in coal $= 4.38/18$	$= 0.244$
From available hydrogen in coal $= 3.50/2.016$	$= 1.738$
Total	$= 2.892$ lb-moles

In boiler-furnace operation it is usually assumed that free hydrogen and hydrocarbons are negligible in the gaseous products and that all the hydrogen introduced into the system leaves as water in the stack gases.

Hydrogen in H_2O of stack gases $= 2.892$ lb-moles

The dew point of the stack gases may now be calculated.

Partial pressure of $H_2O = \dfrac{2.892}{45.1 + 2.892} \times 29.08 = 1.75$ in. Hg

From Table 5, this partial pressure is seen to correspond to a dew point of 36° C or 97° F.

7. Total Volume of Gaseous Products

Basis: 100 lb coal fired.

Moles of wet gas $= 45.1 + 2.892$ $\qquad = 48.0$ lb-moles

Volume at 488° F and 29.08 in. Hg

$= 48.0 \times 359 \times \dfrac{29.92}{29.08} \times \dfrac{948}{492}$ $\qquad = 34{,}150$ cu ft

Summary of Material Balances. To verify the accuracy of experimental data or methods of calculation a summary of all material balances is prepared. The over-all material balance is also indicated on the flow chart of Fig. 109.

Over-all Material Balance

Input			Output	
Coal	100	lb	Refuse	21.2 lb
Dry air (46.6 lb-moles)	1354		Dry gases (45.1 lb-moles)	1367
H₂O in air (0.559 lb-mole)	10.05		H₂O in stack gases	
			(2.892 lb-moles)	52.1
Total	1464	lb	Total	1440 lb

Carbon Balance

In coal	65.93 lb		In gases	63.26 lb
			In refuse	2.67
Total	65.93 lb		Total	65.93 lb

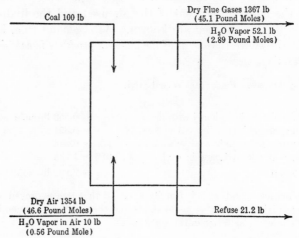

Fig. 109. Material balance of steam generating plant (illustration 4)

Nitrogen Balance

In air (36.85 lb-moles)	1040	lb	In stack gases	
In coal	1.30		(36.90 lb-moles)	1041 lb
Total	1041	lb	Total	1041 lb

Hydrogen Balance

In water vapor of air		In H₂O of stack gases	
(0.559 lb-mole)	1.127 lb	(2.892 lb-moles)	5.83 lb
In combined water of coal			
(0.351 lb-mole)	0.708		
In free moisture of coal			
(0.244 lb-mole)	0.491		
Available hydrogen of coal			
(1.738 lb-moles)	3.50		
Total	5.83 lb	Total	5.83 lb

Ash Balance

In coal	18.58 lb	In refuse	18.58 lb

Oxygen Balance

In combined water of coal		In CO_2 of stack gases	
$6.31 \times 16/18$	5.61 lb	5.26×32	168.5 lb
In free moisture of coal		In O_2 of stack gases	
$4.38 \times 16/18$	3.89	2.94×32	94.2
In dry air		In CO of stack gases	
$36.85 \times 21/79 \times 32$	313.5	$0.01804/2 \times 32$	0.289
In water vapor in air		In H_2O of stack gases	
$0.559/2 \times 32$	8.95	$2.892/2 \times 32$	46.3
Total	332 lb	Total	309 lb

The ash, carbon, nitrogen and hydrogen balances are completed by difference, hence there are no discrepancies between the input and output totals. The over-all balance, however, is not completed by difference, hence the input and output totals normally will not check exactly. In this particular problem, it will be noted that there is a deficit of 24 lb or 1.6% on the output side. The oxygen balance also is not completed by difference, hence it too will show lack of agreement between the input and output totals.

The discrepancies noted above result from errors in the test data, including the analysis of the coal and the analysis of the flue gas. In both instances, the oxygen content is a likely source of error. In coal, the oxygen is determined by difference, and hence is prone to error. In the analysis of the flue gas, the oxygen reported may be low because of the poor absorptive capacity of the alkaline pyragallol usually used in determining the oxygen. Suspended soot and tar, if present and not reported, will be a source of error.

8. Theoretical Amount of Air Required for Combustion

The weight of air theoretically required for complete combustion depends on the chemical composition of the fuel and the stoichiometric relations involved in combustion. Since the one element in common for all combustion reactions is oxygen, the *weight of air required for combustion must be calculated from an oxygen balance.* The oxygen already in the fuel is assumed to be in combination with hydrogen. Hence, only the available hydrogen of the fuel is considered in calculating its oxygen requirement.

Per Cent Excess Air. Basis: 100 lb coal fired.

Oxygen requirements for combustible constituents of coal charged.

		Oxygen Required
Carbon 65.93 lb	= 5.49 lb-atoms	= 5.49 lb-moles
Net hydrogen 3.50 lb	= 1.736 lb-moles	= 0.868
Total		= 6.358 lb-moles
Air required	= 6.358/0.21	= 30.3 lb-moles
Air supplied		= 46.6 lb-moles
Excess air	= 46.6 − 30.3	= 16.3 lb-moles
% excess air	= 16.3/30.3 × 100	= 53.9%

Energy Balance

Since it is conventional to utilize total rather than net heating values in energy balances, the reference state for all water involved in the process is the liquid state; the enthalpies of reactants and products are evaluated on this basis.

Reference temperature: 73° F.
Basis: 100 lb coal fired.

Input

1. **Heating value of the coal** $= (100)(11,670)$ $= 1,167,000$ Btu
2. **Enthalpy of the coal** $=$ 0
3. **Enthalpy of the dry air** $=$ 0
4. **Enthalpy of the water vapor accompanying the dry air.**
   Heat of vaporization $= 18,964$ Btu per lb-mole at 73° F
   Enthalpy $= (0.559)(18,964)$ $=$ 10,600

            Total heat input $= 1,177,600$ Btu

Output

1. **Heating value of refuse**
   Weight of carbon in refuse $= 2.67$ lb
   Heating value $= (2.67)(14,490)$ $=$ 38,690 Btu

2. **Heating value of stack gases**
   Lb-moles of CO in stack gases $= 0.01804$ lb-mole
   Heating value $= (0.01804)(67,636)(1.8)$ $=$ 2190 Btu

3. **Enthalpy of refuse**
   Weight of refuse $= 21.2$ lb
   Enthalpy $= (21.2)(0.23)(255{-}73)$ $=$ 890 Btu

4. **Enthalpy of dry stack gases.** Mean heat capacities between 73 and 488° F from Fig. 62.

Enthalpies:

$$CO_2 = \quad 5.26 \ \ (9.92)(488 - 73) \ = \ 21,650 \text{ Btu}$$
$$CO \ = \quad 0.018 \ (7.05)(488 - 73) \ = \qquad 50$$
$$O_2 \ = \quad 2.94 \ \ (7.24)(488 - 73) \ = \quad 8,830$$
$$N_2 \ = 36.90 \ (7.02)(488 - 73) \ = 107,500$$
$$\text{Total} \qquad\qquad\qquad\quad = 138,030 \text{ Btu}$$

5. **Enthalpy of the water vapor in stack gases**
   Heat of vaporization $= 18,964$ Btu per lb-mole at 73° F
   Heat of vaporization $= (2.892)(18,964)$ $= 54,840$ Btu
   Superheat $= (2.892)(8.20)(488 - 73)$ $= \ 9,840$

            Total enthalpy $= 64,680$ Btu

6. **Heat utilized in generating steam**

   $$\text{Lb of steam generated} = \frac{1,038,400}{119,000} \times 100 = 873 \text{ lb}$$

   Enthalpy of 1 lb steam as produced (151.7 psi absolute, 1.7% moisture), relative to 32° F $= 331.5 + 862.8 - (0.017 \times 862.8) = 1179.6$ Btu per lb

Enthalpy of feed water at 193° F relative to 32° F　　= 161.0
Net heat input into steam produced　　　　　　　　= 1018.6 Btu per lb
Total heat absorbed by steam produced
　= (873)(1018.6)　　　　　　　　　　　　　　　= 889,200 Btu

7. Undetermined losses (by difference)　　　　　= 43,920 Btu

Summarized Energy Balance of Steam Generating Plant

Reference temperature: 73° F
Basis: 100 lb coal fired.

Energy Input	Btu	%
1. Heating value of the coal	1,167,000	99.1
2. Enthalpy of the coal	0	0
3. Enthalpy of dry air	0	0.0
4. Enthalpy of water vapor accompanying dry air	10,600	0.9
Total	1,177,600	100.0

Energy Output	Btu	%
1. Heating value of refuse	38,690	3.3
2. Heating value of stack gases	2,190	0.2
3. Enthalpy of refuse	890	0.1
4. Enthalpy of dry stack gases	138,030	11.7
5. Enthalpy of water vapor in stack gases	64,680	5.5
6. Heat utilized in generating steam	889,200	75.5
7. Undetermined losses (by difference)	43,920	3.7
Total	1,177,600	100.0

Thermal Efficiency and Economy. The thermal efficiency of a boiler furnace may be calculated on the total or the net heating value of the coal. The effectively utilized heat is that which is absorbed in steam generation.

Based on total heating value of coal,

$$\text{Thermal efficiency} = \frac{889,200}{1,167,000} \qquad \text{or } 76.2\%$$

Based on net heating value of coal,

$$\text{Thermal efficiency} = \frac{889,200}{1,167,000 - (42)(1050)} \text{ or } 79.2\%$$

Case 2. Combustion Calculations with Incomplete Ultimate Analysis of Coal.

In the preceding illustration, the calculations were completed without any assumptions being made concerning the composition of the fuel. However, the determination of the hydrogen content of coal is a difficult procedure, to be avoided if possible. For this reason it is frequently necessary to evaluate the material balance of a furnace or gas producer without data on hydrogen content and to calculate this quantity from an oxygen balance. In this illustration an additional complication is introduced in that some of the coal drops through the grate without coking.

Where the hydrogen content of the fuel is calculated from an oxygen balance, great care must be taken in analyzing the flue gases. The sampling and determination of oxygen in hot flue gases are particularly uncertain. For this reason it is frequently preferable to estimate the hydrogen content of the fuel by empirical methods such as those illustrated on page 402 for coal, and page 408, Fig. 108, for petroleum hydrocarbons. The calculations are then carried out as in the preceding illustration.

In the preceding illustration the nitrogen content of the coal was known. In making a complete ultimate analysis it is necessary to determine this element in order that the oxygen content may be obtained by difference. From the results of the previous illustration it is apparent that the nitrogen in the coal might be neglected altogether, assuming all nitrogen in the flue gases to have come from the air. No appreciable error will result from neglecting this nitrogen except in determining the oxygen or combined water content of the coal by difference. For this calculation it is ordinarily sufficient to assume a nitrogen content of 1.7% of the combustible in the coal. This assumption will ordinarily not be in error by more than 0.3% of the weight of the combustible except with anthracite coals. Greater refinement is not justified because of the uncertainty of the sampling of the coal and of other data on which the material balance is based.

Illustration 5. Coal-Fired Boiler Furnace. A furnace is fired with a bituminous coal having the following proximate analysis:

Moisture	2.9%
Volatile matter	33.8
Fixed carbon	53.1
Ash	10.2
	100.0%

The ultimate analysis is known only in part and includes (as-received basis):

Sulfur	1.1%
Carbon	73.8%

The dry refuse from the furnace has the following composition:

Volatile matter	3.1%
Fixed carbon	18.0
Ash	78.9
	100.0%

The Orsat analysis of the flue gases is as follows:

Carbon dioxide	12.1%
Carbon monoxide	0.2
Oxygen	7.2
Nitrogen	80.5
	100.0%

Air enters the furnace at a temperature of 65° F with a percentage humidity of 55%. The barometric pressure is 29.30 in. Hg. The flue gases enter the stack at a pressure equivalent to 1.5 in. of water less than the barometric pressure and at a temperature of 560° F.

Water is fed to the boiler at a temperature of 60° F and vaporized to form wet steam at a gage pressure of 100 psi, quality 98%, at a rate of 790 lb of steam or water per 100 lb of coal charged.

Compute complete material and energy balances, volumes of air and flue gases per 100 lb of coal charged, and percentage of excess air used.

Material Balance

The calculations are similar to illustration 4 with special methods in parts 1, 4, 5.

1. Total carbon content of refuse

Basis: 100 lb of coal charged.

The weight of refuse is calculated from the ash contents of the coal and of the refuse. The weight of ash in 100 lb of coal is 10.2 lb. This weight of ash constitutes but 78.9% of the weight of refuse.

$$\text{Total weight of refuse} = \frac{10.2}{0.789} = 12.9 \text{ lb}$$

Carbon exists in the refuse as fixed carbon and as volatile matter. The volatile matter is due to the dropping of uncoked coal through the grate. The combustible of the uncoked coal in the refuse may be assumed to have the same composition as the combustible of the coal fired. Therefore, the ratio of combustible to volatile matter in the uncoked coal in the refuse will be the same as that in the coal.

$$\text{Ratio of combustible to volatile matter in coal} = \frac{33.8 + 53.1}{33.8} = 2.56$$

Volatile matter in refuse = 12.9×0.031 = 0.40 lb
Unchanged combustible in refuse = 0.40×2.56 = 1.02 lb

Carbon is also present in the refuse as coked coal accompanied by no volatile matter. The amount of carbon as coke is the difference between quantities of total combustible and of unchanged coal combustible in the refuse.

Fixed carbon in refuse = 12.9×0.18 = 2.32 lb
Total combustible in refuse = $2.32 + 0.40$ = 2.72 lb
Coked combustible (carbon) = $2.72 - 1.02$ = 1.70 lb

The total carbon content of the unchanged combustible in the refuse may be determined from the percentage of carbon in the combustible of the original coal.

$$\text{Total carbon content of combustible of coal} = \frac{73.8}{33.8 + 53.1} = 85\%$$

Carbon in uncoked coal in refuse = $(0.85)(1.02)$ = 0.87 lb
Total carbon in refuse = $1.70 + 0.87$ = 2.57 lb

2. Weight of dry flue gases

Carbon Balance. Basis: 100 lb of coal charged.

Carbon gasified = $73.8 - 2.57 = 71.2$ lb or 5.94 lb-atom

1.0 lb-mole of dry flue gases contains:

Carbon dioxide	= 0.121 lb-mole
Carbon monoxide	= 0.002
Total carbon per lb-mole of gas	= 0.123 lb-atom

$$\text{Total dry flue gas} = \frac{5.94}{0.123} = 48.3 \text{ lb-moles}$$

Total dry gases:

CO_2	= 48.3 × 0.121 =	5.85	lb-moles or	× 44	=	258 lb
CO	= 48.3 × 0.002 =	0.096	lb-moles or	× 28	=	3
O_2	= 48.3 × 0.072 =	3.48	lb-moles or	× 32	=	111
N_2	= 48.3 × 0.805 =	38.87	lb-moles or	× 28.2	=	1097
Total		48.30	lb-moles or			1469 lb

3. Weight of air supplied

Nitrogen Balance. Basis: 100 lb of coal charged.

N_2 in flue gases = 38.87 lb-moles

Assuming all N_2 to come from the air:

$$\text{Dry air supplied} = \frac{38.87}{0.79} = 49.2 \text{ lb-moles}$$

$$\text{or } 49.2 \times 29 \qquad = 1430 \text{ lb}$$

From Fig. 20:

Molal humidity of air	= 0.012
Water vapor in air = 0.012 × 49.2	= 0.59 lb-mole
or 0.59 × 18	= 10.6 lb
Total wet air = 49.2 + 0.59	= 49.8 lb-moles
or 1430 + 10.6	= 1440 lb

Volume of air entering at 65° F, 29.3 in. Hg =

$$49.8 \times 359 \times \frac{525}{492} \times \frac{29.92}{29.30} = \qquad\qquad 19{,}450 \text{ cu ft}$$

4. Hydrogen content of coal

Oxygen Balance. Basis: 100 lb of coal charged.

$$\text{Oxygen in dry flue gas} = 5.85 + \frac{0.096}{2} + 3.48 = 9.38 \text{ lb-moles}$$

$$\text{Oxygen entering in dry air} = 49.2 \times 0.21 \qquad = 10.32 \text{ lb-moles}$$

Assuming that the oxygen not accounted for in the dry flue gases was consumed in oxidizing the available hydrogen of the coal:

O_2 oxidizing H_2 = 10.32 − 9.38	= 0.94 lb-mole
H_2 burned = 2 × 0.94 = 1.88 lb-moles or	3.79 lb

The hydrogen burned may be taken as the available hydrogen of the coal, neglecting the small hydrogen content of the uncoked combustible in the refuse.

5. Complete ultimate analysis of coal.
The unknown items of the ultimate analysis are combined water and nitrogen. As pointed out on page 426, the nitrogen

content may be assumed to be $1.7 \times 0.87 = 1.4\%$. The combined water may then be determined as the difference between 100 and the sum of the percentages of moisture, carbon, hydrogen, sulfur, nitrogen, and *corrected* ash.

$$\text{Corrected ash} = 10.2 - 3/8(1.1) \qquad\qquad\qquad = 9.8\%$$
$$\text{Combined } H_2O = 100 - (2.9 + 73.8 + 3.8 + 1.1 + 1.4 + 9.8) = 7.2\%$$

Ultimate analysis:

Moisture	2.9%
Carbon	73.8
Available H	3.8
Sulfur	1.1
Nitrogen	1.4
Corrected ash	9.8
Combined H_2O	7.2
	100.0%

6. Water vapor in flue gases

Hydrogen Balance. Basis: 100 lb of coal charged.

H_2O from air	$= 0.59$ lb-mole
H_2O from coal $= \dfrac{2.9 + 7.2}{18}$	$= 0.56$
H_2O formed from H	$= 1.88$
Total	$= 3.03$ lb-moles or 55 lb

7. Volume of wet flue gases

Moles of wet flue gas $= 48.3 + 3.03 = 51.33$ lb-moles

$$\text{Pressure in flue} = 29.30 - \frac{1.5}{13.6} \qquad = 29.19 \text{ in. Hg}$$

Volume at 560° F, 29.19 in. of Hg

$$= 51.33 \times 359 \times \frac{1020}{492} \times \frac{29.92}{29.19} \qquad = 39,200 \text{ cu ft}$$

8. Complete material balance

Input

Coal	100 lb
Dry air (49.2 lb-moles)	1430
Water vapor in air (0.59 lb-mole)	10
Total	1540 lb

Output

Dry flue gases (48.3 lb-moles)	1469 lb
Refuse	13
Water vapor in flue gas (3.03 lb-moles)	55
Total	1537 lb

This material balance is summarized in Fig. 110.

9. Percentage excess air

Basis: 100 lb of coal charged.

Total carbon in coal charged $= 73.8$ lb $= 6.15$ lb-moles
Available hydrogen in coal charged $= 1.88$ lb-moles

Total O_2 required $= \dfrac{1.88}{2} + 6.15$ $= 7.09$ lb-moles

Air theoretically required $= \dfrac{7.09}{0.21}$ $= 33.8$ lb-moles

Air actually supplied $= 49.2$ lb-moles

Percentage excess air $= \dfrac{49.2 - 33.8}{33.8} = 45.5$, based on that required for complete combustion of all carbon and available hydrogen in the coal charged and neglecting that required for sulfur.

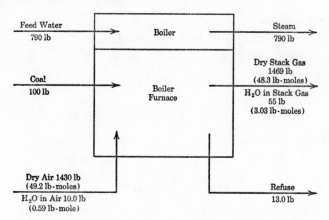

Fig. 110.   Material balance of a boiler furnace (illustration 5)

The percentage excess air may also be calculated directly from the flue-gas analysis:

Basis: 100 lb-moles of flue gas.

Neglecting N_2 from the coal:

O_2 introduced in air $= \dfrac{80.5}{0.79} \times 0.21$ $= 21.4$ lb-moles

O_2 in excess $=$ (surplus present in gases minus that required to complete oxidation of the CO) $= 7.2 - \dfrac{0.2}{2}$ $= 7.1$

O_2 actually consumed $= 14.3$ lb-moles

% excess oxygen $=$ % excess air $= \dfrac{7.1}{14.3}$ $= 49.6.$

This percentage excess is based on the oxygen required for all combustible substances that were *actually burned.*

Energy Balance

The energy balance of this problem is calculated exactly as in illustration 4, estimating the heating value of the coal by the empirical method discussed on page 402.

Effect of Sulfur in Coal. In the preceding illustrations the combustion of the sulfur in the coal has been neglected. This does not introduce appreciable error if the sulfur content is low, 1% or less. It is difficult to take into account the combustion of the sulfur by any rigorous method because of the uncertainty of the forms in which it is present in the coal and the difficulty of determining its distribution in the combustion products. A considerable part of the sulfur that is present in the coal in a combustible or available form will appear as sulfur dioxide in the flue gases. The remainder will be present in the refuse.

The ordinary scheme of flue-gas analysis by the Orsat apparatus, in which the gas sample is confined over water, does not permit determination of sulfur dioxide. Because of the high solubility of sulfur dioxide in water (about 30 times that of carbon dioxide) the bulk of the SO_2 will be absorbed in the water of the sampling apparatus and burette. Any SO_2 that is not removed in this manner will be absorbed and reported as CO_2. The reported analysis will ordinarily represent approximately the composition of the SO_2-free gases. Methods are available by which separate determination may be made of the SO_2, but this is not ordinarily feasible for performance tests of combustion equipment.

Neglect of the combustion of sulfur in calculations of the type carried out in illustration 4, in which the oxygen balance is not used, does not introduce any serious error, even if the sulfur content is relatively high. The calculated total weight of dry flue gases will be low by approximately the weight of SO_2 formed. This error is usually negligible. However, in the method of calculation of illustration 5, in which the net hydrogen in the coal is calculated from an oxygen balance, the errors resulting from neglect of combustion of sulfur may be more serious. In this type of calculation it is assumed that all free oxygen not accounted for as CO_2, CO, or free oxygen in the flue gas was utilized in the oxidation of the net hydrogen of the coal. This assumption neglects the oxygen consumed in oxidation of sulfur and of iron combined with the sulfur as FeS_2. As a result the net hydrogen content calculated by this method will be too high, and the combined water content, calculated by difference, will be too low. These errors are particularly high when the sulfur is originally present as iron pyrites, FeS_2, in which case oxygen is consumed in the oxidation of both the sulfur and the iron. However, because of the relatively high atomic weight of sulfur it is permissible to neglect these errors in dealing with coals containing less than 1% sulfur.

The probable errors involved in neglecting the combustion of the sulfur may be estimated for a typical case by consideration of the data of illustration 5. The coal contained 1.1% sulfur. The maximum error would result if all the sulfur were in the form of iron pyrites. In this case, for each pound-atom of sulfur present 11/8 lb-moles of oxygen would be consumed in producing SO_2 and Fe_2O_3. On the basis of 100 lb of coal charged, 0.048 lb-mole of oxygen would be required for combustion of the sulfur of the coal. Introducing this value into the oxygen balance on page 428 will reduce the unaccounted-for oxygen, used in oxidizing hydrogen, from 0.94 to 0.89 lb-mole. The calculated net hydrogen is correspondingly reduced from 3.8 to 3.6 lb, and the combined water content is increased from 7.2 to 7.4 lb. These errors are not serious and in addition represent maximum errors because actually not all the sulfur will be in the pyritic form and furthermore it may not be completely burned.

Case 3. Method of Calculation Where Sulfur is High and the Carbon and Hydrogen Contents of the Coal Are Unknown. Where sulfur contents of more than about 1% are encountered, it is not ordinarily desirable to use the oxygen balance for computing the net hydrogen content of the coal. In order to develop an accurate oxygen balance it would be necessary to have data on the forms in which the sulfur occurred in the coal and on the sulfur content of the refuse. Without such data the net hydrogen content is better determined analytically or estimated by the empirical method of Uehling, discussed on page 402. This latter method may also be used to estimate the total carbon in the coal from a determination of its heating value.

If the sulfur content of a coal is so high that sulfur dioxide constitutes a considerable part of the flue gas, it is necessary to obtain data from which the amount of sulfur actually burned may be calculated. A determination of either the sulfur in the refuse or of the SO_2 in the flue gases will supply this information. The former determination is more easily carried out. It may then be assumed that the ordinary flue-gas analysis yields the composition of the SO_2-free gases, and the total quantity of gases is then computed on this basis. Direct determination of the SO_2 content of the flue gases is more reliable but frequently unwarranted.

In calculating an energy balance involving a coal of high sulfur content, a sulfur correction should be applied to the heating value directly determined in the oxygen-bomb calorimeter. This correction results from the fact that in the calorimeter the available sulfur is converted almost entirely to aqueous H_2SO_4, whereas in ordinary combustion SO_2 will be formed. The correction may be taken as 1000 cal per gram of sulfur, to be subtracted from the observed heating value.

Case 4. Gas Producers. In the operation of a gas producer, a fuel gas of low calorific value is produced by blowing air, usually accompanied by steam, through a deep incandescent bed of fuel. Carbon monoxide and carbon dioxide are formed by partial combustion of the fuel. Hydrogen and the oxides of carbon are formed from the reduction of water, and volatile combustible matter is distilled from the coal without combustion.

Effect of Soot and Tar. In many combustion processes the gases contain carbon suspended in the forms of soot and tar. These forms of carbon do not appear in an ordinary volumetric gas analysis but must be determined separately by absorption or retention on a weighed filter. The tar can then be separated and analyzed for its hydrogen content although this precision is not usually warranted. It is ordinarily sufficient to assume that the combustible of the suspended tar analyzes 90% carbon and 10% hydrogen and that the combustible of the soot consists of 100% carbon. Frequently refuse also appears suspended in the gaseous products so that the ash content of the suspended material should also be measured.

In addition to the tar and soot suspended in the gases, these products also gradually accumulate as a deposit in the flues. The slight correction necessary for this is usually made by measuring the deposition over a long period of time, for example, over the usual time interval elapsed between successive cleanings of the flues. The tar deposit is removed and weighed, and, since the number of tons of fuel consumed during the time interval is known, a rough approximation can be obtained for the carbon deposited per unit weight of coal charged.

The following illustration of gas-producer operation is of value in that consideration is given to the tar and soot suspended in the producer gas and also to the live steam which is passed into the producer. The problem is of added interest because all experimental data were collected with unusual care. Analyses of gas and coal and measurements of temperatures were made at regular intervals over several days of operation and data carefully weighted to give average results. These data were taken in part from a report by Harrop.[8]

Illustration 6. Air was supplied to a gas producer at 75° F with a percentage humidity of 75% and a barometric pressure of 29.75 in. of Hg. Coal weighing 70,900 lb and having a gross heating value of 11,910 Btu per lb was charged into

[8] C. B. Harrop, *J. Am. Ceramic Soc.*, **1**, 35 (1918).

the producer.   Tar weighing 591 lb was deposited in the flues prior to the point of gas sampling and contained 93% carbon and 7% hydrogen.

The water vapor, suspended tar, and soot in the gases were determined experimentally by withdrawing samples of the gas through an absorption train.   The results are expressed in grains per cubic foot of wet, hot gas measured at 1075° F and 29.75 in. Hg.

Water vapor	3.43 grains per cu ft
Suspended tar	3.31 grains per cu ft
Suspended soot	1.52 grains per cu ft

It is assumed that the suspended tar is 90% carbon and 10% hydrogen and that the soot is 100% carbon.

Saturated steam at a gage pressure of 25.3 psi was introduced at the bottom of the fuel bed.

Analysis of coal as charged:

Carbon	66.31%	Nitrogen	1.52%	Sulfur	1.44%
Available hydrogen	3.53	Total water	23.16	Corrected ash	4.04
					100.00%
Ash (as weighed)					4.58%

Analysis of refuse:

Moisture	1.10%	Fixed carbon	3.08%	Ash	95.82%
					100.00%

Analysis of dry, tar- and soot-free gas, by volume:

CO_2	7.12%	CO	21.85%	H_2	13.65%
O_2	0.90	CH_4	3.25	N_2	53.24
					100.00%

The gases left the flue at 1075° F and 29.75 in. Hg.   The refuse left the producer at 350° F and may be assumed to have a specific heat of 0.22.   The mean specific heat of the tar and soot between 75 and 1075° F may be assumed to be 0.32.   The heating value of the suspended tar was found to be 16,000 Btu per lb, and that of the deposited tar 15,500 Btu per lb.

Calculate complete material and energy balances for the operation of this gas producer.

Solution: Since a majority of the details of this problem are similar to those of illustration 4, full explanations of all steps will not be repeated.   A flow chart of this process is shown in Fig. 111.

Material Balance

Special attention should be given to calculations shown in parts 1 and 3.

1. Weight of gaseous products

Carbon Balance.   Carbon gasified.   *Basis:* 100 lb of coal charged.

Carbon in coal $= 100 \times 0.6631$ $= 66.31$ lb or 5.526 lb-atoms

Carbon in deposited tar $= \dfrac{591}{709} \times 0.93 =$ 0.78 lb or 0.065 lb-atom

Weight of refuse $= 4.58/0.9582$ $= 4.79$ lb

    (calculated on the basis of the ash as weighed in the proximate analysis)

Carbon in refuse = 4.79 × 0.0308   = 0.148 lb or 0.012 lb-atom
Carbon gasified   = 5.526 − (0.065 + 0.012) = 5.449 lb-atoms

Carbon in clean gases. *Basis:* 1.0 lb-mole of dry, tar- and soot-free gas.

    C in CO_2 = 0.0712 lb-atom
    C in CO   = 0.2185
    C in CH_4 = 0.0325
       Total   = 0.3222 lb-atom

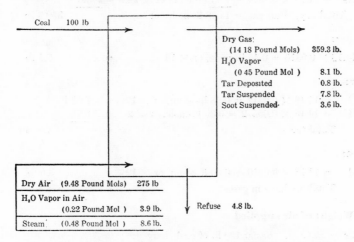

FIG. 111. Material balance of a gas producer (illustration 6)

Water, tar, and soot in gas. *Basis:* 1000 cu ft of wet gases at 1075° F and 29.75 in. Hg.

$$\text{Moles of total gas} = \frac{1000}{359 \times \dfrac{1535}{492} \times \dfrac{29.92}{29.75}} = 0.888 \text{ lb-mole}$$

$$H_2O \text{ present} = \frac{3.43 \times 1000}{7000 \times 18} = 0.0272$$

Moles of dry gas   = 0.8608 lb-mole

Water per mole of dry gas 0.0272/0.8608   = 0.0316 lb-mole

$$\text{Tar per mole of dry gas } \frac{3.31 \times 1000}{7000 \times 0.8608} = 0.550 \text{ lb}$$

$$\text{Soot per mole of dry gas } \frac{1.52 \times 1000}{7000 \times 0.8608} = 0.252 \text{ lb}$$

Total carbon in gases. *Basis:* 1.0 lb-mole of dry, clean gas.

    C in clean gases   = 0.3222 lb-atom
    C in tar = 0.550 × 0.9 = 0.495 lb or 0.0412
    C in soot = 0.252 lb or   0.0210
       Total   = 0.3844 lb-atom

Moles of dry gas per 100 lb of coal = 5.449/0.3844 = 14.18 lb-moles
Total products in gases. *Basis:* 100 lb of coal charged.

CO_2 = 14.18 × 0.0712 = 1.01 lb-moles or × 44 = 44.5 lb
O_2 = 14.18 × 0.0090 = 0.12 lb-mole or × 32 =   3.8
CO = 14.18 × 0.2185 = 3.10 lb-moles or × 28 = 86.8
CH_4 = 14.18 × 0.0325 = 0.46 lb-mole or × 16 =   7.4
H_2 = 14.18 × 0.1365 = 1.94 lb-moles or × 2   =   3.8
N_2 = 14.18 × 0.5324 = 7.55 lb-moles or × 28.2 = 213.0

  Total dry, clean gas = 14.18 lb-moles or   359.3 lb

Water:

  14.18 × 0.0316 = 0.45 lb-mole, or × 18   =   8.1 lb

Tar:

  C = 14.18 × 0.0412 = 0.58 lb-atom, or × 12 =   7.02 lb
  H = 14.18 × 0.055/2 = 0.39 lb-mole, or × 2 =   0.78

     Total tar   =   7.8 lb

Soot:

  C = 14.18 × 0.0210 = 0.30 lb-atom, or × 12 =   3.6 lb

     Total products in gases   =   378.8 lb

2. Weight of air supplied

Nitrogen Balance. *Basis:* 100 lb of coal charged.

  N_2 in gas   =   7.55 lb-moles
  N in coal = 1.52/28   =   0.05

  N_2 from air   =   7.50 lb-moles

  Dry air supplied = 7.50/0.79   =   9.48 lb-moles
    or 9.48 × 29.0   =   275 lb
  Molal humidity of air (Fig. 20)   =   0.023
  Water in air = 9.48 × 0.023   =   0.218 lb-mole
    or 0.218 × 18   =   3.9 lb
       Total wet air = 0.218 + 9.48 =   9.70 lb-moles
         or 275 + 3.9   =   278.9   lb

3. Weight of steam introduced

Hydrogen Balance. *Basis:* 100 lb of coal charged.

Output

H in deposited tar = $\dfrac{591}{709}$ × 0.07 = 0.0585 lb = 0.029 lb-mole

Free H_2 in gas   = 1.94
H_2 in CH_4 in gas = 0.46 × 2   = 0.92
H_2 in H_2O in gas   = 0.45
H in suspended tar   = 0.39

  Total output of H_2 = 3.73 lb-moles

Input

Net H in coal = 3.53/2.02	= 1.75 lb-moles	
H_2 in water in coal = 23.16/18	= 1.28	
H_2 in water in air	= 0.22	
Total input in addition to steam		= 3.25 lb-moles
H_2 from steam = 3.73 − 3.25		= 0.48 lb-mole
Steam introduced = 0.48 lb-mole or		8.64 lb

4. Over-all material balance

Input		Output	
Coal	100 lb	Refuse	4.8 lb
Dry air (9.48 lb-moles)	275	Tar deposited	0.8
Water in air (0.22 lb-mole)	3.9	Soot suspended	3.6
Steam (0.48 lb-mole)	8.6	Tar suspended	7.8
Total	387.5 lb	Dry clean gas (14.18 lb-moles)	359.3
		Water in gas (0.45 lb-moles)	8.1
		Total	384.4 lb

The slight discrepancy in the totals of the material balance results from inaccuracies of the data and neglect of the sulfur content of the coal. The material balance is also summarized in Fig. 111.

5. Gaseous volumes. *Basis:* 100 lb of coal charged.

$$\text{Volume of wet air} = 9.70 \times 359 \times \frac{535}{492} \times \frac{29.92}{29.75} = 3,800 \text{ cu ft}$$

$$\text{Volume of wet gases} = 14.63 \times 359 \times \frac{1535}{492} \times \frac{29.92}{29.75} = 16,500 \text{ cu ft}$$

Energy Balance

Special attention is called to calculations in parts 2, 3, and 5 of output.

Reference temperature: 75° F.

Basis: 100 lb of coal charged.

Input

1. Heating value of coal = 100 × 11,910 = 1,191,000 Btu

2. Sensible enthalpy of coal   =   0

3. Enthalpy of steam

Pressure = 25.3 + 14.6 = 39.9 psi.

From steam tables, the enthalpy of saturated steam at this pressure referred to 32° F is 1170 Btu per lb. Referred to 75° F, this becomes 1170 −43 = 1127 Btu per lb.

Enthalpy of steam = 8.6 × 1127   =   9,700 Btu

4. Enthalpy of dry air   =   0

5. Enthalpy of water vapor in air

Enthalpy = (3.9)(1052)   =   4,100

    Total   = 1,204,800 Btu

Output

1. **Heating value of dry clean producer gas** (calculated from composition)   =   792,200 Btu
2. **Heating value of suspended tar** = $(7.80)(16,000)$ =   124,800
3. **Heating value of soot** = $(0.30)(96,650)(1.8)$ =   52,190
4. **Heating value of carbon in refuse** = $(0.012)(96,650)(1.8)$ =   2,090
5. **Heating value of deposited tar** = $(0.83)(15,500)$ =   12,900
6. **Enthalpy of dry, clean producer gas**

Mean heat capacities between 75 and 1075° F taken from Fig. 62, page 259.

$$
\begin{aligned}
CO_2(1.01)(11.00) &= 11.11 \\
O_2(0.12)(7.60) &= 0.91 \\
CO(3.10)(7.27) &= 22.54 \\
CH_4(0.46)(12.15) &= 5.59 \\
H_2(1.94)(7.01) &= 13.60 \\
N_2(7.55)(7.22) &= \underline{54.51} \\
\text{Total} &= 108.26
\end{aligned}
$$

Enthalpy $= (108.26)(1075 - 75)$ =   108,260

7. **Sensible enthalpy of tar and soot**

Suspended tar and soot $= (11.4)(0.32)(1075 - 75) = 3650$ Btu
Deposited tar and soot $= (0.83)(0.32)(1075 - 75) = \underline{270}$
                          Total $=$ 3920 Btu   3,920

8. **Total enthalpy of water vapor in gases**

Enthalpy $= 0.45 [18,948 + 8.65 (1075 - 75)]$ =   12,420

9. **Sensible enthalpy of refuse** $= 0.22 \times 4.8 (350 - 75)$ = $\underline{290}$
Total energy accounted for   =   1,109,070 Btu

10. **Heat losses, radiation, etc.** $= 1,204,800 - 1,109,070$ =   95,730 Btu

Summary of Energy Balance

Reference temperature: 75° F.

Basis: 100 lb of coal charged.

Input	Btu	%
1. Heating value of coal	1,191,000	98.9
2. Sensible enthalpy of coal	0	0
3. Sensible enthalpy of air	0	0
4. Enthalpy of steam	9,700	0.8
5. Total enthalpy of water in air	4,100	0.3
Total	1,204,800	100.0

Output	Btu	%
1. Heating value of clean gas	792,200	65.8
2. Heating value of suspended tar	124,800	10.4
3. Heating value of suspended soot	52,190	4.3
4. Heating value of carbon in refuse	2,090	0.2
5. Heating value of deposited tar	12,900	1.0
6. Enthalpy of dry, clean gas	108,260	9.0
7. Enthalpy of tar and soot	3,920	0.3
8. Enthalpy of water vapor in gases	12,420	1.0
9. Enthalpy of refuse	290	0.0
10. Heat losses, radiation, etc.	95,730	7.9
Total	1,204,800	100.0

Thermal Efficiency

Cold gas. The effectively utilized energy includes only the heating value of the dry, clean gas.

$$\text{Efficiency} = \frac{792,200}{1,204,800} \times 100 = 65.8\%$$

Hot gas. The effectively utilized energy includes the total sensible enthalpy of all materials in the gases and also the heating value of the suspended tar and soot.

Sensible enthalpy of dry, clean gases	=	108,260 Btu
Sensible enthalpy of water vapor in gases = (0.45)(8.65) (1075 − 75)	=	3,890
Sensible enthalpy of suspended tar and soot	=	3,650
Heating value of dry, clean gas	=	792,200
Heating value of tar and soot	=	176,900
Effectively utilized energy	=	1,084,900 Btu

$$\text{Efficiency} = \frac{1,084,900}{1,204,800} \times 100 = 90.0\%$$

The principal heat losses in the operation of this particular gas producer are by radiation and conduction of heat from the equipment.

Graphical Calculation of Combustion Problems

Where short-cut methods of combustion calculations are desired, a simple graphical solution may be resorted to, provided that the following items are negligible: sulfur content of fuel, combustible content of refuse, water vapor in air, suspended tar and soot in gases, and hydrocarbons in flue gases. The simplified graphical solution of such combustion problems appears in Fig. 112. This method also serves as an approximate solution for other combustion problems where the experimental

data or time required do not warrant precise methods. From Fig. 112
there may be obtained directly the number of pounds of air used and

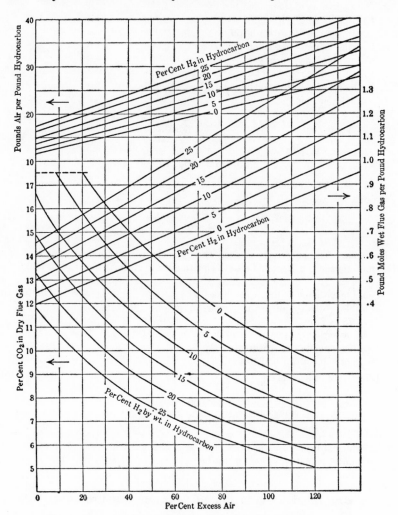

Fig. 112. Combustion chart for fuels

(Reproduced in *CPP Charts*)

pound-moles of wet flue gas produced and the carbon dioxide content
of the dry flue gas on the basis of 1 lb of combustible in oil, coal, gas,
or coke.

Problems

1. A Kentucky coal has the following analysis:

Proximate analysis as received		Ultimate analysis of combustible (corrected ash-free moisture-free basis)	
Moisture	2.97%	C	84.39%
Ash (uncorrected)	2.94	Net hydrogen	4.81
Volatile matter	37.75	N	2.00
Fixed carbon	56.34	S	1.02
		Combined H_2O	7.78
	100.00%		100.00%

The combustible referred to above includes those portions of coal as received that are not classified as moisture or corrected ash.

Determine, on the "as-received" basis:

 (a) Rank of this coal. **Ans.** Bituminous.
 (b) Total heating value by Dulong's formula. **Ans.** 14,372 Btu per lb.
 (c) Total heating value by Uehling's method. **Ans.** 14,268 Btu per lb.
 (d) Estimated available hydrogen content by Uehling's method. **Ans.** 4.30%
 (e) Net heating value by Uehling's method. **Ans.** 13,756 Btu per lb.
 (f) Ultimate analysis (as received, corrected ash).

Ans.

Free water	=	2.97%
Combined H_2O	=	7.35
C	=	79.71
H (available)	=	4.54
N	=	1.89
S	=	0.96
Ash (corrected)	=	2.58
		100.00%

2. In the following table are the analyses of several typical coals, as given in *Bureau of Mines Report of Investigations No.* **3296**, Classification Chart of Typical Coals of the United States. Both the ultimate and proximate analyses are on the "as-received" basis, and the hydrogen given in the ultimate analysis includes the hydrogen of the free moisture in the coal.

	No. 1	No. 2	No. 3	No. 4	No. 5
Proximate Analysis					
Moisture	4.3%	2.0%	2.2%	10.6%	42.6%
Volatile matter	3.0	9.6	21.1	36.4	24.7
Fixed carbon	82.6	77.8	66.2	42.4	27.0
Ash (uncorrected)	10.1	10.6	10.5	10.6	5.7
Ultimate Analysis					
Carbon	82.5%	79.5%	76.1%	62.8%	37.3%
Hydrogen	2.0	3.7	4.8	5.6	7.3
Nitrogen	0.6	0.9	1.2	1.2	0.5
Sulfur	0.5	0.6	2.6	3.3	0.6
Total Heating Value					
Btu per lb	12,650	13,520	13,530	11,300	6,260

No. 1 Anthracite
No. 2 Semianthracite
No. 3 Medium volatile bituminous
No. 4 High volatile bituminous
No. 5 Lignite

For one of the coals, present the following calculations:

(a) Recalculate the ultimate analysis, reporting:

    (1) Carbon.
    (2) Available hydrogen.
    (3) Free H_2O.
    (4) Combined H_2O.
    (5) Sulfur.
    (6) Nitrogen.
    (7) Corrected ash.

(b) Calculate total heating value by Dulong's formula. Compare the result with the experimental value.

(c) Calculate the total heating value by Uehling's method. Compare the result with the experimental value.

(d) Assuming that the proximate analysis and heating value alone are available, calculate the per cent of carbon and of available hydrogen by Uehling's method.

(e) Calculate the net heating value.

3. A gas has the following composition by volume:

Illuminants (C_2H_4 and C_6H_6)	53.6%
O_2	1.6
CH_4	16.9
C_2H_6	24.3
N_2	3.6
	100.0%

The heating value of this gas is 1898 Btu per standard cubic foot. Calculate the percentages of C_2H_4 and C_6H_6 in the gas. **Ans.** 31.7% C_2H_4, 21.9% C_6H_6.

4. Calculate the maximum theoretical flame temperature in degree centigrade when the following gas is burned with the theoretical amount of dry air starting with air and gas at 25° C:

CO	30%
H_2	15
O_2	1
CO_2	5
N_2	49
	100%

5. Calculate the theoretical flame temperature when the above gas is burned with 100% excess air.

6. Calculate the theoretical flame temperature of the above-mentioned gas when burned with the theoretical amount of air and when both gas and air are preheated to 500° C before combustion.

7. Calculate the theoretical flame temperature of the above gas when burned with the theoretical amount of air, the combustion of both CO and H_2 proceeding to only 80% completion. The gas and air are initially at 25° C.

8. A fuel gas has the following analysis:

Carbon dioxide	3.0%
Ethylene	5.8
Benzene	0.8
Oxygen	0.9
Hydrogen	38.8
Carbon monoxide	35.2
Methane	9.5
Ethane	0.7
Nitrogen	5.3
	100.0%

Calculate the theoretical flame temperature if this dry fuel gas is burned with the theoretical amount of dry air. Gas and air are supplied at 25° C. Assume that combustion is complete. Use 1 g-mole of the dry fuel gas as the basis of calculation.

9. For the fuel gas whose analysis is given in problem 8, calculate the theoretical flame temperature if the dry gas is burned with the theoretical amount of dry air. The gas is supplied at 25° C, but the air is preheated to 700° C. Assume that combustion is complete. Use 1 g-mole of the dry fuel gas as the basis of calculation.

10. For the fuel gas whose analysis is given in problem 8, calculate the theoretical flame temperature if the gas is burned with 100% excess air. The gas and air both are dry and are supplied at 25° C. Assume that combustion is complete. Use 1 g-mole of the dry fuel gas as the basis of calculation.

11. The dry fuel gas whose analysis is given in problem 8 is burned with the theoretical amount of dry air. Both air and gas are supplied at 25° C. If carbon monoxide and hydrogen burn to 80% completion, what is the theoretical flame temperature? Assume that the combustion of the gases other than CO and H_2 is complete. Use 1 g-mole of the dry fuel gas as the basis of calculation.

12. A bed of coke having an original weight of 4000 kg and an original temperature of 1400° C has 360 kg of steam blown through it, forming 448 kg CO, 88 kg CO_2, and 40 kg H_2 at an average temperature of 1000° C. The steam is supplied at 120° C, saturated. The radiation loss is 100,000 kcal. Calculate the final temperature of the bed of coke.

13. A fuel gas has the following composition by volume:

CO_2	2.1%
O_2	0.5
$C_{2.5}H_{4.2}$ (illuminants)	7.0
CO	33.8
H_2	40.6
$C_{1.2}H_{4.4}$ (paraffins)	11.2
N_2	4.8
	100.0%

(a) Calculate the analysis of the flue gases formed by burning this gas with 30% excess air, assuming that all combustible components are burned to CO_2 and H_2O. **Ans.** CO_2 11.08%, O_2 4.33%, N_2 71.36%, H_2O 13.23%.

(b) Calculate the heating value in Btu per standard cubic foot. **Ans.** 492 Btu.

14. A fuel oil has a specific gravity of 0.91 and a viscosity of 28 Saybolt Furol seconds at 122° F Estimate the characterization factor, average molecular weight, average boiling point, hydrogen content, specific heat at 150° F, and heating value of this oil.

15. A petroleum fraction has a specific gravity (60°/60° F) of 0.88. When tested in a Saybolt Universal viscometer at 122° F, a reading of 58 sec was obtained. Using the charts in the text, report the following values:

    (a) Characterization factor. **Ans.** 11.9.
    (b) Average molecular weight. **Ans.** 315.
    (c) Average boiling point. **Ans.** 690° F.
    (d) Hydrogen content. **Ans.** 13.2%.
    (e) Specific heat of the liquid oil at 200° F. **Ans.** 0.497.
    (f) Specific heat of the vapor at 800° F. **Ans.** 0.725.
    (g) Heating value. **Ans.** 19,440 Btu per lb.

16. A Pennsylvania bituminous coal has the following composition:

H	4.71%
C	69.80
N	1.42
O_2	7.83
Ash	6.73
H_2O	9.51
	100.00%

Total heating value 6950 kcal per kg

This coal is gasified in a gas producer, using air at 20° C saturated with water vapor. No additional steam or water is admitted into the producer. The barometric pressure is 740 mm. The resulting gas has the following composition:

H_2	0.5%
CO	21.2
N_2	64.5
CH_4	5.8
CO_2	6.2
O_2	1.8
	100%

It may be assumed that no tar or soot is present in the producer gas. During a test period the total coal charged is 10,500 kg.

The dry refuse formed weighs 825 kg and contains 13.3% C.

Temperature of refuse	=	220° C.
Mean specific heat of refuse	=	0.25.
Temperature of outgoing gases	=	450° C.

    (a) Calculate complete material and energy balances of this gas producer on the basis of 100 kg of coal as charged.

    (b) Calculate the thermal efficiency on both the hot and cold bases.

    (c) Calculate the total volume of gases leaving the producer.

    (d) Calculate the heating value of the producer gas in Btu per cubic foot measured at 60° F, 30 in. Hg, saturated with water vapor.

17. A gas producer is charged with bituminous coal having the following composition:

Proximate analysis:

Moisture	2.70%
Volatile matter	25.77
Fixed carbon	62.87
Ash	8.66
	100.00%

Ultimate analysis:

Moisture	2.70%
Carbon	78.55
Net hydrogen	4.13
Nitrogen	1.58
Sulfur	0.69
Corrected ash	8.40
Combined H_2O	3.95
	100.00%

Total heating value = 13,944 Btu per lb

Air is supplied at 75° F with a percentage humidity of 90%. The barometric pressure is 29.65 in. Hg. Dry, saturated steam is supplied at a gage pressure of of 50 psi. The producer gas leaves at a temperature of 1220° F and has the following composition by volume:

CO	25.0%
H_2	22.0
CH_4	3.6
C_2H_4	2.8
CO_2	9.2
N_2	37.4
	100.0%

A sample of gas is withdrawn and cooled to 100° F for determination of suspended tar and soot. The tar and soot content is 10 grains per cubic foot of gas measured at barometric pressure, 100° F, and saturated with water vapor. The tar and soot contain 95% carbon and 5% hydrogen. Its heating value is 17,100 Btu per lb, and its mean specific heat is 0.34. The dew point of the producer gas is 100° F.

The refuse is discharged at 400° F, moisture-free, and containing 4.52% carbon. The specific heat of the refuse is assumed to be 0.23.

Neglecting deposition of tar in the flues and presence of sulfur, calculate:

(a) Complete energy and material balances of the producer, based on 100 lb of coal charged.

(b) Thermal efficiencies on both the hot and cold bases. **Ans.** 100%, 81.1%.

(c) Volume of producer gas, measured at 60° F, 30 in. Hg, saturated with water, formed per 100 lb of coal charged. **Ans.** 5185 cu ft.

(d) Heating value of the producer gas, per standard cubic foot. **Ans.** 229 Btu per cu ft.

The solution of this problem results in a negative radiation loss. What errors in experimental data are most likely responsible for this condition?

18. An Illinois bituminous coal is burned in a boiler furnace with air at 78° F, 92% humidity. The barometric pressure is 29.40 in. Hg. The furnace gases leave at 553° F. The refuse is discharged moisture-free, at 440° F. The refuse as analyzed contains 12.2% carbon as coke and 16.1% moisture. The specific heat of the dry refuse is 0.25. The proximate analysis of coal is:

Fixed carbon	50.34%
Volatile matter	30.68
Moisture	9.61
Ash	9.37
	100.00%

Heating value of coal = 11,900 Btu per lb

The ultimate analysis on the moisture-free basis is:

Carbon	73.70%
Hydrogen	4.75
Oxygen	9.23
Nitrogen	1.58
Sulfur	0.55
Corrected ash	10.19
	100.00%

The flue-gas analysis is as follows:

CO_2	12.2%
CO	0.2
O_2	7.0
N_2	80.6
	100.0%

Calculate:

(a) Total material and energy balances for this process based on 100 lb of coal charged, neglecting the combustion of sulfur. The heat losses and the heat effectively utilized may be considered together as a single item of heat output.

(b) Percentage excess air used in combustion, based on that required for complete combustion of all coal charged.

(c) Dew point of the flue gases.

(d) Actual volumes in cubic feet at the given condition of temperature, humidity, and pressure of the flue gases and air supply, per 100 lb of coal charged.

19. A bituminous coal is burned in a boiler furnace with air at 85° F, 90% humidity. The barometric pressure is 29.20 in. Hg. The furnace gases leave at 572° F. The refuse leaves the furnace moisture-free at 520° F and when analyzed contains 22.3% moisture, 12.3% volatile matter, and 41.4% fixed carbon. The mean specific heat of the refuse is 0.23. The proximate analysis of the coal is:

Fixed carbon	56.34%
Volatile matter	37.75
Moisture	2.97
Ash	2.94
	100.00%

A partial ultimate analysis on the corrected ash-free and moisture-free basis is:

Carbon	84.43%
Nitrogen	2.00
Sulfur	0.82

The total heating value of the coal is 14,139 Btu per lb. Dry, saturated steam at a gage pressure of 150 psi is produced at the rate of 780 lb per 100 lb of coal charged. Water is fed into the boiler at 72° F. The Orsat analysis of the flue gas is:

CO_2	12.0%
CO	1.2
O_2	6.2
N_2	80.6
	100.0%

Calculate:

(a) Net hydrogen content of the coal from an oxygen balance, neglecting combustion of sulfur.

(b) Complete ultimate analysis of the coal.

(c) Complete material balance of the process, based on 100 lb of coal charged.

(d) Complete energy balance of the furnace, based on 100 lb of coal charged.

(e) Thermal efficiencies of the furnace and boiler, based on the total and on the net heating values. **Ans.** 63.7%, 66.5%.

(f) Percentage excess air used, based on the total combustible charged. **Ans.** 27.2%.

20. A heat interchanger, used for heating the oil in a circulating hot oil heating system, is fired with coal having the following proximate analysis:

Moisture	12.38%
Volatile matter	36.88
Fixed carbon	37.50
Ash	13.24
	100.00%

The heating value of the coal is 10,361 Btu per lb, and its sulfur content is 5.1%.

The coal is burned with air at a temperature of 70° F and a percentage humidity of 60. The barometric pressure is 29.3 in. Hg.

The refuse from the furnace is discharged at a temperature of 600° F and contains 16% fixed carbon and 84% ash. The sulfur content of the refuse is 7.8%. Its specific heat may be taken as 0.23.

The flue gases leave the furnace at a temperature of 850° F and have the following composition by volume, on the sulfur- and moisture-free basis.

CO_2	11.50%
CO	0.17
O_2	7.51
N_2	80.82
	100.00%

The oil is circulated at a rate of 3800 lb per 100 lb of coal charged and is heated from 155 to 464° F. The mean specific heat of the oil in this temperature range is 0.55. Calculate:

(a) Complete ultimate analysis of the coal as estimated from the rank and heating value.

(b) Complete material and energy balances of the interchanger, based on 100 lb of coal charged.

(c) Complete analysis, by volume, of the wet flue gases leaving the interchanger.

(d) Percentage excess air, based on the combustible actually burned.

(e) Volume of wet flue gases leaving the interchanger.

(f) Thermal efficiencies of the interchanger, based on both the total and net heating values.

21. A brick kiln, of an intermittent type, is fired with 10,420 lb-moles of dry producer gas. The weight of green ware is 410,000 lb containing 0.52% mechanical water and 3.02% chemically combined water. The gas enters the kiln at 1220° F and a pressure of 29.65 in. Hg, and contains 10 grains of tar (90% C, 10% H) per cubic foot, measured at a pressure of 29.65 in. Hg, 100° F, and saturated with water vapor. The dew point of the producer gas is 100° F.

During the water-smoking period mechanical water is vaporized and leaves the kiln at 300° F, and the chemically combined water leaves at 400° F. During water smoking the saturation temperature of the gases is 150° F. The flue gas leaves at an average temperature of 720° F. The average temperature of the ware at the end of the burn is 2100° F, and its specific heat is 0.23. The producer gas is burned with air at 75° F, 90% humidity, and a pressure of 29.65 in. Hg. The average analysis of gases by volume on the moisture-free basis is:

Producer Gas		Flue Gas	
CO	25.00%	CO_2	12.20%
H_2	22.00	CO	0.16
CH_4	3.60	H_2	0.14
C_2H_4	2.80	O_2	7.10
CO_2	9.20	N_2	80.40
N_2	37.40		100.00%
	100.00%		

Calculate the material balance of the combustion process, the energy balance of the entire unit, and the thermal efficiency. The heat absorbed by the kiln structure may be included with the other undetermined heat losses as a single item.

22. Limestone is burned in a continuous vertical kiln which is heated by coal burned on an external grate located beside the bottom of the kiln shaft. The limestone is charged at the top of the shaft at atmospheric temperature and gradually descends in contact with a rising stream of the flue gases from the grate. The burned lime is discharged from the bottom of the shaft at a temperature of 950° F. The flue gases, mixed with all gases and vapors evolved by the charge, leave the top of the shaft at 560° F. For each 100 lb of coal burned, 161 lb of burned lime are produced. The limestone charged has the following composition:

CaO	51.0%
MgO	2.0
CO_2	42.2
Al_2O_3	1.5
SiO_2	1.2
H_2O	2.1
	100.0%

The ultimate analysis of the coal is as follows:

Moisture	10.69%
C	66.62
Net H	3.18
N	1.57
S	1.91
Corrected ash	6.41
Combined H_2O	9.62
	100.00%

The total heating value of the coal is 11,805 Btu per lb. The gases leaving the top of the kiln have the following composition by volume:

CO_2	16.4%
N_2	76.8
O_2	6.8
	100.0%

The coal is burned with air at a temperature of 70° F having a humidity of 80% The barometric pressure is 29.4 in. Hg.

The refuse from the grate contains 4.2% fixed carbon and 95.8% ash. Its sulfur content is 3.1%.

It may be assumed that in the burning process all CO_2 and water are driven from the limestone. The heat of wetting of granular limestone is negligible. It may be assumed that the sulfur burned forms SO_2 which is further oxidized and absorbed by the lime to form $CaSO_4$. The mean specific heat of the burned lime is 0.21. The enthalpy of the refuse may be neglected.

Calculate the complete energy and material balances of the grate and kiln on the basis of 100 lb of coal fired.

Calculate the thermal efficiency of the process, considering the effectively utilized heat to be consumed in the decomposition of the limestone.

23. A 12-hr test was conducted on a steam-generating plant with four of the boilers in operation. The data for the 12-hr test are as follows:

Proximate Analysis of Coal:

Moisture	4.38%	Fixed carbon	48.98%
Volatile matter	29.93	Ash (uncorrected)	16.71
			100.00%

Half of the sulfur of the coal appears in the volatile matter.

Ultimate Analysis of Coal:

Carbon	65.93%	Combined water	6.31%
Available hydrogen	3.50	Free moisture	4.38
Nitrogen	1.30	Ash (corrected, sulfur free)	15.59
Sulfur	2.99		100.00%

Heating value of coal as fired (total)	11,670 Btu per lb
Weight of coal fired	119,000 lb
Temperature of coal	65° F

Proximate Analysis of Refuse:

Moisture	4.77%	Fixed carbon	12.51%
Volatile matter	2.08	Ash (uncorrected)	80.64
			100.00%

Flue Gas:

CO_2	11.66%	CO	0.04%
O_2	6.52	N_2	81.78
			100.00%

Temperature	488° F

Air:

Dry-bulb temperature	73° F
Wet-bulb temperature	59.4° F
Barometer	29.08 in. Hg

Steam:

Feed water	193° F
Water evaporated	1,038,400 lb
Steam pressure	137.4 psi gage
Quality	98.3%

Calculate the complete material and energy balances for this steam-generating plant.

24. Calculate the complete energy balances for illustration 5.

25. A steam superheater is to be designed to heat 10,000 lb per hr of low-pressure process steam from 250 to 700° F. The fuel is a residual cracked oil having an API gravity of 8.5 and a viscosity of 200 Saybolt Furol seconds at 122° F. The design bases are as follows:

% excess air	40
Air temperature, ° F	70
% relative humidity of air	60
Stack temperature, ° F	600

Assuming a radiation loss of 5% of the heating value of the fuel burned, calculate complete material and energy balances for 1.0 hr of the combustion operation and the thermal efficiency based on the total heating value of the fuel. Figure 112 may be used for establishing the material balance of the combustion.

26. The wet felting removed from the sheet-making machine of a shingle mill is partially dried in an adiabatic tunnel drier. The felt is passed countercurrent to a stream of hot gases resulting from the combustion of coke.

Analysis of flue gas: CO_2 14.35%, CO 2.05%, O_2 3.79%, N_2 79.81%.

Gas is fed to dryer at 300° F, 31 in. Hg, at a wet-bulb temperature of 118° F, and leaves at 160° F, 31 in. Hg, at a rate of 350,000 cu ft per min at these conditions.

Calculate the weight of water evaporated in pounds per hour.

27. A rotary drier is heated by burning liquefied bottle gas containing 30% propane and 70% n-butane by weight. The liquid is vaporized and injected into the drier at a rate of 50 lb per hr, 80° F. Air at 80° F and 29.5 in. Hg is fed at a rate of 10% in excess of the amount theoretically required for complete combustion. The combustion of hydrogen is complete. The carbon is oxidized 90% to CO_2 and 10% to CO. The stack gases leave at 600° F and 29.3 in. Hg.

Calculate complete material and energy balances of combined vaporization and combustion processes on the basis of 1 hr performance.

28. The unit for producing the protective atmosphere for a steel heat-treating furnace consists of a combustion chamber which is fed with a fuel gas and a limited amount of air. A gas high in carbon monoxide and nitrogen and low in carbon dioxide and oxygen is thus produced. In order to reduce the water vapor content of the gas, it is cooled to as low a temperature as feasible by the available water supply. The gas emerging from the water-spray cooler is fed to the heat-treating furnace, where it flows around the steel being heated, thus protecting the steel against surface oxidation.

Operating data on a particular unit are as follows:

Fuel Gas:

Volumetric Analysis, Dry Basis:

CH_4	90%
C_2H_6	5
N_2	5
	100%

90° F, 29.70 in. Hg, 30% humidity.
Air: 70° F, 29.50 in. Hg, 50% humidity.

Protective Gas Formed

Volumetric Analysis, Dry Basis:

CO	12.5%
CO_2	1.1
O_2	0.1
N_2	86.3
	100.0%

Temperature leaving combustion unit and entering spray chamber	= 1525° F
Temperature leaving spray chamber	= 60° F
Humidity leaving spray chamber	= 100%

The cooling water enters at 50° F and leaves at 90° F.

Calculate the following, on the basis of 100 cu ft of gas supplied:

(1) Complete material balance.
(2) Complete energy balance. Assume heat loss from the water-spray cooler is negligible.

12

Chemical, Metallurgical, and Petroleum Processes

The methods applied in calculating material and energy balances are alike in principle for all industrial processes, differing only in detail. Three illustrative material and energy balances are presented in this chapter, representing the procedures applicable to chemical processes in general. Typical illustrations in the chemical, metallurgical, and petroleum industries have been selected. The chemical and metallurgical illustrations summarize the principles involved in dealing with complex chemical reactions where the intermediate reactions are unknown. The petroleum process illustrates the use of an energy balance to establish a material balance and the principles of recycling to increase yield and effect temperature control. The chemical and metallurgical processes illustrate the analysis of experimental data from existing plants while the petroleum process is presented as a problem in the design of a new plant.

Chamber Sulfuric Acid Plant

The material and energy balances of a chamber process sulfuric acid plant are selected for examination as representative of the chemical industries. The operating conditions have been taken from data published by Kaltenbach.[1] In this particular problem iron pyrites is burned with air in a shelf burner. The burner gases consisting of sulfur dioxide, oxygen, and nitrogen pass through a dust chamber where suspended matter is removed, and then into the Glover tower. In the Glover tower the hot gases meet a descending stream of acids from the Gay-Lussac tower and chambers. Oxides of nitrogen lost in the system are replaced by nitric acid introduced at the top of the tower. In passing through the Glover tower the hot gases are cooled, and some conversion of SO_2 to sulfuric acid takes place, the chamber acid is concentrated, and the oxides of nitrogen are evolved from the acids for recirculation. The mixed gases leaving the Glover tower pass into a series of large lead chambers where conversion to H_2SO_4 is completed.

[1] M. Kaltenbach, *Chimie & Industrie*, 3, 407 (1920); *Chem. Age*, 28, 295 (1920).

Finally, the oxides of nitrogen are recovered in the Gay-Lussac tower by passing the spent gases from the chamber countercurrent to a stream of cold sulfuric acid from the Glover tower. These different steps in the manufacture and the material and energy balances of the burner, Glover tower, chambers, and Gay-Lussac tower are each discussed separately. The balances for each unit are based on 100 kg of dry pyrites charged into the burner. The reference temperature for the energy balances is taken at 25° C.

The concentration of a strong aqueous solution of sulfuric acid is usually determined by measurement of its specific gravity. Concentrations are usually expressed in terms of specific gravities or of degrees Baumé rather than in percentages.

Illustration 1. Material and Energy Balances of a Chamber Sulfuric Acid Plant. Pyrites, containing 85.3% FeS_2, 2% H_2O, and 12.7% inert gangue, is burned in a shelf furnace yielding a gas containing 8.5% SO_2, 10.0% O_2, and 81.5% N_2 by volume. The pyrites is charged at 25° C and the air is supplied at the same temperature with a humidity of 31% and at a pressure of 750 mm Hg. The cinder leaves the burner at 400° C containing 0.42% sulfur as unburned pyrites. The pyrites burned forms Fe_2O_3 and SO_2, and the gangue passes into the cinder unchanged. The mean specific heat of the cinder is 0.16. The gases from the burner, after passing through the dust chamber, enter the Glover tower at 450° C and leave at 91° C. In the Glover tower 16% of the SO_2 in the gas is converted to H_2SO_4. There are sprayed into the top of the Glover tower, per 100 kg of moisture-free pyrites charged:

(a) 182 kg of aqueous sulfuric acid at 25° C from the chambers, containing 64.0% H_2SO_4 and 36% H_2O.

(b) 580 kg of mixed acid at 25° C from the Gay-Lussac tower, containing 77% H_2SO_4, 22.1% H_2O, and 0.885% N_2O_3.

(c) 1.31 kg of aqueous nitric acid at 25° C, containing 36% HNO_3 and 64% H_2O.

Acid leaves the bottom of the Glover tower, free from oxides of nitrogen, at a temperature of 125° C, and containing 78.0% H_2SO_4 and 22% H_2O. This acid is cooled to 25° C, part of it is returned to the top of the Gay-Lussac tower, and the remainder is withdrawn as the final product of the plant. The gases leaving the Glover tower are passed through a series of four chambers and finally enter the Gay-Lussac tower, at 40° C. Spray water is introduced at the tops of the various chambers at 25° C. The acid formed in the chambers is withdrawn from the first chamber at 68° C, containing 64.0% H_2SO_4. This acid is cooled to 25° C and is all fed into the top of the Glover tower. Part of the Glover acid after cooling to 25° C is returned to the top of the Gay-Lussac tower. The Gay-Lussac acid leaves the bottom of the tower at 27° C and is all fed to the top of the Glover tower at 25° C. The spent gases leave the top of the Gay-Lussac tower at 30° C. The flow chart of the entire process is shown diagrammatically in Fig. 113.

Calculate material and energy balances of the entire plant and of each of the following units, all based on 100 kg of moisture-free pyrites charged:

(a) The burner.
(b) The Glover tower.
(c) The four chambers as a single unit.
(d) The Gay-Lussac tower.

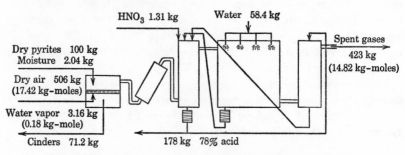

FIG. 113.   Material balance of an entire sulfuric acid plant

Material Balance of Entire Plant

Before discussing the material and energy balances of the individual units in the sulfuric acid plant it is desirable to calculate the material balance of the entire plant in order to have in mind a perspective of the whole process. This balance is represented by the following entries:

	Input		Output
Dry ore	Moisture in air		Acid produced
Moisture in ore	Nitric acid		Dry spent gases
Dry air	Spray water		Cinder

1. Weight of cinder formed

Basis: 100 kg of moisture-free pyrites charged.

Weight of pyrites as charged $= \dfrac{100}{0.98}$ $= 102$ kg

Weight of FeS_2 charged $= 102 \times 0.853$ $= 87$ kg
     or $87/120$ $= 0.726$ kg-mole
Weight of gangue charged $= 102 \times 0.127 = 13$ kg
% S in cinder $= 0.42\%$

% FeS_2 in cinder $= \dfrac{0.42}{32} \times \dfrac{1}{2} \times 120$ $= 0.78\%$

Let $x =$ kilograms of FeS_2 in cinder.

Weight of FeS_2 oxidized $= (87 - x)$ kg or $(87 - x)/120$ kg-moles

Fe_2O_3 formed $= \dfrac{87 - x}{120} \times \dfrac{1}{2} \times 159.7 = (58 - 0.667x)$ kg

Weight of cinder $= 13.0 + x + (58 - 0.667x) = (71 + 0.333x)$ kg
Weight of FeS_2 in cinder $= x = 0.0078(71 + 0.333x)$ or $x = 0.56$ kg
     or $= 0.56/120 = 0.0048$ kg-mole
Fe_2O_3 in cinder $= 58 - (0.667 \times 0.56)$ $= 57.63$ kg
     or $= 57.63/159.7 = 0.361$ kg-mole
Weight of cinder $= \overline{71.2}$ kg

2. Weight of dry burner gases

Sulfur Balance.   Basis: 100 kg of dry pyrites charged.

$$FeS_2 \text{ burned} = 0.726 - 0.005 = \quad 0.721 \text{ kg-mole}$$
$$S \text{ burned} = 0.721 \times 2 \qquad = \quad 1.442 \text{ kg-atoms}$$
$$S \text{ per kg-mole of burner gas} = \quad 0.085 \text{ kg-atom}$$
$$\text{Dry burner gas} = \frac{1.442}{0.085} \qquad = \quad 16.95 \;\; \text{kg-moles}$$

$$SO_2 = 16.95 \times 0.085 = \;\; 1.441 \text{ kg-moles or} \times 64 \quad = \quad 92.2 \text{ kg}$$
$$O_2 \; = 16.95 \times 0.100 = \;\; 1.70 \;\; \text{kg-moles or} \times 32 \quad = \quad 54.4$$
$$N_2 \; = 16.95 \times 0.815 = \underline{13.81} \;\; \text{kg-moles or} \times 28.2 = \underline{389.8}$$

$$\text{Total dry gases} \qquad = \overline{16.95} \;\; \text{kg-moles or} \qquad \overline{536.4 \text{ kg}}$$

3. Weight of dry air used

Nitrogen Balance

$$\text{Nitrogen in burner gas} \; = 13.81 \text{ kg-moles}$$
$$\text{Air introduced} = \frac{13.81}{0.79} = 17.5 \text{ kg-moles}$$
$$\text{or } 17.5 \times 29 \qquad \quad = 506 \text{ kg}$$

4. Weight of water vapor in dry air. The air supply enters at 25° C, 31% humidity, and 750 mm pressure.   From Fig. 20 the molal humidity is 0.0101.

$$\text{Water vapor in air} = 0.0101 \times 17.5 \; = 0.176 \text{ kg-mole}$$
$$\text{or} \qquad\qquad\qquad 0.176 \times 18 \quad = 3.16 \text{ kg}$$

5. H_2SO_4 produced in system. Sulfuric acid is formed only in the Glover tower and in the chambers.

$$SO_2 \text{ entering Glover tower} \qquad\qquad\qquad = \quad 1.44 \;\; \text{kg-moles}$$
$$SO_2 \text{ converted to acid in Glover tower} = 1.44 \times 0.16 = \quad 0.230 \text{ kg-mole}$$
$$H_2SO_4 \text{ formed in Glover tower} = 0.230 \times 98.1 \quad = \quad 22.6 \quad\;\; \text{kg}$$
$$H_2SO_4 \text{ in acid from chambers} = 182 \times 0.64 \quad = \underline{116.5}$$
$$\text{Total } H_2SO_4 \text{ formed} \qquad\qquad\qquad\qquad = \quad 139.1 \quad\;\; \text{kg}$$
$$\text{or} = 139.1/98 \qquad\qquad\qquad\qquad\qquad = \quad 1.42 \;\; \text{kg-moles}$$

$$\text{Total product of aqueous acid, } 78\% \; H_2SO_4 = 139.1/0.78 = 178 \text{ kg}$$

6. Weight of spray water in chambers

Water Balance

Output

The small amount of water vapor in the gases from the Gay-Lussac tower may be neglected.   On this basis all water leaves the process in the 78% acid product.

$$H_2O \text{ used in forming } H_2SO_4 \qquad\qquad = 1.42 \text{ kg-moles}$$
$$\text{or } 1.42 \times 18 \qquad\qquad\qquad\qquad\quad = 25.5 \;\; \text{kg}$$
$$H_2O \text{ in aqueous acid} = 178 \times 0.22 \qquad = \underline{39.0}$$
$$\text{Total } H_2O \text{ output} \qquad\qquad\qquad\quad = 64.5 \;\; \text{kg}$$

Input

$$H_2O \text{ in ore} = 102 \times 0.02 \qquad\qquad\quad = \quad 2.04 \text{ kg}$$
$$H_2O \text{ in air} \qquad\qquad\qquad\qquad\qquad\quad = \quad 3.16$$
$$H_2O \text{ from aqueous nitric acid} = 1.31 \times 0.64 \; = \quad 0.84$$

It is assumed that the HNO_3 introduced is completely decomposed into H_2O, NO, and O_2.

HNO_3 introduced = 1.31 × 0.36	=	0.47 kg
or 0.47/63	=	0.0075 kg-mole
H_2O formed from HNO_3 = 0.0075/2	=	0.0037 kg-mole
or	=	0.067
Total H_2O input accounted for	=	6.11 kg
Water supplied by sprays = 64.5 − 6.1	=	58.4 kg

7. Gases leaving Gay-Lussac tower. The gases leaving the acid plant consist of SO_2, O_2, and N_2 from the burner and the oxides of nitrogen which are supplied by the nitric acid and lost from the system. Most of the SO_2 and a corresponding amount of oxygen are removed from the burner gases to form H_2SO_4. It is assumed that no water vapor leaves the Gay-Lussac tower.

SO_2 from burner	=	1.441 kg-moles
SO_2 forming H_2SO_4	=	1.42
SO_2 in gases leaving	=	0.021 kg-mole = 1.34 kg
O_2 in burner gases	=	1.70 kg-moles
O_2 used in oxidizing SO_2 = 1.42/2	=	0.71
O_2 from burner in gases leaving	=	0.99 kg-mole = 31.9 kg
N_2 in gases from burner	=	13.81 kg-moles = 389.8 kg
HNO_3 decomposed	=	0.0075 kg-mole

According to the reaction,

$$2HNO_3 = H_2O + 2NO + 3/2\ O_2$$

O_2 from HNO_3 = 0.0075(3/4)	=	0.0056 kg-mole = 0.18 kg
NO from HNO_3	=	0.0075 kg-mole = 0.22 kg

Total gases leaving:

SO_2	0.021 kg-mole	1.3 kg
O_2	0.995	31.9
NO	0.0075	0.22
N_2	13.81	389.8
Total	14.82 kg-moles	423.2 kg

Material Balance of Entire Plant

Input		Output	
Dry pyrites	100.0 kg	Cinder	71.2 kg
H_2O in pyrites	2.04	Spent gases (14.82 kg-moles)	423.2
Dry air (17.42 kg-moles)	506.00	Acid (78% H_2SO_4)	178.0
H_2O in air (0.176 kg-mole)	3.16	Total	672.4 kg
Nitric acid	1.31		
Spray water	58.4		
Total	670.9 kg		

This material balance is summarized in Fig 113.

Material Balance of Burner

Input			Output	
Dry ore	100	kg	Cinder	71.2 kg
Moisture in ore	2.04		Dry gases	536.4
Dry air (17.42 kg-moles)	506.		Water vapor (3.16 + 2.04) =	5.2
Moisture in air	3.16		Total	612.8 kg
Total	611.2	kg		

This material balance is summarized in Fig. 114.

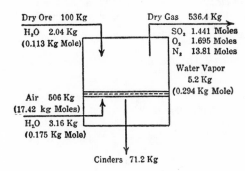

Dry Ore   100 Kg

H₂O   2.04 Kg
(0.113 Kg Mole)

Air   506 Kg
(17.42 kg Moles)

H₂O   3.16 Kg
(0.175 Kg Mole)

Dry Gas   536.4 Kg

SO₂   1.441 Moles
O₂   1.695 Moles
N₂   13.81 Moles

Water Vapor
5.2 Kg
(0.294 Kg Mole)

Cinders 71.2 Kg

FIG. 114. Material balance of pyrites burner

Energy Balance of Burner

The energy balance of the burner includes the heat of combustion of FeS_2 to Fe_2O_3 and SO_2 and the enthalpy of the water vapor in the air as the only important sources of heat. The energy output is distributed as sensible enthalpy of the outgoing cinders and dry gases and as enthalpy of outgoing water vapor. The heat losses include radiation from the dust chamber and flues up to the entrance of the Glover tower where the temperature of the burner gases, 450° C, was measured.

1. Heat evolved in the combustion of pyrites. The reaction involved in the combustion of pyrites and the corresponding standard heat of reaction are as follows:

$$4FeS_2 + 11O_2 = 2Fe_2O_3 + 8SO_2$$
$$\Delta H_{25} = 2(-196,500) + 8(-70,960) - 4(-42,520)$$
$$\Delta H_{25} = -790,600$$

FeS_2 actually burned = 0.721 kg-mole

Heat evolved in combustion of FeS_2 burned = $0.721 \times \dfrac{790,600}{4} = 142,510$ kcal

2. Enthalpy of water vapor in air

Heat of vaporization at 25° C = 10,513 kcal per kg-mole
Enthalpy = 0.176 (10,513) = 1850 kcal

3. Enthalpy of cinder = (71.2)(0.16)(450 − 25) = 4840 kcal.

4. Enthalpy of dry burner gas. Mean heat capacities between 25 and 450° C taken from Table 19, page 258.

$$SO_2 = (1.44)(11.08) = 15.96$$
$$O_2 = (1.70)(7.461) = 12.68$$
$$N_2 = (13.81)(7.124) = \underline{98.38}$$
$$Total = 127.0$$

Enthalpy $= (127.0)(450 - 25) = 53,980$ kcal

5. Enthalpy of water vapor in burner gases

Water vapor present $= 5.2/18 = 0.29$ kg-mole
Heat of vaporization at $25°$ C $= 10,513$ kcal per kg-mole
Mean molal heat capacity of water vapor $(25°$ C $- 450°$ C$) = 8.474$

Enthalpy $= 0.29[10,513 + 8.474(450 - 25)] = 4090$ kcal

Summary of Energy Balance of Burner

Input

	Kcal	%
1. Heat evolved in combustion of pyrites	142,510	98.7
2. Enthalpy of water vapor in air	1,850	1.3
Total input	144,360	100.0

Output

	Kcal	%
1. Enthalpy of cinder	4,840	3.4
2. Enthalpy of dry burner gas	53,980	37.4
3. Enthalpy of water vapor in gases	4,090	2.8
4. Heat losses (by difference)	81,450	56.4
Total output	144,360	100.0

Material Balance of Glover Tower

The functions of the Glover tower are as follows:

1. Cooling of burner gases before being blown into the chambers.
2. Conversion of about 16% of SO_2 in burner gas to H_2SO_4.
3. Mixing of Gay-Lussac and chamber acids and nitric acid.
4. Evolution of oxides of nitrogen from Gay-Lussac acid and from nitric acid.
5. Concentration of chamber acid.

At the top of the tower, mixing of the Gay-Lussac and chamber acids takes place with the release of the oxides of nitrogen from the Gay-Lussac acid when this acid is heated and diluted. Nitric acid is also added to make up for the losses of the oxides of nitrogen through incomplete absorption in the Gay-Lussac tower. The oxides of nitrogen are present as a mixture of NO and NO_2, but, since the equilibrium mixture gradually changes owing to variation in temperature and in oxygen content, it is assumed that the oxides of nitrogen leave the Glover tower as NO. The released oxides of nitrogen react in the vapor state with SO_2, oxygen, and water vapor, forming liquid nitrosyl–sulfuric acid

$$2NO + \tfrac{3}{2}O_2 + 2SO_2 + H_2O \rightarrow 2(NO_2SO_2OH)$$

Because of the high concentration of SO_2 in the Glover tower, the decomposition of this nitrosyl–sulfuric acid is complete, forming sulfuric acid according to the equation

$$2H_2O + 2(NO_2SO_2OH) + SO_2 = 3H_2SO_4 + 2NO$$

The second reaction proceeds much more rapidly than the first in moles per unit volume of space, chiefly because it is a reaction that proceeds in the liquid phase. The high concentration in the liquid state permits a more rapid rate of reaction, other conditions being the same. The final concentration of the Glover acid takes place at the hottest zone near the bottom of the tower, and the acid finally leaves as 78% H_2SO_4.

The input of the material balance includes the gases from the burner, the chamber acid, the make-up nitric acid, and the Gay-Lussac acid. The output includes the Glover acid and the gases leaving to enter the chambers.

Input to Glover Tower

Basis: 100 kg of moisture-free pyrites charged.

1. Dry burner gases = 16.95 kg-moles		=	536.4 kg
2. Water vapor in burner gases = 0.29 kg-mole		=	5.2
3. Chamber acid (64% H_2SO_4; 36% H_2O)		=	182
4. Nitric acid (36% HNO_3)		=	1.31
5. Gay-Lussac acid (77% H_2SO_4; 22.1% H_2O; 0.885% N_2O_3)		=	580
Total input		=	1304.9 kg

Output

1. Weight of Glover acid. The H_2SO_4 leaving the Glover tower as 78% acid includes that from the chamber and Gay-Lussac acids and that formed by conversion of SO_2 to H_2SO_4 in the tower already calculated to be 0.230 kg-mole.

H_2SO_4 from chambers = 182 × 0.64   = 116.4 kg
H_2SO_4 from Gay-Lussac acid = 580 × 0.77 = 446
H_2SO_4 formed in tower = 0.230 × 98   = 22.6
       Total H_2SO_4 = 585.0 kg

Weight of 78% acid leaving Glover tower $= \dfrac{585}{0.78} = 750$ kg

2. Weight of dry gases leaving the Glover tower. The weight and composition of the dry burner gases in passing through the Glover tower are changed, owing to the disappearance of some SO_2 and the corresponding amount of oxygen in the formation of SO_3 and to the evolution of NO and O_2 due to the decomposition of nitric acid and the release of the oxides of nitrogen from the Gay-Lussac acid.

Dry burner gases entering tower:

SO_2	=	1.44 kg-moles
O_2	=	1.70
N_2	=	13.81
Total	=	16.95 kg-moles

SO_2 converted to H_2SO_4 in tower   = 0.230 kg-mole
SO_2 leaving tower = 1.44 − 0.230   = 1.21 kg-moles
O_2 used in forming SO_3 = 0.230/2   = 0.115 kg-mole
O_2 remaining from burner gases = 1.70 − 0.115 = 1.58 kg-moles

Nitric acid decomposed $= 1.31 \times 0.36 = 0.472$ kg or   0.0075 kg-mole
$2HNO_3 = H_2O + 2NO + 1\frac{1}{2}O_2$
O_2 from $HNO_3 = 0.0075 \times \frac{3}{4}$ $= 0.0056$ kg-mole
NO from HNO_3 $= 0.0075$ kg-mole
H_2O from $HNO_3 = 0.0075/2$ $= 0.0037$ kg-mole

N_2O_3 from Gay-Lussac acid $= 580 \times 0.00885 = 5.13$ kg or $= 0.0675$ kg-mole
$N_2O_3 = 2NO + \frac{1}{2}O_2$
O_2 from $N_2O_3 = 0.0675/2$ $= 0.0338$ kg-mole
NO from N_2O_3 $= 0.1350$ kg-mole
Total O_2 leaving $= 1.58 + 0.0056 + 0.0338$ $= 1.62$ kg-moles
Total NO leaving $= 0.0075 + 0.1350$ $= 0.143$ kg-mole

Total dry gases leaving

SO_2 =	1.21 kg-moles or	$\times$ 64	=	77.5 kg
O_2 =	1.62 kg-moles or	$\times$ 32	=	51.8
NO =	0.143 kg-mole or	$\times$ 30	=	4.29
N_2 =	13.81 kg-moles or	$\times$ 28.2	=	390.0
Total	16.78 kg-moles or		=	523.6 kg

Total weight of dry gases leaving Glover tower $= 523.6$ kg

3. Water vapor in the gases leaving the Glover tower. The weight of water vapor in the gases leaving the Glover tower is calculated on the basis of a water balance. Water enters as vapor in the gases, as water in the Gay-Lussac and chamber acids, and is associated with the nitric acid as HNO_3 and as water. The 0.230 kg-mole of H_2SO_4 formed in the tower requires 0.230 kg-mole of H_2O or 4.15 kg. The 78% acid leaving requires $(750)(0.22) = 165$ kg water.

The 1.31 kg of 36% nitric acid charged yields upon dehydration and decomposition $0.84 + 0.07 = 0.91$ kg H_2O.

Water Balance

Input

In gas (0.289 kg-mole)	=	5.2 kg
In chamber acid (182) (0.36)	=	65.5
In Gay-Lussac acid 580×0.221	=	129.0
From nitric acid	=	0.9
Total	=	200.6 kg

Output

In 78% acid	165.0 kg
In formation of H_2SO_4	4.1
In water vapor (by difference)	31.5
Total	200.6 kg

Total water vapor in gases leaving $= 31.5$ kg
        or 31.5/18 $= 1.75$ kg-moles

Molal humidity of gases leaving $= \dfrac{1.75}{16.78} = 0.103$

Dew point of gases leaving (from Fig. 19) $= 113°$ F or $45°$ C

Summary of Material Balance of Glover Tower

Input

Dry gases (16.95 kg-moles)	536.4 kg
Water vapor (0.29 kg-mole)	5.2
Chamber acid (64% H_2SO_4)	182.0
Nitric acid (36% HNO_3)	1.31
Gay-Lussac acid	580.0
Total	1304.9 kg

Output

Dry gases	523.6 kg
Water vapor	31.5
Glover acid (78% H_2SO_4)	750.0
Total	1305.1 kg

This material balance is summarized in Fig. 115.

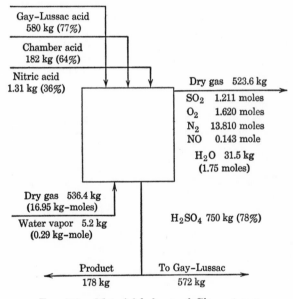

FIG. 115.   Material balance of Glover tower

Energy Balance of Glover Tower

In addition to the enthalpy of all materials entering and leaving, the input side of the energy balance of the Glover tower includes heat evolved in the formation of H_2SO_4 within the tower.   This acid goes into solution, but at the same time some water leaves the solution, the net effect being one of concentration and hence requiring the input of heat.   In addition, heat is required to remove the oxides of nitrogen from solution and to decompose them into NO and O_2.   It will be assumed that all oxides of nitrogen, both from the Gay-Lussac and the make-up nitric acids, are decomposed and leave the Glover tower as NO.

1. Heat evolved in formation of H₂SO₄. Sulfuric acid is formed from SO_2 gas, liquid H_2O, and oxygen. Actually the conversion takes place in two steps with intermediate formation of nitrosyl-sulfuric acid. However, the net effect is the same as though the reaction proceeded as follows:

$$SO_2(g) + \tfrac{1}{2}O_2(g) + H_2O(l) = H_2SO_4(l)$$

$$\Delta H_{25} = (-193{,}910) - [(-70{,}960) + (-68{,}317)]$$
$$\Delta H_{25} = -54{,}633 \text{ kcal}$$

Heat evolved in formation of $H_2SO_4 = 0.230(54{,}633) = 12{,}570$ kcal

2. Heat absorbed in concentrating acid. The sulfuric acid formed in the tower is diluted by the acid already present to form 78% acid leaving. The chamber acid entering at 64% is concentrated to 78%. The Gay-Lussac acid yields aqueous sulfuric acid containing 77/0.991 or 77.7% H_2SO_4. This is concentrated to 78% acid. The net result of these changes is a concentrating effect.

The net heat effect of the concentration changes in the Glover tower may be calculated from integral heat of solution data by the method discussed on page 319. The heat evolved is the difference between the total heat evolved in forming each of the entering acid solutions from H_2SO_4 and H_2O and that evolved in forming the solution leaving from H_2SO_4 and H_2O. These thermal effects may be calculated from the integral heat of solution data plotted in Fig. 116, page 463. This curve was plotted from data of the *International Critical Tables*, Vol. V, with permission. Molal integral heats of solution of H_2SO_4:

$$64\% \ H_2SO_4 \ = -11{,}8000 \text{ kcal per kg-mole}$$
$$77.7\% \ H_2SO_4 = - \ 8{,}700 \ \text{ kcal per kg-mole}$$
$$78\% \ H_2SO_4 \ = - \ 8{,}600 \ \text{ kcal per kg-mole}$$

Heat of solution of chamber acid $= \dfrac{182 \times 0.64}{98} \times -11{,}800 \quad = -14{,}000$ kcal

Heat of solution of Gay-Lussac acid $= \dfrac{580 \times 0.77}{98} \times -8700 \quad = -39{,}700$

Heat of solution of entering acids $\qquad\qquad\qquad\qquad = \overline{-53{,}700}$ kcal

Heat of solution of Glover acid leaving $= \dfrac{750 \times 0.78}{98} \times -8600 = -51{,}300$

Heat absorbed in concentration of acid $= 53{,}700 - 51{,}300 \qquad = \overline{+2{,}400}$ kcal

The net heat absorbed in concentration is 2400 kcal, which should be placed on the output side of the energy balance. The large amount of heat required for the concentration of chamber acid is offset by the heat evolved when the H_2SO_4 formed in the tower is dissolved.

3. Heat absorbed in release and decomposition of N₂O₃ from Gay-Lussac acid. The oxides of nitrogen enter the Glover tower as recovered nitrosyl–sulfuric acid from the Gay-Lussac tower and as make-up nitric acid. The Gay-Lussac acid may be considered to consist of 0.0677 kg-mole of N_2O_3 dissolved in 129 kg (7.15 kg-moles) of water. This concentration corresponds to 106 moles of water per mole of N_2O_3. The thermal effects in the evolution of the N_2O_3 from solution and its decomposition may be calculated from the following thermochemical data:

Formula	State	Heat of Formation, kcal per kg-mole	Moles of Water	Heat of Solution, kcal
NO	g	+ 21,600		
N_2O_3	g	+ 20,000	100	− 28,900
NO_2	g	+ 8,091		
N_2O_4	g	+ 2,309		
N_2O_5	g	+ 3,600		
HNO_3	l	− 41,404	6.22	−7,000

Heat *absorbed* in evolution of N_2O_3 from solution in the Gay-Lussac acid
= 0.0677 × 28,900 = 1960 kcal

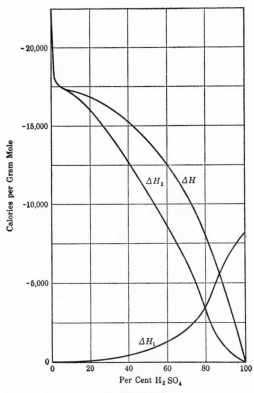

ΔH = integral heat of solution of H_2SO_4, cal per g-mole
ΔH_1 = differential heat of solution of H_2O, cal per g-mole
ΔH_2 = differential heat of solution of H_2SO_4, cal per g-mole

Fig. 116.   Differential and integral heats of solution of
aqueous solutions of sulfuric acid at 25° C

This entry is not exact since it has been assumed that the heat of solution of N_2O_3
is the same in 77.7% H_2SO_4 acid as it is in water, which is a poor approximation.

The N_2O_3 is assumed to break down entirely to NO and O_2; this is also an approximation. However, in view of the relatively small heat effects involved, these approximations seem justified.

$$N_2O_3 = 2NO + \tfrac{1}{2}O_2$$

$$\Delta H_{25} = 2(21,600) - (20,000) = 23,200 \text{ kcal}$$

Heat absorbed in decomposition of $N_2O_3 = 0.0677 \times 23,200 = 1570$ kcal
Total heat absorbed in release and decomposition of $N_2O_3 = 1960 + 1570$
    $= 3530$ kcal

4. Heat absorbed in decomposition of nitric acid. The nitric acid consists of 0.0075 kg-mole of HNO_3 dissolved in 0.0466 kg-mole of water, corresponding to 6.22 moles of water per mole of HNO_3. The acid may be considered to be separated into its liquid components, HNO_3, and water, and then decomposed according to the following reaction:

$$HNO_3(l) = \tfrac{1}{2}H_2O(l) + NO(g) + \tfrac{3}{4}O_2(g)$$

$$\Delta H_{25} = [\tfrac{1}{2}(-68,317) + (21,600)] - (-41,404) = 28,845 \text{ kcal}$$

Heat absorbed in separating HNO_3 from solution $= 0.0075 \times 7000 \quad = \quad 52$ kcal
Heat absorbed in decomposition of $HNO_3 = 0.0075 \,(28,845) \qquad = 216$
    Total heat absorbed $\qquad\qquad\qquad\qquad\qquad\qquad = \overline{268}$ kcal

The decomposition of the oxides of nitrogen is of particular interest because of their endothermic heats of formation.

5. Heat input in burner gases. This has already been calculated as the heat output of the burner (page 458) $= 58,070$ kcal
  6. Enthalpy of chamber acid   0 kcal
  7. Enthalpy of Gay-Lussac acid   0 kcal
  8. Enthalpy of nitric acid   0 kcal
  9. Enthalpy of dry gases leaving. Mean heat capacities between 25 and 91° C taken from Table 19, page 258.

$$
\begin{array}{llll}
SO_2 & (1.21)(9.81) & = & 11.87 \\
N_2 & (13.81)(6.97) & = & 96.26 \\
NO & (0.14)(7.14) & = & 1.00 \\
O_2 & (1.62)(7.08) & = & \underline{11.70} \\
& & & 120.8
\end{array}
$$

Enthalpy $= 120.8(91 - 25) = 7970$ kcal

10. Total enthalpy of water vapor leaving Glover tower

Heat of vaporization at 25° C $= 10,513$ kcal per kg-mole
Mean molal heat capacity of vapor between 25 and 91° C $= 8.077$

Enthalpy $= 1.75[10,513 + 8.077(91 - 25)] = 19,330$ kcal

11. Heat absorbed in cooling outgoing acid

Heat capacity (Fig. 117) $= 0.44$ kcal per kg per C°, neglecting the temperature coefficient of the heat capacity
Heat absorbed $= 750 \times 0.44(125 - 25) = 33,000$ kcal

12. Enthalpy of outgoing acid   0 kcal

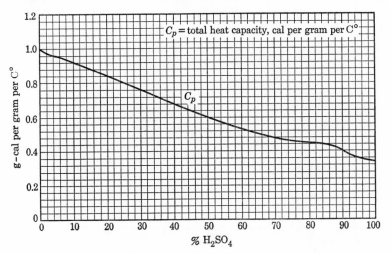

Fig. 117. Heat capacity of aqueous solutions of sulfuric acid at 20° C

Summary of Energy Balance of Glover Tower

Input

	kcal	%
Enthalpy of dry burner gases	53,980	76.4
Enthalpy of water vapor from burner	4,090	5.8
Enthalpy of chamber acid	0	0.0
Enthalpy of Gay-Lussac acid	0	0.0
Enthalpy of nitric acid	0	0.0
Heat evolved in formation of H_2SO_4	12,570	17.8
Total	70,640	100.0

Output

Enthalpy of dry gases	7,970	11.3
Enthalpy of water vapor in gases	19,330	27.4
Enthalpy of acid leaving cooler	0	0.0
Heat absorbed in concentration of acid	2,400	3.4
Heat absorbed in decomposition of nitric acid	270	0.4
Heat absorbed in release and decomposition of N_2O_3 from Gay-Lussac acid	3,530	5.0
Heat absorbed by outgoing acid cooler	33,000	46.6
Heat losses (by difference)	4,140	5.9
Total	70,640	100.0

It will be seen that over one half of the available energy in the Glover tower is absorbed by the concentrated acid. This acid must be cooled before it can be used for absorption of gases in the Gay-Lussac tower and before storage. The cooling of this acid represents one of the difficult problems in acid manufacture, in the development of a heat interchanger which will withstand hot concentrated sulfuric acid and at the same time permit a rapid transfer of heat.

Material Balance of Chambers

It would be proper to consider the material and energy balances of each chamber separately, but to avoid needless repetition all four chambers will be considered as a unit. In the chambers, H_2O, SO_2, and O_2 are removed from the gases to form H_2SO_4.

Basis: 100 kg of dry pyrites charged.

H_2SO_4 formed in chambers = 116.5 kg or 1.19 kg-moles

	Gas Entering		Gas Removed		Dry Gas Leaving
SO_2	1.21	kg-moles	1.19	kg-moles	0.021 kg-mole
O_2	1.62		0.595		1.025
NO	0.143				0.143
N_2	13.81				13.81
H_2O	1.75		1.19		
	Total				15.00 kg-moles

Water vapor leaving. The water entering with the gases and from the sprays is used in the formation of H_2SO_4 and dilution of the H_2SO_4 to form the chamber acid.

Water Input

H_2O from gases	31.5 kg
H_2O from sprays	58.4
Total	89.9 kg

Water Output

H_2O to form H_2SO_4 = 1.19 × 18	= 21.4	kg
H_2O in chamber acid = 182 × 0.36	= 65.5	
Output accounted for	= 86.9	kg
H_2O in gases leaving = 89.9 − 86.9	= 3.0	kg
or 3.0/18	= 0.167	kg-mole
Moles of dry gases leaving	= 15.0	kg-moles
Molal humidity = 0.167/15.0	= 0.011	
Dew point (Fig. 20)	= 46° F or 8° C	

Summary of Material Balance of Chambers

Entering			Leaving		
SO_2 (1.21 kg-moles)	77.5 kg		SO_2 (0.021 kg-mole)	1.3 kg	
O_2 (1.62 kg-moles)	51.8		O_2 (1.025 kg-moles)	32.8	
N_2 (13.81 kg-moles)	390.0		N_2 (13.81 kg-moles)	390.0	
H_2O (1.75 kg-moles)	31.5		H_2O (0.168 kg-mole)	3.0	
NO (0.143 kg-mole)	4.3		NO (0.143 kg-mole)	4.3	
Spray water	58.4		Acid	182.0	
Total	613.5 kg		Total	613.4 kg	

This material balance is summarized in Fig. 118.

Energy Balance of Chambers

The first reaction proceeding in the chambers is a gaseous reaction among the SO_2, H_2O, O_2, and NO gases in contact with the water spray forming nitrosyl–sulfuric acid.

$$2SO_2 + 2NO + 1\tfrac{1}{2}O_2 + H_2O \rightarrow 2(NO_2SO_2OH)$$

This reaction is favored by high concentrations of SO_2 and NO. The acid spray is swept against the side walls of the chamber where the spray is condensed owing to cooling, and by dilution with water H_2SO_4 is formed with the release of the oxides of nitrogen.

$$2(NO_2SO_2OH) + H_2O = 2H_2SO_4 + N_2O_3$$
$$N_2O_3 = 2NO + \tfrac{1}{2}O_2$$

The first of these reactions proceeds in the liquid phase and is favored by a high concentration of water brought in by the spray and condensed upon the side walls.

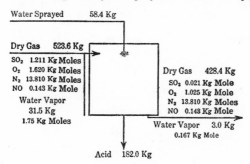

FIG. 118. Material balance of chambers

In calculating the heat evolved in the chamber reactions, only the final, net effects need be considered. It will be assumed that the oxides of nitrogen ultimately leave the chambers in the same form in which they entered as NO. This assumption is not exact because some oxidation of NO to N_2O_3 probably takes place at the relatively low temperatures of the last chamber. The ultimate effects of the reactions in the chambers are then the production of H_2SO_4 from SO_2, O_2, and H_2O and the dissolution of this acid to form an aqueous solution containing 64% H_2SO_4.

Reference temperature: 25° C.

Basis: 100 kg of dry pyrites charged.

1. Heat evolved in formation of H_2SO_4. From item 1 of the energy balance of the Glover tower, the heat evolved is 54,633 kcal per kg-mole of H_2SO_4 formed.

H_2SO_4 formed in chambers   = 1.19 kg-moles
Heat evolved = 1.19 × 54,633 = 65,010 kcal

2. Heat evolved in dissolving H_2SO_4. Integral heat of solution (Fig. 116) at a concentration of 64% H_2SO_4 = 11,800 kcal per kg-mole.

Heat evolved in dissolution = 11,800 × 1.19 = 14,040 kcal

3. Enthalpy of dry gases and water vapor entering. This has already been calculated as part of the heat output of the Glover tower as 27,300 kcal.

4. Enthalpy of spray water. Since the spray water enters at the reference temperature, 25° C, its enthalpy is equal to 0 kcal.

5. Enthalpy of dry gases leaving. Mean heat capacities between 25 and 40° C taken from Table 19, page 258.

$$SO_2 \quad (0.021)(9.60) \quad = \quad 0.2$$
$$O_2 \quad (1.025)(7.03) \quad = \quad 7.2$$
$$N_2 \quad (13.81)(6.96) \quad = \quad 96.1$$
$$NO \quad (0.14)(7.14) \quad = \quad \underline{1.0}$$
$$Total \quad\quad\quad = \quad 104.4$$

Enthalpy of dry gases $= (104.4)(40 - 25) = 1570$ kcal

6. Enthalpy of water vapor leaving

Heat of vaporization at $25°$ C (Fig. 21) $= 10,513$ kcal per kg-mole.
Total enthalpy $= 0.167[10,513 + 8.036(40 - 25)] = 1780$ kcal

7. Heat absorbed in cooling the acid leaving.
Heat capacity (Fig. 117) of acid containing 64% $H_2SO_4 = 0.50$ kcal per kg per ° C.

Enthalpy $= 0.50 \times 182(68 - 25) = 3910$ kcal

8. Enthalpy of acid leaving the cooler $= 0$ kcal.

Summary of Energy Balance of Chambers

Input

	Kcal	%
Enthalpy of dry gases from Glover tower	7,970	7.5
Enthalpy of water vapor in gases entering	19,330	18.2
Enthalpy of spray water		0.0
Heat evolved in forming H_2SO_4	65,010	61.1
Heat evolved in dissolving H_2SO_4	14,040	13.2
Total	106,350	100.0

Output

	Kcal	%
Enthalpy of dry gases leaving	1,570	1.5
Enthalpy of water vapor in gases	1,780	1.7
Enthalpy of acid leaving the cooler	0	0.0
Heat absorbed by cooler	3,910	3.7
Heat loss from chambers (by difference)	99,090	93.1
Total	106,350	100.0

It will be seen that nearly the entire source of energy (74.3%) comes from the formation of H_2SO_4 and its dilution, whereas nearly the entire energy input is lost by radiation from the extensive lead surfaces of the chambers. More recent developments in the chamber process have provided for more rapid means of heat removal with much less floor space and size of equipment by rapid circulation of both gases and acid in packed towers.

Material Balance of Gay-Lussac Tower

The purpose of the Gay-Lussac tower is to recover the oxides of nitrogen from the chamber gases. These oxides are then returned to the system in the Glover tower. The water remaining in the chamber gases is also absorbed. The conditions favorable for absorption of the oxides of nitrogen are a high concentration of acid, a low temperature, and a low concentration of SO_2 in the residual gas. A high water content in the gas from the chambers causes objectionable dilution of the acid. Small amounts of SO_2 will cause decomposition of nitrosyl–sulfuric acid with release and loss of the oxides of nitrogen. The presence of oxygen is essential to effect the oxidation of NO to N_2O_3 and its absorption in the acid.

The loss of oxides of nitrogen in the gases from the Gay-Lussac tower may be assumed to be equivalent to the make-up nitric acid introduced. It will be assumed that these oxides leave in the form of NO.

1. Input. The input to the Gay-Lussac tower consists of the wet gases from the chambers and the Glover acid which is introduced. All the gases pass through the tower unchanged with the exception of the NO and O_2.

2. NO in gas leaving = 0.008 kg-mole or = 0.2 kg

3. O_2 in gases leaving:

NO oxidized to N_2O_3 = 0.143 − 0.008   = 0.135 kg-mole
O_2 consumed = 0.135/4   = 0.034 kg-mole
O_2 leaving = 1.025 − 0.034 = 0.991 kg-mole   = 31.7   kg

Summary of Material Balance of Gay-Lussac Tower

Input			Output		
SO_2	(0.021 kg-mole)	1.3 kg	SO_2	(0.021 kg-mole)	1.3 kg
O_2	(1.025 kg-moles)	32.8	O_2	(0.991 kg-mole)	31.7
N_2	(13.81 kg-moles)	390.0	N_2	(13.81 kg-moles)	390.0
NO	(0.143 kg-mole)	4.3	Acid leaving		580.0
H_2O	(0.167 kg-mole)	3.0	NO (0.008 kg-mole)		0.2
Glover acid		572.0	Total		1003.2 kg
	Total	1003.4 kg			

This material balance is summarized diagrammatically in Fig. 119.

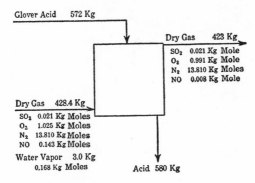

FIG. 119. Material balance of Gay-Lussac tower

Energy Balance of Gay-Lussac Tower

1. Heat evolved in forming and dissolving N_2O_3. The N_2O_3 released from the Gay-Lussac acid and decomposed in the Glover tower is re-formed and recovered in the Gay-Lussac tower evolving 3530 kcal calculated as part of the energy balance of the Glover tower, page 465.

2. Heat evolved in the dissolution of the water vapor absorbed. The water vapor leaving the last chamber is completely absorbed in the Gay-Lussac tower. Since the resulting concentration change in this absorption is negligible, it is necessary to determine the heat evolved in the dissolution of the water from data on the differential heat of solution of water in sulfuric acid solutions. From Fig. 116, the differential molal heat of solution of water in a sulfuric acid solution containing 77.7% H_2SO_4 is 3100 kcal per kg-mole. This value neglects the effect of the dissolved oxides of nitrogen.

$$\text{Heat evolved} = 3100 \times 0.167 = 520 \text{ kcal}$$

3. Enthalpy of Glover acid introduced $= 0$ kcal

4. Enthalpy of entering gases. Already calculated as output items in the energy balance of the chambers $= 3350$ kcal.

5. Enthalpy of gases leaving. Mean heat capacities between 25 and 30° C taken from Table 19, page 258.

$$
\begin{array}{lll}
SO_2 & (0.021)(9.56) = & 0.12 \\
O_2 & (0.99)(7.02) = & 6.95 \\
N_2 & (13.81)(6.96) = & 96.12 \\
NO & (0.01)(7.13) = & \underline{0.07} \\
& & 103.3
\end{array}
$$

Enthalpy $= (103.3)(30 - 25) = 520$ kcal

6. Heat lost in pipe lines between Gay-Lussac tower and Glover tower

$580 \times 0.45(27 - 25) = 520$ kcal

Summary of Energy Balance of Gay-Lussac Tower
(including pipe line to Glover tower)

Input	Kcal	%
Enthalpy of dry gases entering	1570	21.2
Enthalpy of water vapor entering	1780	24.1
Enthalpy of Glover acid entering	0	0.0
Heat evolved in forming N_2O_3	1570	21.2
Heat evolved in dissolving N_2O_3	1960	26.5
Heat evolved in dissolving H_2O	520	7.0
Total	7400	100.0

Output	Kcal	%
Enthalpy of gases leaving	520	7.0
Heat losses from pipe line	520	7.0
Enthalpy of acid to Glover tower	0	0.0
Heat loss (by difference)	6360	86.0
Total	7400	100.0

Summarized Energy Balance for Entire Plant

The summarized energy balance for the entire plant is obtained from the net input and output energy items taken from the energy balances of the separate parts without additional calculations.

Reference temperature: 25° C

Basis: 100 kg of dry pyrites charged.

Input

	Kcal	%
1. Heat evolved in combustion of pyrites	142,510	60.2
2. Enthalpy of water vapor in air	1,850	0.8
3. Heat evolved in formation of H_2SO_4 in Glover tower	12,570	5.3
4. Sensible enthalpy of nitric acid	0	0.0
5. Enthalpy of water spray	0	0.0
6. Heat evolved in formation of H_2SO_4 in chamber	65,010	27.5
7. Heat evolved in solution of H_2SO_4 in chamber	14,040	6.0
8. Heat evolved in solution of H_2O in Gay-Lussac tower	520	0.2
Total	236,500	100.0

Output

	Kcal	%
1. Enthalpy of cinder	4,840	2.0
2. Net heat absorbed in concentrating acid in Glover tower	2,400	1.0
3. Concentration and decomposition of nitric acid	270	0.1
4. Radiation losses from burners	81,450	34.4
Radiation losses from Glover tower	4,140	1.8
Radiation losses from chamber	99,090	41.9
Radiation losses from Gay-Lussac tower	6,360	2.7
5. Cooling of Glover acid	33,000	14.0
Cooling of chamber acid	3,910	1.7
Cooling of Gay-Lussac acid in pipe line	520	0.2
6. Enthalpy of acid product	0	0.0
7. Enthalpy of spent gases leaving Gay-Lussac tower	520	0.2
Total	236,500	100.0

Material and Energy Balances of a Blast Furnace

A blast furnace is essentially a huge gas producer where, in conjunction with the partial combustion and distillation of a carbonaceous fuel, the reduction of ore and the formation of slag occur simultaneously. The charge, consisting of iron ore, coke, and limestone in proper proportions, is fed into the top of the blast furnace. Preheated air, preferably free from water vapor, is blown through the tuyères near the bottom of the furnace into the descending stream of materials. This results in combustion of the coke to carbon dioxide. The carbon dioxide gas in the presence of excess coke is reduced at the high prevailing temperature to carbon monoxide. A great many chemical reactions occur within the furnace. As the charge descends the shaft and as its temperature is gradually increased, dehydration of the ore, coke, and limestone takes place, followed by distillation of the remaining volatile matter of the coke, calcination of magnesium and calcium carbonates present in the limestone or ore, and reduction of the higher oxides of manganese and iron to manganous and ferrous oxides by the rising stream of reducing gases. The carbon dioxide formed by reduction of the ore with carbon monoxide is reduced in the presence of excess coke. As the temperature of the descending charge becomes still higher, the lower oxides of iron and manganese are reduced to the metallic state.

At the highest temperature of the tuyères, part of the silica present is reduced to the metallic state and is dissolved in the molten iron. The excess silica and alumina of the charge are fluxed by reaction with the metallic bases present, resulting in the formation of a fusible slag consisting of complex silicates and aluminates of calcium, magnesium, and iron. The high temperature at the tuyères produces a fluid slag and molten metal which readily flow through the solid reacting charge, which separate into two layers at the bottom of the furnace, and which are periodically run out in two separate streams as molten pig iron and as molten slag. A high temperature at the tuyères favors a ready separation of the slag and removal, as CaS in the slag, of much sulfur which was originally present in the coke. The temperature of the blast furnace is too low and insufficient coke is present to reduce the oxides of calcium, magnesium, and aluminum and the silicates. Hence, these compounds pass into the slag. Silica is used as a flux in a few exceptional cases where the alkaline earths and alumina predominate in the gangue present in the ore.

The purpose of preheating the air used in combustion of the coke is to permit the attainment of the high temperatures necessary for the final reduction of the ore and the fusion of the pig iron and slag. Any water vapor present in the incoming air will lower the temperature in the fusion zone on account of the heat absorbed in its reduction to hydrogen and carbon monoxide. For this reason it is desirable to use a blast of dried air.

The products of the blast furnace consist of molten pig iron, slag, and blast-furnace gas. The outgoing gas consists essentially of nitrogen, carbon monoxide, carbon dioxide, and water vapor, with small amounts of hydrogen and methane, and also carries in suspension a considerable amount of dust. The heating value of this gas is low because of its high content of nitrogen and the small amount of volatile matter in the coke which is used for reduction. The free moisture in the incoming charge is distilled off near the top of the furnace and escapes into the blast-furnace gas without reduction.

The slag contains all the lime, magnesia, alumina, and alkalies originally present in the ore and flux, together with most of the silica and some ferrous and manganous oxides. The exact mineralogical compositions of ore, flux, and slag are not known completely, so that some uncertainty exists concerning the exact thermal energy involved in reduction and chemical transformations. The molten pig iron contains, in addition to iron, some carbon present in solution and lesser amounts of silicon and manganese. In the solid pig iron the carbon normally exists partly as free graphite and partly in the combined form as Fe_3C.

In order to establish the energy balance of a blast furnace it is necessary to know the masses and chemical compositions of the ore, flux, dust, and pig iron, and the analysis of the dry blast-furnace gas. The masses of slag, air, and water vapor can then be calculated.

The material balance includes:

Input	Output
Iron ore	Dry gases
Flux	Water vapor in gases
Coke	Pig iron
Air	Dust
Water vapor	Slag

As an illustration of the calculations involved in the material and energy balance of a blast furnace, the data for the reduction of a basic iron ore with charcoal and an acid flux will be given. An example of the more usual operation with a limestone flux is given in the problems at the end of this chapter. The balances are worked out on a basis of 100 kg of pig iron produced.

Illustration 2. A blast furnace is charged with 212.7 kg of ore, 110.0 kg of charcoal, and 13.9 kg of flux per 100 kg of pig iron produced. The compositions of these materials are as follows:

Ore (212.7 kg)		Charcoal (110.0 kg)	
Fe_2O_3	54.93%	C	86.89%
FeO	8.48	O	3.15
CaO	9.58	H	0.45
Mn_3O_4	4.97	N	0.51
Al_2O_3	3.00	H_2O	7.00
MgO	1.83	Ash	2.00
SiO_2	4.92		100.00%
H_2O	4.48		
CO_2	7.81		
	100.00%		

Flux	(13.9 kg)		Pig iron	(100.0 kg)
SiO_2	78.38%		C	3.12%
Al_2O_3	13.99		Si	1.52
CaO	0.53		Mn	2.22
Fe_2O_3	3.90		Fe	93.14
H_2O	3.20			100.00%
	100.00%			

The total heating value of the charcoal is 7035 kcal per kg.

The clean gas produced has the following composition by volume on the moisture-free basis:

CO_2	12.62%
CO	25.56
CH_4	0.69
H_2	1.34
N_2	59.79
	100.00%

The ore, flux, and charcoal are charged to the furnace at an average temperature of 25° C. The air blast is dried and enters the tuyères at a temperature of 300° C and moisture-free.

The gases leave the furnace at a temperature of 173° C and contain only negligible quantities of dust.

The slag and pig iron are poured at an average temperature of 1360° C.

In order to cool the outside of the bosh of the furnace and thereby protect the refractories from excessive heating, water is circulated in a pipe passing around the circumference of the bosh. On the basis of 100 kg of pig iron produced, 576 kg of water are circulated and heated through a temperature rise of 13° C.

Calculate the complete material and energy balances of this furnace.

Material Balance

The mineralogical composition of the ore is not given, although it is customary to assume that the carbon dioxide present is combined with the lime and the magnesia present as limestone. This limestone is calcined at about 900° C to CaO, MgO, and CO_2. The silica is present chiefly as silicates of aluminum and magnesium. Any free moisture present in the ore is driven off at the top of the furnace without reduction. However, the chemically combined water, as in the minerals kaolinite and limonite, will be retained until the ore reaches a hotter zone, where it will be partly reduced by coke to hydrogen and carbon monoxide. It will be assumed that all the elements of the carbon dioxide and water of the ore leave the blast furnace in the gases.

The alumina, lime, and magnesia pass into the slag without reduction but combine with the silica of the flux to form complex silicates of calcium, magnesium, and aluminum. The exact thermal energy of this latter change is unknown since the mineralogical composition of the slag as well as the heats of formation of complex silicates are unknown.

The higher oxides of iron are reduced at a relatively low temperature to the lower oxides. In the hot zone of the furnace the oxides are further reduced to the metallic state to supply the iron requirements of the pig iron. It is assumed that the remaining iron, as ferrous oxide, passes into the slag forming ferrous silicate. This assumption is inexact because it is known that a part of the iron exists as metallic particles included in the slag.

Sufficient silica is reduced to metallic silicon to supply the silicon content of the pig iron. The remainder combines with the basic oxides to form silicates in the slag.

The Mn_3O_4 of the ore is in part reduced to metallic manganese, supplying that present in the pig iron. The remainder of the manganese is assumed to enter the slag as MnO, forming silicates.

The oxygen given up in the reduction of the oxides of iron, silicon, and manganese will be present in the gases as CO, CO_2, or H_2O. The gases also contain the CO_2 and H_2O of the ore.

Basis: 100 kg pig iron produced.

1. Weight of slag formed. The weight of slag may be determined by setting up material balances for Fe, Mn, Al_2O_3, CaO, and MgO. Before setting up the detailed material balances, it will prove convenient to compute molal quantities of certain substances that are involved not only in the material balances pertaining to the evaluation of slag weight, but in subsequent ones as well.

$$Fe_2O_3 \text{ in ore } = \frac{212.7 \times 0.5493}{159.70} = 0.73160 \text{ kg-mole}$$

$$Fe_2O_3 \text{ in flux } = \frac{13.9 \times 0.0390}{159.70} = 0.003394 \text{ kg-mole}$$

$$Mn_3O_4 \text{ in ore } = \frac{212.7 \times 0.0497}{228.79} = 0.04620 \text{ kg-mole}$$

(a) *Iron Balance* (kg-atoms Fe or kg-moles FeO)

Input	Kg-atoms	Output	Kg-atoms
1. From Fe_2O_3 of ore: 0.73160 × 2 = 1.4632		1. To pig iron: $\frac{100 \times 0.9314}{55.85}$ = 1.6677	
2. From FeO of ore: $\frac{212.7 \times 0.0848}{71.85}$ = 0.2510		2. To slag (by difference) = 0.0533	
3. From Fe_2O_3 of flux: 0.003394 × 2 = 0.0068			
	1.7210		1.7210

Weight of FeO in slag = 0.0533 × 71.85 = 3.83 kg

(b) *Manganese Balance* (kg-atoms Mn or kg-moles MnO)

Input	Kg-atoms	Output	Kg-atoms
1. From Mn_3O_4 of ore: 0.04620 × 3 = 0.1386		1. To pig iron: $\frac{100 \times 0.0222}{54.93}$ = 0.0404	
		2. To slag (by difference) = 0.0982	
	0.1386		0.1386

Weight of MnO in slag = 0.0982 × 70.93 = 6.97 kg

(c) *Alumina Balance* (kg Al_2O_3)

Input	Kg	Output	Kg
1. From ore:		1. To slag (by dif-	
212.7 × 0.0300	= 6.38	ference)	= 8.32
2. From flux:			
13.9 × 0.1399	= 1.94		
	8.32		8.32

(d) *Silicon Balance* (kg-atoms Si)

Input	Kg-atoms	Output	Kg-atoms
1. From SiO_2 of ore:		1. To pig iron:	
$\dfrac{212.7 \times 0.0492}{60.06}$	= 0.1742	$\dfrac{100 \times 0.0152}{28.06}$	= 0.0542
2. From SiO_2 of flux:		2. To slag (by dif-	
$\dfrac{13.9 \times 0.7838}{60.06}$	= 0.1814	ference)	= 0.3014
	0.3556		0.3556

Weight of SiO_2 in slag = 0.3014 × 60.06 = 18.10 kg

(e) *Calcium Oxide Balance* (kg CaO)

Input	Kg	Output	Kg
1. From ore:		1. To slag (by dif-	
212.7 × 0.0958	= 20.38	ference)	= 20.45
2. From flux:			
13.9 × 0.0053	= 0.07		
	20.45		20.45

(f) *Magnesium Oxide Balance* (kg MgO)

Input	Kg	Output	Kg
1. From ore:		1. To slag (by dif-	
212.7 × 0.0183	= 3.89	ference)	= 3.89

Besides the materials derived from the ore and the flux, the slag contains the ash originally present in the charcoal.

In the following tabulation the weights of the various oxides that formed the slag are given. It should, however, be understood that the slag is not a simple mixture of the pure oxides but rather is a mixture of complex compounds formed from the simple oxides.

Component	Weight, kg	%
FeO	3.83	6.0
MnO	6.97	10.9
Al_2O_3	8.32	13.0
SiO_2	18.10	28.4
CaO	20.45	32.1
MgO	3.89	6.1
Ash 110.0 × 0.0200	2.20	3.5
	63.76	100.0

2. Weight of dry blast-furnace gas formed. A carbon balance will serve to establish the weight of dry blast-furnace gas.

Carbon Balance (kg-atoms carbon)

Input	Kg-atoms	Output	Kg-atoms
(a) From charcoal:		(a) To pig iron:	
$\dfrac{110.0 \times 0.8689}{12.01} = 7.958$		$\dfrac{100 \times 0.0312}{12.01} = 0.260$	
(b) From ore:		(b) To gases (by difference) $= 8.075$	
$\dfrac{212.7 \times 0.0781}{44.01} = 0.377$			
	8.335		8.335

Carbon in 1 kg-mole dry blast-furnace gas
$$= 0.1262 + 0.2556 + 0.0069 = 0.3887 \text{ kg-atoms}$$
$$\text{Dry blast-furnace gas} = \frac{8.075}{0.3887} = 20.774 \text{ kg-moles}$$

Constituent	Kg-moles		Kg	
CO_2	$20.774 \times 0.1262 =$	2.6217	2.622×44.01	$= 115.39$
CO	$20.774 \times 0.2556 =$	5.3098	5.310×28.01	$= 148.73$
CH_4	$20.774 \times 0.0069 =$	0.1433	0.143×16.04	$= 2.29$
H_2	$20.774 \times 0.0134 =$	0.2784	0.278×2.016	$= 0.56$
N_2	$20.774 \times 0.5979 =$	12.4208	12.421×28.2	$= 350.27$
		20.774		617.24

$$\text{Molecular weight} = \frac{617.24}{20.774} = 29.71$$

3. Weight of air supplied. A nitrogen balance will establish the weight of dry air supplied.

Nitrogen Balance (kg-moles N_2)

Input	Kg-moles	Output	Kg-moles
1. From charcoal:		1. To dry blast-furnace	
$\dfrac{110 \times 0.0051}{28} = 0.0200$		gas $= 12.4208$	
2. From air (by difference) $= 12.4008$			
	12.4208		12.4208

$$\text{Dry air} = \frac{12.4008}{0.79} = 15.6972 \text{ kg-moles}$$
$$\text{or } 15.6972 \times 29.0 = 455.22 \text{ kg}$$

The air was dried to a low humidity before being admitted to the preheaters that supply the blast furnace with air. The moisture content of the air is therefore negligible.

4. Weight of water vapor in the blast-furnace gas. A hydrogen balance will give the weight of the water vapor appearing in the gas emerging from the top of the blast furnace.

Hydrogen Balance (kg-moles H_2)

Input	Kg-moles
1. From H_2O of ore:	
$\dfrac{212.7 \times 0.0448}{18.016} = 0.5289$	
2. From H_2O of charcoal:	
$\dfrac{110.0 \times 0.0700}{18.016} = 0.4274$	
3. From available hydrogen of charcoal:	
$\dfrac{110.0 \times 0.0045}{2.016} = 0.2455$	
4. From H_2O of flux:	
$\dfrac{13.9 \times 0.0320}{18.016} = 0.0247$	
	1.2265

Output	Kg-moles
1. To dry blast-furance gas:	
(a) In CH_4:	
0.1433×2	= 0.2866
(b) Free hydrogen	= 0.2784
2. To H_2O in blast furnace gas (by difference)	= 0.6615
	1.2265

Weight H_2O in blast-furnace gas = $0.6615 \times 18.016 = 11.92$ kg

5. Over-all material balance

Input	Kg	Output	Kg
1. Ore	212.7	1. Pig iron	100.0
2. Charcoal	110.0	2. Slag	63.8
3. Flux	13.9	3. Dry Gas	617.2
4. Air	455.2	4. Water vapor in gas	11.9
	791.8		792.9

This particular set of operating data gives a close check between the input and output tabulations, the difference being only 1.1 kg, or 0.14%. Experimental error in the data frequently produces discrepancies of several per cent between the input and output totals.

It will be noted that no oxygen balance was employed in the foregoing calculations. Since all of the material balances were completed by difference, all of the errors will accumulate in the oxygen balance, and the error will be the same as the difference between the input and output totals of the over-all material balance. In this particular case, the error in the oxygen balance will be small, as the input and output totals for the over-all material balance are in close agreement. Since the sources of oxygen and the utilization of oxygen are of some interest, an oxygen balance is included below. In setting up this balance, compounds that go to the slag unchanged in oxygen content (most of the SiO_2, some FeO, all Al_2O_3, all CaO, all MgO) are not included in the tabulation.

6. Oxygen balance (kg-atoms oxygen)

Input	Kg-atoms
(a) From ore:	
1. Reduction of Fe_2O_3:	
$0.7316 \times 3 = 2.1948$	

Output	Kg-atoms
(a) To blast-furnace gas:	
1. In CO_2:	
2.6217×2	= 5.2434

Input	Kg-atoms	Output	Kg-atoms
2. Reduction of FeO:		2. In CO	= 5.3098
$0.2510 - 0.0533 = 0.1977$		3. In H_2O	= 0.6615

3. Reduction of Mn_3O_4:
   $(0.04620 \times 4) - 0.0982 = 0.0866$

4. Reduction of SiO_2:
   $0.0542 \times 2 = 0.1084$

5. From H_2O:
   $$\frac{212.7 \times 0.0448}{18.016} = 0.5289$$

6. From CO_2:
   $$\frac{212.7 \times 0.0781}{44.01} \times 2 = 0.7549$$

(b) From charcoal:

1. Oxygen:
   $$\frac{110.0 \times 0.0315}{16} = 0.2166$$

2. From H_2O:
   $$\frac{110.0 \times 0.0700}{18.016} = 0.4274$$

(c) From flux:

1. Reduction of
   Fe_2O_3:
   $0.003394 \times 3 \quad = \quad 0.0102$

2. From H_2O:
   $$\frac{13.9 \times 0.0320}{18.016} = 0.0247$$

(d) From air:
   $15.6972 \times 0.21 \times 2 = \quad \underline{6.5928}$

                                                     11.1430   11.2147

The output side of the oxygen balance is greater than the input side by 0.0717 kg-mole, or 1.15 kg, which closely checks the discrepancy between the output and input sides of the over-all material balance. The material balance is summarized in Fig. 120.

Energy Balance

An energy balance might be established by considering the enthalpies and heats of formation of all components of the charge and all components of the slag, pig iron, and furnace gas together with the heat loss by radiation. Such an energy balance would be disproportionate since the chemical energies of formation of the oxides and silicates which pass through the process unchanged, such as the oxides of aluminum, silicon, calcium, and magnesium, are of no interest.

During the course of reduction many intermediate chemical reactions take place, each accompanied by a certain thermal change. Examples are the progressive reduction of the higher oxides of iron and manganese, the oxidation of carbon at the tuyères and its subsequent reduction by coke to carbon monoxide, the reduction of metallic oxides by carbon and by carbon monoxide, and the reaction of water to carbon monoxide, hydrogen, and carbon dioxide. However, in any chemical process the total change in energy is dependent only on the initial and final states of chemical constitution, temperature, pressure, and state of aggregation and is independent of any intermediate state. Hence, in calculating the energy balance of a blast furnace, the numerous intermediate reactions involved need not be considered. It is sufficient to know the temperature, state of aggregation, and composition of each material charged and each product leaving, without knowing how the various components of the products are actually produced.

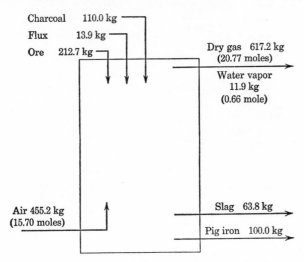

FIG. 120.   Material balance of a blast furnace

The energy balance must be set up in such a way that the fundamental equation

$$H_R = H_P + \Delta H_{25} + Q$$

is satisfied. In this equation,

H_R = enthalpy of all entering streams of material, relative to 25° C and the selected reference states for the energy balance

H_P = enthalpy of all leaving streams of material, relative to 25° C and the selected reference states for the energy balance

ΔH_{25} = the over-all standard heat of reaction for the process

Q = heat losses

There is little trouble in deciding what items to include for H_R, H_P, and Q in the detailed balance, and in evaluating each from the operating data. The chief difficulty arises in handling the ΔH_{25} item properly. The best method of handling this part of the energy-balance calculations is to break down the over-all process into several steps, which, if summated, would be equivalent to the actual over-all process. This breakdown is as follows:

Decomposition Processes. For purposes of calculation, it will be assumed that the oxides decompose into metals, lower oxides, and oxygen. This does not actually occur; the removal of oxygen from the oxides is really accomplished by reaction with carbon monoxide. This, however, does not make it inadmissible to assume a decomposition reaction, provided that subsequent steps are assumed which yield the same materials as does the actual process.

The decomposition processes can be shown as follows, it being understood that a chemical symbol does not represent one mole of the material in question, but rather the total mass of the substance in question, which either enters or leaves the system.

(Fe_2O_3 in ore and flux) + (FeO in ore)
$$\rightarrow \text{(Fe in pig iron)} + (O_2 \text{ that later reacts with C or H)}$$

(Mn_3O_4 in ore)
$$\rightarrow \text{(Mn in pig iron)} + \text{(MnO in slag)} + (O_2 \text{ that later reacts with C or H)}$$

(SiO_2 in ore and flux)
$$\rightarrow \text{(Si in pig iron)} + (SiO_2 \text{ in slag}) + (O_2 \text{ that later reacts with C or H)}$$

For each decomposition process, ΔH_{25} may be evaluated, using heat of formation data.

Since there is no reduction of Al_2O_3, CaO, and MgO to the metallic states or to lower oxides, there are no decomposition equations written for these oxides.

Calcination Processes

$$(CaCO_3 \text{ in ore}) \rightarrow (CaO \text{ in slag}) + (CO_2 \text{ in gases)}$$
$$(MgCO_3 \text{ in ore}) \rightarrow (MgO \text{ in slag}) + (CO_2 \text{ in gases)}$$

For each of these calcination processes, ΔH_{25} may be evaluated through the use of heat of formation data.

Combustion Process

(Charcoal) + (dry air) + (O_2 from decomposition processes)
$$+ (CO_2 \text{ from calcination processes)}$$
$$+ (H_2O \text{ from ore, charcoal, flux)}$$
$$\longrightarrow \text{(dry blast-furnace gas)} + (H_2O \text{ in blast-furnace gas)} + \text{(ash)}$$
$$+ (C \text{ that later goes to pig iron)}$$

For this process, ΔH_{25} is best evaluated using heat of combustion data, remembering that the heats of combustion of H_2O, CO_2, and ash are zero.

Formation of Fe_3C. Since the data on the enthalpy of pig iron shown in Fig. 121 were obtained under such conditions that all of the carbon was in combined form as Fe_3C and none existed as free graphite when the pig iron was at 25° C, the reference state for the carbon in the pig iron will be indicated as carbon combined in Fe_3C. In view of this circumstance, it will be necessary to consider the following process in setting up the energy balance.

(C from combustion process) + (Fe from decomposition reactions)
$$\rightarrow (Fe_3C \text{ in pig iron)}$$

ΔH_{25} for this process may readily be evaluated from the data on heat of formation of Fe_3C.

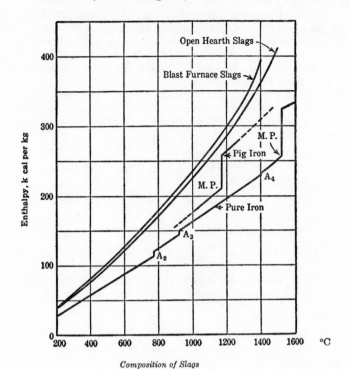

Composition of Slags

	Blast-Furnace Slags			Open-Hearth Slags		
	1	2	3	4	5	6
SiO₂	34.50	37.26	34.22	18.20	18.28	20.28
FeO	1.58	0.82	0.74	13.45	10.27	10.47
Fe₂O₃	0.29	0.84	0.20	2.40	3.63	2.50
CaO	40.92	43.15	41.80	42.63	43.55	44.20
MgO	3.90	1.80	4.56	9.14	11.84	10.47
P	Trace	Trace	Trace	0.33	0.25	0.25
S	0.98	0.11	0.90	0.54	0.45	0.40
MnO	2.24	1.82	1.88	6.97	6.60	6.60
Al₂O₃	15.48	13.15	15.60	5.00	4.68	4.79
Apparent sp. gr. 20° C	2.80	2.94	2.97	3.46	3.56	3.12

Composition of Pig Iron

C 4.31%
Si 1.11%
Mn 0.53%
P 0.12%
S 0.022%
Cu 0.21%

Heat of fusion = 46.63 kcal per kg

Thermal Data on Pure Iron

Heat of transition A_2 (α to β)	6.5 kcal per kg
Heat of transition A_3 (β to γ)	5.6 kcal per kg
Heat of transition A_4 (γ to δ)	1.9 kcal per kg
Heat of fusion	65.6 kcal per kg

FIG. 121. Enthalpies of iron and slags, referred to 25° C (Data taken from S. Umino, *Science Repts. Tôhoku Imp. Univ.*, **17**, 985, 1928, for slag; **16**, 575, 1927, for pig iron; **18**, 91, 1929, for pure iron)

Slag-Forming Process. The slag is formed by reaction among FeO, MnO, SiO_2, Al_2O_3, CaO, MgO, and the ash from the charcoal to form a complex mixture of various unknown compounds. An accurate calculation of ΔH_{25} for this process cannot be made owing to lack of data. However, an approximation will be made by assuming that CaO and SiO_2 combine to form $CaSiO_3$ and by assuming that other heat effects in forming the slag are negligible.

Energy Balance

Reference Conditions: 25° C
              H_2O in liquid state
              C in pig iron as Fe_3C

Input Items

1. Enthalpy of ore. The ore enters at the reference temperature for the energy balance; hence its enthalpy is zero.   0 kcal

2. Enthalpy of charcoal. Its enthalpy is zero, as it enters at the reference temperature for the energy balance.   0

3. Enthalpy of flux. Since it enters at 25° C, its enthalpy is zero.   0

4. Enthalpy of hot blast $15.6972 \times 7.073 \times (300 - 25)$ =   30,532

5. Heating value of charcoal 110.0×7035 = 773,850

6. Heat evolved in forming slag

     Kg-moles SiO_2 = 0.3014
     Kg-moles CaO = 0.3647

The SiO_2 is thus the limiting reactant in forming $CaSiO_3$.

Heat evolved = $0.3014 \times 21{,}320$ =   6,420

     Total input   = 810,802 kcal

Output Items

1. Enthalpy of dry blast-furnace gas. Mean heat-capacity values between 173 and 25° C were taken from Table 19, page 258, for the following calculations:

	Kg-moles		c_{pm}		
CO_2	2.6217	×	9.580	=	25.12
CO	5.3098	×	7.008	=	37.21
CH_4	0.1433	×	9.45	=	1.35
H_2	0.2784	×	6.948	=	1.93
N_2	12.4208	×	6.990	=	86.82
					152.43

Total enthalpy = $152.43 (173 - 25) = 22{,}560$ kcal

2. Enthalpy of water vapor in blast-furnace gas

     $0.6615 [10{,}513 + 8.152 (173 - 25)] = 7{,}752$ kcal

3. Enthalpy of slag. Umino[2] determined the relative enthalpies of various blast-furnace and open-hearth furnace slags and found that the enthalpies of these two types of slags are nearly the same when measured at the same temperature. These values are shown graphically in Fig. 121. It will be seen that there is no sudden break in the enthalpy-temperature curve, thus indicating that no sudden transformations take place in cooling and that the slag exists essentially in the form

[1] S. Umino, *Science Repts. Tôhoku Imp. Univ.*, **17**, 985 (1928).

of a glass. From Fig. 121 it will be seen that at a temperature of 1360° C the enthalpy of the slag is 365 kcal per kg.

Total enthalpy of slag $= 63.76 \times 365 = 23,272$ kcal

4. Enthalpy of pig iron. The average enthalpy of molten pig iron from blast furnaces has also been determined by Umino[3] and is shown in Fig. 121. Although the composition of the pig iron used in his experiments was not identical with the one under discussion, nevertheless the compositions are sufficiently alike to justify the use of his enthalpy values. It may also be pointed out that his experimental conditions were such that in the samples cooled to room temperature all carbon existed in combination as Fe_3C, with no graphite present. At a pouring temperature of 1360° C the enthalpy is 300 kcal per kg, and this includes the sensible enthalpy of the solid and liquid states, the latent heat of fusion, and the heat of solution of the Fe_3C.

Total enthalpy of pig iron $= 100 \times 300 = 30,000$ kcal

5. Heat absorbed in decomposing iron oxides

Fe_2O_3 in ore	0.73160 kg-mole
Fe_2O_3 in flux	0.00339
	0.73499 kg-mole
FeO in ore	0.2510 kg-mole
FeO going to slag	0.0533
FeO decomposed	0.1977 kg-mole

Heat absorbed in decomposing $Fe_2O_3 = 0.73499 \times 196,500 = 144,426$ kcal
Heat absorbed in decomposing FeO $\quad = 0.1977 \times \quad 64,300 = \underline{\quad 12,712}$
$\qquad\qquad\qquad\qquad\qquad\qquad\qquad\qquad\qquad\qquad 157,138$ kcal

6. Heat absorbed in decomposing Mn_3O_4

Mn_3O_4 in ore	0.04620 kg-mole
MnO going to slag	0.0982 kg-mole

Net heat absorbed $= (0.04620 \times 331,400) - (0.0982 \times 92,000)$
$\qquad\qquad\qquad\quad = 15,311 - 9034 = 6,277$ kcal

7. Heat absorbed in decomposing SiO_2

Total SiO_2 input	0.3556 kg-mole
SiO_2 to slag	0.3014
SiO_2 decomposed	0.0542 kg-mole

Heat absorbed in decomposing $SiO_2 = 0.0542 \times 205,400 = 11,133$ kcal

8. Heat absorbed in calcining carbonates

CO_2 in ore			$= 0.3775$ kg-mole
CaO in ore	20.38 kg		
CaO in flux	0.07		
	20.45 kg	or	$\dfrac{20.45}{56.08} = 0.3647$ kg-moles
MgO in ore	3.89 kg	or	$\dfrac{3.89}{40.32} = 0.0965$
			0.4612 kg-mole

[3] S. Umino, *Science Repts. Tôhoku Imp. Univ.*, **16**, 575 (1927).

It will be noted that there is insufficient CO_2 to combine with all of the CaO and MgO present. It will be assumed that all CaO is combined with CO_2 to form $CaCO_3$, and that any remaining CO_2 is combined with MgO to form $MgCO_3$.

$CaCO_3$ = 0.3647 kg-moles
$MgCO_3$ = (0.3775 − 0.3647) = 0.0128 kg-moles
$CaCO_3 \rightarrow CaO + CO_2$
ΔH_{25} = (−151,900 − 94,052) − (−288,450) = −42,498 kcal
$MgCO_3 \rightarrow MgO + CO_2$
ΔH_{25} = (−143,840 − 94,052) − (−266,000) = −28,108 kcal

Heat absorbed calcining $CaCO_3$ = 0.3647 × 42,498 = 15,499 kcal
Heat absorbed calcining $MgCO_3$ = 0.0128 × 28,108 = ___360___
15,859 kcal

9. Heating value of blast-furnace gas

Heating value of CO = 5.3098 × 67,636 = 359,134 kcal
Heating value of CH_4 = 0.1433 × 212,798 = 30,494
Heating value of H_2 = 0.2784 × 68,317 = _19,019_
408,647 kcal

10. Heating value of carbon appearing in pig iron. In the previous discussion on the method of evaluating ΔH_{25} for the over-all blast-furnace process, it was suggested that the complex process be broken down into several simpler steps, one of which was designated as the combustion process. In this combustion process, one of the terms on the right side of the process equation is carbon going to the pig iron. Accordingly, even though the carbon in the final pig iron does not occur in the free state, an entry must be made for the heat of combustion of this carbon.

Heating value of carbon appearing in pig iron = 0.260×94,052 = 24,454 kcal

11. Heat absorbed in forming Fe_3C. This item takes account of the fact that the carbon in the pig iron does not exist in the free state.

Heat absorbed forming Fe_3C = 0.260 × 5000   1,300 kcal

12. Heat absorbed by cooling water = 576 × 13   7,488 kcal
13. Heat losses (by difference)   _94,922 kcal_
Total output   810,802 kcal

Summary of Energy Balance

Basis: 100 kg pig iron.

Input

	Kcal	%
1. Enthalpy of ore	0	0.00
2. Enthalpy of charcoal	0	0.00
3. Enthalpy of flux	0	0.00
4. Enthalpy of hot blast	30,532	3.77
5. Heating value of charcoal	773,850	95.44
6. Heat evolved forming slag	6,420	0.79
Total	810,802	100.00

Output

	Kcal	%
1. Enthalpy of dry blast-furnace gas	22,560	2.78
2. Enthalpy of water vapor in blast-furnace gas	7,752	0.96
3. Enthalpy of slag	23,272	2.87
4. Enthalpy of pig iron	30,000	3.70
5. Heat absorbed in decomposing iron oxides	157,138	19.38
6. Heat absorbed in decomposing Mn_3O_4	6,277	0.77
7. Heat absorbed in decomposing SiO_2	11,133	1.37
8. Heat absorbed in calcining carbonates	15,859	1.96
9. Heating value of blast-furnace gas	408,647	50.40
10. Heating value of carbon appearing in pig iron	24,454	3.02
11. Heat absorbed forming Fe_3C	1,300	0.16
12. Heat absorbed by cooling water	7,488	0.92
13. Heat losses (by difference)	94,922	11.71
Total	810,802	100.00

Petroleum Cracking Process

In a so-called "vapor phase cracking" process a clean, well-fractionated gas-oil cut from petroleum, containing no material of gasoline boiling range, may be decomposed to form gasoline and gas by heating in the tubes of a furnace designed to provide the necessary reaction time at elevated temperatures. In order to arrest the reaction and minimize coke formation, the hot vapor mixture of gas, gasoline, oil, and tar from the furnace is "quenched" by a cooler stream of oil. The resulting mixture passes to an evaporator where further cooling and rough fractionation is accomplished by means of a reflux stream of cool oil sprayed in at the top of the vessel. The quantity of reflux is regulated to obtain such a temperature in the evaporator as to produce the desired quality of "tar" which is withdrawn from the bottom of the evaporator and generally sold as heavy fuel oil.

The vapors from the evaporator pass to a fractionating tower, the lower section of which is utilized for preheating the fresh charge by direct heat exchange. This tower is operated to produce well-fractionated gasoline and gas as the overhead product. All partially decomposed feed in the boiling range between gasoline and the tar is condensed as a bottom product. This "recycle stock" mixes in the bottom section of the tower with the fresh feed to form the "combined feed," part of which is charged to the furnace, part used for quenching the hot vapors leaving the furnace, and part used as a reflux in the evaporator. It may be assumed that the mixture leaving the furnace is completely vaporized and that negligible condensation is produced by the quench. It may also be assumed that all the oil used for quench and reflux is vaporized under the conditions of the evaporator, forming no tar.

Illustration 3. The flow diagram of such a process is shown in Fig. 122, on which are indicated significant temperatures and characteristics of the streams. The furnace is of the gas-fired type, provided with both radiant and convection heating sections. An air preheater transfers heat from the stack gases to the air used for combustion.

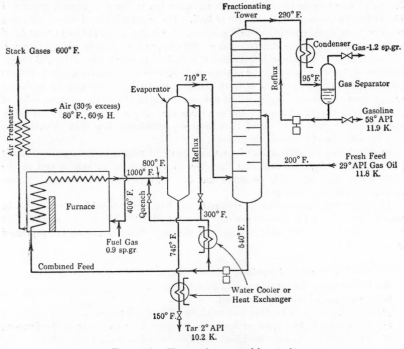

FIG. 122. Vapor-phase cracking unit

The gas used for heating the furnace is supplied at 65° F and has the following composition by volume:

CH₄	14.6%
C₂H₆	77.3
CO	1.2
H₂	6.1
N₂	0.8
	100 %

The temperatures shown on the flow diagram are either arbitrarily set as bases for design or are derived from previous pilot-plant or commercial experience indicating the temperatures necessary for the desired reaction rates and separations.

Pilot-plant tests indicate that in cracking a gas oil having a gravity of 29° API and a characterization factor of 11.8 the following yields are obtained, expressed as percentage by liquid volume of the fresh feed:

Gasoline (58° API, K = 11.9)	61.0%
Tar (2° API, K = 10.2)	24.5%

It may be assumed that the operation is conducted with 100% material recovery and that the yield of gas is determined by difference.

Previous commercial experience has indicated that such an operation may be carried out satisfactorily under reaction conditions which result in the conversion per pass into gasoline and gas of 22.0 weight per cent of the combined feed entering the furnace. The recycle stock from such an operation is found to have a gravity of 19° API and a characterization factor of 10.7. The standard heat of the endothermic reaction from liquid charge to liquid gasoline, recycle stock, and tar, and gas at 60° F and atmospheric pressure may be taken as 600 Btu per lb of gasoline plus gas formed.

On the basis of the above information it is desired to develop material and energy balances of the separate parts and of the entire plant for the design of a unit to process 5000 barrels (42 gal) per day of fresh charge. Radiation losses may be neglected except from the furnace where a loss corresponding to 5% of the heating value of the fuel is assumed. Although the pressures throughout will be somewhat above atmospheric, it will be assumed that they are sufficiently low that all enthalpies may be taken as at atmospheric pressure.

In calculating the enthalpies of mixtures, heats of mixing in both gas and liquid phases may be neglected and also the effect of pressure on enthalpies and on the heat of cracking. It should be noted that the characterization factors and degrees API are additive on a weight basis.

The following information is required: All flow rates should be expressed in pounds per hour and in barrels per day (at 60° F) for the liquids and in cubic feet (at 60° F, 30 in. Hg, saturated) per hour for the gas. Heat rates should be expressed in Btu per hour.

(a) Production rates of gasoline, gas, and tar.

(b) Flow rate of the combined feed to the furnace.

(c) Properties of the combined feed.

(d) Heat absorbed in heating and cracking oil in the furnace.

(e) Flow rate of the combined feed used for quenching vapors from the furnace.

(f) Flow rate of the combined feed used for the reflux in the evaporator.

(g) Reflux rate of gasoline in the fractionating column.

(h) Temperature of flue gases leaving the convection section of the furnace.

(i) Fuel gas burned in the furnace.

(j) Thermal efficiency of the combined furnace and preheater.

(k) Heat-transfer duties of the condenser and coolers.

(l) Cooling water required by the condenser and coolers in gallons per minute. (Assume a 20° F temperature rise of the water.)

Solution

For ready reference the physical and thermal properties of the various petroleum fractions are presented in Tables 38 and 39. The characterization factors and degrees API of a mixture are additive properties on a weight basis. The average molecular weights and boiling points are obtained from values of API and K by use of Fig. 106. Latent heats of vaporization are calculated from Fig. 72, page 280. The mean specific heats of liquids and vapors are obtained from Figs. 66 and 61. Enthalpies at various temperatures are then calculated from the above data, using 65° F as the reference temperature.

All calculations are based upon 1 hr of operation.

TABLE 38. PHYSICAL PROPERTIES OF OIL FRACTIONS

	°API	K	G, 60°/60°	Avg. Molecular Weight	Avg. Boiling Point, °F
Gasoline	58.0	11.9	0.747	109	240
Tar	2.0	10.2	1.06	320	800
Fresh feed	29.0	11.8	0.88	300	675
Recycle	19.0	10.7	0.940	220	560
Gas			1.22 (air)		
Combined feed (calculated)	22.1	11.04	0.922	240	585

1. Rates of production. Rates of production are calculated from the yield statement, as follows:

	Charge	Gasoline	Tar	Gas
Bbl per day	5000	3050	1225	....
Gal per hr	8750	5340	2145	....
Sp. gravity	0.882	0.747	1.06	....
Lb per hr	64,290	33,210	18,930	12,150
M cu ft per day	....	....	....	3230

2. Flow rate of combined feed to furnace. Of the combined feed entering the furnace, 22% by weight is converted into gas plus gasoline.

Gas plus gasoline, lb per hr = (33,210 + 12,150) = 45,360

Combined feed to furnace, lb per hr = $\dfrac{45,360}{0.22}$ = 206,170

Recycle stock produced in furnace, lb per hr = 206,170 −
(33,210 + 18,930 + 12,150) = 141,880

3. Properties of combined feed

	Recycle Stock	Fresh Feed
Gravity, °API	19°	29°
Characterization factor K	10.7	11.8
Lb per hr	141,880	64,290
% by weight	68.8	31.2

K of combined feed = (10.7)(0.688) + (11.8)(0.312) = 11.04
°API of combined feed = (19)(0.688) + (29)(0.312) = 22.1
From Fig. 106 average molecular weight = 240

Average boiling point = 585° F

The thermal properties are tabulated in Table 39.

4. Heat absorbed in heating and cracking oil in furnace

Input

Enthalpy of combined feed at 540° F = (206,170)(248.4)   =   51,212,600 Btu
Heat supplied to oil from combustion of fuel (by difference) = 111,318,600
                                                                162,531,200 Btu

TABLE 39.   THERMAL PROPERTIES OF OIL FRACTIONS

	Boiling Point °F	Heat of Vaporization Btu/lb	Mean Specific Heat of Liquids (65° to t°F)		Mean Specific Heat of Vapors (240° to t°F)		Enthalpies 65°F reference	
			t°F	C_p	t°F	C_p	t°F	Btu/lb
Gasoline	240	134.8	95	0.475	290	0.464	95(l)	14.3
			240	0.518	710	0.574	240(l)	90.7
					800	0.596	240(v)	225.5
					1000	0.640	290(v)	248.7
							710	495.2
							800	559.0
							1000	711.7
Tar	800	100			(800°F to t°F)		150(l)	43.4
			150	0.377	800	0.620	745(l)	353.6
			745	0.520	1000	0.660	800(l)	388.1
			800	0.528			800(v)	488.1
							1000(v)	620.3
Gas					(65°F to t°F)			
					95	0.444	95	13.3
					290	0.498	290	112.1
					710	0.608	710	392.3
					800	0.630	800	463.0
					1000	0.675	1000	631.3
Fresh Feed	675	88	200	0.464	200		200(l)	62.6
Combined Feed	585	104.4			(585°F to t°F)			
			300	0.462	710	0.610	300(l)	108.6
			540	0.523	800	0.630	540(l)	248.4
			585	0.534			585(l)	277.7
							585(v)	382.1
							710(v)	458.5
							800(v)	517.4
Recycle	560	110.6	560	0.519	710	0.586	560(l)	256.9
					800	0.605	560(v)	367.5
					1000	0.644	710(v)	455.3
							800(v)	512.7
							1000(v)	650.7

Output

Enthalpy of gas = (12,150)(631.3)	=	7,670,000 Btu
Enthalpy of gasoline = (33,210)(711.7)	=	23,635,600
Enthalpy of tar = (18,930)(620.3)	=	11,742,300
Enthalpy of recycle = (141,800)(650.7)	=	92,269,300
Energy absorbed in cracking oil (600)(45,360)	=	27,214,000
Total	=	162,531,200 Btu

5. Weight of combined feed used for quenching hot vapors from furnace.
The energy balance about the quench point is shown diagrammatically in Fig. 123.
The required quantity of quench is fixed by this balance.
Let x = lb quench supplied to vapor from furnace.

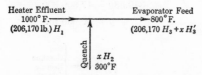

Fig. 123.   Flow chart about quench point

Input

Enthalpy of vapors at 1000° F = 135,317,200 Btu
Enthalpy of quench at 300° F = $\underline{}$ 108.6x
  Total = 135,317,200 + 108.6x

Output

Enthalpy of gas = (12,150)(463.)	=	5,625,100
Enthalpy of gasoline = (33,210)(559)	=	18,564,400
Enthalpy of tar vapor = (18,930)(488.1)	=	9,239,300
Enthalpy of recycle vapors = (141,880)(512.7)	=	72,741,900
Enthalpy of combined feed quench = (x)(517.42)	=	517.42x
Total	=	106,170,700 + 517.42x

From the above balance:

$$106,170,700 + 517.42x = 135,317,200 + 108.6x$$
$$408.82x = 29,146,500$$
$$x = 71,294 \text{ lb of combined feed for quench}$$

6. Flow rate of combined feed used for reflux in the evaporator.
Let y = number of pounds of combined feed used for reflux in evaporator.

The reflux rate required in the evaporator is determined by an energy balance
following the flow chart shown in Fig. 124.

Energy Balance of Evaporator

Input

Enthalpy of vapors at 800° F
  = 106,170,700 + (71,294)(517.42) = 143,059,600
Enthalpy of combined feed for reflux = $\underline{}$ 108.6y
  Total = 143,059,600 + 108.6y

Output

Enthalpy of gas = (12,150)(392.29)   =   4,766,300
Enthalpy of gasoline = (33,210)(495.23)   =   16,468,800
Enthalpy of recycle = (141,880)(455.34)   =   64,636,400
Enthalpy of combined feed quench = (71,294)(458.45) = 32,684,700
Enthalpy of combined feed reflux = $(y)(458.45)$ = $+ 458.45y$
Enthalpy of tar at 745° F = (18,930)(353.6)   =   6,693,600
  Total   = 125,249,800 + 458.45y

From the energy balance:

$$125,249,800 + 458.45y = 143,059,600 + 108.6y$$
$$349.85y = 17,809,800$$
$$y = 50,906 \text{ lb.}$$

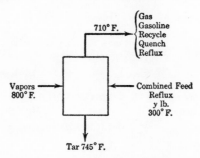

710° F.

{Gas
Gasoline
Recycle
Quench
Reflux

Vapors
800° F.

Combined Feed
Reflux
y lb.
300° F.

Tar 745° F.

Fig. 124. Flow chart of evaporator

7. Rate of gasoline reflux in fractioning tower. The reflux rate required in the fractionating tower is determined from an energy balance following the flow chart shown in Fig. 125.

Energy Balance. Let z = reflux, pounds per hour.

Input

Enthalpy of gas   =   4,766,300 Btu
Enthalpy of gasoline   =   16,468,800
Enthalpy of recycle   =   64,636,400
Enthalpy of combined feed recycle and quench
 (122,200)(458.45)   =   56,022,600
Enthalpy of fresh feed = (64,290)(62.57)   =   4,022,600
Enthalpy of gasoline reflux   =   14.25z
  Total   = 145,916,700 + 14.25z

Output

Enthalpy of gas = (12,150)(112.05)   =   1,361,400 Btu
Enthalpy of gasoline = $(33,210 + z)(248.65)$ =   8,257,700 + 248.65z
Enthalpy of combined feed = (328,370)(248.42)   =   81,573,700
  Total   =   91,192,800 + 248.65z

From the energy balance:

$$91,192,800 + 248.65z = 145,916,700 + 14.25z$$
$$234.4z = 54,723,900$$
$$z = 233,463 \text{ lb of gasoline reflux per hr}$$

8. Temperature of stack gases leaving furnace. The temperature of the stack gases leaving the furnace is calculated from an energy balance of the air preheater, based on the combustion of 100 lb-moles of fuel gas with 30% excess air.

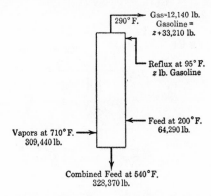

Gas=12,140 lb.
Gasoline =
z +33,210 lb.

290° F.

Reflux at 95° F.
z lb. Gasoline

Feed at 200° F.
64,290 lb.

Vapors at 710° F. →
309,440 lb.

Combined Feed at 540° F.
328,370 lb.

FIG. 125.   Flow chart of fractionating tower

Basis: 100 lb-moles of fuel gas.

	Moles	Mol Wt	Lb	O_2 Required, lb-moles	CO_2, lb-moles	H_2O, lb-moles	Heating Value, Btu per lb-mole	Heating value $\times$ Mole Fraction
CH_4	14.6	16	2.34	29.2	14.6	29.2	383,036	55,920
C_2H_6	77.3	30	23.19	270.55	154.6	231.9	671,076	518,740
CO	1.2	28	0.34	0.6	1.2		121,745	1,460
H_2	6.1	2	0.12	3.05		6.1	122,971	7,500
N_2	0.8	28	0.22					
			26.21	303.40	170.4	267.2		583,620 Btu per lb-mole

Oxygen theoretically required	= 303.40 lb-moles
Oxygen actually supplied = (303.40)(1.3)	= 394.4 lb-moles
Nitrogen actually supplied = $394.4 \times \dfrac{0.79}{0.21}$	= 1483 lb-moles
Total moles air supplied per 100 moles fuel	= 1877 lb-moles
Molal humidity, lb-moles water per lb-mole dry gas (Fig. 20)	= 0.022
Water from air = (0.022)(1877) lb-moles	= 41.3 lb-moles

Products of combustion:

Carbon dioxide   = 170.4 lb-moles
Water = 267.2 + 41.3   = 308.5
Oxygen = 394.4 − 303.40 = 91.0
Nitrogen = 0.8 + 1483   = 1483.8
                                 2053.7 lb-moles

Enthalpy of water vapor in gas streams:

	Btu per lb-mole
In entering air, 80° F	19,165
In leaving air, 400° F	21,785
In waste gases, 600° F	23,457

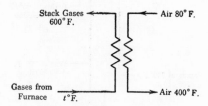

FIG. 126.   Flow chart of air preheater

The temperature of the gases from the furnace is determined by an energy balance of the air preheater, illustrated in Fig. 126. As a first approximation it may be assumed that the temperature drop of the gases in the preheater approximately equals the temperature rise of the air. Mean heat capacities are based on this assumed temperature, which may be corrected by a second approximation if the energy balance shows it to be seriously in error.

Energy Balance of Preheater

Basis: 1.0 lb-mole of fuel gas.

Input

Enthalpy of dry air $(18.77)(80 − 65)$
  (6.97) = 1,962 Btu
Enthalpy of water vapor in air (0.413)
  $(19,165)$ = 7,915
Enthalpy of hot waste gases:
  CO_2 $1.704(t − 65)(c_{pm})_{CO_2}$
  H_2O Latent enthalpy 3.085
        × 19,045   = 58,754
  H_2O Sensible enthalpy $3.085(t − 65)(c_{pm})_{H_2O}$
  O_2 $0.910(t − 65)(c_{pm})_{O_2}$
  N_2 $14.838(t − 65)(c_{pm})_{N_2}$
                                    68,631 Btu

Total input $= 68,631 + [1.704(c_{pm})_{CO_2} + 3.085(c_{pm})_{H_2O} + 0.910(c_{pm})_{O_2}$
$$+ 14.838(c_{pm})_{N_2}](t − 65)$$

Output

Enthalpy of waste gases leaving at 600° F:

CO_2 = (1.704)(600 − 65)(10.17) = 9,271 Btu
H_2O = (3.085)(23,457)   = 72,365
O_2 = (0.910)(600 − 65)(7.31) = 3,559
N_2 = (14.838)(600 − 65)(7.04) = 55,886

                                           141,081 Btu

Enthalpy of air at 400° F:

Dry air = (18.77)(400 − 65)(7.02) = 44,141 Btu
Water vapor = (0.413)(21,785)   = 8,997

                                             53,138 Btu

        Total Output   194,219 Btu

If input and output are equated, the following equation results:

$$[1.704(c_{pm})_{CO_2} + 3.085(c_{pm})_{H_2O} + 0.910(c_{pm})_{O_2} + 14.838(c_{pm})_{N_2}](t - 65) = 125,588$$

If this equation is solved by the use of mean heat-capacity data as demonstrated on page 353, $t = 865°$ F.

9. Fuel gas burned in furnace per hour. The amount of fuel gas burned in the furnace may be calculated from an energy balance either of the furnace or the furnace plus preheater based on one hour of operation. The latter balance is illustrated in Fig. 127.

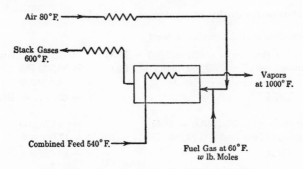

FIG. 127.   Flow chart of furnace and preheater

Let w = number of pound-moles of fuel gas burned per hour.

Energy Balance of Furnace plus Preheater

                    Input

Enthalpy of dry air at 80° F =   1,962w Btu
Enthalpy of H_2O in air   =   7,915w
Enthalpy combined feed   =   + 51,212,600
Heating value of fuel   = 583,620w
        Total   = 593,497w + 51,212,600 Btu per hr

Output

Enthalpy of vapors at 1000° F = 135,317,200
Enthalpy of waste gases   =   141,081w
Heat absorbed in cracking   = 27,214,000
Heat losses = (0.05)(583,620)w =   29,181w
         Total   = 162,531,200 + 170,262w

From the energy balance,

$$593,497w + 51,212,600 = 162,531,200 + 170,262w$$

$$423,235w = 111,318,600$$

$$w = 263.0 \text{ lb-moles dry fuel gas per hr}$$

Cu ft per hr of gas, at 60° F and 30 in. Hg saturated with H_2O = (263.0)(385.2) = 101,308 cu ft per hr

10. Thermal efficiency of furnace. From part 4,

Heat supplied to oil = 111,318,600 Btu

Total heating value of fuel gas (583,620)(263.0) = 153,492,060 Btu

Thermal efficiency based on heating value of gas $\dfrac{111,318,600}{153,492,060} \times 100 = 72.5\%$

11. Heat-transfer duties of condenser and coolers

1. Condenser on fractionation tower:
   Gasoline product = (33,210)(248.7 − 14.3)   =   7,784,420 Btu per hr
   Gasoline reflux = (233,463)(248.7 − 14.3)   = 54,723,730
   Gas = (12,150)(112.1 − 13.3)   =   1,199,690
            Total   = 63,707,840 Btu per hr

2. Tar cooler (18,930)(253.6 − 43.4)   =   5,873,200
3. Combined feed cooler (122,200)(248.4 − 108.6) = 17,086,000
            Total   = 86,667,040 Btu per hr

12. Cooling water required by condenser and cooler

Heat absorbed per gal of water = 20 × 8.33 = 166.6 Btu

1. Condenser on fractionating tower = $\dfrac{63,707,840}{(166.6)(60)}$ = 6373 gpm

2. Tar cooler = $\dfrac{5,873,200}{(166.6)(60)}$ = 587

3. Combined feed cooler = $\dfrac{17,086,000}{(166.6)(60)}$ = 1709

         Total cooling water required   = 8669 gpm

Problems

1. In a plant for the manufacture of sulfuric acid by the chamber process pyrites is burned in a shelf burner. The gases from the burner enter the Glover tower at 480° C and leave this tower at 105° C, entering the first chamber. The gases leave the last chamber at 42° C and finally leave the Gay-Lussac tower at 21° C. On the basis of 100 kg of pyrites, as charged, there are charged into the Glover tower 175 kg of chamber acid, 65.2% H_2SO_4 (51.8° Bé) at 30° C; 610 kg of Gay-Lussac acid at 23° C; and 1.30 kg of 40% nitric acid at 20° C. The Gay-Lussac acid contains 78.0% H_2SO_4 (60.0° Bé), 0.984% N_2O_3 in solution, and 21.0% H_2O.

The analyses of the pyrites, cinder, and the moisture-free gases leaving the burner are as follows:

Pyrites		Cinder		Gases (by Volume)	
FeS_2	90.00%	Fe_2O_3	89.80%	O_2	9.32%
SiO_2	4.80	FeS_2	1.65	N_2	82.38
H_2O	5.20	SO_3	1.93	SO_2	8.00
	100.00%	SiO_2	6.62	SO_3	0.30
			100.00%		100.00%

The pyrites is charged to the burner at 20° C. The air enters at 20° C, under a barometric pressure of 722 mm Hg and with a percentage humidity of 40. The cinder is withdrawn at 320° C.

For each 100 kg of pyrites as charged, 768 kg of acid containing 79.4% H_2SO_4 leave the Glover tower at 100° C and are cooled to 23° C. The chamber acid leaves the first chamber at 65° C and is cooled to 30° C before entering the Glover tower. The acid leaves the Gay-Lussac tower at 30° C and is cooled to 23° C for recirculation. The spray water enters the chambers at 20° C.

From the flow chart and assumptions of illustration 1, calculate individual material and energy balances, on the basis of 100 kg of pyrites as fired, of

    (*a*) The burner.
    (*b*) The Glover tower.
    (*c*) The chambers.
    (*d*) The Gay-Lussac tower.
    (*e*) The entire plant.

2. The charge delivered to a blast furnace, on the basis of 1000 lb of pig iron, consists of 1810 lb of ore, 361 lb of limestone, and 892 lb of coke. The analyses of various components of the charge are as follows:

Ore (1810 lb)		Limestone (361 lb)		Coke (892 lb)	
Fe_2O_3	62.10%	CaO	51.12%	Carbon	88.20%
FeO	19.07	MgO	2.10	Hydrogen	2.00
Mn_3O_4	2.12	SiO_2	2.89	Fe_2O_3	2.10
Al_2O_3	2.89	Al_2O_3	4.12	SiO_2	1.98
SiO_2	8.62	Fe_2O_3	0.52	CaO	2.32
H_2O	5.20	CO_2	35.05	MgO	1.10
	100.00%	H_2O	4.20	S	0.20
			100.00%	H_2O	2.10
					100.00%

The total heating value of the coke is 14,200 Btu per lb.

On the basis of 1000 lb of pig iron produced, 51 lb of dust are collected from the gases leaving the furnace. The analyses of the products are as follows:

Pig Iron (1000 lb)		Flue Dust (51 lb)		Gas Analysis (by Volume)	
Fe	92.28%	FeO	83.2%	CH_4	0.80%
Si	2.10	C	10.1	CO_2	12.10
Mn	1.38	CaO	3.1	CO	29.30
S	0.03	SiO_2	3.6	H_2	2.12
C	4.21		100.0%	O_2	0.20
	100.00%			N_2	55.48
					100.00%

The surrounding air is at 70° F, 40% humidity and a barometric pressure of 29.2 in. Hg. This air is heated and supplied to the tuyères at 850° F.

The ore, flux, and coke are charged at an average temperature of 65° F.

The gases leave the furnace at a temperature of 422° F. The molten slag and pig iron are tapped from the furnace at a temperature of 2500° F. The sensible enthalpy of the flue dust is negligible.

Calculate the complete material and energy balances of this furnace, using the assumptions of illustration 2.

3. In producing 1 ton (2000 lb) of steel in an open-hearth furnace, the following charge was supplied:

Hot metal from blast furnace (2400° F)	814 lb
Cold scrap iron	1250 lb
Limestone (95.5% $CaCO_3$, 4.5% H_2O)	118 lb
Iron ore (94% Fe_2O_3, 6% H_2O)	56 lb
Fuel oil	28.2 gal

Air was supplied to the regenerators at 80° F, 60% relative humidity, atmospheric pressure. The air (85% of total supply) was preheated to 2000° F in the regenerators. The hot gaseous products of combustion left the hearth at 2860° F, entered the regenerators at 2560° F, and entered the stack at 1000° F. The average analysis of the flue gases measured over the 9-hr run was as follows:

$$CO_2 = 17.0\%$$
$$O_2 = 0.8$$
$$N_2 = 82.2$$
$$100.0\%$$

The metals analyzed as follows:

	Hot Metal	Cold Scrap	Steel
Carbon	4.25%	0.15%	0.15%
Silicon	1.92	0.02	0.02
Manganese	0.32	0.50	0.25
Phosphorus	0.65	0.065	0.02
Iron	92.86	99.265	99.56

The steel and slag were poured at 2800° F. From experience it is known that 15% of the air used in the furnace leaks in through the doors and brickwork of the hearth.

The fuel oil had the following properties:

Characterization factor   11.1
API gravity   12.0

The following information is desired on the basis of 1 ton of steel produced (use 80° F as basis of enthalpies):

1. Material balance of the hearth proper.
2. Weight of flue gas.
3. Weight of dry air used.
4. Weight of water vapor in air supply and in flue gas.
5. Over-all material balance of combustion and refining processes.
6. Air theoretically required for combustion of the fuel.
7. Percentage excess air used, based on requirements of fuel oil.
8. Over-all energy balance of process.
9. Energy balance of reactions on hearth.

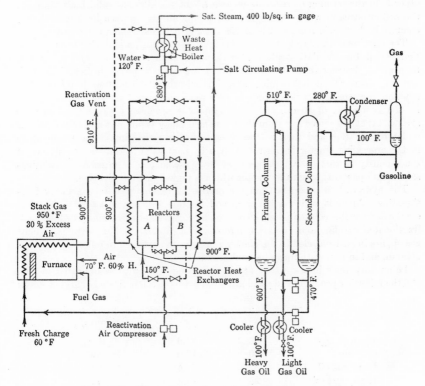

FIG. 128.   Flow chart of catalytic cracking process

4. It is desired to prepare a preliminary engineering design study of a proposed catalytic unit for the "cracking" of higher-boiling petroleum fractions into gasoline and gases rich in recoverable olefins.   A flow diagram is shown in Fig. 128.

The process illustrated in Fig. 128 is not in commercial application as shown but represents a possible combination of features described in the many patents relating to this subject.

The charge, a well-fractionated gas oil, is pumped through a heater where its temperature is raised to 900° F at a pressure somewhat above atmospheric and passed to a reactor B where it is contacted with a refractory catalyst in the form of small pellets. The reactor is provided with a heat-exchange system through which a molten salt is circulated for temperature control and is so designed that heat-exchange surface is uniformly distributed throughout the entire catalyst bed. The decomposition reaction is endothermic, requiring the supply of heat from the salt system in order to maintain isothermal conditions.

As the cracking reaction progresses, a carbonaceous deposit accumulates on the catalyst which reduces its activity. In order to maintain continuous operation an alternate reactor A is provided in which the hydrocarbon stream may be processed while the catalyst deposit is removed frum reactor B by oxidation with air. Thus, one reactor system is at all times in the *process period* of the operation while the other is in the burning or *reactivation period*. In Fig. 128 the solid lines represent the various streams while reactor B is processing and the broken lines indicate the changes in flow when reactor A is processing and B is undergoing reactivation.

The burning of the catalyst deposit is highly exothermic, and during the reactivation period the circulating-salt heat-exchange system serves to cool the catalyst to prevent destructive overheating. Since the heat liberated in reactivation is generally greater than that required to maintain isothermal conditions in the cracking reaction, a waste heat boiler is provided for removing heat from the circulating salt by the generation of steam. The salt system is provided with switch valves so arranged that the cooled salt from the waste heat boiler is pumped first to the reactor under reactivation and then to the reactor that is processing. The temperature in the processing reactor is controlled by a by-pass valve permitting regulation of the quantity of salt passed through the heat exchanger. The entire circulating-salt stream passes through the reactor which is undergoing reactivation.

The hydrocarbon products from the processing reactor pass to a primary fractionating column in which a heavy gas-oil fraction is removed as bottoms. The overhead from this column passes to a secondary fractionating column in which well-fractionated gasoline and gases are the overhead products. The bottoms, a light gas oil, are in part recycled to the cracking process, in part used to reflux the primary column, and in part withdrawn as final product.

Laboratory tests indicate that at the operating temperatures indicated in Fig. 128 the following products and yield are obtained from the indicated charging stock:

	° API	K	% by Volume of Charge
Fresh charge	30	11.9	100
Gasoline	60	12.0	49
Light gas oil	32	11.4	24
Heavy gas oil	25	11.5	18

The gas has a specific gravity of 1.6 and a molal heat capacity corresponding to that of propane.

The catalyst deposit is found to correspond to 3% by weight of the fresh charge and contains 4% hydrogen and 96% carbon by weight.

The hydrocarbon combined feed, consisting of fresh charge and recycled light gas oil, is passed through the processing reactor at a rate of 1.2 volumes of oil, measured at 60° F, per hour per volume of catalyst. This method of expressing reactor feed rates is commonly used in catalytic processes and is termed the *liquid hourly space velocity*. The catalyst volume includes the actual volume of the catalyst pellets and the voids between the pellets in the catalyst bed. The density of the catalyst bed is 55 lb per cu ft.

At the above operating conditions it is found that 42% by weight of the combined feed is decomposed into gasoline and gas in a single pass through the unit.

The average standard heat of reaction at 60° F and one atmosphere from liquid combined feed to liquid gasoline and gas oils and gas is found to be +220 Btu per lb of gasoline plus gas formed, corresponding to an endothermic reaction.

The fuel gas burned in the furnace has the following composition:

	Mole %
Hydrogen	22.4
Methane	26.0
Ethylene	6.8
Ethane	7.2
Propylene	29.5
Propane	8.1
	100.0

In accordance with the process information given above, develop the following design factors, and evaluate complete material and energy balances for a plant to charge 10,000 barrels (42 gal) per day. Base the balances on one hour of operation and express rates in barrels per day and pounds per hour for liquids and pounds per hour and thousands of cubic feet per day for gases. Heat losses may be neglected except from the furnace.

(*a*) Production rates of all net hydrocarbon products and the catalyst deposit.

(*b*) Rates and properties of the combined feed to the heater and the light gas-oil recycle stream.

(*c*) Heat absorbed by the oil in the furnace, assuming complete vaporization but no decomposition in the heater.

(*d*) Rate of fuel consumption in the furnace, assuming complete combustion and a radiation loss of 5% of the heating value of the fuel burned.

(*e*) Thermal efficiency of the furnace.

(*f*) Reflux rates to the primary and secondary fractionating columns.

(*g*) Heat-transfer duties of the gasoline condenser and the light and heavy gas-oil coolers.

(*h*) Cooling water requirements of the plant in gallons per minute with a 30° F temperature rise for the water.

(*i*) Volume and weight of the catalyst required in each reactor.

(*j*) Length of the process period in minutes if it is desired to limit the deposit on the catalyst to a maximum of 2.5% by weight of the catalyst.

(*k*) Rate at which air must be supplied in order that reactivation may be completed in the time of the process period. It may be assumed that the oxygen of the air is 100% utilized, going to carbon dioxide and water under the catalytic combustion conditions.

(*l*) Rate at which heat must be removed by the circulating salt from a reactor under reactivation, neglecting changes in the enthalpy of the catalyst bed.

(*m*) Rate at which salt must be circulated through a reactivating reactor heat exchanger. The heat capacity of the molten salt may be taken as 0.25 Btu per lb per ° F.

(*n*) Temperature of the salt entering the waste heat boiler.

(*o*) Quantity of steam generated in the waste heat boiler, in thousand pounds per hour.

5. In the *fluid* catalytic-cracking process heavy petroleum fractions are decomposed into gasoline, gas, and furnace oil by contact in the vapor state with a powdered catalyst which is maintained in a fluidized or partially suspended state by upward flow of the oil vapors and the products of cracking. Such a plant is shown in diagrammatic form in Fig. 129.

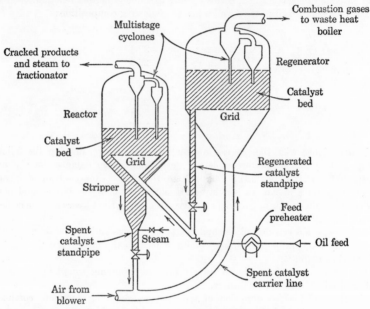

FIG. 129. Fluid catalytic-cracking unit

Oil to be cracked is fed through a feed preheater where it undergoes heat exchange with hot liquid products of the process. The heated feed then enters the regenerated catalyst carrier line where it contacts hot regenerated catalyst powder which is introduced at a rate controlled by a slide valve. The heat content of the catalyst vaporizes the oil, producing a velocity in the carrier line sufficient to carry the catalyst upward into the reactor in a highly dispersed state. The mixture of catalyst and oil enters the reactor through an entrance cone, on top of which is a grid perforated with relatively large holes to produce approximately 2% open area. The purpose of the grid is to produce a pressure drop to assist in the uniform distribution of the catalyst-oil mixture.

The reactor catalyst bed is supported by the grid in a relatively dense fluidized state which produces vigorous turbulence and mixing, with the result that temperatures and compositions are essentially uniform throughout the bed. The oil vapors and cracked products emerge from this bed into a disengaging zone in which a major portion of the entrained catalyst powder settles down to the bed. The remainder of the catalyst is removed from the vapors by passage through a multi-stage cyclone separator which returns the separated catalyst to the bed. The clean vapors pass to the fractionating tower where they are separated into the desired products.

To offset the quantity of catalyst which is constantly introduced with the oil feed, it is necessary to withdraw spent catalyst continuously from the reactor bed. This catalyst is coated with a carbonaceous deposit formed as a by-product of the cracking reaction. The spent catalyst flows downward around the entrance cone into a stripper section where it is contacted with steam to vaporize adsorbed oil vapors. The stripped catalyst passes into a short spent-catalyst standpipe from which it is fed, at a rate controlled by a slide valve, into a stream of air in the spent-catalyst carrier line.

Contact with air immediately starts burning off the deposit from the spent catalyst and raises the temperature of the air as the mixture passes up the carrier line into the regenerator. The catalyst bed in the regenerator is supported on a grid similar to that in the reactor and is of sufficient size to permit substantially complete combustion of the catalyst deposit. As a result of the turbulence of the bed, the temperature is uniform throughout, and the gases leave at the average temperature of the bed. Regenerated catalyst flows from the regenerator bed into the regenerated-catalyst standpipe and thence through a regulating slide valve into the carrier line, completing the circuit.

The transport of the catalyst in a fluid catalytic-cracking plant is energized entirely by the energy contents of the oil vapors and the air in the carrier lines. Circulation is maintained as a result of the differences in density of the fluidized catalyst at different points, which is determined largely by the velocity of the fluidizing or carrying medium. Typical densities are as follows:

Reactor and regenerator beds	25–35 lb per cu ft
Standpipes	35–45 lb per cu ft
Carrier lines	1–2 lb per cu ft

It is evident that the high densities in the beds and standpipes compared to those in the carrier lines result in the equivalent of a hydrostatic head which causes circulation. This head must be sufficient to overcome friction losses and the pressure drops through the grids and the slide valves. This requirement dictates the height of the unit which is typically of the order of 150 ft.

It is desired to prepare a process design study of a plant of the foregoing type to crack catalytically 18,000 barrels (42 gal) per operating day of an oil that laboratory cracking tests have shown will yield the following products:

	° API	K	% by Volume of Charge
Charge	27	11.6	100
Gasoline	59	12.0	35
Furnace oil	28	11.7	20
Heavy gas-oil recycle stock	23	11.4	34

The catalyst deposit corresponds to 5.5% by weight of the charge and contains 6% hydrogen and 94% carbon and may be assumed to have a total heating value of 17,000 Btu per lb, and a specific heat equal to that of graphite. The remainder of the charge goes to form hydrocarbon gases having an average molecular weight of 37, a heating value of 2100 Btu per cu ft (60° F, 30 in. Hg, sat.), and a mean specific heat of 0.634 between 70° F and 900° F.

The design study is to be based on the following operating conditions:

Reactor temperature	900° F
Regenerator temperature	1080° F
Stripping steam (saturated, dry, 100 psi gage)	
rate, lb per bbl of charge	0.7
Regeneration air, temperature	70° F
Regeneration air, % humidity	40 %
Oxygen content of dry combustion gases	3 %
Carbon on regenerated catalyst, % by wt.	0.4%

Laboratory burning tests have indicated that with the catalyst employed 43% of the carbon burned from the catalyst will form CO while the remainder forms CO_2 at the conditions of the regenerator. The hydrogen is burned to water, and it may be assumed that the hydrogen content of the deposit on the regenerated catalyst is the same as that of the deposit on the spent catalyst. The effects of pressure on enthalpies may be neglected. Radiation losses may also be neglected, as well as the enthalpies of the steam and air streams which may be introduced at various points in such a unit to assist fluidization. It may be assumed that no reaction occurs in the stripper and that the heat of hydration of the catalyst and the heat of vaporization of the stripped oil are negligible at the conditions of the stripper.

(a) Calculate the standard heat of reaction of the cracking reaction.

(b) From an over-all energy balance of the unit, calculate the temperature to which the oil feed must be preheated without vaporization in order to maintain a heat balance.

(c) Assuming that the specific heat of the catalyst is 0.23, calculate the rate of catalyst circulation required to maintain the reactor temperature, the temperature of the catalyst leaving the stripper, and also the carbon content of the spent catalyst.

(d) Prepare a complete energy and material balance of the reactor-regenerator unit.

Appendix

Table A. International Atomic Weights—1952

Element	Symbol	Atomic Number	Atomic Weight	Element	Symbol	Atomic Number	Atomic Weight
Actinium	Ac	89	227	Mercury	Hg	80	200.61
Aluminum	Al	13	26.98	Molybdenum	Mo	42	95.95
Americium	Am	95	[243]	Neodymium	Nd	60	144.27
Antimony	Sb	51	121.76	Neptunium	Np	93	[237]
Argon	A	18	39.944	Neon	Ne	10	20.183
Arsenic	As	33	74.91	Nickel	Ni	28	58.69
Astatine	At	85	[210]	Niobium	Nb	41	92.91
Barium	Ba	56	137.36	Nitrogen	N	7	14.008
Berkelium	Bk	97	[245]	Osmium	Os	76	190.2
Beryllium	Be	4	9.013	Oxygen	O	8	16.0000
Bismuth	Bi	83	209.00	Palladium	Pd	46	106.7
Boron	B	5	10.82	Phosphorus	P	15	30.975
Bromine	Br	35	79.916	Platinum	Pt	78	195.23
Cadmium	Cd	48	112.41	Plutonium	Pu	94	[242]
Calcium	Ca	20	40.08	Polonium	Po	84	210
Californium	Cf	98	[246]	Potassium	K	19	39.100
Carbon	C	6	12.010	Praseodymium	Pr	59	140.92
Cerium	Ce	58	140.13	Promethium	Pm	61	[145]
Cesium	Cs	55	132.91	Protactinium	Pa	91	231
Chlorine	Cl	17	35.457	Radium	Ra	88	226.05
Chromium	Cr	24	52.01	Radon	Rn	86	222
Cobalt	Co	27	58.94	Rhenium	Re	75	186.31
Columbium (see Niobium)				Rhodium	Rh	45	102.91
				Rubidium	Rb	37	85.48
Copper	Cu	29	63.54	Ruthenium	Ru	44	101.7
Curium	Cm	96	[243]	Samarium	Sm	62	150.43
Dysprosium	Dy	66	162.46	Scandium	Sc	21	44.96
Erbium	Er	68	167.2	Selenium	Se	34	78.96
Europium	Eu	63	152.0	Silicon	Si	14	28.09
Fluorine	F	9	19.00	Silver	Ag	47	107.880
Francium	Fr	87	[223]	Sodium	Na	11	22.997
Gadolinium	Gd	64	156.9	Strontium	Sr	38	87.63
Gallium	Ga	31	69.72	Sulfur	S	16	32.066
Germanium	Ge	32	72.60	Tantalum	Ta	73	180.88
Gold	Au	79	197.2	Technetium	Tc	43	[99]
Hafnium	Hf	72	178.6	Tellurium	Te	52	127.61
Helium	He	2	4.003	Terbium	Tb	65	159.2
Holmium	Ho	67	164.94	Thallium	Tl	81	204.39
Hydrogen	H	1	1.0080	Thorium	Th	90	232.12
Indium	In	49	114.76	Thulium	Tm	69	169.4
Iodine	I	53	126.91	Tin	Sn	50	118.70
Iridium	Ir	77	193.1	Titanium	Ti	22	47.90
Iron	Fe	26	55.85	Tungsten	W	74	183.92
Krypton	Kr	36	83.80	Uranium	U	92	238.07
Lanthanum	La	57	138.92	Vanadium	V	23	50.95
Lead	Pb	82	207.21	Xenon	Xe	54	131.3
Lithium	Li	3	6.940	Ytterbium	Yb	70	173.04
Lutetium	Lu	71	174.99	Yttrium	Y	39	88.92
Magnesium	Mg	12	24.32	Zinc	Zn	30	65.38
Manganese	Mn	25	54.93	Zirconium	Zr	40	91.22

A value given in brackets denotes the mass number of the most stable known isotope.

TABLE B. UNITS OF ENERGY*

Carnegie Institute of Technology American Petroleum Institute Research Project 44 Pittsburgh, Pa.

To convert the numerical value of a property expressed in one of the units in the left-hand column of the table to the numerical value of the same property expressed in one of the units in the top row of the table, multiply the former value by the factor in the block common to both units.

Units	G-mass (energy equiv)	Joule	Int joule	Cal	I. T. cal	Btu	Kw-hr	Hp-hr	Ft-lb(wt)	Cu ft-lb(wt)/sq.in.	Liter-atm
1 g-mass (energy equiv) =	1	8.987416×10^{13}	8.985933×10^{13}	2.148044×10^{13}	2.146640×10^{13}	8.518554×10^{10}	2.496505×10^{7}	3.347861×10^{7}	6.628764×10^{13}	4.603308×10^{11}	8.869642×10^{11}
1 joule =	1.112667×10^{-14}	1	0.999835	0.239006	0.238849	0.947831×10^{-3}	2.777778×10^{-7}	3.72505×10^{-7}	0.737561	5.12195×10^{-3}	9.86896×10^{-3}
1 int joule =	1.112850×10^{-14}	1.000165	1	0.239045	0.238889	0.947988×10^{-3}	2.778236×10^{-7}	3.72567×10^{-7}	0.737682	5.12279×10^{-3}	9.87058×10^{-3}
1 cal =	4.655398×10^{-14}	4.1840	4.18331	1	0.999346	3.96573×10^{-3}	1.162222×10^{-6}	1.558562×10^{-6}	3.08595	2.14302×10^{-2}	4.12917×10^{-2}
1 I. T. cal =	4.658444×10^{-14}	4.18674	4.18605	1.000654	1	3.96832×10^{-3}	1.162983×10^{-6}	1.559582×10^{-6}	3.08797	2.14443×10^{-2}	4.13187×10^{-2}
1 Btu =	1.173908×10^{-11}	1055.040	1054.866	252.161	251.996	1	2.930667×10^{-4}	3.93008×10^{-4}	778.156	5.40386	10.41215
1 kw-hr =	4.005601×10^{-8}	3,600,000	3,599,406	860,421	859,858	3412.19	1	1.341019	2,655,218	18439.01	35528.2
1 hp-hr =	2.986982×10^{-8}	2,684,525	2,684,082	641,617	641,197	2544.48	0.745701	1	1,980,000	13750.	26493.5
1 ft-lb (wt) =	1.508577×10^{-14}	1.355821	1.355597	0.324049	0.323837	1.285089×10^{-3}	3.766169×10^{-7}	5.05051×10^{-7}	1	6.94444×10^{-3}	1.338054×10^{-2}
1 cu ft-lb (wt)/sq in. =	2.172351×10^{-12}	195.2382	195.2060	46.6630	46.6325	0.1850529	5.423283×10^{-5}	7.27273×10^{-5}	144.	1	1.926797
1 liter-atm =	1.127441×10^{-12}	101.3278	101.3111	24.2179	24.2021	0.0960417	2.814662×10^{-5}	3.77452×10^{-5}	74.7354	0.518996	1

* The electrical units are those in terms of which certification of standard cells, standard resistances, etc., is made by the National Bureau of Standards. Unless otherwise indicated, all electrical units are absolute.

TABLE C. CONVERSION FACTORS AND CONSTANTS

Data on Air

1. Average Dry Analysis by Volume
International Critical Tables, Vol. 1, p. 393 (1926).

N_2	78.03%
A	0.94
O_2	20.99
	99.96%
CO_2	0.03
H_2, Ne, He, Kr, Xe	0.01
	100.00%

2. Values to Use in Combustion Calculations. In combustion calculations, the 0.04% of CO_2, H_2, and rare gases may be ignored. Furthermore, the argon may be lumped with the nitrogen; this is referred to as atmospheric nitrogen.

	% by Volume	% by Weight	Molecular Weight
Atmospheric nitrogen	79.00	76.80	28.16
Oxygen	21.00	23.20	32.00
	100.00	100.00	

Molecular weight = 28.97

Physical Constants

The Gas-Law Constant R

Numerical Value	Units
1.987	g-cal/(g-mole)(K°)
1.987	Btu/(lb-mole)(R°)
82.06	$(cm^3)(atm)/(g-mole)(K°)$
0.08205	(liter)(atm)/(g-mole)(K°)
10.731	$(ft^3)(lb_f)/(in.)^2(lb-mole)(R°)$
0.7302	$(ft)^3(atm)/(lb-mole)(R°)$

1 faraday = 96,493.1 (abs. coulomb)/(g-equivalent)
Avogadro constant
= 6.02380×10^{23} atoms per gram-atom or molecules per gram-mole

Density

1 g-mole of an ideal gas at 0° C, 760 mm Hg	=	22.4140 liters
	=	22,414.6 cc
1 lb-mole of an ideal gas at 0° C, 760 mm Hg	=	359.05 cu ft
Density of dry air at 0° C and 760 mm Hg	=	1.2929 g per liter
	=	0.080711 lb per cu ft

1 gram per cc = 62.43 lb per cu ft
1 gram per cc = 8.345 lb per U. S. gal

Length

1 in.	2.540 cm
1 micron	10^{-6} meter
1 Ångstrom	10^{-10} meter

Appendix

Mass

1 lb (avoirdupois)	16 oz
1 lb (avoirdupois)	7000 grains
1 lb (avoirdupois)	453.6 grams
1 ton (short)	2000 lb (Av.)
1 ton (long)	2240 lb (Av.)
1 gram	15.43 grains
1 kilogram	2.2046 lb (Av.)

Mathematical Constants

e	2.7183
π	3.1416
$\ln N$	$2.303 \log N$

Power

1 kw	56.87 Btu per min
1 kw	1.341 hp
1 hp	550 ft-lb per sec
1 watt	44.25 ft-lb per min
1 watt	14.34 g-cal per min

Pressure

1 psi	2.036 in. Hg at 0° C
1 psi	2.311 ft water at 70° F
1 atm	14.696 psi
1 atm	760 mm Hg at 0° C
1 atm	29.921 in. Hg at 0° C

Temperature Scales

Degrees Fahrenheit = 1.8 (degrees centigrade) + 32
Degrees Kelvin = degrees centigrade + 273.16
Degrees Rankine = degrees Fahrenheit + 459.69

Volume

1 cu in.	16.39 cc
1 liter	61.03 cu in.
1 liter	1000.028 cc
1 cu ft	28.32 liters
1 cu meter	1.308 cu yd
1 cu meter	1000 liters
1 U. S. gal	4 qt
1 U. S. gal	3.785 liters
1 U. S. gal	231 cu in.
1 British gal	277.42 cu in.
1 British gal	1.20094 U. S. gal
1 cu ft	7.481 U. S. gal
1 liter	1.057 U. S. qt
1 U. S. fluid oz	29.57 cc

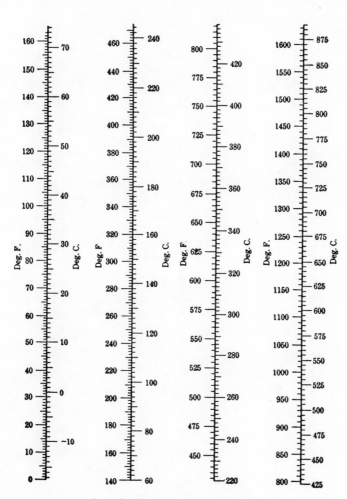

Fig. A. Temperature conversions

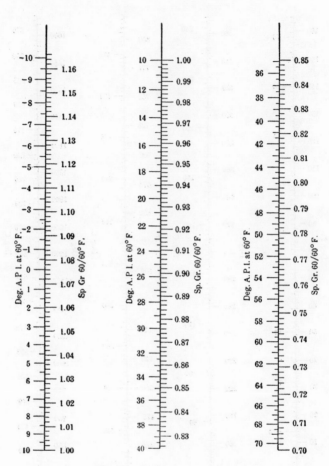

FIG. B.   Gravity conversions

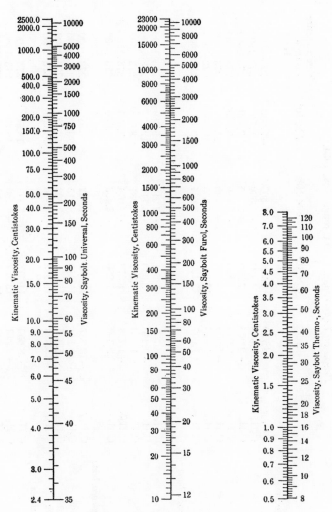

FIG. C.   Viscosity conversions

TABLE C. MOLAL HEAT CAPACITIES OF GASES AT ZERO PRESSURE *

$$c_p° = a + bT + cT^2 + dT^3; \quad (T = °K)$$

By K. A. Kobe and associates, "Thermochemistry for the Petrochemical Industry," Petroleum Refiner, Jan. 1949 through Nov. 1954.

		a	$b \times 10^2$	$c \times 10^5$	$d \times 10^9$	Temperature Range, °K	Error Max. %	Error Avg. %
Paraffinic Hydrocarbons								
Methane	CH_4	4.750	1.200	0.3030	−2.630	273–1500	1.33	0.57
Ethane	C_2H_6	1.648	4.124	−1.530	1.740	273–1500	0.83	0.28
Propane	C_3H_8	−0.966	7.279	−3.755	7.580	273–1500	0.40	0.12
n–Butane	C_4H_{10}	0.945	8.873	−4.380	8.360	273–1500	0.54	0.24
i–Butane	C_4H_{10}	−1.890	9.936	−5.495	11.92	273–1500	0.25	0.13
n–Pentane	C_5H_{12}	1.618	10.85	−5.365	10.10	273–1500	0.56	0.21
n–Hexane	C_6H_{14}	1.657	13.19	−6.844	13.78	273–1500	0.72	0.20
Monoolefinic Hydrocarbons								
Ethylene	C_2H_4	0.944	3.735	−1.993	4.220	273–1500	0.54	0.13
Propylene	C_3H_6	0.753	5.691	−2.910	5.880	273–1500	0.73	0.17
1–Butene	C_4H_8	−0.240	8.650	−5.110	12.07	273–1500	0.25	0.18
i–Butene	C_4H_8	1.650	7.702	−3.981	8.020	273–1500	0.11	0.06
cis-2-Butene	C_4H_8	−1.778	8.078	−4.074	7.890	273–1500	0.78	0.14
trans-2-Butene	C_4H_8	2.340	7.220	−3.403	6.070	273–1500	0.54	0.12
Cycloparaffinic Hydrocarbons								
Cyclopentane	C_5H_{10}	−12.957	13.087	−7.447	16.41	273–1500	1.00	0.25
Methylcyclopentane	C_6H_{12}	−12.114	15.380	−8.915	20.03	273–1500	0.86	0.23
Cyclohexane	C_6H_{12}	−15.935	16.454	−9.203	19.27	273–1500	1.57	0.37
Methylcyclohexane	C_7H_{14}	−15.070	18.972	−10.989	24.09	273–1500	0.92	0.22
Aromatic Hydrocarbons								
Benzene	C_6H_6	−8.650	11.578	−7.540	18.54	273–1500	0.34	0.20
Toluene	C_7H_8	−8.213	13.357	−8.230	19.20	273–1500	0.29	0.18
Ethylbenzene	C_8H_{10}	−8.398	15.935	−10.003	23.95	273–1500	0.34	0.19
Styrene	C_8H_8	−5.968	14.354	−9.150	22.03	273–1500	0.37	0.23
Cumene	C_9H_{12}	−9.452	18.686	−11.869	28.80	273–1500	0.36	0.17

Acetylenes and Diolefins

Acetylene	C_2H_2	5.21	2.2008	−1.559	4.349	273–1500	1.46	0.59
Methylacetylene	C_3H_4	4.21	4.073	−2.192	4.713	273–1500	0.36	0.13
Dimethylacetylene	C_4H_6	3.54	5.838	−2.760	4.974	273–1500	0.70	0.16
Propadiene	C_3H_4	2.43	4.693	−2.781	6.484	273–1500	0.37	0.19
1,3-Butadiene	C_4H_6	−1.29	8.350	−5.582	14.24	273–1500	0.91	0.47
Isoprene	C_5H_8	−0.44	10.418	−6.762	16.93	273–1500	0.99	0.43

Combustion Gases (Low Range)

Nitrogen	N_2	6.903	−0.03753	0.1930	−0.6861	273–1800	0.59	0.34
Oxygen	O_2	6.085	0.3631	−0.1709	0.3133	273–1800	1.19	0.28
Air		6.713	0.04697	0.1147	−0.4696	273–1800	0.72	0.33
Hydrogen	H_2	6.952	−0.04576	0.09563	−0.2079	273–1800	1.01	0.26
Carbon monoxide	CO	6.726	0.04001	0.1283	−0.5307	273–1800	0.89	0.37
Carbon dioxide	CO_2	5.316	1.4285	−0.8362	1.784	273–1800	0.67	0.22
Water vapor	H_2O	7.700	0.04594	0.2521	−0.8587	273–1800	0.53	0.24

Combustion Gases (High Range)

Nitrogen	N_2	6.529	0.1488	−0.02271	—	273–3800	2.05	0.72
Oxygen	O_2	6.732	0.1505	−0.01791	—	273–3800	3.24	1.20
Air		6.557	0.1477	−0.02148	—	273–3800	1.64	0.70
Hydrogen	H_2	6.424	0.1039	−0.007804	—	273–3800	2.14	0.79
Carbon dioxide	CO_2	See footnote † for special equation.				273–3800	2.65	0.54
Carbon monoxide	CO	6.480	0.1566	−0.02387	—	273–3800	1.86	1.01
Water vapor	H_2O	6.970	0.3464	−0.04833	—	273–3800	2.03	0.66

Sulfur Compounds

Sulfur	S_2	6.499	0.5298	−0.3888	0.9520	273–1800	0.99	0.38
Sulfur dioxide	SO_2	6.157	1.384	−0.9103	2.057	273–1800	0.45	0.24
Sulfur trioxide	SO_3	3.918	3.483	−2.675	7.744	273–1300	0.29	0.13
Hydrogen sulfide	H_2S	7.070	0.3128	0.1364	−0.7867	273–1800	0.74	0.37
Carbon disulfide	CS_2	7.390	1.489	−1.096	2.760	273–1800	0.76	0.47
Carbonyl sulfide	COS	6.222	1.536	−1.058	2.560	273–1800	0.94	0.49

* Reprinted with permission. In the original article, constants are also given for T in degrees centigrade, degrees Fahrenheit, and degrees Rankine.

† Equation for CO_2, 273 to 3800 °K: $c_p° = 18.036 − 0.000044747T − 158.08/\sqrt{T}$.

TABLE C. MOLAL HEAT CAPACITIES OF GASES AT ZERO PRESSURE (*Continued*)

		a	$b \times 10^2$	$c \times 10^5$	$d \times 10^9$	Temperature Range, °K	Error Max. %	Error Avg. %
Halogens and Halogen Acids								
Fluorine	F_2	6.115	0.5864	-0.4186	0.9797	273-2000	0.78	0.45
Chlorine	Cl_2	6.8214	0.57095	-0.5107	1.547	273-1500	0.50	0.23
Bromine	Br_2	8.051	0.2462	-0.2128	0.6406	273-1500	0.43	0.15
Iodine	I_2	8.504	0.13135	-0.10684	0.3125	273-1800	0.11	0.06
Hydrogen fluoride	HF	7.201	-0.1178	0.1576	-0.3760	273-2000	0.37	0.09
Hydrogen chloride	HCl	7.244	-0.1820	0.3170	-1.036	273-1500	0.22	0.08
Hydrogen bromide	HBr	7.169	-0.1604	0.3314	-1.161	273-1500	0.27	0.12
Hydrogen iodide	HI	6.702	0.04546	0.1216	-0.4813	273-1900	0.92	0.39
Chloromethanes								
Methyl chloride	CH_3Cl	3.05	2.596	-1.244	2.300	273-1500	0.75	0.16
Methylene chloride	CH_2Cl_2	4.20	3.419	-2.3500	6.068	273-1500	0.67	0.30
Chloroform	$CHCl_3$	7.61	3.461	-2.668	7.344	273-1500	0.92	0.42
Carbon tetrachloride	CCl_4	12.24	3.400	-2.995	8.828	273-1500	1.21	0.57
Phosgene	$COCl_2$	10.35	1.653	-0.8408	—	273-1000	0.97	0.46
Thiophosgene	$CSCl_2$	10.80	1.859	-1.045	—	273-1000	0.98	0.71
Cyanogens								
Cyanogen	$(CN)_2$	9.82	1.4858	-0.6571	—	273-1000	0.69	0.42
Hydrogen cyanide	HCN	6.34	0.8375	-0.2611	—	273-1500	1.42	0.76
Cyanogen chloride	CNCl	7.97	1.0745	-0.5265	—	273-1000	0.97	0.58
Cyanogen bromide	CNBr	8.82	0.9084	-0.4367	—	273-1000	0.85	0.54
Cyanogen iodide	CNI	9.69	0.7213	-0.3265	—	273-1000	0.75	0.37
Acetonitrile	CH_3CN	5.09	2.7634	-0.9111	—	273-1200	0.45	0.26
Acrylic nitrile	CH_2CHCN	4.55	4.1039	-1.6939	—	273-1000	0.63	0.41
Oxides of Nitrogen								
Nitric oxide	NO	6.461	0.2358	-0.07705	0.08729	273-3800	2.23	0.54
Nitric oxide	NO	7.008	-0.02244	0.2328	-1.000	273-1500	0.97	0.36
Nitrous oxide	N_2O	5.758	1.4004	-0.8508	2.526	278-1500	0.59	0.26
Nitrogen dioxide	NO_2	5.48	1.365	-0.841	1.88	273-1500	0.46	0.18
Nitrogen tetroxide	N_2O_4	7.9	4.46	-2.71	—	273-600	0.97	0.36

Oxygenated Hydrocarbons								
Formaldehyde	CH_2O	5.447	0.9739	0.1703	−2.078	273–1500	1.41	0.62
Acetaldehyde	C_2H_4O	4.19	3.164	−0.515	−3.800	273–1000	0.40	0.17
Methanol	CH_4O	4.55	2.186	−0.291	−1.92	273–1000	0.18	0.08
Ethanol	C_2H_6O	4.75	5.006	−2.479	4.790	273–1500	0.40	0.22
Ethylene oxide	C_2H_4O	−1.12	4.925	−2.389	3.149	273–1000	0.36	0.14
Ketene	C_2H_2O	4.11	2.966	−1.793	4.22	273–1500	0.48	0.17
Miscellaneous Hydrocarbons								
Cyclopropane	C_3H_6	−6.481	8.206	−5.577	15.61	273–1000	0.94	0.35
Isopentane	C_5H_{12}	−2.273	12.434	−7.097	15.86	273–1500	0.34	0.14
Neopentane	C_5H_{12}	−3.865	13.305	−8.018	18.83	273–1500	0.27	0.12
o-Xylene	C_8H_{10}	−3.789	14.291	−8.354	18.80	273–1500	0.52	0.15
m-Xylene	C_8H_{10}	−6.533	14.905	−8.831	20.05	273–1500	0.67	0.16
p-Xylene	C_8H_{10}	−5.334	14.220	−7.984	17.03	273–1500	0.56	0.18
C₃ Oxygenated Hydrocarbons								
Carbon suboxide	C_3O_2	8.203	3.073	−2.081	5.182	273–1000	1.22	0.31
Acetone	C_3H_6O	1.625	6.661	−3.737	8.307	273–1500	0.56	0.10
i-Propyl alcohol	C_3H_8O	0.7936	8.502	−5.016	11.56	273–1500	0.35	0.18
n-Propyl alcohol	C_3H_8O	−1.307	9.235	−5.800	14.14	273–1500	0.90	0.30
Allyl alcohol	C_3H_6O	0.5203	7.122	−4.259	9.948	273–1500	0.23	0.14
Chloroethenes								
Chloroethene	C_2H_3Cl	2.401	4.270	−2.751	6.797	273–1500	0.46	0.22
1,1-Dichloroethene	$C_2H_2Cl_2$	5.899	4.383	−3.182	8.516	273–1500	0.81	0.40
cis-1,2-Dichloroethene	$C_2H_2Cl_2$	4.336	4.691	−3.397	9.010	273–1500	0.70	0.39
trans-1,2-Dichloroethene	$C_2H_2Cl_2$	5.661	4.295	−3.022	7.891	273–1500	0.50	0.27
Trichloroethene	C_2HCl_3	9.200	4.517	−3.600	10.10	273–1500	1.17	0.50
Tetrachloroethene	C_2Cl_4	15.11	3.799	−3.179	9.089	273–1500	0.94	0.42
Nitrogen Compounds								
Ammonia	NH_3	6.5846	0.61251	0.23663	−1.5981	273–1500	0.91	0.36
Hydrazine	N_2H_4	3.890	3.554	−2.304	5.990	273–1500	1.80	0.50
Methylamine	CH_5N	2.9956	3.6101	−1.6446	2.9505	273–1500	0.59	0.07
Dimethylamine	C_2H_7N	−0.275	6.6152	−3.4826	7.1510	273–1500	0.96	0.15
Trimethylamine	C_3H_9N	−2.098	9.6187	−5.5488	12.432	273–1500	0.91	0.18

Author Index

Subject Index

Subject Index

Subject Index

Subject Index

Subject Index

Subject Index

Subject Index